21st International Symposium on Aviation Psychology (ISAP 2021)

Online
18-21 May 2021

ISBN: 978-1-7138-3082-5

Printed from e-media with permission by:

Curran Associates, Inc.
57 Morehouse Lane
Red Hook, NY 12571

CURRAN ASSOCIATES INC. proceedings.com

Some format issues inherent in the e-media version may also appear in this print version.

Wright State University – Conferences & Events
3640 Colonel Glenn Highway
Dayton, Ohio 45435
USA

Phone: (937) 775-1000

http://www.wright.edu

Additional copies of this publication are available from:

Curran Associates, Inc.
57 Morehouse Lane
Red Hook, NY 12571 USA
Phone: 845-758-0400
Fax: 845-758-2633
Email: curran@proceedings.com
Web: www.proceedings.com

21st International Symposium on Aviation Psychology

18–21 May, 2021

Photo

Katharine and Orville Wright sitting in Wright Model H Flyer, created 1914. Used courtesy of Special Collections and Archives, Wright Brothers Collection, Wright State University Libraries. https://corescholar.libraries.wright.edu/special_ms1_photographs/1088/

Description. Katharine Wright, wearing flying goggles, and Orville Wright sitting in a Wright Model H Flyer. Note on back, "Note vertical radiator made out of speaking tubes. Also note brake against wheel."

ISAP

International Symposium on Aviation Psychology

https://doi.org/10.5399/osu/1148

TABLE OF CONTENTS

ATTENTION, SEARCH, AND SCANNING

AVIATION INTERFACE DESIGN

EDUCATION

EXPLORING METHODS TO COLLECT AND ANALYZE HUMAN CONTRIBUTIONS TO SAFETY

FLIGHT DECK DISPLAYS AND DECISION MAKING

HEALTH AND WELL-BEING

HUMAN-MACHINE TEAMING

ORGANIZATIONAL SAFETY, CULTURE, AND CLIMATE

PERCEPTION AND PERFORMANCE

POSTERS

PROCEDURES IN PRACTICE AND TRAINING

RISK AND HAZARD ANALYSIS

SAFETY

SKILLS DEGRADATION AND ASSESSMENT

SKILLS TRAINING

WHAT WE KNOW ABOUT TEAMWORK AND MULTITEAM COORDINATION

WORKLOAD AND SITUATION AWARENESS

AN ANALYSIS OF INFORMATION REQUIREMENTS FOR PASSENGERS OF (AUTONOMOUS) URBAN AIR MOBILITY VEHICLES

Dennis B. Beringer
(Unaffiliated)
Yukon, OK

Much effort has been put into examining control/monitoring strategies for semi-autonomous/autonomous urban air mobility vehicles (UAMVs). Less has been done to define information requirements for passengers to facilitate their cognitive comfort. Similarities and differences between driverless automobiles (and transport-category aircraft) and UAMVs will both affect what information is needed and what operational factors influence that need, including; perceived locus of control, shared fate, ambient visibility, familiarity with the area to be traversed, and operational status of the vehicle. Information impacted includes route/progress (location, estimated time of arrival), phase of flight, and system status as well as communications between passenger and vehicle operator/monitor. Some intermediate level of information less than that required for orientation and control will likely suffice to achieve passenger acceptance, and that the level of information required is likely negatively correlated with both the visibility in the external environment and with the perceived safety/reliability of the vehicle/system.

There are, on the near horizon (metaphorically and literally), several interesting means by which people could be conveyed from point to point. We have seen testing of driverless automobiles and use of same to capture street-level photographs for use by Google Maps. We are also now seeing efforts to take this type of approach to the airborne environment in the context of autonomous urban-air-taxi operations. As an example, an electrically powered autonomous passenger-carrying air vehicle has been fielded in China by Ehang (https://www.ehang.com/ehangaav). This vehicle is in advanced testing and has carried occupants on flights. Airbus is testing a similar vehicle in Oregon (Banse, 2019), and Boeing also has activity ongoing, but the FAA is working with as many as 30 manufacturers pursuing vehicle certification (Reichmann, 2021). This type of operation has been discussed at various meetings with emphasis being on (1) system designs to allow operation within safe bounds, (2) infrastructure needed for integration into the existing airspace (or restructure present airspace), and (3) the business case for profitable operations (NASA, 2018). There has been discussion of user acceptance (Edwards & Price, 2020; survey of 2500 potential users) but this has largely been related to safety features and ride quality that would enhance acceptability (see Edwards & Price for full list of issues). There is a separate issue that has not received much attention, and that is what one might label as the "*cognitive comfort*" of the passenger (vaguely labeled "psychological factors" by one source).

Definitions and Analysis

If we consider an operational definition of "cognitive comfort," it might be the condition of the passenger that results in minimal feelings of anxiety. Anxiety can be generated by a number of factors in this situation, which could include but not be limited to (1) perceived (not necessarily actual) safety, (2) degree of perceived control, (3) predictability of vehicle behavior, (4) familiarity with the vehicle environment, etc. For the purposes of this examination, we can restrict the consideration to information that is available to the passenger that allows them to (1) assess vehicle state (normal/abnormal), (2) determine progress towards the destination, and (3) initiate communication with the operator/monitor of the vehicle. In doing this it is assumed that the most important concern of potential passengers, safety (Edwards & Price, 2020), can and will be addressed elsewhere (both the necessary hardware and software systems to achieve a high level of safety and the necessary user education/briefing materials). It should be noted that at this point in time there is little in the regulations that can be used to specify many of the details of what is necessary (trend towards performance-based criteria, attainable in a number of nonspecific ways).

It may not be appropriate to use existing systems as sources from which to generalize passenger needs as there are some aspects of the airborne on-demand system concept that differ from other transportation systems used in the past or presently. The two most relevant ones appear to be that (1) the vehicle is airborne as opposed to being a ground vehicle, ground vehicles being the most common experience for the majority of the population, and (2) there is no onboard operator/pilot (no shared fate). However, it may be appropriate to reuse information sources from those vehicles as they will be familiar to the potential passengers, and thus *familiarity may breed cognitive comfort rather than contempt* (Johnny Carson: "These ARE the jokes…").

Shift of Locus of Control, Shared Fate, and Airborne versus Ground

A feature of autonomous systems that has been a point of issue for potential passengers has been the lack of an onboard pilot or operator (Edwards & Price, 2020, indicate that 75% of those queried had reservations about using autonomous aircraft, and 25% said they would not use such services if they became available). This has been proposed previously for transport-category aircraft, but was not widely accepted by the public. If we look at the ground-vehicle environment and the efforts towards autonomous passenger-carrying vehicles, we collectively have some illusions regarding locus of control. It has been established that drivers of automobiles are not as aware of the true risks involved in operating a car in comparison with flying in an aircraft because, in some sense, they believe that the locus of control in their automobile is internal; they control their vehicle and hence their exposure to risk. However, most do not appreciate their lack of control of the myriad other vehicles that pose hazards. In pubic ground transportation, passengers forfeit control over the vehicle (e.g., bus) to the operator but the environment is similar to that with which they are familiar and in which they believe they have control. Thus, there is the familiarity with the environment, and that may also work to the benefit of autonomous ground vehicles (may have a back-up onboard control option). The notion of a small autonomous air taxi is not something that people in general have acquaintance or familiarity with, and thus there are additional hurdles to overcome to generate cognitive comfort.

People often associate aircraft with more hazard exposure than that in ground vehicles because aircraft leave the ground and it is perceived that they can come back into contact with the ground in an unpleasant manner. Second, in piloted aircraft, the locus of control is, again, transferred to the pilot(s) and is not with the passengers. We all have that experience when we ride in a car with someone else, and the degree of our discomfort is frequently correlated with how closely we are related to the driver (positive correlation?). The difference between this situation and an aircraft is that most people have experience driving a car and thus may compare the driver's actions with what they would have done as the driver. This is not the same in an aircraft as most of the passengers are not pilots, cannot (except when a passenger in a small General Aviation aircraft) see displays showing aircraft performance indices (one can usually see the speedometer in a car even when not driving), and cannot thus make direct comparisons of personal expectations with pilot performance. However, there is the perception that there is an operator on the aircraft who is experiencing the same things as are the passengers, is in the same "boat" so to speak, and also has a vested interest, at the most basic level (survival), in getting the aircraft back on the ground safely (shared fate). In contrast, there is no such pilot/operator onboard an autonomous or remotely-operated aircraft who has a shared fate with the passengers, and thus the remote operator/monitor could be perceived as largely dissociated from the passenger.

Passenger information needs

In discussing this new environment with a number of potential users, and with UAV researchers Kevin Williams and Anthony Tvaryanas at the Civil Aerospace Medical Institute, it was apparent that desired information included (1) indications that the aircraft was continuing to operate safely and within expected parameters, (2) indications that the flight was making progress towards the destination (location, destination, estimated time of arrival; these are not regularly scheduled flights as are scheduled carriers), and (3) indications of deviations from the intended route/destination or schedule. Associated with (1) and (3) was a desire for a means of communicating with the operator/monitor of the flight (to both deliver and receive information). These categories of information were identified as those that would make the potential users feel more comfortable with the vehicle and the flight (again, assuming that all safety issues have been satisfied). Entertainment was not a top issue, particularly as flight durations appear limited by stated ranges of vehicles (power limited: see Table 1). To set a context, three of the four eVTOL (electric Vertical Take Off and Landing) vehicles listed by Bellamy (2021) that had no pilot onboard were limited to 20 minutes or less of operation as a function of stated range and cruise speed.

Table 1. Range and speed, maximum duration of flight based upon range and speed, and number of passengers for 4 eVTOL aircraft without pilots onboard (summarized from Bellamy, 2021).

Name	Company	Range (miles)	Cruise speed (mph)	Calculated max duration, (hrs:mins) @ cruise*	Passengers
The Cora	Wisk	25	100	0:15	2
City Airbus	Airbus	80	75	1:00	4
EH216	Ehang	21	83	0:15	2
Volocity	Volocopter	22	68	0:19	2

*Actual flight durations should be shorter due to power requirements to climb and reach cruise speed

Further examination of the variables that are likely to affect the passenger's need for information can begin with the simplest case where much of the information is received by direct viewing of the contact environment (out-the-window view). We can then look at best and worst possible cases of the environment, and the one case of *transmitting* information from the passenger to the operator/monitor. These should shape the approach to how much and what kind of information needs to be given to the passenger.

1) Communications with operator/monitor (labeled "connectivity with the ground" by Edwards & Price, 2020). Need for this information/function may be influenced by -
    a. Knowledge of how many vehicles the "monitor" is tracking
        i. If the passenger knows it is 1 operator/many vehicles, the passenger may want a dedicated system onboard that is guaranteed, by minimal passenger action, to reach the operator/monitor (may be influenced by experience of having problems contacting someone on 911 in a large metroplex area).
        ii. If the passenger knows it is 1 operator/one vehicle, they may accept a cell-phone option or similar means of communication.
    b. Desire to actually "see" an operator and converse in real time (can interface with aft-cabin crew in scheduled-carrier operations)
2) Geographic Orientation (Where am I? Am I making progress towards my destination? When will I arrive?) Need for this information is likely influenced by -
    a. Level of landmark detail that can be seen out the window (NOTE: presently there appears to be a belief amongst the potential operators that these vehicles are not likely to use airspace at altitudes greater than 3,000 feet AGL regularly; as such, terrain and cultural features would be visible as long as there were no impediments to visual acquisition)
        i. Geographic operating environment
            1. Need lower in areas with prominent cultural or geographic features
            2. Need higher in areas without prominent terrain or with few cultural features
        ii. Meteorological operating environment
            1. Need lower when visibility is high (daytime, no atmospheric obscuration)
            2. Need higher when visibility is low (nighttime, precipitation, cloud cover, haze, fog)
    b. Familiarity of passenger with flight area (terrain, cultural features)
        i. Need lower in areas with which passenger is familiar
        ii. Need higher in areas with which passenger is unfamiliar
3) Visibility of vehicle to other vehicles/operators (this is really connected with "safety" to a large degree)
    a. Can other operators of similar aircraft and of piloted aircraft see/detect this one?
        i. Visual-detection-enhancement devices on aircraft (lights; presence, strobes, beacons)
        ii. Electronic presence (tracking) indicators (ADS-B, radar, etc.)
4) Operational status of the vehicle
    a. Knowledge of the current state of the vehicle by category
        i. Operating within expected parameters
        ii. Operating safely but with limitations on performance/duration.
        iii. Deviating to "safe" landing site to avoid exceeding operational limits (or partial failure)

From these we can construct a "worst-possible-case" situation. Table 2 presents the worst-possible factor levels with the required information required under those conditions and possible means of information presentation. The design approaches one could follow would be (a) tailor the interface to automatically provide the minimum information required under the established conditions or (b) provide information at all times that supports worst-possible-case conditions. Selection of means/formats should likely lean towards formats that would be most familiar to the user/passenger and thus more likely to support cognitive comfort as a function of their familiarity and little need for learning how to interpret the presentation.

Table 2. Environmental/flight condition, required information, and possible means of presentation.

Condition	Information/function required	Possible means of presentation*
Many-to-one vehicle-to-monitor configuration	Direct link with flight monitor	Call button, in-vehicle two-way audio-video; cell-phone-based video (e.g. FaceTime)
Minimal topographic and cultural features in area	Vehicle location and reference points	Plan-view electronic map; voice messaging with progress/location data+
Low visibility (night or meteorological obscuration)	View of outside world	Forward-looking synthetic vision (dash-mounted display or head-worn appliance); sensor-based forward view; plan-view electronic map; voice messaging with progress/location information
Geographical area unfamiliar to passenger	Vehicle location and reference points	Plan-view electronic map; voice messaging delivering progress/location information
Vehicle entering "recovery" mode, nearest landing site	Status of vehicle and relative locations of vehicle and landing site	Plan-view electronic map; voice messaging delivering progress/location information; status indicator showing phase/mode of flight

*Listing most likely and user-familiar means; list is not all-inclusive or exhaustive

+ Consistent with current PA announcements by crew in scheduled carriers

Information display/communication link

It is also of interest to providers to determine how best to bring this information into the vehicle. Ehang advertises that they are using 4G-5G networks to provide communication with and control of their vehicles. This suggests that an in-vehicle display would make sense for providing information regarding status and progress as well as audio-video communication with the flight monitor, and Ehang's document (Xu, 2020) depicts an in-vehicle display (Figure 1). The display appears to have attitude and airspeed indications in the upper left quadrant, some kind of vehicle status information in the upper-right quadrant, other undecipherable (at this resolution) vehicle related information and temperature (C) in the lower left quadrant, and an apparently forward-looking-camera image in the lower-left quadrant (does not appear to be synthet-

Figure 1. Example in-vehicle display (from Xu, 2020)

ic vision). Lacking a detailed explanation in the source document as to intent/use of this format, it is difficult to conclude much about this set of displays. However, it does illustrate that there is sufficient graphical-display real estate to present information of passenger interest. Unfortunately, there is little information available at this time for any of the proposed vehicles regarding interior displays or communication systems for passengers. It is also possible that a Bluetooth or USB connection could be provided by which a passenger could connect their personal electronic device, use an app provided by the service provider, and have these same data/functions provided on their cell phone or tablet. The disadvantage of the latter is that the passenger loses this information if the device battery goes down unless power can be drawn from the aircraft (USB; communications AND device charging at the same time). Ehang is proposing an app for scheduling rides (the Uber model), but Ehang shows the in-vehicle display.

Conclusion & Prospectus

This effort was originally intended to be an analysis/survey/simulator experiment study, but the latter phases were temporarily delayed by Covid 19 and by retirement, both of which occasioned a reduction in resources and loss of simulator-study participants. As such, we have to be temporarily satisfied with suggesting what information may be useful to and needed by passengers of autonomous airborne vehicles. The survey portion, to validate the analysis/identification of factors, is in progress now. However, it may remain for a later time, hopefully soon, to conduct the simulation experiment with appropriate displays and manipulation of levels/formats of information display and environmental conditions to determine if potential users' stated preferences align with their actual responses in a quasi-realistic environment (the oft-observed dissociation between preference and performance). The limitation of the approach is, of course, that the vehicle will be "simulated," and that the passenger will not actually be potentially exposed to any real hazard, and thus the potential difficulty of observing any true variation in perceived anxiety. Hopefully validation of the analysis will produce reasonable recommendations for creating a harmonious environment for future passengers in these systems.

References

Banse, T. (2019). In the Skies Of Eastern Oregon, An Autonomous Robo Taxi Takes Flight. Retrieved from https://www.opb.org/news/article/pacific-northwest-oregon-drone-air-taxi-airbus-test-flights/

Bellamy, W., III (2021). 10 eVTOL Development Programs to Watch in 2021. Aviation Today, Febr4uary-March 2021. Retrieved from: file:///Volumes/NO%20NAME/ISAP_2021/First%20FAA%20Type%20Certification%20for%20Advanced%20Air%20Mobility%20Aircraft%20Could%20Come%20This%20Year%20-%20Aviation%20Today.htm.

Edwards, T. and Price, G. (2020). eVTOL Passenger Acceptance. Alexandria, VA: National Technical Information Service, NASA/CR—2020–220460. (also available at http://ntrs.nasa.gov)

NASA (2018). Urban air mobility market study. Retrieved from https://www.nasa.gov/sites/default/files/atoms/files/uam-market-study-executive-summary-v2.pdf

Reichman, K. (2021). First FAA type certification for Advanced Air Mobility Aircraft could come this year. Aviation Today, January 26, 2021. Retrieved from: file:///Volumes/NO%20NAME/ISAP_2021/First%20FAA%20Type%20Certification%20for%20Advanced%20Air%20Mobility%20Aircraft%20Could%20Come%20This%20Year%20-%20Aviation%20Today.htm.

Xu, E. (H.) (2020). The Future of Transportation: White Paper on Urban Air Mobility Systems. Ehang. Retrieved from https://www.ehang.com/ehangaav.

HUMAN FACTORS CONSIDERATIONS FOR URBAN AIR MOBILITY

Bhoomin B Chauhan
Florida Institute of Technology
Melbourne, Florida, U.S.A
Dr. Meredith Carroll
Florida Institute of Technology
Melbourne, Florida, U.S.A

Before urban air mobility (UAM) flights are safely integrated into the current airspace system, it is necessary to identify and address human factors issues associated with UAM. Various industry and academic institutions are currently exploring a range of different aspects of UAM, such as vehicle concepts, airspace integration, and ground infrastructure, all of which have human factors implications. These human factors issues, which will heavily influence how UAM operations will evolve with growth in demand and autonomous technology, are in need of research. Potential human factors issues include UAM pilot's trust in automation, situational awareness, visual scanning, decision-making capabilities, as well as workload and stress of pilots, air traffic controllers, and ground personnel, to name a few. This paper aims to examine UAM's current research and identify potential human factors issues in need of future research.

Introduction

The idea of using manned flying vehicles for inter-city transport first emerged in the 1940s with the advent of helicopters, which provided vertical take-off and landing capability without requiring extensive and costly infrastructure, such as the case with fixed-wing aircraft (Straubinger et al., 2020). New technological advances in the aviation industry, such as electric propulsion systems, efficient battery technology, and UAM concept aircraft, have laid the foundation for the development of UAM (Straubinger et al., 2020). Many academic institutions and industry stakeholders are currently working on understanding how components of UAM, like vehicle design, airspace corridors, operational models, and infrastructure, need to be developed. It is crucial to identify and examine potential human factors issues associated with these UAM components. With UAM, there will be an entirely new class of aircraft, new cockpit designs, new operational procedures, and new infrastructure. It is likely to affect the pilot's situational awareness, decision-making capabilities, trust in automation, stress, and workload. UAM also has the potential to significantly impact the general public's trust in automation, air traffic controllers (ATC) interactions with the pilots, and the comfort of the passengers riding in the vehicles. This paper aims to examine UAM's current research and identify potential human factors issues and areas in need of future research.

UAM Concept of Operations

The National Aeronautics and Space Administration (NASA) has proposed a three-phase UAM maturity level (UML) scale to measure UAM growth over the coming years, based on air traffic density, operational complexity, and automation (Patterson et al., n.d). The initial stage of

UAM integration into the National Airspace System (NAS) will involve developing UAM aircraft, determining the UAM aircraft certification process, and establishing UAM corridors in controlled airspace to identify the level of changes necessary for safe integration (Pinto Neto et al., 2019). During early UAM integration, it is critical to understand the effect of increased traffic in the NAS on air traffic controllers' workload and their ability to manage two potentially different types of separation techniques. Workload is related to the difficulty of the task at hand, aircraft count, cluttering, and use of restricted airspace (Stein, 1985). It is necessary to empirically study controller capabilities in this performance context to set safe and effective performance standards.

The Intermediate state represents low density, medium complexity operations. In this state, operations are tested with more scalable and weather-tolerant designs and consideration for local regulations (Pinto Neto et al., 2019). There will be an increase in the use of automation for low and medium-density UAM operations, which will require development of collaborative and robust automated systems. Research has shown that low-level automation may increase pilot workload, but high-level automation may result in a loss of pilot awareness of the state of aircraft or airspace, which can lead to errors and reduced performance (Gill et al., 2012). Further, research has shown that as level of automation increases and pilots move towards a supervisory position, the trust in automation will vary based on the number of aircraft under supervision, decision-making capabilities, and ability to identify automation failures (Ruff et al., 2002; Dikmen & Burns, 2017). Not only will levels of automation increase as UAM operations mature, but pilot-in-command distance will increase as pilots transition from onboard piloting to remote piloting and monitoring of multiple aircraft. Research is needed to understand the impact on pilot trust in and use of automation, and ultimately on performance and safety.

The mature state is associated with high density and highly complex operations with fully autonomous systems, including a large-scale, widely distributed UAM flight network (Pinto Neto et al., 2019. In this state it is assumed that the majority of UAM flights will be remotely operated, where the pilot will be given the task of supervising multiple flights. Due to the remote and supervisory nature of their responsibilities, research has shown that it can result in potential loss of situational awareness of aircraft state because of absence of visual, auditory, proprioceptive, and olfactory sensations during remote operations making it more difficult for the pilot to maintain an awareness of the aircraft's state (Hobbs & Lyall, 2016). This is complicated by the added roles of monitoring multiple aircraft simultaneously. Empirical research is needed to understand how many aircraft an operator can effectively supervise at once and under what conditions performance and safety can be optimized.

UAM Vehicle Concepts

Aircraft with VTOL capability are the primary vehicles under consideration, including three basic conceptual models: Quadrotor, Side-by-Side Helicopter, and Lift + Cruise VTOL aircraft. These designs will be prone to localized turbulence and poor visibility due to large infrastructure near the landing and take-off facilities (Price et al., 2020). Research has shown that the combined effect of degraded visual environment and turbulence can lead to workload exceeding the pilot's capability and lead to a significant decrease in pilot's ability to maintain situational awareness and control of their aircraft (Hoh, 1990; Ji et al., 2021). Therefore, there is a need to empirically study UAM pilot performance under these circumstances.

UAM pilots will be required to maintain separation and aircraft stability while operating at low altitudes, the time in flight in which there is typically the highest workload and the greatest levels of risk. With increased task demand, anxiety can influence the pilot's ability to perform adequately, as stress and anxiety will use up the necessary cognitive resources, causing the performance to deteriorate (Dismukes et al., 2015). A potential research gap is investigating the impact pilot workload, and stress levels will have on the pilot's ability to operate in the UAM environment.

New cockpit designs also introduces human factors issues. Research has shown that if the outputs provided to the pilots are less predictable, unexpected automation surprises can compromise the pilot's situational awareness and degrade performance (Dorneich et al., 2012). It is, therefore, necessary to examine how emerging UAM cockpit designs can effectively support pilots in maintaining situation awareness and making effective decisions. As UAM moves to fully automated operations, the ground control station interface will also have to be designed to facilitate situational awareness and support for aeronautical decision-making (Williams et al., 2001).

As UAM operations transition to automated flights, it is important to identify designs that will ensure appropriate levels of trust in automation (e.g., transparency). Research has shown that human-automation interaction can lead to unbalanced mental workload, reduced situational awareness, decision biases, mistrust, overreliance, and complacency (De Visser & Parasuraman, 2011). It is necessary to study how emerging UAM automation interfaces will impact pilot trust in and use of automation.

UAM Infrastructure

The ground infrastructure for electric VTOL flights needs to be designed to handle different levels of traffic from single flights to multiple flights taking off simultaneously. One of the key elements of infrastructure being proposed are vertiports, a type of airport that is designed explicitly for aircraft capable of vertical take-off and landing, passenger embarking and disembarking, pre-and post-flight checks, aircraft battery charging, and general day-to-day maintenance of the aircraft systems (Taylor et al., 2020). These vertiports will either be located at ground level or positioned on the top of high buildings, and ground staff will be tasked with performing their task under sometimes severe environmental conditions, including at great heights where winds can be high. The total area available to handle multiple aircraft at vertiports will be less than traditional airports or helipads. Research has shown that as the number of aircraft increases, it results in more complex operations and faster turnaround time, resulting in pilot and ground personnel error, workload, and misjudgment (Cardosi & Yost, 2001). Also, due to smaller space available, the number of ground personnel would also be limited. As ground operations increase, ground personnel will have less time to complete a task which can result in increased levels of stress (Sun & Chiou, 2010). Further, ground staff working hours, physical work, and lack of rest contribute to fatigue (Rosskam et al., 2009), a precursor to unsafe acts and accidents. There is a need to study how UAM infrastructure constraints will impact ground worker's workload, stress, and performance.

UAM Roles and Responsibilities

According to the Federal Aviation Administration (2020), UAM pilot/operator, state/local/federal authorities, service providers, aerodrome facilities, NAS users, and public interest stakeholders will hold prominent roles and responsibility in developing UAM. ATC's responsibilities in overseeing UAM operations will be extensive, including setting up the UAM airspace corridor based on the functional design of flights, time of the day, departure & approach paths, location, availability of vertiports, and separation with other UAM flights and manned flights. Airspace can accommodate more aircrafts if they are flying under visual flight rules (VFR) (Holcombe, 2018). This principle can be applied to UAM flights; however, the effect of attention and distraction for controllers and pilots, as well as UAM pilot's visual scanning performance in the highly congested conditions of this context, needs to be studied. As air traffic will increase, there is the potential for an increase in the air traffic controller's perceived workload (Hah et al., 2006). Controllers will have to monitor a greater number of aircraft, and research has shown that inattentional blindness, attentional blink, and working-memory capacity are top contributing factors for ATC operational errors (Xing & Bailey, 2005). Therefore, there is a need to study these phenomena in a UAM performance context.

It is also important to consider public perception and the passenger ride quality. Factors like vehicle inputs (manual or automatic maneuvering capabilities), vehicle characteristics (e.g., aircraft motion, vibration, noise, seat geometry, temperature, ambient lighting conditions), passenger motivation and willingness, cost, flight routes, schedule, and convenience (Edwards & Price, 2020) can influence passenger perception. These factors need be studied to ensure optimal ride quality and UAM acceptance.

Conclusions

Like any new concept, there is a need to identify the primary research areas that will help develop successful UAM operations. This paper aimed to identify key components of UAM and the associated human factors issues. Key human factors areas in need of future research include UAM pilot's trust in automation, situational awareness, visual scanning performance, decision-making capabilities; pilot, controller, ground personnel's workload and stress; and passenger ride quality and public perception. Understanding these areas of research will not only help the aviation community better understand how to implement UAM successfully but will also help the UAM stakeholders develop standards and procedures that keep "the human" in mind.

References

Cardosi, K., & Yost, A. (2001). Controller and Pilot Error in Airport Operations: A Review of Previous Research and Analysis of Safety Data (DOT/FAA/AR 00/51). U.S. Department of Transportation, Office of Aviation Research.

De Visser, E., & Parasuraman, R. (2011). Adaptive aiding of human-robot teaming. *Journal of Cognitive Engineering and Decision Making*, *5*(2), 209-231. https://doi.org/10.1177/1555343411410160

Dikmen, M., & Burns, C. (2017). Trust in autonomous vehicles: The case of Tesla autopilot and summon. *2017 IEEE International Conference on Systems, Man, and Cybernetics (SMC)*. https://doi.org/10.1109/smc.2017.8122757

Dismukes, R., Goldsmith, T., & Kochan, J. (2015). Effects of Acute Stress on Aircrew Performance: Literature Review and Analysis of Operational Aspects (NASA/TM—2015–218930). NASA.

Dorneich, M. C., Rogers, W., Whitlow, S. D., & DeMers, R. (2012). Analysis of the risks and benefits of flight deck adaptive systems. *PsycEXTRA Dataset*. https://doi.org/10.1037/e572172013-017

Federal Aviation Administration. (2020). Urban Air Mobility. NASA. https://nari.arc.nasa.gov/sites/default/files/attachments/UAM_ConOps_v1.0.pdf

Edwards, T., & Price, G. (2020). *eVTOL passenger acceptance* (NASA/CR—2020–220460). NASA. https://ntrs.nasa.gov/citations/20200000532

Gill, G., Kaber, D., Kaufmann, K., & Kim, S. (2012). Effects of modes of cockpit automation on pilot performance and workload in a next generation flight concept of operation. *Human Factors and Ergonomics in Manufacturing & Service Industries*, *22*(5), 395-406. https://doi.org/10.1002/hfm.20377

Hah, S., Willems, B., & Phillips, R. (2006). The effect of air traffic increase on controller workload. Proceedings of the Human Factors and Ergonomics Society 50th Annual Meeting. https://doi.org/10.1037/e577552012-011

Hobbs, A., & Lyall, B. (2016). *Human Factors Guidelines for Remotely Piloted Aircraft System Remote Pilot Stations* (TN–34128). National Aeronautics and Space Administration.

Hoh, R. H. (1990). *The Effect of Degraded Visual Cueing and Divided Attention on Obstruction Avoidance in Rotorcraft* (DOT/FAA/RD-90/40). Federal Aviation Administration.

Holcombe, R. (2018). Integrating drones into the us air traffic control system. SSRN Electronic Journal. https://doi.org/10.2139/ssrn.3191352

Ji, H., Chen, R., Lu, L., & White, M. D. (2021). Pilot workload investigation for rotorcraft operation in low-altitude atmospheric turbulence. *Aerospace Science and Technology*, *111*, 106567. https://doi.org/10.1016/j.ast.2021.106567

Patterson, M. D., Medonca, N. L., Neogi, N. A., Metcalfe, M., Hill, B. P., Wiggins, S., & Metts, C. (n.d.). An Initial Concept for Future Urban Air Mobility Operations.

Pinto Neto, E. C., Baum, D. M., Almeida Jr., J. R., Camargo Jr., J. B., & Cugnasca, P. S. (2019). Trajectory-Based Urban Air Mobility (UAM) Operations Simulator (TUS). arXiv. https://arxiv.org/ftp/arxiv/papers/1908/1908.08651.pdf

Price, G., Helton, D., Jenkins, K., Kvicala, M., Wolfe, R., & Parker, S. (2020). Urban Air Mobility Operational Concept (OpsCon) Passenger-Carrying Operations (NASA/CR—2020–5001587). NASA.

Rosskam, E., Greiner, B., Mateski, M., McCarthy, V., Siegrist, J., Smith, S., Wege, N., & Zsoldos, L. (2009). Stressed and fatigued on the ground and in the sky. International Transport Workers' Federation. Civil Aviation Section.

Ruff, H. A., Narayanan, S., & Draper, M. H. (2002). Human interaction with levels of automation and decision-aid fidelity in the supervisory control of multiple simulated unmanned air vehicles. *Presence: Teleoperators and Virtual Environments*, *11*(4), 335-351. https://doi.org/10.1162/105474602760204264

Stein, E. S. (1985). *Air traffic controller workload: An examination of workload probe* (DOT/FAA/CT-TN84/24). Federal Aviation Administration Technical Center. https://hf.tc.faa.gov/publications/1985-air-traffic-controller-workload/

Straubinger, A., Rothfeld, R., Shamiyeh, M., Büchter, K., Kaiser, J., & Plötner, K. O. (2020). An overview of current research and developments in urban air mobility – Setting the scene for UAM introduction. Journal of Air Transport Management, 87, 101852. https://doi.org/10.1016/j.jairtraman.2020.101852

Sun, K., & Chiou, H. (2010). Aviation ground crews: Occupational stresses and work performance. *African Journal of Business Management*, *5*(7), 2865-2873.

Taylor, M., Saldanli, A., & Park, A. (2020). Design of a Vertiport design tool. 2020 Integrated Communications Navigation and Surveillance Conference (ICNS). https://doi.org/10.1109/icns50378.2020.9222989

Williams, D. M., Waller, M. C., Koelling, J. H., Burdette, D. W., Capron, W. R., Barry, J. S., . . . Doyle, T. M. (2001). Concept of operations for commercial and business aircraft synthetic vision systems. Hampton, VA: National Aeronautics and Space Administration.

Xing, J., & Bailey, L. (2005). Attention and memory in air traffic control tasks. *Journal of Vision*, *5*(8), 427. https://doi.org/10.1167/5.8.427

VALIDATING STRESS TOLERANCE EXERCISE WITH GAME-BASED ASSESSMENT AND STRESS CHECKLISTS

Jóhann Wium

Isavia Air Navigation Services

Iceland

Stress tolerance is an important attribute for air traffic controllers. Assessing an individual's stress tolerance should therefore be considered when selecting student air traffic controllers. Unfortunately, measures of stress tolerance are often self-reports and as such are subject to "faking good". This paper details the validation of an observation-based stress tolerance exercise and its convergent validity with behavioral signs of stress (from stress checklists) and results of game-based assessments of Emotional Stability and Performance Under Pressure. The resulting data suggests that ratings from the exercise are valid and that game-based assessments can be used to predict ratings made from observable behavior in candidates.

A high attrition rate is a persistent problem in air traffic controller (ATCO) training (Broach, 2017). Previous job analyses of the role of ATCOs emphasize being able to perform in high stress environments (Goeters et al, 2004; Nickles et at 1995; Suresh et al, 2012) so stress tolerance is an issue in ATCO selection. Stress tolerance is defined as being capable of reacting with a problem-solving approach rather than an emotional approach when faced with adversity and remaining calm, even-tempered, and composed in stressful situations (Nickles et at, 1995).

Some studies indicated that stress tolerance can affect student ATCO success (see e.g. Chapelle et al, 2015; Roe et al, 2012; Collins et al, 1989) while others found no significant connection (Geven et al, 2008; Luuk et al, 2009; Oakes et, 2001; Schroeder et al 1993). One possible explanation is that these studies used self-report questionnaires when measuring stress tolerance and they can be susceptible to "faking-good". Griffith and Converse (2012) estimate that 30% (+/-10%) of job applicants will give overly positive answers about their capabilities and that "faking good" is more likely in 'high stakes' selection (Ellingsen, 2012; Griffith et al, 2007).

To counter this bias, an assessment center exercise was developed to assess stress tolerance under pressure. The exercise was validated by comparing it to behavioral observations by two raters and scores from a game-based assessment (GBA). The behavioral observations were in the form of a checklist score of observed behaviors connected to stress (i.e. stress checklist). The game-based assessment is a novel method to assess various applicant characteristics, such as performance under pressure and emotional stability. Applicants had completed the GBA one month earlier as part of a multiple hurdle student ATCO selection process. While extensive, that process did not specifically select for stress tolerance or similar traits until the assessment center.

Inter-rater reliability between the two assessors was also calculated to ensure consistency of ratings.

Game-based assessments

Convergent validity for GBA's had been demonstrated with cognitive tests before in Icelandic student ATCO selection (Boardman, 2017), but this is the first study to use behavioral assessments. A GBA uses methods of psychological testing embedded in a gamified interface, collecting game performance data, both overtly (e.g. candidate choices in the game) and covertly (e.g. reaction times), to assess cognitive and personality factors (Arctic Shores, 2017). GBA's are also resistant to "faking-good" (Armstrong et al, 2016) making them ideal for this comparison.

GBA's combine research from I/O, neuropsychology (Ferreira-Brito et al, 2019), behavioral economics, and education (Reiners & Woods, 2015) to assess applicants. Reaction time, for example, can assess Neuroticism (Robinson & Tamar, 2005) or self-confidence (Wichmann et al, 2016), time on task can assess persistence (Ventura & Shute, 2013), time logs can assess collaboration (Mislevy at al, 2015), and so on.

The GBA used was Skyrise City from Arctic Shores. It measures several psychological constructs including Emotional Stability (ES), consistency of affect, i.e., stressful situations are dealt with in a calm and even-tempered manner, and Performance under Pressure (PP), maintaining goal-oriented behavior while subjected to negative stressors (Arctic Shores, 2017).

Stress tolerance assessment exercise

An applicant entered a room with two assessors and sat at a table with a countdown clock, 12 puzzle pieces, and a rectangular frame. A standardized briefing was given stating that the exercise would be 10 minutes long (counted down on the clock) and that the goal was to fit all puzzle pieces inside the rectangular frame. The briefing stated that warnings would be given when there were 5 minutes and 1 minute left. The applicant was not explicitly told that stress tolerance was being assessed, and, if asked, the standard answer was that assessment was based on task performance.

The puzzle was a reproduction of an assembly puzzle called Calibron-12 created by Theodore Edison in 1933. While it is theoretically possible to finish within the allotted time, it is exceptionally unlikely (estimated minimum time is about 4 hours). The puzzle, however, looks like it should be easily solved within the allotted time (Creative Crafthouse, 2020). Because an applicant is unlikely to solve the puzzle, the performance measure was stress tolerance, not completion time.

The stress in this exercise is created by the fact that the exercise is part of a high-stakes selection process and occurs under time pressure while the candidate is being observed by two assessors. Additionally, the candidate is faced with a frustrating task, which may engender a sense of impending failure. The capacity to continue to perform, while remaining unaffected by these stressors, is an indication of stress tolerance.

Once 10 minutes had passed and the applicant had left, each rater gave an independent rating (IR) of estimated stress tolerance. The score ranged from 1 (Very poor stress tolerance) to 6 (Very good stress tolerance). Scores from the raters was then averaged to give a final stress tolerance (ST) score for the exercise.

Second, each rater completed a checklist for observed signs of stress and tallied the number of signs to give a checklist score. The score ranged from 0-9 and the signs included were: shaking, fidgeting, stiffness, defensive behavior, avoiding behavior, inadvertent sounds, flushed skin, forceful movements and slow movements. Assessors received detailed descriptions of stress signs as part of their training.

Methods and results

Participants

Thirty participants were assessed in an assessment center for Icelandic student ATCO applicants. Internal data protection policy dictates that identifiable personal information is removed from research datasets so information on age and gender is not available.

Assessor measures

The independent scores (IR) for stress tolerance as given by the assessors, demonstrated an inter-rater reliability of $r(30) = 0.853$, $p>0.001$. For the checklist of observed signs of stress, assessors had a 68% agreement rate.

The total number of stress signs observed by each assessor was counted from the checklists. The correlation between the total number of stress signs and independent rating of stress tolerance (IR) was $r(60) = -0.582$, $p>0.001$ (with a lower IR meant that more signs were observed and vice versa). This supports the hypothesis that IR and the checklist score (observable signs of stress) demonstrate convergent validity.

Game-based assessment

The correlation between final stress tolerance score (ST) and Emotional Stability (ES) was $r(30) = 0.444$, $p>0.05$. The correlation between final stress tolerance score (ST) and Performance Under Pressure (PP) was $r(30) = 0.380$, $p>0.05$. The GBA measures also reached significance when compared to the independent scores of the raters (IR). Between IR and ES the correlation was $r(60) = 0.348$, $p>0.01$, and between IR and PP it was $r(60) = 0.295$, $p>0.05$. Correlation between checklist scores and ES or PP did not reach significance.

Discussion

The results suggest that the exercise is valid. It has reliability as evidenced by high inter-rater reliability of the independent ratings. The independent ratings and checklist scores demonstrate convergent validity as there is a significant negative correlation between observable stress signs and stress tolerance. The assessment also demonstrates convergent validity in that the GBA measures of Emotional Stability (ES) and Performance Under Pressure (PP) both correlate significantly with stress tolerance (ST).

While this study suffers from a potential lack of generalizability due to its small sample size it provides two useful insights for practitioners. First, the exercise described shows an observational method to rate stress tolerance free from applicants "faking good". Second, as game-based assessments are a new approach in selection it is important to note that they can show significant correlations to more traditional ratings made by behavioral observations.

References

Arctic Shores (2017). Skyrise City: Technical manual. Game version 1.26, Scoring version 1.5.

Armstrong, M., Landers, R., & Collmus, A. (2016). Gamifying Recruitment, Selection, Training, and Performance Management: Game-Thinking in Human Resource Management. In Gangadharbatla, H., and Davis, D.Z. (Eds.), *Emerging Research and Trends in Gamification* (pp. 140-165). doi: 10.4018/978-1-4666-8651-9.ch007.

Boardman, E.C. (2017). Convergent and Predictive Validity of a Game-based Assessment in Air Traffic Control Recruitment. United Kingdoms: Arctic Shores

Broach, D. (2017). Attrition in U.S. Air Traffic Control Specialist (ATCS) Training: A Review of 50 Years of Data. 19th International Symposium on Aviation Psychology, 542-547.

Chappelle, W., Thompson, W., Goodman, T., Bryan, C., & Reardon, L. (2015). The Utility of Testing Noncognitive Aptitudes as Additional Predictors of Graduation From US Air Force Air Traffic Controller Training. *Aviation Psychology and Applied Human Factors. 5.* 93-103. doi: 10.1027/2192-0923/a000082

Collins, W. C., Schroeder, D. J., & Nye, L. G. (1989). Relationship of anxiety scores to academy and field training performance of air traffic control specialists. DOT-FAA-AM-89-7. Washington, DC: Federal Aviation Administration Office of Aviation Medicine.

Creative Crafthouse (2020, October 10). Calibron 12 – A Very Difficult Wood Puzzle Brain Teaser. Retrieved from https://www.creativecrafthouse.com/calibron-12-very-difficult-design-from-1933-now-with-cover.html

Ellingson, J. E. (2012). People fake only when they need to fake. In Ziegler, M., MacCann, C., and Roberts, R. (Eds.), *New Perspectives on Faking in Personality Assessment*, (pp. 19-33). New York: Oxford University Press.

Ferreira-Brito, F., Fialho, M., Virgolino, A., Neves, I., Miranda, A., Sousa-Santos, N., Caneiras, C., Carriço, L., Verdelho, A., & Santos, O. (2019). Game-based interventions for neuropsychological assessment, training and rehabilitation: Which game-elements to use? A systematic review. *Journal of Biomedical Informatics. 98*. doi: 103287. 10.1016/j.jbi.2019. 103287.

Geven, E.., Oprins, E., Veldhuijzen, E., and Roe, R.A. (2008). Using personality scales in ATC selection: Predictive validities for ATC task performance in Initial Training. Proceedings of the 28th European Association for Aviation Psychology Conference.

Goeters, K.M., Maschke, P., and Eißfeldt, H. (2004). Ability Requirements in Core Aviation Professions: Job Analyses of Airline Pilots and Air Traffic Controllers. In K. M. Goeters (Ed.), *Aviation psychology: Practice and Research* (pp. 135-140). Aldershot: Ashgate Publishing Limited.

Griffith, R. L., Chmielowski, T., and Yoshita, Y. (2007). Do applicants fake? An examination of the frequency of applicant faking behaviour. *Personnel Review*, *36*(3), 341-355. doi: 10.1108/00483480710731310

Griffith, R. L., and Converse, P. D. (2012). The rules of evidence and the prevalence of applicant faking. In Ziegler, M., MacCann, C., and Roberts, R. (Eds.), *New Perspectives on Faking in Personality Assessment* (pp. 34-52). New York: Oxford University Press.

King, R.E., Retzlaff, P.D., Detwiler, C.A., Schroeder, D.J., Broach, D. (2003). Use of personality assessment measures in the selection of air traffic control specialists. DOT/FAA/AM-03/20. Washington, DC: Federal Aviation Administration Office of Aviation Medicine

Luuk, K., Luuk, A., & Aluoja, A. (2009). Predicting professional success of air traffic control personnel from their personality profile at admission to Ab Initio training. *The International Journal of Aviation Psychology, 19(3),* 235–251. doi:10.1080/10508410902983896

Mislevy, R.J., Corrigan, S., Oranje, A., DiCerbo, K., Bauer, M.I., von Davier, A., & John, M. (2015). Psychometrics and Game-Based Assessment, in F. Drasgow (Ed.), *Technology and Testing: Improved Educational and Psychological Measurement*. United Kingdoms: Routledge.

Nickles, B.J., Bobko, M.D., Sands, W.A., & Tartak, E.L. (1995). Separation and Control Hiring Assessment (SACHA) final job analysis report, Bethesda, MD: University Research Corporation

Reiners, T., & Wood, L.C. (2015). Gamification in Education and Business. Springer International Publishing. doi: 10.1007/978-3-319-10208-5

Robinson, M., & Tamir, M. (2005). Neuroticism as mental noise: A relation between neuroticism and reaction time standard deviations. *Journal of Personality and Social Psychology, 89(1),* 107– 114. doi: 10.1037/0022-3514.89.1.107

Roe, R.A., Oprins, E., and Geven, E. (2012). The predictive validity of personality tests in air traffic controller selection. Proceedings of the 30th European Association for Aviation Psychology Conference.

Schroeder, D. J., Broach, D., & Young, W. C. (1993). Contribution of personality to the prediction of success in initial air traffic control specialist training. DOT/FAA/AM93/4. Washington, DC: FAA Office of Aviation Medicine.

Suresh, A., Ramachandran, K., & Srivastava, A. (2012). Personality Based Job Analysis of Air Traffic Controller. *Indian Journal of Aerospace Medicine. 56.* 21-31.

Ventura, M. & Shute, V. (2013). The validity of a game-based assessment of persistence. *Computers in Human Behavior. 29.* 2568-2572. doi: 10.1016/j.chb.2013.06.033.

Wichmann, T., Buchheim, A., Menning, H., Schenk, I., George, C., & Pokorny, D. (2016). A reaction time experiment on adult attachment: The development of a measure of neuro-psychological settings. *Frontiers in Human Neuroscience.* doi: 10.3389/ fnhum.2016. 00548

THE ROLE OF A GROUP ASSESSMENT CENTRE IN THE SELECTION OF AB INITIO AIR TRAFFIC CONTROLLERS

Jennifer Eaglestone
Netherlands Aerospace Centre (NLR) previously Air Traffic Control the Netherlands (LVNL)
Amsterdam, The Netherlands
Alexander W. J. Arnold
Air Traffic Control the Netherlands (LVNL)
Schiphol, The Netherlands

The selection of Air Traffic Controllers (ATCOs) is known to be extensive and highly selective. Air Traffic Control the Netherlands (LVNL) has a six-stage procedure for ab initio applicants. It includes two rounds of cognitive ability testing, personality testing, two job sample tests, an interview, and an assessment centre (AC). This study examines the construct- and predictive validity of the AC using 15 dimension-scores as predictor variables, with the final score on a high-fidelity job sample test as the dependent variable (criterion). A Principal Component Analysis (PCA) of the 15 dimension-scores resulted in two components, one (inter)personal behaviour component and one performance component. The performance component was found to have a stronger correlation with the cognitive ability tests than the (inter)personal behaviour component. Eight of the 15 AC predictor variables had a significant positive correlation with the high-fidelity job sample test.

Air Navigation Service Providers (ANSPs) are known to use extensive selection procedures before deciding which applicants can start training to become an Air Traffic Controller (ATCO). Training for ATCOs is long and expensive and, as with all such professions, any reduction in attrition rates means costs will be significantly reduced (Martinussen et al. 2000). Psychological assessment for Air Traffic Control (ATC) applicants has taken place since the early 60s, however up until the 90s it focussed mainly on cognitive attributes (Hättig, 1991). Job analyses have however underlined the importance of personality-related worker attributes in ATC with ATCO performance not only depending on technical knowledge and cognitive ability, but also on characteristics related to personality (Wium & Eaglestone, 2021).

Personality related factors are most often assessed using personality questionnaires and interviews. Damitz et al. (2003) however pointed out that these methods probably only assess personality related aspects of performance by evaluating signs of behaviour, as opposed to assessing actual behaviour. It assumes that how a candidate describes him or herself is predictive of the way that he or she will actually act. A broader approach, in which actual behaviour is also observed, would complement evaluation using a questionnaire and/or interview.

Assessment centres have been used since the 1950s to assess behaviour (Cascio and Silbey, 1979) and by the mid-70s assessment centres were, according to Mitchel (1975), being used widely within industry for the selection of managers. The use of assessment centers in selection in aviation is, however, comparatively new. In 1994 the German Aerospace Centre (DLR) developed one of the first assessment centres for pilot selection (Hörmann et al., 1997) and in 1996 an assessment centre was also added to the ATCO selection at DLR (Pecena, 2003).

In 2005 Air Traffic Control the Netherlands (LVNL) started using an assessment centre (AC) as part of their selection procedure for ab initio ATCOs. The AC in its current form has been used since 2014 and is the fourth stage of ATCO selection. In total candidates take part in six selection stages (hurdled approach), comprising of two rounds of cognitive ability testing, a low-fidelity job sample, the AC, an interview combined with a personality questionnaire and lastly a high-fidelity job sample test.

Aim and hypothesis

This study aims to undertake a step in the validation (predictive and construct validity) of the assessment centre as a predictive selection tool for the selection of ab initio ATCOs at LVNL. Even though assessment centres can be valid predictors of job success (e.g., Hermelin et al., 2007) the construct validity of assessment centres has often been questioned (e.g., Sackett & Dreher, 1982). In 1990, Shore suggested that construct validity might be increased if single rating items were grouped into wider categories (Shore et al., 1990). He felt that assessors could probably only distinguish between interpersonal related behaviour and performance related behaviour and not between a vast number of competences or behaviours.

More recently Damitz et al. (2003) also stated that ratings of single behavioural dimensions have only low convergent and discriminant validity. While validating an assessment centre for pilot selection, Damitz et al. found that a Principal Component Analysis (PCA) clearly showed an interpersonal component and a performance related component in the assessment centre ratings. In this study we therefore expect the 15 Assessment Centre (AC) dimensions to consist of two components, an interpersonal and a performance related component (hypothesis 1a). Those AC dimensions in the performance (task-related) component should be more reliant on cognitive ability, whereas interpersonal behaviour is considered distinct from cognitive ability. Therefore, the AC dimensions in the performance component are expected to have a higher correlation with the cognitive aptitude tests than the AC dimensions in the interpersonal component (hypothesis 1b).

A lack of standardised training results that could be used as a criterion (due to multiple changes in training) fuelled the decision to use the LVNL high-fidelity job sample test (ACT) as the dependent measure when assessing predictive validity in this study. Not only does this job-sample have a high resemblance to actual ATCO training (taking place in the training simulator) but it also an expensive and time-consuming selection round, making it a necessity to only allow candidates with a high chance of passing to take part. The ACT job sample test is taken by applicants in the last stage of selection and candidates are taught to control traffic in an approach simulator for three days. After the three days they are scored by ATCO instructors on both behaviour and aptitude in the simulator. The 15 assessment centre dimension scores are expected to have a positive significant correlation with the final score on the ACT job sample test (hypothesis 2).

Method

Sample

This study was conducted using data from ab initio applicants for air traffic controller training at LVNL. The sample contained 1158 candidates who took part in the AC between October 2014 and July 2019. The average age is 23.4 years ($SD = 3.4$) with a range from 17 years to 31 years. A total of 18.4% of the sample is female. Of these 1158 candidates, 160 took part in the ACT job sample test. These were candidates that had passed all previous selection stages including the assessment centre. Average age for this group is 24.2 (*SD* = 3.3) years with a range from 17 years to 30 years (16.4% of the sample is female). ACT job sample data was collected between October 2016 and July 2019. This sample was used for the analysis of predictive validity.

Predictors and Criteria

The LVNL assessment centre consists of three exercises: two group discussions (in groups of 4 candidates) and one individual exercise. Each of which is scored using the same 15 item score-form. All items are scored on a 4-point scale by a psychologist, whereby a maximum of 2 exercises per candidate are scored by the same psychologist. The three scores for each of the items were summed to create 15 dimension-scores (predictor variables). An overall score was also calculated for the assessment centre.

Performance on the LVNL high fidelity Job Sample test (ACT) is the dependent variable (criterion) in this study. Candidate performance during the ACT job sample was scored by ATCOs using 19 items (for example, planning, decisiveness, information processing) on a 6-point scale. The scores on the 19 items are summed to create the ACT final score. The dependent variable consisted of this final score.

In conjunction with Hypothesis 1b, scores from cognitive ability tests (round 2 of selection) were used. This selection round uses four subtests of the First European ATC Selection Test (FEAST) battery (Rathje, 2004) namely Heading and Range (SAHR), ATC Planning (SAP), Sorting Ability (SORT) and Visualisation (FOLD). These tests are administered with a cut-off per test and all subtests must be passed to be able to take part in the assessment centre.

Analyses

In order to investigate the construct validity of the AC, a PCA was used with varimax rotation and a Kaiser normalization. The analysis was conducted using the 15 final dimension scores. Furthermore, the dimensions in each of the components resulting from the PCA were summed to create a competence score and then correlated with the raw FEAST subtest scores from the second round of selection. To assess the predictive validity, the 15 assessment dimension-scores were correlated with the ACT job sample final score. SPSS v25 was used for all analyses.

Results

To assess hypothesis 1, a PCA with varimax rotation and Kaiser normalisation was used. A two factor solution was chosen based on an inspection of the scree plot and eigenvalues (> 1.0) accounting for 49.9% of the variance. The results are presented in Table 1. Component 1 is composed of not only dimensions that would be categorised as interpersonal but seems to be somewhat broader and focussed on personal behaviour in general. Component 2 contains the more task-related dimensions and can indeed be classified as a performance component.

Processing speed however does not fit this pattern entirely. It scores relatively high on both components, with the highest loading on the (inter)personal component.

Table 1
PCA of Assessment Centre dimensions

	Component	
Assessment Centre dimensions	1	2
Presence	.87	
Initiative	.85	
Assertive	.75	
Decisiveness	.68	
Flexibility	.48	
Accountability	.40	
Teamwork	.39	
Composure	.44	
Concentration	.45	.58
Argument effectiveness		.66
Processing speed	.55	.43
Solution effectiveness		.66
Prioritising		.67
Accuracy		.78
Structure		.75

Note: Factor loadings < 0.35 have been omitted for clarity.

To assess hypothesis 1b, two final assessment centre ratings were calculated for the two PCA components by summing the related dimension scores, creating an (inter)personal competence score and a performance competence score. These were correlated with the four cognitive ability subtests to analyse the relationship between cognitive ability and assessment centre performance. As the assessment centre comes after the cognitive ability tests in the selection procedure the sample was severely restricted in range. Table 2 shows the (uncorrected) correlations. For the SORT and FOLD subtests the correlation is significantly higher for the performance competence than for the (inter)personal competence.

Table 2
Correlations of assessment centre scores with the FEAST cognitive ability subtests

	FEAST cognitive ability subtests			
	SAHR	SAP	SORT	FOLD
(Inter)personal behaviour competence (1)	.04	.02	.03	-.07*
Performance competence (2)	.08*	.08*	.13**	.08*

Note: $n = 832$. *$p < .05$,**$p < .01$. Difference $r_{1,\text{ subtest}}$ and $r_{2,\text{subtest}}$ is sig. for SORT* and FOLD**

To study the predictive validity of the assessment centre (hypothesis 2) the overall AC rating and an (Inter)personal and Performance rating were correlated with the dependent variable – the ACT job sample final score. The three overall ratings all correlated significantly with the ACT final score. Furthermore eight of the 15 AC dimensions showed a significant correlation with the ACT job sample final score. Results are presented in table 3.

Table 3
Correlations between Predictor and Criterion Measures (ACT job sample final score)

	r (uncorr.)		r (uncorr.)
Overall ratings		*Dimension ratings*	
Overall Assessment centre rating	.33**	Presence	.16*
(Inter)personal	.17*	Initiative	.09
Performance	.32**	Assertive	.09
		Decisiveness	.08
		Teamwork	.08
		Flexibility	.07
		Accountability	.06
		Concentration	.27**
		Composure	.20**
		Argument effectiveness	.25**
		Processing speed	.09
		Solution effectiveness	.24**
		Prioritising	.29**
		Accuracy	.25**
		Structure	.20**

Note: $n = 155$, $*p < .05$, $**p < .01$(two-tailed test).

Discussion

The results of this study support earlier findings (Damitz et al. 2003, Shore et al. 1990) that assessors can distinguish between interpersonal behaviour and performance behaviour, although the interpersonal component here was somewhat broader and also encompassed worker-related personal behaviour such as composure, accountability and initiative that may not be considered *inter*personal. The exception in this study was the dimension 'processing speed' which had a higher loading on the (inter)personal component than on the performance component, while one could argue that it is a more task/performance related dimension. This could be a sign of the difficulty with operationalisation of this dimension, making it difficult to evaluate. What assessors seem to be observing here is something more personality related. Correlations between the cognitive ability tests and the performance and (inter)personal behaviour dimensions have provided us with some evidence of construct validity.

Predictive validity was shown by the significant correlation between the assessment centre scores and the ACT final score. Although the separate (inter)personal behaviour dimensions do not correlate significantly, the combined score of the (inter)personal competence does show a significant correlation with the ACT final score. Restriction of range here is also greater than for the performance competence as the assessment centre is followed by an interview rating very similar (inter)personal dimensions.

Altogether it seems that the assessment centre is a promising tool in the selection of ab initio ATCOs at LVNL, however it is important that a full validation is carried out using training and job performance criteria.

References

Cascio, W. F., & Silbey, V. (1979). Utility of the assessment center as a selection device. *Journal of Applied Psychology, 64*(2), 107–118. doi: 10.1037/0021-9010.64.2.107

Damitz, M., Manzey, D., Kleinmann, M. & Severin, K. (2003), Assessment Center for Pilot Selection: Construct and Criterion Validity and the Impact of Assessor Type. *Applied Psychology*, 52, 193-212. doi: 10.1111/1464-0597.00131

Hättig, H. J. (1991). Selection of air traffic control cadets. In R. Galand, and A. D. Mangelsdorff (Eds.), *Handbook of Military Psychology* (p. 115-129). New York, NY: John Wiley & Sons.

Hermelin, E., Lievens, F. & Robertson, I. T. (2007). The validity of assessment centres for the prediction of supervisory performance ratings: A meta-analysis. *International Journal of Selection and Assessment, 15*, 405-411. doi: 10.1111/j.1468-2389.2007.00399.x

Hörmann, H.-J., Manzey, D., Maschke, P., & Pecena, Y. (1997). Behavior-oriented assessment of interpersonal skills in pilot selection. In R.S. Jensen (Ed.), *Proceedings of the ninth international symposium on aviation psychology: Vol. 2* (pp. 1110-1115). Columbus, OH: The Ohio State University.

Martinussen, M., Jenssen, M., & Joner, A. (2000). Selection of air traffic controllers: Some preliminary findings from a meta- analysis of validation studies. Paper presented at the 24th EAAP conference (European Association for Aviation Psychology), Crieff, Scotland.

Mitchel, J. O. (1975). Assessment center validity: A longitudinal study. *Journal of Applied Psychology, 60*(5), 573–579. doi: 10.1037/0021-9010.60.5.573

Pecena, Y. (2003). An assessment-center approach to ATCO selection – An evaluation study. In Proceedings of the second EUROCONTROL Selection Seminar (pp. 201-208). Luxembourg, 16.-18.Nov. 2003.

Rathje, H., Golany, Z., and Eißfeldt, H. (2004). Pan-European Selection Test Battery for Air Traffic Control Applicants. In K. Goeters (Ed.), *Aviation Psychology: Practice and Research* (p. 171 – 201). Aldershot: Ashgate

Sackett, P. R., & Dreher, G. F. (1982). Constructs and assessment center dimensions: Some troubling empirical findings. *Journal of Applied Psychology, 67*(4), 401-410. doi: 10.1037/0021-9010.67.4.401

Shore, T. H., Thornton, G. C., & McFarlane Shore, L. M. (1990). Construct validity of two assessment center dimension ratings. *Personnel Psychology*, *43*, 101-116.

Wium, J., & Eaglestone, J. (2021). Job analysis for air traffic control: A review of the literature from 1949-2020. Unpublished manuscript.

MEASURING MOTIVATION IN PILOT TRAINING APPLICANTS

John D. Trent
United States Air Force
Air Force Personnel Center
San Antonio, Texas, USA

Mark R. Rose
United States Air Force
Air Education and Training Command
San Antonio, Texas, USA

This study evaluated applicant motivation in pilot selection. Traditional measures of motivation have moderate to low predictive validity in flight training and often show large subgroup differences. The purpose of this study was to evaluate a self-report measure of motivation by examining 1) dimensionality, 2) reliability, 3) predictive validity, 4) construct validity, 5) validity based on known groups, and 6) subgroup differences. In a sample of 16,911 pilot applicants, a composite score correlated $r = .37$ with success in flight training and provided predictive validity beyond current measures of motivation. Differences between subgroups were small to medium. Thus, a self-report measure may be a better indicator of motivation than measures that come at a substantial financial cost (e.g., flight hours). As a result, it may be possible to improve pilot selection, decrease training costs, and make training more accessible to a wide range of applicants.

Motivation has been theorized to be one of three essential determinants of performance, along with declarative and procedural knowledge (Campbell, Gasser, & Oswald, 1996). Consistent with the importance of the concept, there has been a profusion of work motivation theories over the past 50 years (Muchinsky & Howes, 2019), and vast research documenting positive relations of motivation to performance in educational (Conti, 2000; Tanaka & Yamauchi, 2001) and work (Joo et al., 2010; Li et al., 2015) settings.

Nonetheless, as it pertains to training for a pilot career, there have been few empirical attempts to understand motivation. One exception is research that examined whether students who were more internally motivated would be more successful in flight training (Frederick-Recascino & Hall, 2003). The study, which evaluated 193 archival student records, found that student motivation operationalized as number of times a student cancelled their flight lessons, accounted for a significant amount of variance in flight performance, measured through number of lessons required to graduate, performance on written exams, and grade-point average (Frederick-Recascino & Hall, 2003).

Other exceptions have been attribute rating and biodata studies. For example, achievement motivation was ranked highest out of 27 cognitive and non-cognitive attributes by Air Force fighter pilots for relevance to major tasks (Carretta et al., 1993). Also, biodata has had a long history of success in predicting pass/fail in flight training. For example, Henry (1966, as cited in Hough, 1988) reported that the item "Did you ever build a model airplane that flew?" was almost as good a predictor of pilot training success as the entire Air Force test battery. Use of biodata has apparently been curtailed in US pilot selection due to issues such as differential prediction by sex (Damos, 2011).

Finally it has been proposed that measures of aviation knowledge and flying hours may function in part as indicators of general interest and motivation in aviation. For example, because the U.S. Air Force's aviation knowledge test is administered pre-accession, before undergoing any required training, it may measure variance that can be attributed to motivation. Applicants with high motivation for a pilot career may be more likely to actively pursue opportunities to learn about motivation and dedicate time to self-study in preparation for the test (Barron, Carretta, & Rose, 2016).

The purpose of the current study is to examine preliminary evidence for validity of a measure of motivation to become a pilot by examining 1) evidence of dimensionality in a group of pilot applicants, 2) reliability estimates by dimension, 3) evidence for predictive criterion-related validity, 4) evidence for convergent and discriminant validity, 5) evidence for validity based on known groups, and 6) subgroup group differences for males and females as compared to measures sometimes assumed to be indicators of motivation – aviation knowledge and flying hours. Increased insight into the construct validity and impact of motivation for a pilot career can help to improve the validity and fairness of selection systems used for pilots, critical factors given the substantial cost of pilot training and widely recognized need for increased diversity among pilots.

Method and Results

Participants in this study were 16,911 applicants being considered for at least one of four U.S. Air Force flying careers, who had completed a survey designed to measure motivation for a pilot career. Participants had varying amounts of data on outcome and other measures used to evaluate reliability and validity of the measure, and subgroup differences. The source of items of motivation was the 48 item Work Interest Inventory, a survey initially comprised of 37 items intended to measure motivation for unmanned aircraft pilot careers (Paullin et al., 2011) and later expanded with intentions to measure motivation for manned aircraft pilot careers (Barron et al., 2015).

Initial analyses focused on identifying underlying dimensions of the measure using exploratory factor analysis, and conducting internal consistency reliability analyses of the dimensions. Results showed that the Work Interest Inventory measures seven dimensions with coefficient alpha reliability estimates that ranged from .61 to .91 (Table 1).

Table 1.
Factor Analysis of Work Interest Inventory Items.

Prompt: Rate how important each job characteristic is to your ***ideal job***.
Scale: 1 = *This is something I would actively try to avoid in a job* to 5 = *This is something I would actively seek out as part of a job.*

Dimension	Description
Performance Under Pressure ($\alpha = .91$)	Seeks out work that involves multitasking and working under high stress conditions.
RPA-Specific Working Conditions ($\alpha = .81$)	Favors a work context and tasks typical for RPA operators (e.g., focused on reconnaissance).
Manned-Aircraft-Pilot-Specific Tasks ($\alpha = .82$)	Seeks a career that involves using skills of a manned aircraft pilot.
Competitive/Independent ($\alpha = .66$)	Seeks a career that provides opportunities to compete with other and make independent decisions.
Sociable ($\alpha = .75$)	Seeks out work that allows for interaction with coworkers.
Cutting Edge Technology to Protect ($\alpha = .61$)	Seeks out work that involves using cutting edge technology to help others.
Lethal Action ($\alpha = .65$)	Seeks a career that involves application of lethal force.

We then conducted logistic regression analysis, regressing manned aircraft pilot training success on the extracted dimensions. Dimensions with theoretical and empirically stronger relationships to training success were used to compose the preliminary measure of motivation for a manned aircraft pilot career. We next evaluated the measure's convergent and discriminant validity, using other components of the Air Force pilot selection test battery, and examined subgroup differences.

Validity analyses showed that three dimensions in particular, labeled Manned-Aircraft-Pilot-Specific Tasks, Sociable, and RPA-Specific Working Conditions, were significant predictors and a unit weighted composite score correlated .37 with pilot training success. This composite score also correlated .40 with a one-item measure of interest in a manned aircraft pilot career, .46 with pilot training success when combined with the interest measure (Table 2), and predicted success beyond flying hours and aviation information.

Table 2.
Means, Standard Deviations, and Correlations of Variables.

Variable	Mean	*SD*	1	2	3	4	5	6
1. Pilot Motivation (PM)	6.36	8.67	1.00					
2. Career Interest (CI)	0.87	0.34	.40**	1.00				
3. Flying Hours	2.40	3.15	.25**	.21**	1.00			
4. Aviation Knowledge	11.91	4.17	.23**	.15**	.56**	1.00		
5. PM + CI	1.37	1.78	.84**	.83**	.28**	.23**	1.00	
6. Training P-F	0.87	0.33	.37**	.39**	.25**	.29**	.46**	1.00

Finally, subgroup differences for groups based on gender, race, and ethnicity were small to medium (Table 3).

Table 3.
Cohen's d Values for Variables by Subgroup.

Variable	Male - Female	Black/African American - White	Hispanic-Non-Hispanic
Pilot Motivation	.16	.42	.12
Flight Hour Code	.23	.16	.21
Aviation Knowledge	.66	.53	.30

Discussion

These results have several implications. First, the measure of motivation appears to provide substantial incremental validity beyond other motivation-relevant U.S. Air Force selection tools. Similarly, a relatively simple self-report measure of motivation may be a better indicator of motivation than measures that come at a substantial financial cost to applicants (e.g., flying hours), especially when paired with a simple measure of interest in a pilot career. As a result, it may be possible to improve selection of applicants into training, decreasing costs related to attrition and making training more accessible to a wide range of applicants. Further study is needed to determine the extent to which these results generalize to future applicants.

Acknowledgements

The views expressed in this paper are those of the authors and are not necessarily those of the U.S. Government, Department of Defense, or the U.S. Air Force.

References

Barron, L. G., Carretta, T. R., & Rose, M. R. (2016). Aptitude and trait predictors of manned and unmanned aircraft pilot job performance. *Military Psychology, 28*, 1-16.

Barron, L., Rose, M. R., & Burns, G. N. (2015). *Initial validation of the Work Interest Inventory for Remotely Piloted Aircraft (RPA) career fields.* In J. Wood (Chair), UAV Selection and Assessment. Symposium presented at the International Symposium on Aviation Psychology, Dayton, OH.

Carretta, T. R., Rodgers, M. N., & Hansen, I. (1993). *The identification of ability requirements and selection instruments for fighter pilot training.* (AL/HR-TP-1993-0016). Brooks AFB, TX: Armstrong Laboratory.

Campbell, J. P., Gasser, M. B., & Oswald, F. L. (1996). The substantive nature of job performance variability. In K. R. Murphy (Ed.), *Individual differences and behavior in organizations* (pp. 258-299). Jossey-Bass.

Conti, R. (2000). College goals: Do self-determined and carefully considered goals predict intrinsic motivation, academic performance and adjustment during the first semester? *Social Psychology of Education, 4,* 189–211.

Damos, D. (2011). *KSAOs for military pilot selection: A review of the literature* (Rep. No. AFCAPS-FR-2011-0003). Randolph AFB, TX: HQ AFPC/DSYX Strategic Research and Assessment Branch.

Frederick-Recascino, C. M., & Hall, S. (2003). Pilot motivation and performance: Theoretical and empirical relationships. *The International Journal of Aviation Psychology, 13*, 401-414.

Hough, L. M. (1988). *Literature review: Utility of temperament, biodata, and interest assessment for predicting job performance.* (ARI Research Note 88-02). U.S. Army Research Institute for the behavioral and Social Sciences.

Joo, B.K., Jeung, C.W., & Yoon, H. J. (2010). Investigating the influences of core self-evaluations, job autonomy, and intrinsic motivation on in-role job performance. *Human Resource Development Quarterly, 21*, 353-371.

Li, Y., Wei, F., Ren, S. and Di, Y. (2015). Locus of control, psychological empowerment and intrinsic motivation relation to performance, *Journal of Managerial Psychology, 30*, 422-438.

Muchinsky, P., & Howes, S. (2019). *Psychology applied to work* (12th ed.). Hypergraphic Press.

Paullin, C., Ingerick, M., Trippe, D. M., & Wasko, L. (2011). *Identifying best bet entry-level selection measures for US Air Force remotely piloted aircraft (RPA) and sensor operator occupations* (FR-11-64). HQ AFPC/DSYX Strategic Research and Assessment Branch.

Tanaka, A., & Yamauchi, H. (2001). A model of achievement motives, goal orientations, intrinsic interest and academic achievement. *Psychological Reports, 88,* 123–125.

USING A PERCEPTUAL SPEED TEST TO PREDICT FLIGHT TRAINING PERFORMANCE: NEW FINDINGS

Diane L Damos, Ph.D.
Damos Aviation Services, Inc.
Gurnee, IL
Alvah C Bittner, Ph.D., CPE
Bittner and Associates
Kent, WA
Mary Niemczyk, Ph.D.
Arizona State University
Mesa, AZ

Perceptual speed tests have been successfully used for US military pilot selection for 80 years yet are rarely used by other militaries or for civilian pilot selection. This paper describes a study examining the predictive validity of a perceptual speed test, the Tabular Speed Test (TST), for civilian flight training. The TST was administered to university students as they entered a professional pilot curriculum. The number correct (COR) correlated significantly with time to solo ($N = 119$, $r = -.15$, $p = .05$) and with time to private pilot's certificate (N = 51, $r = -.34$, $p = 0.01$) but not with graduation/elimination from the pilot curriculum. The number of incorrect responses (WR) did not correlate with any performance measure. Average test-retest reliability from 5 to 17 months was $r = .604$ for COR and $r = .431$ for WR, $p < .001$ for both.

The first use of a perceptual speed test for pilot selection occurred in December 1942 when the Dial Reading and Table Reading Tests were included in the US Army Air Forces aircrew classification battery (Guilford & Lacey, 1947). The Dial Reading Test was eventually dropped, leaving the Table Reading Test as the only perceptual speed test. This test is still in the US Air Force pilot selection battery (Carretta & Ree, 1995; Johnson, Barron, Carretta, & Rose, 2017). Despite the well-documented usefulness of a perceptual speed test for over 75 years, only the Norwegian Air Force includes a perceptual speed test in its pilot selection battery (Martinussen & Torjussen, 2004).

Like their military counterparts, civilian pilot selection batteries rarely include a perceptual speed test. One exception to this is the German Aerospace Center (DLR) selection battery that was developed for civilian ab initio pilot selection (Goeters, Hoermann, & Maschke, 1989). Another exception is the selection battery for entrance into the professional flying curriculum of the Artic University of Norway (M. Martinussen, personal communication, February 22, 2021).

In 2008, Mount, Oh, and Burns (2008) demonstrated that, for perceptual speed tests, the number correct (COR) and the number wrong (WR) assess different attributes. They suggested that, whereas COR assesses speed of processing, WR reflects an inability to focus on the task at hand and a tendency towards "distractibility, carelessness, recklessness, or apathy on the job (p.118)". They suggested that WR may predict problems with rules compliance, which would be manifested in "accidents, safety violations, tardiness, and use of alcohol or drugs on the job (p. 118)." If this were true, then pilot selection batteries should include a perceptual speed test, and both WR and COR should be included in any selection decision.

This study had two major goals. The first was to confirm the Mount et al. results that COR and WR measure different attributes. We sought to confirm Mount et al.'s results in three ways. First, we obtained test-retest data at three different intervals. If COR and WR measure stable attributes, they should demonstrate significant reliabilities with little decrease over time. Second, assuming that the COR and WR measure different attributes, their correlation should be low. Third, again assuming that COR and WR assess different attributes, they should correlate with different behaviors.

The second goal was to determine the predictive validity of COR and WR. US civilian ab initio training has three major milestones: time to solo, time to the private pilot certificate (PPC), and program completion. This study evaluated the predictive validity of COR and WR for all three milestones.

Approach

A perceptual speed test, the Tabular Speed Test (TST), was administered each semester beginning with the fall semester of 2005 through the spring semester of 2008 to all students enrolled in the aviation program of a large western university. The college offered three majors: professional flight (pilot), air traffic control, and aviation management. To obtain retest data, the test also was administered once during this 3.5-year period in an advanced course typically taken by third- and fourth-year professional flight and aviation management students and once during a departmental safety meeting that was mandatory for all flight students.

Mekhail, Niemczyk, Ulrich, and Karp (2010) obtained the time to solo and time to PPC data presented in this study. Our data analysis differs in that we limited our analyses to those students with known outcomes, i.e., those definitely identified as either having dropped out of the curriculum or graduated from the curriculum. We also analyzed WR, which they did not.

Mekhail et al. (2010) observed that the completion times reported by some students appeared to be estimates. During the time of our study, the department kept a student's flight records for only two years after the student either graduated or left the program. Thus, it was not possible to obtain a more accurate measure of time to solo or time to PPC for many of the students. Additionally, not all data were available for all students, i.e. a student might report time to PPC but not time to solo. These estimates and the unknown cause for dropping out of the flight curriculum resulted in greater random error than in a more controlled study. We adopted a one-tailed $p < .10$ level of statistical significance rather than the more common, two-tailed $p < .05$ (Wickens, 2015) for testing the three milestones because of the greater level of random error and because prior studies provided directional expectations concerning the relation between the three milestones and the two dependent measures.

Methods

Subjects

The TST was administered to all students taking the two classes described earlier and attending the departmental flight safety meeting. Participation was voluntary; students who did not want to participate did not complete the test form. Students were given no financial incentives to participate nor did they receive any additional course credit.

Tabular Speed Test

The TST is a paper-and-pencil, 50 question, multiple-choice test. The test has a 9-minute time limit. The TST has two parallel versions (equivalent means, standard deviations, and distributions), both of which were used.

Administration

The TST forms were distributed at the beginning of class or at the beginning of the safety meeting. Approximately equal numbers of each version were distributed during each testing session. The test administrator started and stopped the testing period but provided no information on the time remaining during the test. None of the test administration classrooms had clocks. The test administrator did not control which version of the TST a testee received for the retest. All test administrations were conducted by the third author.

Results

Although all students enrolled in the aviation program took the introductory class, only data from students enrolled in the professional flight curriculum were analyzed. Both COR, time to solo, and time to PPC had normal distributions. WR, however, was skewed and leptokurtic. A *ln* (WR+1) transformation resulted in acceptable distributions for all of the analyses. Unless otherwise indicated, all analyses were conducted using Excel 365.

COR Versus WR

Test-retest. Because of differences in student schedules, the retest interval varied. Three intervals had sufficient data to allow reliability calculations: 5 months, 12 months, and 17 months. Data for the 5-month interval was obtained from students who took the retest the following semester during the departmental safety meeting or who repeated the introductory class because they dropped the class or failed it. Data for the 12- and 17-month interval came from students who took the retest during the safety meeting or during the advanced course. The results shown in Table 1 are Pearson Correlations.

Table 1.
Test-retest reliabilities as a function of testing interval

Testing interval in months	N	COR	Ln(WR+1)
5	41	.70****	.29
12	24	.45*	.53**
17	23	.64***	.45*

** **p<.0001, *** p< .001. ** p< .01, * p< .05

If COR and *ln* (WR+1) test-retest reliabilities represent attributes that are stable across time, then their respective reliabilities should be essentially constant over time. If they assess attributes that are affected by experience, then their reliabilities should decrease over time. Table 1 does not show either pattern clearly, though there is a slight suggestion of a break after 5 months. Consequently, we sequentially tested (Lenhard & Lenhard, 2014) both the Olkin-Pratt averaged 12-17 month COR correlation against the 5 month correlation after we tested the 12 versus 17 month correlation; neither contrast was significant ($p > 0.10$ for both). This process was repeated for the *ln* (WR +1) correlations, and again neither contrast was significant ($p > 0.10$). Finally, we averaged the respective correlations using the Olkin and Pratt (1958) weightings. The averaged reliabilities were .604 for COR (weighted N estimate = 86, $p < 0.00001$) and .431 for *ln* (WR+1) ($p < 0.0001$).

Correlation between COR and WR. To calculate the correlation between COR and WR, first-administration data were used. The correlation was $r = -.19$, $N = 144$, $p = 01$.

Performance Measures

Time to solo. The first major milestone in ab initio flight training is time to solo. We performed directional tests on COR and WR because we wished to determine if time to solo decreased as COR increased (negative relation) and increased as WR increased (see Bittner, Bittner, Winn, & Lundy, 2004 for a discussion of the use of directional testing). The correlation between COR and time to solo was $r = -0.15$, $N = 119$, $p = 0.05$. The corresponding value for *ln* (WR+1) was $r = 0.01$, $N = 119$, $p = 0.46$.

Time to PPC. The second major milestone is time to PPC. Again, we used directional testing to determine if time to PPC decreased with increasing COR (negative relation) and increased with increasing WR. For this milestone, the correlation between time to PPC and COR was $r = -.34$, $N = 51$, $p = .01$. For *ln* (WR+1) the corresponding values were $r = -.06$, $N = 51$, $p = .34$.

Completion of program. The completion status of each student (graduated from the professional flight curriculum or dropped out) was obtained by examining the university's records. We obtained graduation data on 144 students. Of these, 94 completed the professional flight curriculum and 50 did not. We performed two directional (one-tailed) comparisons (Bittner et al., 2004). The first determined if COR for those who completed flight training was significantly greater than for those who dropped out. The second determined if WR was greater for those who dropped out versus those who completed the curriculum. *T*-tests for unequal variances conducted on COR and *ln* (WR +1) were both nonsignificant (t (90) = 0 .98, p = .16 and t (90) = .58, p = .28, respectively).

Discussion

This study had two goals. The first of these was to confirm Mount et al's (2008) finding that COR and WR measure different attributes. We used three different methods to confirm their results. The first was to demonstrate that COR and WR have significant and stable test-retest reliabilities over extended time intervals. Table 1 shows that COR had significant reliabilities up to 17 months after initial testing. WR showed a similar pattern although the reliabilities were generally lower but still statistically significant. The second method concerned the correlation between COR and WR. Mount et al. found a significant correlation of -.46 between these two variables. We used only the first administration data to make our results comparable to those of Mount et al. and found a significant correlation of -.19. The third was to contrast the behaviors predicted by COR versus WR. Mount et al. (2008) argued that COR measures performance, whereas WR measures compliance behaviors. In this study COR correlated significantly with two of the three performance measures, time to solo and time to PPL; WR correlated with none of the three performance measures. Because this study had no measures of compliance behavior, we cannot determine which attributes may be measured by WR. Consequently, we were only partially successful in supporting Mount et al.

The second goal of this paper was to determine the predictive validities of COR and WR for three milestones in ab initio student training: time to solo, time to PPC, and program completion. COR successfully predicted time to solo and time to PPC but failed to predict program completion. The department had no information on why a student dropped out of the professional flying program. Thus, we could not distinguish between those students who were failing, found that they disliked flying, or dropped out for financial reasons. The last administration of the TST occurred during the spring 2008 semester. By this time, the economic recession of 2008 had begun, which may have caused more students than usual to drop out for financial reasons.

In summary, COR predicted both time to solo and time to PPC, which supports the use of perceptual speed tests in civilian, as well as military, pilot selection batteries. The correlation between COR and WR and the constant test-retest reliabilities provide good support for Mount et al's (2008) findings that COR and WR assess different attributes. COR clearly assesses performance, but the attributes assessed by WR must await further research.

References

Bittner, A. C. J., Bittner, R. C., Winn, F. J., & Lundy, N. C. (2004). Theory-based directional testing: Revisiting a primary option for efficient human factors/ergonomics (HF/E) studies. In *Human Factors and Ergonomics Society 48th Annual Meeting* (pp. 2466-2470). Washington, DC: Human Factors and Ergonomics Society.

Carretta, T. R., & Ree, M. J. (1995). Air force officer qualifying test validity for predicting pilot training performance. *Journal of Business and Psychology, 9*(4), 379-388.

Goeters, K.-M., Hoermann, H.-J., & Maschke, P. (1989). The DLR test system ab initio pilot selection. In *Fifth International Symposium on Aviation Psychology* (pp. 663-668). Columbus, OH: Department of Aviation, the Ohio State University.

Guilford, J. P., & Lacey, J. I. (Eds.). (1947). *Printed classification tests Report No. 5*. Washington, D.C.: U.S. Government Printing Office.

Johnson, J. F., Barron, L. G., Carretta, T. R., & Rose, M. R. (2017). Predictive validity of spatial ability and perceptual speed tests for aviator training. *International Journal of Aerospace Psychology, 27*(3-4), 109-120.

Lenhard, W., & Lenhard, A. (2014). *Hypothesis Tests for Comparing Correlations*. Bibergau, Germany: Psychometrica. doi:DOI: 10.13140/RG.2.1.2954.1367.

Martinussen, M., & Torjussen, T. M. (2004). Initial validation of a computer-based assessment battery for pilot selection in the Norwegian Air Force. *Human Factors and Aerospace Safety, 4*(3), 233-243.

Mekhail, A., Niemczyk, M., Ulrich, J., & Karp, M. (2010). Using the Table Reading Test as an Indicator for Success in Pilot Training. *Collegiate Aviation Review, 28*(1), 101-114.

Mount, M. K., Oh, I., & Burns, M. (2008). Incremental validity of perceptual speed and accuracy over general mental ability. *Personnel Psychology, 61*, 113-139.

Olkin, I., & Pratt, J. W. (1958). Unbiased estimation of certain correlaton coefficients. *Annals of Mathematical Statistics, 29*, 201-211.

Wickens, C. D. (2015). Statistical errors in aviation psychology: Commonsense statistics in aviation safety research. In *18th International Symposium on Aviation Psychology*. Dayton, OH: Wright State University.

DECISION SUPPORT TOOLS
FOR THE COLLABORATIVE TRAJECTORY OPTIONS PROGRAM

Philip J. Smith, Amy Spencer and Mark Evans
The Ohio State University
Columbus OH

Bob Hoffman, Bert Hackney and Rafal Kicinger
Metron Aviation
Herndon, VA

The FAA currently makes frequent use of Flow Constrained Areas (FCAs) to thin traffic through some region of airspace by assigning departure delay to flights filed to fly through that airspace. An important potential future use of such FCAs is their integrated application within a Collaborative Trajectory Options Program (CTOP) Traffic Management Initiative (TMI). This paper reports the results of cognitive walkthroughs completed with ten recently retired traffic managers. These walkthroughs were designed to evaluate information and information display requirements, as well as other decision support requirements, for software to enable the creation of FCAs for a CTOP TMI.

In this paper we provide the results of a cognitive walkthrough designed to provide guidance on the design of an effective interface for access to the information and decision support tools necessary to plan a CTOP TMI. (See Smith et al., 2019 for a description of CTOP.) Broadly speaking, each FCA in CTOP allows traffic managers to control traffic rates to constrain the volume of aircraft through a certain airspace region. Through the use of multiple (typically adjacent) FCAs within a given CTOP (see Figure 1 at the end of this paper), the traffic managers can manage the traffic flows through these FCAs in a coordinated manner. The definition of a specific FCA includes filters that can limit the included flights by departure and arrival centers, time frame, altitude range, etc., as well as the geographic airspace transited.

To evaluate the information, information display and functional requirements necessary to support effective decision making when using a CTOP, this study focused on completion of a cognitive walkthrough focused on planning a CTOP TMI for a given day. For the cognitive walkthrough, the participants in a group that included experienced traffic managers from the relevant Enroute Centers viewed a storyboard together, evaluating a prototype interface design in terms of the following user interface elements: Filters, FCAs (location and types) and throughputs for five FCAs/FEAs that were included in this CTOP. The walkthroughs were completed over the internet, with participants viewing the storyboard using a shared display.

Step 1. Review of Weather on Date of Interest

The storyboard begins at 1500Z. To begin the walkthrough, the color codings used in the images for the weather displays (actual Vertically Integrated Liquid (VIL) and echo tops; forecast VIL and echo tops) were described. The traffic managers were told that, following a review of the actual and forecast weather, they would be asked to critique a presentation indicating how a hypothetical traffic manager could use the prototype design to manage a CTOP. They were further told that, when providing their critiques, they should assume that certain other TMIs were being used to move much of the other traffic out of this area. These other TMIs, which consisted of the actual TMIs in use on this date, were presented to them. They were then shown the forecast for the VIL and echo tops from 1500-2100Z in order to understand the forecast weather pattern that required some form of traffic management. (See Figures 1 and 2 at the end of this paper.)

Step 2. Review of CTOP

The traffic managers (who were already familiar with the use of FCAs from their past work experience) were then told that, for the next step, their goal was to evaluate the use of a CTOP TMI. They then were presented with a review describing the nature and design of CTOP initiatives.

Step 3. Walkthrough of Traffic Manager using CTOP

The participants were told that they would would be asked to critique the use of a CTOP by a hypothetical traffic manager to deal with the weather forecast from 1500-2100Z, as illustrated in Figures 1 and 2 at the end of this paper, in order to manage the traffic departing ZDC, ZJX, ZMA and ZTL and arriving ZAB, ZAU, ZDV, ZFW, ZHU, ZID, ZKC, ZLA, ZLC, ZME, ZMP, ZOA, ZOB and ZSE (i.e., all Centers except ZNY and ZBW). The traffic managers were reminded that they had already reviewed the plays and other required reroutes actually used on this day to move some of the traffic around the forecast weather. They were then presented with a walkthrough of a prototype illustrating the use of CTOP to manage the remaining traffic.

Participants

The primary group studied consisted of five recently retired traffic managers, one each from ZDC (Washington), ZID (Indianpolis), ZME (Memphis), ZTL (Atlanta) and ATCSCC (Air Traffic Control Systems Command Center). They had 12, 4, 29, 13 and 22 years of experience respectively as traffic managers in these facilities. The intention was to run a second session with five other traffic managers from the same facilities. However, due to scheduling and technical (internet) difficulties, it was not possible to conduct this session. We did, however, conduct the walkthrough on two additional dates, with other traffic managers representing the key facilities. One group (Group 2) had a traffic manager from each of ZID, ZME and ZTL with 2, 1 and 12 years of experience respectively; the other (Group 3) had traffic managers from ZDC and ATCSCC, with 14 and 10 years of experience respectively. Since these sessions did not represent the full set of relevant expertise, we simply report results from those sessions that provided insights that were significantly different from those of the Primary Group and that did not appear to be affected by the fact that the full complement of relevant traffic managers was not present.

Results and Discussion

Following the review of CTOP, the participants were shown the prototype design and its components were described. The left pane was used to specify filters for the FCAs; the upper right to show the current and forecast weather (VIL or echo tops); and the lower right to set and display the rates through FCAs in a CTOP. They also were shown the CTOP (with FCA locations and filters) as set up by a hypothetical traffic manager to manage the traffic flows for flights departing ZDC, ZJX, ZMA and ZTL and arriving ZAB, ZAU, ZDV, ZFW, ZHU, ZID, ZKC, ZLA, ZLC, ZME, ZMP, ZOA, ZOB and ZSE (i.e., all Centers except ZNY and ZBW).

Assessment of Filters

After reviewing the left pane for entering parameters for the filters, during the walkthrough the participants were each asked *individually* to respond to the following questions: Is this a reasonable set of filters? What would you add or delete from this set? Why? They were then given an opportunity to respond to what they heard from the other traffic managers.

Primary Group.

ZID: "The only thing I'd add is that you might want different tops for different FCAs. If a Citation can go to 41,000 over the tops, then you should let them."

ZTL: "The European traffic to Hartsfield needs to be picked up in the total count even though they are exempt from delay with these filters."

Groups 2 and 3.

ATCSCC: "I might leave ZDC out of the filter for departure centers and focus on the traffic to the Midwest with the CTOP. You could handle the rest of the traffic separately."

ZDC: "I'm ok with ZDC being included. There will be crossing flows, but that's the reality of what we live with. In addition, I'm going to be running EDC or TBFM for Chicago. The flights are going to get whacked twice, for CTOP and EDC or TBFM."

"The forecast doesn't show major input west of New York. I'm not worried about them. New York to Detroit or Chicago won't file into DC."

Assessment of FCA Locations and Types

Regarding the FCA locations and types, the interface to draw or edit the FCAs using the FCA Locations tab was not shown. Instead, the traffic managers were shown a pre-defined set of FCAs. Four FCAs (solid colored lines) and one monitoring Flow Evaluation Area or FEA (a dashed line segment) are shown in Figures 1 and 2. Note that the directional FCAs and FEA used for this example were designed so that they each capture different major flows focusing on flights from the Southeast to the Midwest, Northwest, West and Southwest that were likely to be used as initial routes or as alternative reroutes (FEA1A captures traffic routed through MEM; FCA1B traffic through BNA; FCA1C through IIU and FLM; FCA1D through the vicinity of BKW; and FCA1E for traffic crossing the line of weather further to the north, with each placed to capture traffic slightly north and east of the preceding one).

After viewing the forecast for a given one hour period, each traffic manager was asked to individually critique the FCAs shown relative to the forecast for that hour. Responses are indicated below for the 1600Z hour. (Results for other forecast times are in a full report.)

Primary Group:

ZID: "They look pretty good. Maybe move 1D a little northwest. Move it closer to the boundary, align it more north/south."

ZTL: "I don't have a problem. The FCAs run along my northern boundary. I think it's gonna give me some good numbers to look at. I'd make 1A an FCA right away because of the volume we'll want to move to the west of the weather."

ZDC: "The goal is to get rid of structured routes. The way 1D and 1E are set up, they're not in a good position. They're not capturing J6 New York traffic. New York/Boston can file through this weather. You need to draw 1D at the Indy/Cleveland/Washington Center boundary." "I would absolutely use a CTOP. You just have to define the FCAs right. CTOP won't solve the whole problem. It might be more efficient in some scenarios than others." "It gives you breathing room to solve problems more granularly and efficiently at the facilities."

ZME: "1A needs to be an FCA. Most of the airplanes aren't going to go through 1C. They will go around the edges. I'd extend 1A further south down into Fort Worth and then put an FEA down from there to the coast line. Every CTOP ought to have 1-2 monitoring FEAs at each end."

Group 2 and 3:

ATCSCC: "Don't filter by direction."

ZDC: "I'd put 1D along the Indy line and 1E along the Cleveland line."

ATCSCC: "Using Center or sector boundaries is a good idea because jet routes don't normally go along Center or sector boundaries."

ATCSCC: "I don't believe Memphis will buy off on an FEA. The planes will go around as tightly as they can and there is some weather in 1A."

ATCSCC: "We treated 1A as an FEA. It needs to be an FCA. We are going to hurt them badly. And we need an FEA at each end."

FCA Rates. The traffic managers again were shown the actual weather (VIL and echo tops) for 1500Z and the forecast weather for 1600Z and 1700Z (see Figures 1 and 2 for the one hour forecast of VIL and echo tops at 1600Z). They also were again shown the actual weather at 1500Z for the entire U.S. Then, they were each asked to individually write down the rates (% reduction relative to maximum capacity) that they would recommend for the second hour (1600-1700Z) based on the one and two hour forecasts (1600 and 1700Z). The results are shown in Figure 3 at the end of this paper, with the bold type indicating traffic managers with the most expertise for a particular FEA/FCA. (Similar results are available in a full report for the 3 hour and 5 hour forecasts.)

Note especially the differences in some of the recommended rates for traffic managers with expertise covering the same airspace (e.g., up to a 50% difference for the two traffic managers from ZDC for FCA1E).

Conclusion

At an abstract level, the feedback from the traffic managers generally indicated that, at a conceptual level, the traffic managers were comfortable with the use of the filters and the number and general locations of the FCAs presented and the use of a directional filter. Their feedback regarding FCA designs did indicate possible refinements in the locations of the FCAs, a desire to turn the FEA1A into an FCA and a desire to add FEAs at both ends of the CTOP. Important exceptions raised by some of the traffic managers, however, included recommendations to:

- Add required reroutes for specific FCAs.
- Draw FCAs on Center boundaries when possible to enhance coordination and communication within and across Centers.
- Include flights traversing the FCAs from north to south as well as from south to north to address southbound traffic.
- Use polygons or boxes instead of line FCAs to capture traffic in any direction.
- Use moving FCAs to deal with weather movement over time.
- Move the FCAs after 3-4 hours to deal with weather movement over time.

The individual differences in rate estimations of up to 50%, however, are a significant concern regarding the use of dynamic FCAs in a CTOP. Further research is needed to understand how to best determine rate reductions due to convective weather.

Acknowledgements

This research was funded by NASA Ames Research Center under the direction of Deepak Kulkarni and support from Paul Lee and Nancy Smith. We also are grateful to the traffic managers who participated in this study.

References

Smith, P.J., Evans, M., Spencer, A., Hoffman, R., Myers, T. & Hackney, B. (2019). Integrated Application of the Collaborative Trajectory Options Program. Proceedings of the 2019 Digital Avionics Systems Conference (DASC).

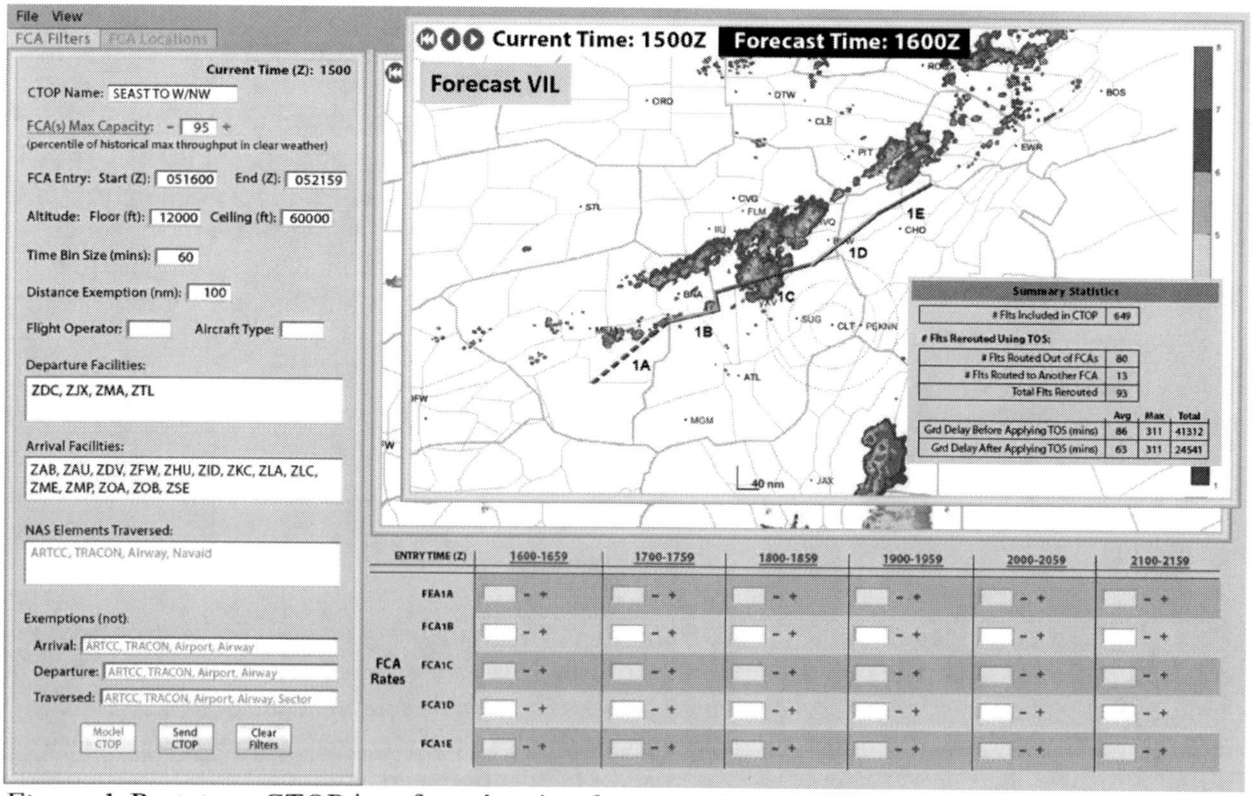

Figure 1. Prototype CTOP interface showing forecast VIL at 1600Z.

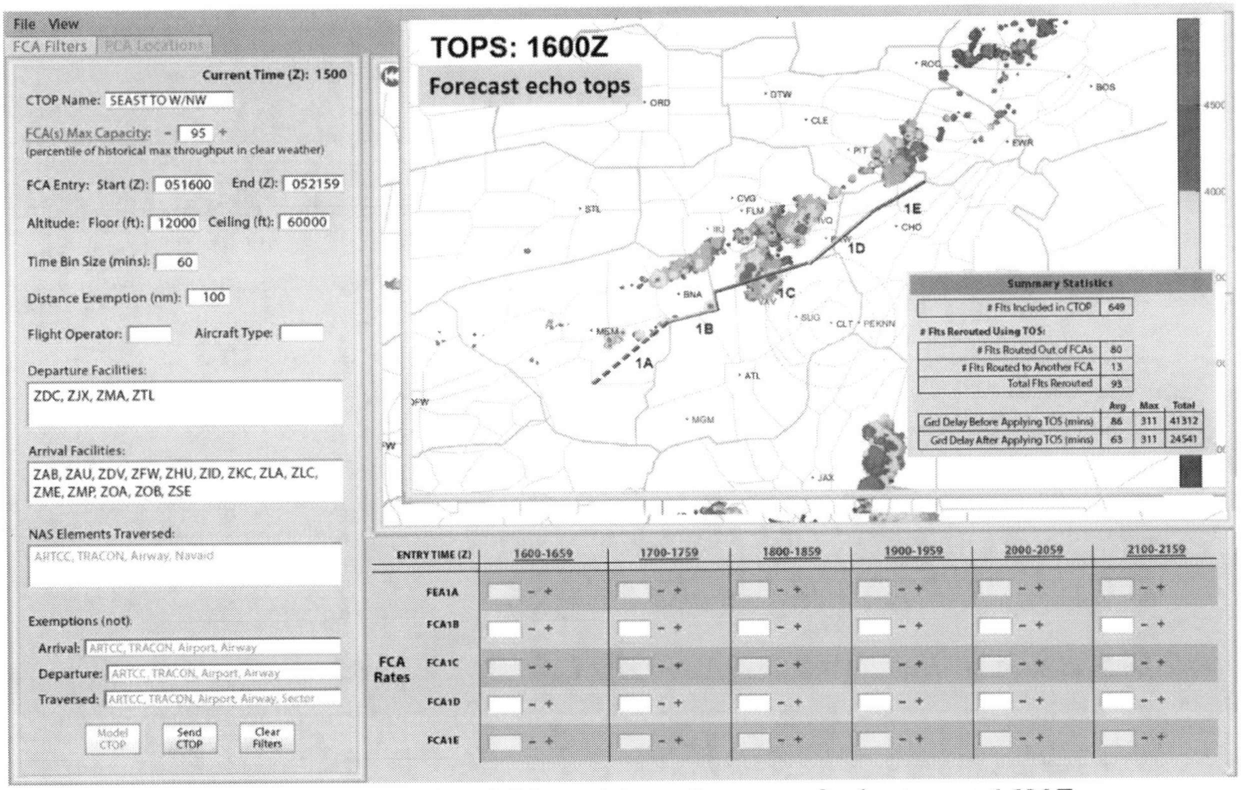

Figure 2. CTOP filters and associated FCAs: 1 hour forecast of echo tops at 1600Z.

Results

Set rates for FCAs for 1600Z hour (view 1 hour forecast and then write down individually on form). Then discuss.

Primary Group:

1600-1659Z:	FEA1A	FCA1B	FCA1C	FCA1D	FCA1E
ZDC	60%	70%	50%	**70%**	**40%**
ZID	40%	40%	**35%**	**50%**	60%
ZME	**80%**	**70%**	40%	40%	60%
ZTL	80%	**80%**	**40%**	50%	60%
DCC	80%	80%	50%	50%	80%

- ZTL: "Because of deviations, we don't want to go over 100% for 1A."
- ZDC: "1E is normally a departure corridor. It will be affected more because you've got flights lower in altitude."

Group 2:

1600-1659Z:	FEA1A	FCA1B	FCA1C	FCA1D	FCA1E
ZID	60%	40%	**30%**	**10%**	30%
ZME	**80%**	**50%**	50%	80%	80%
ZTL	80%	**75%**	**60%**	80%	70%

- *ZTL: "For FCA 1C, Atlanta and Indy would work together to find a number that works for both of us."*

Group 3:

1600-1659Z:	FEA1A	FCA1B	FCA1C	FCA1D	FCA1E
ZDC	90%	90%	65%	**65%**	**90%**
DCC	90%	80%	50%	40%	80%

- *And for 1A you easily have 30 miles for deviations that won't cause a workload issues, that won't require point outs."*
- *DCC: "I'm looking at those tops. It's not a solid line at 40,000. I'm allowing more room to pick through. There's so much room to handle some minor deviations."*

Figure 3. Rates generated by individual traffic managers (% reduction in throughput relative to maximum capacity) that they would recommend considering the actual weather at 1500Z, the one hour forecast weather at 1600Z and the two hour forecast (VIL and echo tops) for 1700Z.

AIRSPACE COMPLEXITY FOR PILOTS OPERATING IN HIGH-DENSITY TERMINAL AIRSPACE: NEW YORK CASE STUDY

Divya C. Chandra, Andrea Sparko, Andrew Kendra, and Janeen Kochan
USDOT Volpe Center
Cambridge, MA

Performance-based navigation (PBN) has been implemented in the redesign of terminal airspace across the National Airspace System (NAS). However, some locations, such as the New York metro area (NY), have not adopted PBN widely. Modernizing flight operations at high-density terminal airspace like NY is challenging, but also has the potential for significant operational benefits if successful. This research aims to understand the flight deck perspective on flying in high-density terminal airspace. We analyzed 73 events from the Aviation Safety Reporting System (ASRS) to assess flight operations at four major NY airports before COVID-19. We defined and explored the concept of airspace complexity for pilots operating in the terminal airspace. Our concept is comprised of four types of external threats related to flight path management: air traffic control interactions, autoflight systems on the flight deck, airspace and flight procedures, and environment. Our findings paint a picture of flight operations in NY.

The Federal Aviation Administration (FAA) has made significant progress modernizing the National Airspace System (NAS) through new technologies and procedures for pilots and controllers under the Next Generation Air Transportation System (NextGen) program. Performance-based navigation (PBN) is a cornerstone of NextGen. It is based upon Area Navigation (RNAV) and Required Navigation Performance (RNP), which allow aircraft to fly more precise lateral routes using satellite navigation and/or other aircraft navigation systems. NextGen leverages PBN for the design of new instrument flight procedures (IFPs) that define routes in and out of terminal airspace, including arrival, departure, and approach procedures.

A PBN NAS strategy report describes the benefits of a PBN-centric NAS (FAA, 2016). One benefit is an improvement in system-wide efficiency by increasing the homogeneity of NAS operations across the country. Another benefit is that NextGen paves the way toward future air traffic management capabilities. And, a PBN-centric NAS allows stakeholders to take advantage of investments in advanced navigation capabilities. In order to realize the benefits of NextGen, however, it is important to increase the utilization of PBN procedures. One of the areas that has been slow to adopt PBN is the New York metro region (NY), which has four busy airports in close proximity: John F Kennedy International (KJFK), La Guardia (KLGA), Newark International (KEWR), and Teterboro (KTEB). This is high-density terminal airspace, with multiple airports and a large number of flights. Here we explore what makes NY a challenging area from the pilot's perspective.

Background

NY has a complex terminal airspace, in part, because of the close physical proximity of its major airports. KJFK is 18 miles east of KTEB and KEWR, and just 9 miles southeast of KLGA. Their relative locations and runway configurations constrain the arrival, departure, and approach procedures that can be assigned to aircraft while keeping them safely separated. In addition, their close proximity necessitates coordinated changes to the airport runway-use configuration. Operations for all the core NY airports are controlled by a Terminal Radar Approach Control (TRACON) that is known for its fast-paced communications and strong expectations of pilot responsiveness to assigned headings, altitudes, and speeds, especially during visual meteorological conditions when arrival and departure rates peak. Pilots familiar with the area told us that prior experience with NY Air Traffic Control (ATC) and airspace procedures makes the flight operations manageable.

Our concept of terminal airspace complexity for pilots was informed by two research strands. The first strand was done to support the Free Flight concept (RTCA, 1995). The goal of Free Flight was to allow pilots more freedom to select optimal flight routes with the ability to self-separate from other air traffic under some conditions. Free Flight operations were focused on enroute airspace, where flows are more structured and traffic density is lower. Free Flight spurred research on airspace complexity from a controller perspective because it was a way to understand "the effect of changing airspace configurations and traffic patterns on the workload of air traffic controllers" (Sridhar et al., 1998). A key parameter of interest was "dynamic density," an idea first mentioned in the 1995 RTCA report. Dynamic density takes into account not just the number of aircraft, but their relative positions and how those positions (and geometries) are changing over time (cf. Kopardekar, et al., 2007; Histon, et al., 2002).

Riley and others studied airspace complexity for pilots for the task of strategic conflict avoidance (Riley, et al., 2003; Riley, et al., 2004). Riley et al. (2004) point out that the concept of airspace complexity is relevant to other pilot tasks, not just to flight-deck decision aids for conflict resolution. They recognized that the definition of airspace complexity should be expanded to include real-world aspects such as weather, restricted airspace, and terrain, especially when fast-changing weather could constrain future aircraft maneuvering.

The second research strand we built upon was work on PBN flight operations, their associated charting, and design of IFPs. Chandra and Markunas (2017) studied line pilot perspectives on the complexity of IFPs, aeronautical charts, and flight path management. Complexity associated with the design of individual IFPs includes factors such as the energy profile, altitude and speed constraints, transitions (i.e., branches) in the route, restricted airspace, and even explanatory text notes.

Chandra & Markunas (2017) also defined five sources of "operational complexity," which occurs in normal operations. These are: ATC interventions, aircraft equipment and performance, environment, crew, and operator factors. Operational complexity factors vary day to day in real-time (e.g., ATC clearance amendments). Controllers and pilots work together to resolve operational issues because these cannot be mitigated in advance through IFP design. Chandra, et al. (2020) found that although PBN can complicate the situation, operational complexity exists even without PBN. Environment factors within operational complexity include terrain, traffic, weather and prohibited airspace; these same factors were mentioned by Riley et al. (2004) in terms of airspace complexity for pilots.

IFP design is distinct from operational complexity, but it is related to airspace design and traffic flows. Arrival, departure, and approach IFPs that pilots expect to fly are proposed in their flight plans and assigned via ATC clearances. Published IFPs are selectable within the aircraft's navigation database. The terminal airspace contains multiple published IFPs that may cross in three-dimensions.

Airspace Complexity for Pilots

Our concept of airspace complexity for pilots includes four types of factors: air traffic control interactions, autoflight systems on the flight deck, airspace design and IFPs, and environment. We chose to focus on external factors that affect the pilot's ability to manage their flight path under normal operating conditions in the terminal airspace. We do not consider emergencies or non-normal conditions. We also excluded internal pilot factors such as fatigue and training. The concept does consider the entire airspace design, not just a single IFP.

This concept of airspace complexity for pilots combines aspects of the concept of airspace complexity for controllers (e.g., traffic and airspace geometry) with IFP design factors and operational complexity factors. As with airspace complexity for controllers, we expect that airspace complexity for pilots will vary in time. We also expect that airspace complexity for pilots will vary by airspace, traffic density, and traffic geometries. Airspace complexity for pilots is also impacted by the capabilities of the aircraft autoflight systems. This is a difference between the pilot and controller views of airspace complexity; controllers are generally unaware of the autoflight system capabilities. The design of IFPs, and use of PBN, is factored into this concept through the airspace design factors.

Method

We reviewed 100 events from the National Aeronautics and Space Administration (NASA) Aviation Safety Reporting System (ASRS) that occurred in the NY terminal area between October and December 2019 (before the impacts of COVID-19). From these, we selected a dataset of 73 events that were from one of the four major NY airports and were relevant to airspace complexity for pilots. These events involved interactions between ATC and pilots and had narratives from the pilot's perspective.

The limitations of ASRS reports are well known. The events are self-reported, subjective, and written from memory. The narratives can be incomplete and difficult to interpret. They can also be biased because of difficulty in observing one's own behavior. The frequency of events in the database may not represent the frequency of occurrence in actual operations. Also, ASRS reports are typically filed when there is an undesired outcome, so findings tend to be framed in terms of negatives rather than positives.

We developed a coding rubric to classify each event. The rubric included a synopsis of the event, factual information (e.g., where the event occurred and who reported it), the outcome, threat(s), context, and an explanation of the coding for internal use. We also recorded whether pilots hand-flew during the event or used the flight management system (FMS). Two researchers reviewed each event and resolved any discrepancies. Table 1 lists the threats we recorded, which were elements from our concept of airspace complexity for pilots.

Table 1.
Threats related to Airspace Complexity for Pilots in High-Density Terminal Airspace.

Threat Type	Threats	Examples
ATC interactions	(Lack of) clarity of communications	Confusing phraseology
	Unpublished restrictions assigned	ATC assigned speed
	Changing instructions	Clearance amendments
	Time-pressure	Difficulty reaching ATC
Flight deck equipment	Unexpected behavior of automated system	Trouble resolving a route discontinuity
	Time-pressured setup or configuration	Managing airspeed on descent
	Aircraft performance requires attention	Use of speed brakes
Airspace	(Complex) design of IFPs	Multiple constraints along an IFP
	High density terminal airspace design	Multiple IFPs, airport interactions
	Large amount of information to brief/know, impacting pilot tasks	Difficulty interpreting charts
Environment	Weather (of all types) that requires attention	Low visibility or shifting winds
	(High) traffic	Mix of aircraft types

Results and Discussion

Of the 73 events, 31 occurred at KEWR, 29 at KLGA, 8 at KTEB, and 5 at KJFK. Thirteen events occurred on departure, 16 on arrival, 20 on approach, and 24 occurred while connecting from the arrival to the approach. Most of the events (58, or 79%) were reported by a Part 121 operator (scheduled air carrier). There was one event each from a Part 135 (charter) and Part 91 (general aviation) operator. The type of flight operation was not specified for 13 events. We ascertained that pilots flew with the FMS in at least 28 events (38%) and hand-flew the aircraft in at least 26 events (36%). Pilots may have only flown a portion of an event with either method. For example, sometimes pilots disconnected the autopilot and hand-flew the aircraft to quickly resolve a traffic conflict. Thirteen events (18%) involved wind-related issues.

The most common outcomes (occurring in at least 10 events each) were Vertical Deviations, Unstable Approaches, Traffic Alert and Collision Avoidance System (TCAS) Resolution Advisories (RA), and Lateral Deviations. Note that a single event might have had more than one outcome (e.g., both a lateral deviation and a TCAS RA). Speed Management Issues, Misconfigurations, Go-Arounds, Terrain Alerts, Vectors, TCAS Traffic Advisories, and Losses of Separation each occurred in fewer than 10 events. We also identified 24 "Other" outcomes. Examples of these included exceeding 250 knots below 10,000 feet (which is generally not authorized under federal regulations), landing without a clearance, and losing sight of the runway while flying a charted visual approach. KLGA has two charted visual approaches, the River visual and the Expressway visual, which may be difficult for pilots to fly if they are not familiar. These charted visuals are generally hand-flown, which can be especially challenging with crosswinds.

Figure 1 shows how often each threat occurred as a percent of the 73 events in the dataset. Note that a single event might have multiple associated threats. Threats related to ATC Interactions occurred most often. There was at least one ATC Interaction factor present in 49 events (67%). This relatively high number confirms what we learned anecdotally, that flight operations in NY are demanding. For example, ATC issued unpublished restrictions in 14 events; 12 of these were higher than preferred speeds during descent or approach. Two were altitude constraints, one of which was assigned, atypically, for a visual approach.

Airspace threats were present in just 11 of the 73 NY events (15%). Interestingly, Complex Design of IFPs was mentioned by pilots in only 4% of the NY reports. In contrast, Chandra et al. (2020), found that Complex Design of IFPs was coded in 35% of 148 events analyzed at locations that had PBN IFPs. It appears that the pilots are more aware of complex IFP designs when PBN is implemented. At NY, the tactical nature of ATC may make the airspace complexity less visible to pilots. The downside of this tactical approach is that it creates time-pressure, which then creates the potential for other undesirable outcomes. For example, pilots might miss the clearance due to frequency congestion, they may not have time to clarify an instruction, or they may run out of time to verify their automation set up, setting up future errors.

Environment factors were recorded in 35 of the 73 NY events (48%), whereas they were present in 36% of events in Chandra et al. (2020). The difference between these two sources of data may be the relative volumes of air traffic. With four major airports, there is more air traffic at NY than at the locations evaluated in the 2020 PBN-related study; those locations, on average, have lower traffic volumes than NY. The rate of Flight Deck Equipment issues reported in Chandra et al. (2020), 32% of 148 events, was similar in these results (30%).

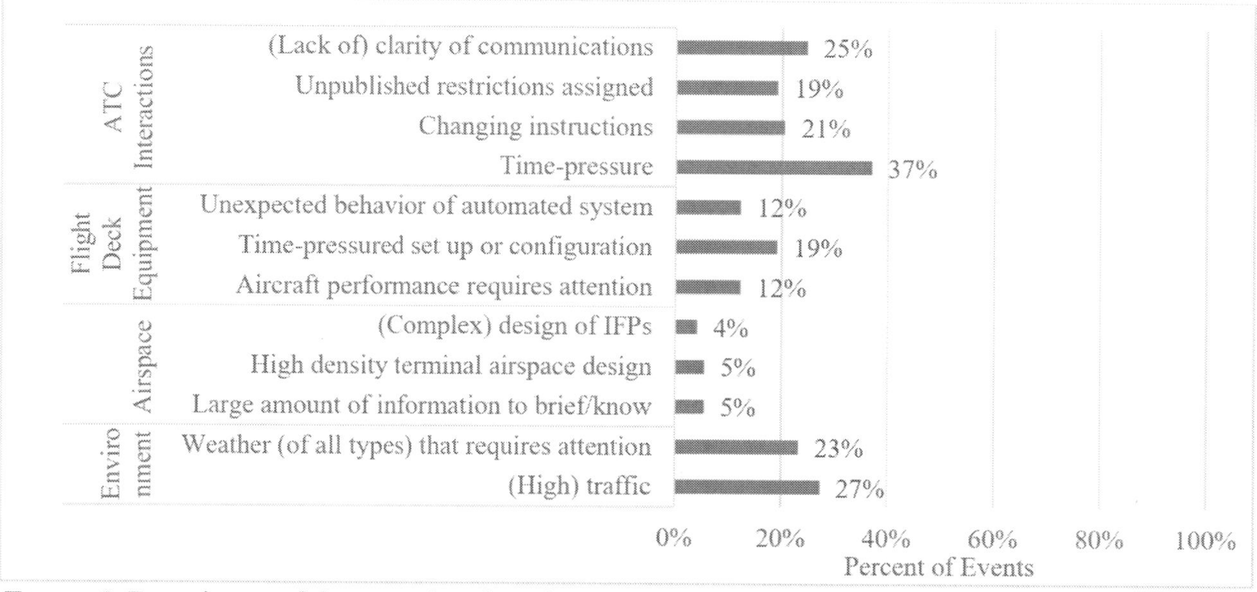

Figure 1. Prevalence of threats related to airspace complexity for pilots in the dataset.

Summary and Next Steps

We explored the pilot perspective on the challenges of operating in a high-density terminal airspace, using NY as a case study. ATC Interactions are high in this airspace confirming that flight operations are uniquely demanding at NY. PBN is not used often at NY, and it appears that the tactical nature of ATC at NY makes the structure of the airspace and procedures less visible to pilots. Out next planned step is to compare these findings from ASRS data with data from discussions with professional pilots who operate in NY. Discussions with pilots may give us further insights into the challenges of flying in NY, and of the potential impacts of greater adoption of PBN IFPs.

Acknowledgements

This document was prepared for the FAA with funding from the NextGen Human Factors Division (ANG-C1). This paper was prepared for the FF01 Interagency Agreement (NextGen NAS and Flight Crew Procedures). We thank our technical sponsors, Kathy Abbott and Jeff Kerr, and our program manager, Victor Quach. We would also like to thank the NASA ASRS Program Office for their assistance in gathering the ASRS data. The views expressed herein are those of the authors and do not necessarily reflect the views of the Volpe National Transportation Systems Center or the United States Department of Transportation.

References

Chandra, D.C., & Markunas, R. (2017). *Line Pilot Perspectives on Complexity of Terminal Instrument Flight Procedures*, (DOT-VNTSC-FAA-17-06). Cambridge, MA, U.S.DOT Volpe National Transportations Systems Center. Retrieved from https://www.volpe.dot.gov.

Chandra, D.C., Sparko, A., Kendra, A., & Kochan, J. (2020). *Operational Complexity in Performance-Based Navigation (PBN) Arrival and Approach Instrument Flight Procedures (IFPs),* DOT-VNTSC-FAA-20-02. U.S.DOT Volpe National Transportation Systems Center. https://rosap.ntl.bts.gov/view/dot/43835

FAA (2016). *Performance-based navigation (PBN) NAS Navigation strategy*. Retrieved from NextGen – Resources (faa.gov).

Histon, J.M., Hansman, R.J., Aigoin, G., Dalahaye, D. & Puechmorel, S. (2002). Introducing Structural Considerations into Complexity Metrics. https://doi.org/10.2514/atcq.10.2.115. *Air Traffic Control Quarterly*. 10(2), 2002, pp. 115-130.

Kopardekar, P., Schwartz, A., Magyarits, S., & Rhodes, J. (2007). Airspace complexity measurement: An air traffic control simulation analysis. *7th USA/Europe Air Traffic Management R&D Seminar*. Barcelona, Spain, 2-5 July 2007. http://atmseminar.org/7th-seminar/papers/

Riley, V., Sierra, E., Mogford, R., Johnson, W., Kopardekar, P., Chatterji, G., & Lawton, G. (2003) Pilot perceptions of airspace complexity. *22nd IEEE Digital Avionics Systems Conference*. 12-16 October 2003. Indianapolis, IN. https://doi.org/10.1109/DASC.2003.1245867

Riley, V., Chatterji, G., Johnson, W., Mogford, R., Kopardekar, P., Sierra, E., & Lawton, G. (2004) Pilot perceptions of airspace complexity. Part 2, *23rd IEEE Digital Avionics Systems Conference.* 24-28 October 2004. Salt Lake City, UT.

RTCA. (1995). *Task Force 3 Final Report on Free Flight Implementation.* RTCA, Inc., Washington, DC.

Sridhar, B., Sheth, K.S., & Grabbe, S. (1998). Airspace complexity and its application in air traffic management. *2nd USA/Europe Air Traffic Management R&D Seminar.* Orlando, Florida 1-4 December 1998. http://atmseminar.org/2nd-seminar/papers/

MODELING CONTINGENCY MANAGEMENT IN UNMANNED AIRCRAFT SYSTEMS TRAFFIC MANAGEMENT

Emily Barrett and Martijn IJtsma
The Ohio State University
Columbus, OH

Contingency management in future Unmanned Aerial Vehicles (UAVs) Traffic Management (UTM) requires a variety of distributed and interdependent functions and services—such as flight tracking and conformance monitoring, weather detection and prediction, and ground-based detection and avoidance—that need to be coordinated across multiple roles and organizations. This paper describes a combination of cognitive walkthroughs and computational modeling of work to analyze edge case scenarios and assess resiliency in future UTM operations. We discuss how the walkthrough and modeling inform each other and present early results. The ultimate goal of this work is to identify requirements for robust and resilient system responses in future UTM contingency management.

Unmanned Aircraft System Traffic Management (UTM) is an envisioned concept of operation for lower-altitude airspaces with a mix of unmanned and manned capabilities (National Aeronautics and Space Administration (NASA), 2019). UTM operations rely on effective information sharing and coordination among a number of interdependent roles and organizations, including and facilitated by automated services. To assure efficiency and safety of the operations, the system needs to be robust and resilient against anticipated and unanticipated contingencies.

This paper discusses early work on exploring robustness and resilience in contingency management (CM) in UTM operations. We conducted several cognitive walkthroughs and developed a computational model of CM operations for a variety of edge case scenarios. The approach demonstrates how cognitive walkthroughs and computational work modeling can inform each other and provide early results from a computational experiment testing two different types of CM automation.

Background

Figure 1 shows a notional architecture for the UTM system (see NASA, 2019 for a detailed description of the architecture). Actors in the system include Remote Pilots in Command (RPIC), UAS Service Suppliers (USS) and/or Supplemental Data Service Providers (SDSS). At the heart of the system is the UTM Operations Center, tasked with supervising the UTM system and managing the airspace. Information sharing is handled through an Unmanned Traffic Information Management System. The UTM system also interfaces with Air Traffic Control (ATC) in the area via the Flight Information Management System (FIMS).

Robustness and resilience describe a system's ability to adapt and maintain performance under anticipated and unanticipated disruptions, respectively. Resilient CM requires fast-paced responses with interaction and coordination across various roles, see the example procedural information flows in Figure 1. Earlier research on coordination and adaptation for resilient behavior used edge case scenarios and cognitive walkthroughs with subject-matter experts (SMEs) to assess the system's response at the boundary of performance envelopes (Bisantz & Roth, 2007; Woods & Balkin, 2018).

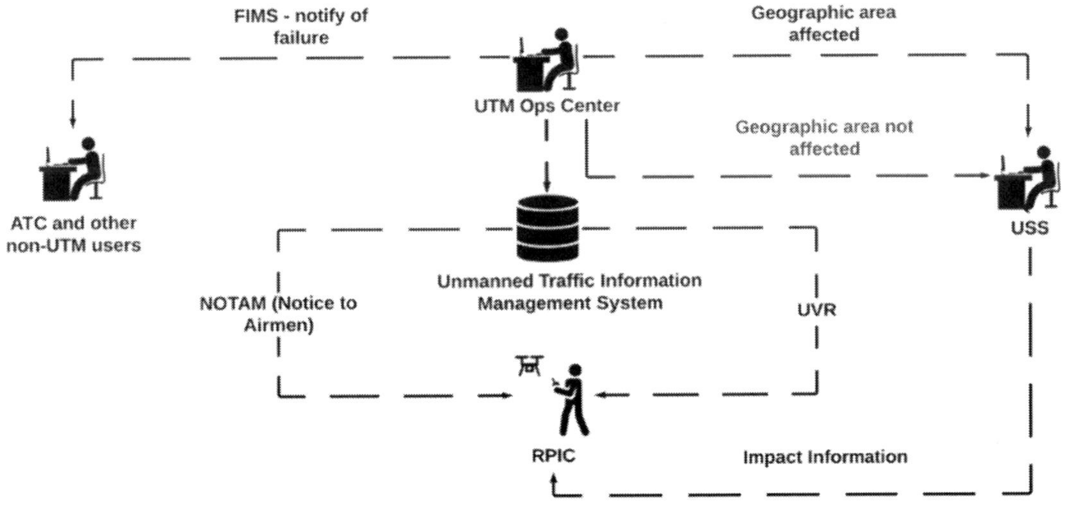

Figure 1. Information Flow Diagram for Component Failure Contingency

Modeling of cognitive work can support assessment of robustness and resilience. Work in complex work domains like UTM is driven by constraints and dynamics in the work environment that can be identified and codified (Vicente, 1999). Once codified, models can be simulated to evaluate the dynamics of such work (Pritchett, Bhattacharyya, & IJtsma, 2016; Pritchett, Feigh, Kim, & Kannan, 2014). We argue that for assessing resilience in future UTM operations, knowledge-elicitation and modeling can be part of a formative and iterative cycle in which exploration of system characteristics and responses support identification of design requirements, similar to Vicente (1999) and Woods & Roth (1994). In this paper, we combine cognitive walkthroughs and computational modeling and simulation of edge case scenarios to perform model-based exploration of a UTM system's robustness and resilience.

Edge Case Scenarios & Cognitive Walkthroughs

We conducted cognitive walkthroughs with SMEs to explore how actors in future UTM operations would need to respond and coordinate during CM. A document review was conducted to learn about the envisioned UTM system at hand, including the various types of contingencies that could take place and disrupt the nominal flow of operation. Five classes of contingencies were created that span a range of disruptions to the system's nominal operations, see Table 1.

For each of these classes, we developed narratives with a representative traffic situation and a set of probing questions for the SMEs. The probing questions were targeted at discovering how actors, as part of the bigger UTM system, would adapt and coordinate to respond to disruptions and at testing the validity of the scenarios and envisioned procedures. All interviewees were subject matter experts in the field of aviation who have experience in UAV operations, air traffic management, and/or resilience engineering.

Table 1.
Contingency classes and descriptions

Contingency	Description
Component failure	Failure of a component or system critical to the operations (e.g., radar, ping station)
Loss of link	UTM is not receiving telemetry data and/or cannot communicate with a UAV
Weather event	Weather front moves through area, and/or micro-weather conditions deteriorate
External emergency	An external event (e.g., fire, police activity) requires unanticipated airspace changes
Unidentified actor	UAV is not conforming to the expected flight plan or uncontrollable moving objects

As an example of an edge case scenario, the component failure narrative involved two RPICs, pilots of a commercial flight, one UTM Supervisor, and one USS. The traffic situation consisted of three vehicles flying west of Columbus, Ohio: a high priority UAV flight transporting a liver transplant, a law enforcement UAV surveying a crime scene, and a commercial airline flight landing at the John Glenn Columbus International Airport (CMH). When the traffic is nearing closest-points-of-approach, a radar fails unexpectedly, resulting in a loss of sensing capability for the UTM system and a need for to reconfigure the airspace.

The walkthroughs revealed various complicating factors to CM, such as constraints, goal conflicts, time pressures, and the need to coordination between actors, particularly between the RPICs and UTM supervisor. For example, interviewees noted trade-offs between closing the airspace for all current traffic (requiring rerouting) or allowing existing flights to continue, with the ability to monitor the separation as a determining factor. The findings from the document review and walkthroughs were aggregated into an abstraction hierarchy for the overall UTM system (Vicente, 1999), see Figure 2.

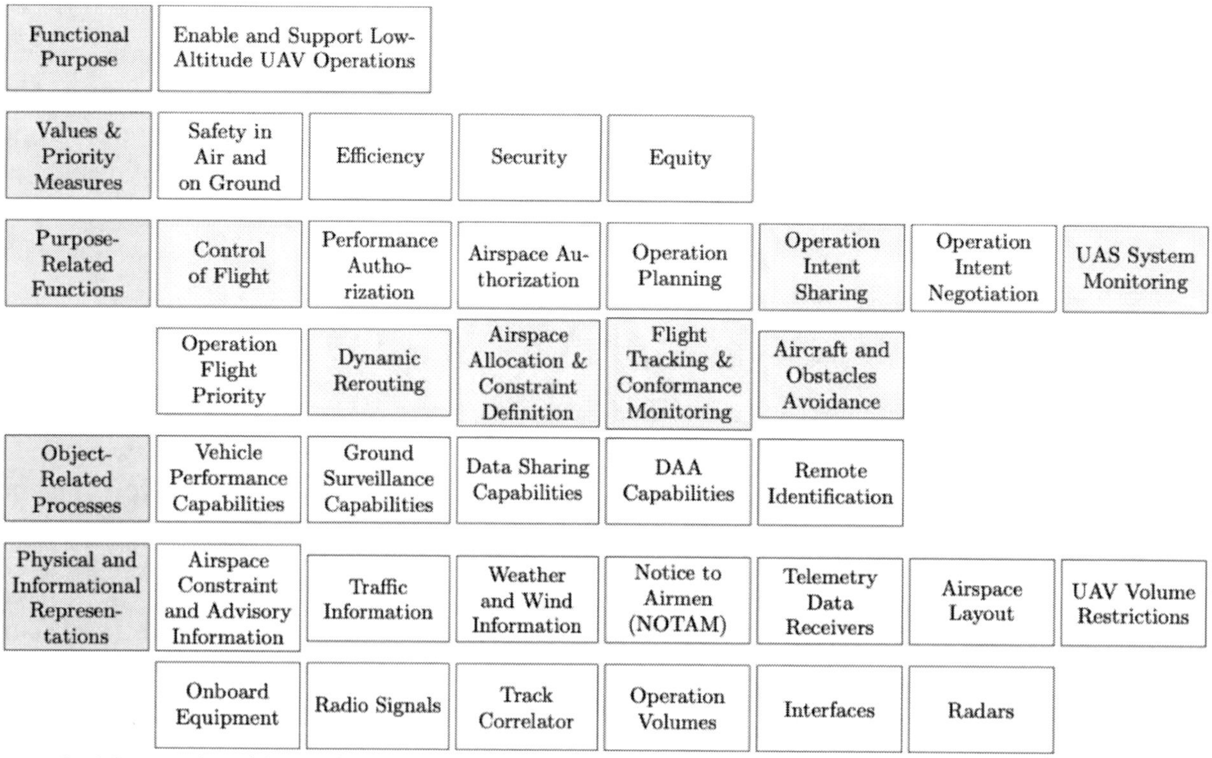

Figure 2. Abstraction hierarchy for a UTM system.

Work Models that Compute

In parallel with the walkthroughs, we developed a computational model of the work involved in UTM CM. Work Models that Compute (WMC) is a computational modeling and simulation framework for analyzing situated work (Pritchett et al., 2014), used before to analyze work allocation in air traffic management and space operations. Through models of resources, actions, and agents, WMC can make quantitative predictions of system performance given different system configurations.

The first two columns of Table 2 show the actions that were modeled in WMC for what were deemed the purpose-related functions most critical to CM, see the highlighting in Figure 2. Furthermore, the flight dynamics of the aircraft are deemed an important driver of the UTM system's dynamics, determining much of the actors' timing of activity to keep pace with disturbances. Thus, the computational work model includes a model of the flight dynamics for a generic UAV, with parameters that can be changed to simulate a variety of vehicle classes (e.g., a small quadrotor UAV or a large package delivery drone).

Results from the cognitive walkthroughs directly informed the modeling, with the system's response captured primarily in the first three and last rows of Table 2. The work model also includes descriptions of the information resources (such as geographic location, altitude, and radar status) that are shared amongst the actors. As an example of how the walkthrough informed the modeling, several SMEs noted their decisions about the impact of the radar failure depended on the vertical separation between aircraft. Thus, the "assess impact" action is modeled to compare the difference in altitudes of the two vehicles, then assigning High, Medium, or Low to the Impact resource that is shared with the other actors in the system.

Table 2.
Work model actions with two allocations of authority (A) and responsibility (R) (format: A/R)

Purpose-Related Function	Work Model Action(s)	Allocation 1	Allocation 2
Airspace Allocation and Constraint Definition	Generate UVR	Supervisor/Supervisor	Automation/Supervisor
UAS System monitoring	Assess impact, monitor system integrity	Supervisor/Supervisor	Automation/Supervisor
Operation Intent Sharing	Communicate via NOTAM	Supervisor/Supervisor	Automation/Supervisor
Flight Tracking and Conformance Monitoring	Track flight, manage waypoint progress	RPIC/RPIC	RPIC/RPIC
Control of Flight	Change altitude, change speed, change heading, takeoff, land, direct to waypoint, distance to next waypoint, flight dynamics	UAV/RPIC	UAV/RPIC
Aircraft and Obstacle Avoidance	Avoid conflict, detect conflict	RPIC/RPIC	RPIC/RPIC
Dynamic Rerouting	Reroute flight	RPIC/Supervisor	Automation/Supervisor

A WMC run simulates the detailed interaction between actions and the work environment (as captured in resources), including how activity of actors in the system is interconnected through dynamics and information. WMC provides quantitative data on the dynamics of activity, such how often and when actions are performed, and how often and what information is shared amongst actors. In addition, WMC can be used to evaluate effect of system design choices, such as the allocation of authority and responsibility between human operators and various autonomous capabilities. Here, authority denotes the agent that will be executing an action, and responsibility denotes who is held accountable for the outcome of an action.

To demonstrate, we conducted simulation runs with two types of automated capabilities, see the last two columns of Table 2: Allocation 1 with a UTM supervisor performing the majority of the work manually, and Allocation 2 in which a majority of the CM is automated, with the UTM supervisor monitoring the automated response. In the latter case, the UTM supervisor is still responsible for the outcome. In these instances of mismatching authority and responsibility, WMC automatically engenders a monitoring action for the authorized agent (i.e., UTM supervisor), executed in parallel with the automation's actions (Pritchett et al., 2016).

Figure 3 and 4 show early results from simulation runs. Figure 3 illustrates when each actor is performing an action. Because actions related to control of flight (executed by the UAV) are updated relatively frequently, and the CM actions are of primary concern to this analysis, these actions are omitted from the figure. The figure shows when human actors need to monitor automation agents due to authority-responsibility mismatches (shown as 'teamwork'), clearly indicating that more autonomous capabilities lead to higher monitoring loads. Time pressure was an important concern during the walkthroughs, and data like this can provide estimates for how quickly UTM supervisors and RPICs need to coordinate a response to a radar failure.

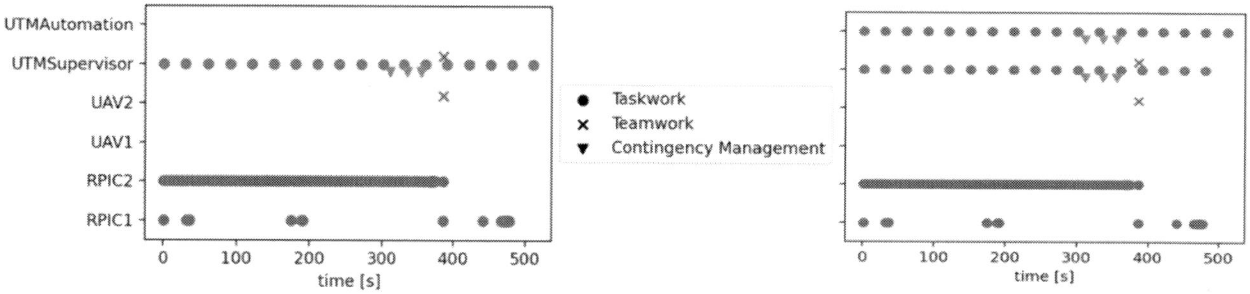

Figure 3. Plots for every instance an agent performs an action for allocation 1 (left) and allocation 2 (right).

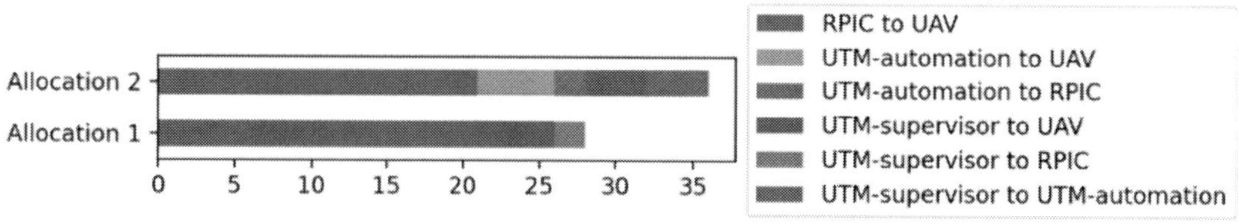

Figure 4. Information exchange requirements for various work allocations

Figure 4 shows data on the total number of information exchange requirements for each simulation run, categorized by the agents that are involved in the exchange. Every time an action is carried out, the simulation logs what information is needed and who last updated that information. Allocation 2 shows more information exchange requirements, particularly due to increased requirements for human-automation information exchange. These data provide insight into who needs to communicate with whom and how often, addressing a theme from the walkthroughs related to communication across the various actors in the system.

Conclusion and Future Work

We used a combination of cognitive walkthroughs and computational modeling and simulation of edge case scenarios to analyze CM in future UTM operations. The scenario development, cognitive walkthrough, and computational modeling occurred in an iterative process and highlighted how insights from interviewing SMEs can be used to inform computational modeling. The walkthroughs provide data and insights for the modeling effort, and the computational models of work afford a thorough analysis of the system's dynamic response. Future work includes more detailed modeling of other classes of contingencies and performing larger-scale analysis using the computational models. Ultimately, with extended modeling capabilities and testing of various system architectural characteristics, the aim is to identify specific requirements for robustness and resilience in the UTM system.

Acknowledgments

This work is sponsored in part by a subcontract from CAL Analytics, funded by the Ohio Federal Research Network (OFRN) and a scholarship from the Eddowes fund. The authors would like to thank the CAL team, Maggie Miles, Jonathan Dowling, and Renske Nijveldt for their support in preparing and running the cognitive walkthroughs.

References

Bisantz, A., & Roth, E. (2007). Analysis of Cognitive Work. *Reviews of Human Factors and Ergonomics*, *3*(1), 1–43. https://doi.org/10.1518/155723408x299825

Naikar, N. (2017). Cognitive work analysis: An influential legacy extending beyond human factors and engineering. *Applied Ergonomics*, *59*, 528–540. https://doi.org/10.1016/j.apergo.2016.06.001

National Aeronautics and Space Administration (NASA). (2019). Unmanned Aircraft System (UAS) Traffic Management (UTM). In *Enabling Civilian Low-altitude Airspace and Unmanned Aircraft System Operations*.

Pritchett, A. R., Bhattacharyya, R. P., & IJtsma, M. (2016). Computational Assessment of Authority and Responsibility in Air Traffic Concepts of Operation. *Journal of Air Transportation*, *24*(3), 93–102. https://doi.org/10.2514/1.D0024

Pritchett, A. R., Feigh, K. M., Kim, S. Y., & Kannan, S. K. (2014). Work Models that Compute to describe multiagent concepts of operation: Part 1. *Journal of Aerospace Information Systems*, *11*(10), 610–622. https://doi.org/10.2514/1.I010146

Vicente, K. (1999). *Cognitive Work Analysis: Toward Safe, Productive, and Healthy Computer-Based Work*. CRC Press.

Woods, D. D., & Balkin, A. E. (2018). *Resiliency Trade Space Study: The Interaction of Degraded C2 Link and Detect and Avoid Autonomy on Unmanned Aircraft*.

Woods, D. D., & Roth, E. M. (1994). Symbolic AI Computer Simulations as Tools for Investigating the Dynamics of Joint Cognitive Systems. In P. C. Cacciabue, E. Hollnagel, P. C. Cacciabue, & J.-M. Hoc (Eds.), *Expertise and Technology* (pp. 75–90). Psychology Press.

UAS SAFETY ZONES: A MODEL FOR ADDRESSING INCREASED AIR TRAFFIC CONTROLLER WORKLOAD

Pratik Jadhav
Dr. Damon Lercel
Dr. Sarah Hubbard
Purdue University
West Lafayette, IN

This paper presents a model to address the increased workload for air traffic controllers (ATC) due to the integration of unmanned aircraft systems (UAS) into the airport ecosystem. The FAA established small UAS operational regulations in 2016. Already, UAS pilots comprise over 20% of the total pilot population, and UAS account for 89% of the total aircraft registrations. Given current FAA resource constraints and the forecast growth of air traffic, innovative safety management solutions are required to address the increase in controller workload and the associated safety risks. This research presents one perspective regarding the impact of UAS operations on air traffic controller workload and a conceptual model to address the increase in UAS operations at airports. The model proposes designating predetermined UAS safety zones, or routes, inside the airport. By limiting UAS operations to designated zones, controller resources may be more focused while maintaining an acceptable level of safety.

Air traffic controllers (ATC) are responsible for air traffic management in the national airspace system (NAS). ATC workload and ergonomics is a wide-ranging topic due to the multiple responsibilities that encompass equipment, operations, communications, and management (Loura, 2014). ATC workload is expected to increase due to projected increases in air traffic (after aviation recovers from the pandemic). The impact of this increase will be exacerbated when coupled with the number of UAS currently flying in the NAS and the continued growth of UAS activities

In 2012 the Federal Aviation Administration (FAA) introduced the Modernization and Reform Act that actively seeks to integrate civil UAS into the NAS (FAA, 2013). The FAA asserted that UAS standards shall mimic traditional "manned aircraft training standards to the maximum extent possible" (FAA, p. 28, 2013). This proliferation of UAS technologies requires ATC not only to undergo additional training but also maintain situational awareness across an increasingly complex airspace system, such as multiple aircraft types at different altitudes, emergency situations, and manned-unmanned interoperability. Ultimately, ATC will maintain overall responsibility for air traffic separation based on airspace class and the type of UAS without necessarily maintaining a direct link to the UAS. The FAA has also called for various entities in air traffic management to cluster and work towards air traffic interoperability, which in the current aviation scenario can create increased workload on existing ATC personnel. One example of this workload increase is the difference between UAS and manned aircraft flight plans (Semanek and Kamienski, 2015).

Integration of UAS in the NAS has certainly impacted the ATC system, which includes controller job tasks and responsibilities. For example, in 2017, the FAA launched the Low Altitude Authorization and Notification Capability (LAANC), which allows both civil and public UAS operators to expeditiously obtain FAA authorization to fly in controlled airspace near airports. LAANC automates the application and approval process for airspace authorizations through the use of internet-based technologies. LAANC allows pilots to receive their authorization in near-real time and is currently available at 538 air traffic facilities and 731 airports, and the FAA continues to add locations (FAA, 2020a). Another option is the FAA Drone Zone, which allows UAS operators to request airspace waivers; these waivers enable longer term access to controlled airspace and may allow for UAS operations on the airport property itself (FAA, 2020b). The maturation of these regulatory policies and processes have enabled more UAS operations near airports and at airports. These activities also require additional responsibility and/or tasks for ATC controllers.

According to Semanek and Kamienski (2015), the impact on ATC must be considered and the potential increase in ATC workload may be concerning. UAS issues include "UAS flight plan, UAS communication link, UAS types, and training for ATC regarding UAS" (p. 1049). UAS potentially increases ATC workload in terms of flight plan issues and control and communication (C2) link (Semanek and Kamienski, 2015). Traffic management of small UAS may be a strenuous task as the lack of UAS identification and tracking capabilities continue to persist. Vu et al. (2014) suggest UAS in the NAS negatively affects safety of the NAS, stating "[the] increasing number of UAS did negatively impact the ATC's performance to some degree" (Vu et al., p. 6, 2014). This increase in workload may adversely affect the performance of ATC controllers, such as increases in work-related stress and fatigue.

In response to these challenges, this research presents the use of UAS safety zones, which provide an operational area that has been predetermined to increase the safety of UAS operations within the airport area. These systems allow for ongoing hazard assessment through strategic use of technology that may only require monitoring by controllers, minimizing direct controller interdiction and communication with UAS operators. This model provides one way to simplify UAS operations at airports and help address the safety of the projected increase in UAS operations to support airport activities while minimizing the effect of this additional traffic on ATC workload. There is significant interest in UAS for airport operations due to its ease of use, mobility, and low cost. UAS can be used for scheduled maintenance of airport navigation aids such as the instrument landing system (ILS) and VHF omnidirectional radio range (VOR) system (Bredemeyer & Schrader, 2018). The mobility aspect of UAS allows "critical areas to be accessible only with a flying platform" (Bredemeyer & Schrader, p. 279, 2018). As the UAS is a relatively small machine, it is possible to record measurements in unsafe areas with more precision, cost-effectiveness, and optimal time (Bredemeyer & Schrader, 2018).

Lawrence and Mackie (2019) of Woolpert Inc. detailed the practical use of UAS at Savannah Airport. The team successfully integrated UAS into the airport ecosystem to support wildlife mitigation, first responder services, pavement inspections, and UAS based security services (Lawrence and Mackie 2019). Integrating UAS operations with traditional airport traffic may be achieved through ATC coordination, robust safety systems, coordinated flight systems, and compliance with all FAA guidance in Class C airspace (Lawrence and Mackie 2019).

Additional benefits of UAS were elucidated by Hubbard et al. (2017) regarding the applications of UAS at airports such as airport obstruction analysis, runway and taxiway pavement surveys, wildlife mitigation, emergence response services, airport construction aid, and post-snow runway inspections. Advances in UAS technology and the benefits it may provide to airport stakeholders has led to the exploration of UAS airport applications along with techniques to safely integrate UAS with manned aircraft operations.

The initial review of literature found that ATC workload may be impacted with the integration of UAS in the NAS, whereas UAS operations on and near airports is expected to accelerate. Looking beyond the current use cases, the accelerated development of Urban Air Mobility has now moved well beyond the concept phase and presents further challenges to controlling air traffic in the NAS (FAA, 2020c). This paper explores the issue of ATC workload and attempts to provide a working concept of UAS at airports to aid day-to-day operations.

UAS Safety Zones – A Theoretical Model

The concept of designated operating areas for air traffic is not new. The FAA has designated air routes across the United States with one of the primary benefits being more efficient traffic management, increased safety, and reduced air traffic control workload. An intuitive way of facilitating the operation of UAS in the airport environment is by identifying dedicated corridors with appropriate safety zones for UAS. Zones that enable and restrict movement of UAS through certain waypoints in the airspace. This concept is similar to the FAA's military training routes (MTR), which are specific routes designated to separate military activities below 10,000 feet and speeds above 250 knots (FAA, 2016). In terms of flight corridors inside urban cities, utilization of a robust UAS traffic management (UTM) system, safety systems, and aligned interests between government bodies and UAS entities may promote the logistical use of UAS (Ronczka, 2018). The recent development of a 50-mile UAS corridor in the state of New York as a test site for UAS operations beyond visual line of sight (BVLOS) provides an example of the method to advance toward an ultimate goal of full interoperability between manned and unmanned aircraft.

Referring to Figure 1 below, UAS safety zones are identified on the airport map, which will facilitate UAS operations. Two safety zones: Zone X1 – X2 and Zone Y1 – Y2. UAS will be operated along a fixed designated route inside these zones. The UAS will be monitored by a ground controller to facilitate effortless movements. Points defined below, X1 and Y1 represent UAS-stations near the airport terminal whereas points X2 and Y2 are UAS-stations across the runways 05L and 05R respectively. Figure 1 also shows safety zones that run perpendicular to points X2 and Y2, illustrated by red arrows. Such zones support UAS transport inside the airport at some routes that may not overlap with a taxiway or runway and facilitate the formation of UAS safety zone linkages. These safety zones will extend across runways and taxiways to form a transportation route.

Figure 1. A theoretical example of UAS safety zones at Indianapolis International Airport.

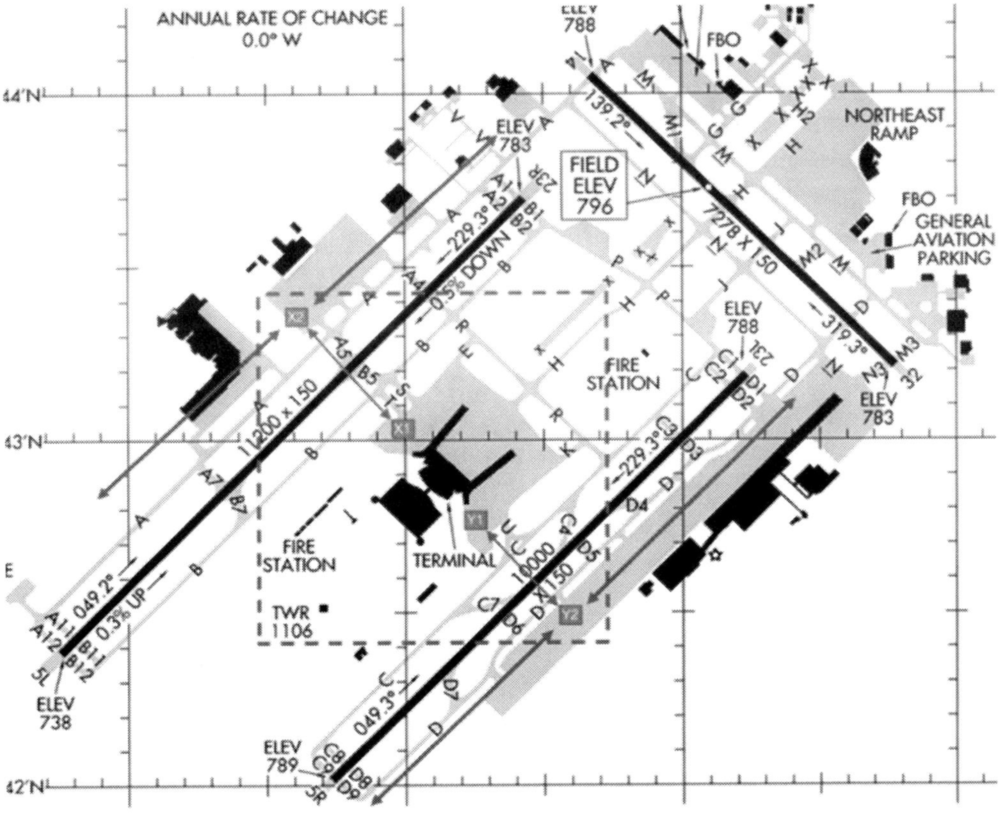

Note. Airport diagram obtained from FAA website (for representation purpose only).

UAS flight paths through these safety zones will be defined based on fixed waypoints. Waypoints will be used by the UAS to fly specified paths across airport runways under a controller's supervision. These waypoints will also be developed with consideration for manned aircraft operations at airports. Designated flight times for UAS through these zones (especially zones that fly over runways) can be determined based on characteristics of manned aircraft activities. Due to its mobility feature, UAS may allow transport across runways in reduced times. Integration with the airport operations ecosystem will be the most essential feature of this model. UAS operations may be coordinated with manned aircraft operations to support simultaneous operations of manned-unmanned airport operations.

To provide clarity about the course of UAS, guidance paths can be setup on a virtual display that enables the controller to maintain a visual line of sight (VLOS) with the UAS. As seen in Figure 1, the location of the control tower (TWR 1106) is in line of sight of the proposed UAS safety zones, which are Zone X1 – X2 and Zone Y1 – Y2. These virtual paths will complement airport lights and taxiways to enhance UAS traffic management. A designated, identifiable UAS path on a virtual display will aid in better management of UAS operationally and visually.

Use of Technology

UAS integration may present a hazard during emergency situations at airports such as manned aircraft runway incursions and excursions, obstacle collisions, ground vehicle collisions, and wildlife strikes. To mitigate such hazards, geofence technology may be used. Geofence technology inside airport boundaries may prohibit the UAS from initiating launch. Ongoing emergency situations may require unobstructed airspace; therefore, restricting UAS flights. Inflight UAS will be forced to land at a designated location within these safety zones, such as those designated as X1, X2, Y1, and Y2 in Figure 1. Geofence activation may be easily initiated by controllers when responding to an emergency.

Safety is also enhanced by UAS sense and avoid technology. This technology uses onboard sensors to detect, sense, and avoid any obstructions in the UAS flight path. The use of sense and avoid technology may further increase airspace safety in terms of manned-unmanned interactions. The ability of an UAS to avoid dangers with the assistance of sensors may aid ground controllers when flying beyond visual line of sight.

Summary

Coupling UAS into daily airport operations presents some challenges but this paper presents one approach to support the safe integration of UAS at airports, manned-unmanned flight coordination, UAS strategic flight-path planning, and UAS safety zones. The proposed safety zones may also reduce ATC's UAS related workload. Initiation of UAS safety zones may support a decrease in the number of ground vehicle movements while reducing the time associated with such movements. Ultimately, the use of these UAS safety zones may promote efficient management of controller workload, while supporting safe UAS airport operations.

Future Research

Further research needs to be done to identify the technical aspects of integrating the UAS into the airport ecosystem using such proposed safety zones. This includes operational details, financial analysis, and regulatory issues. Financial analysis should be used to quantify the cost of using UAS and the benefits to the airport and airport tenants as well as benefits to ATC. Regulatory issues include compliance with FAA regulations for air traffic control as well as ground operations at the airport.

References

Bredemeyer, J., & Schrader, T. (2018, April). Employing UAS to perform low altitude navaids measurements. In *International Flight Inspection Symposium (IFIS), Monterey, California*. http://www.icasc.co/sites/faa/uploads/documents/20th_IFIS_Papers/Papers/IFIS18-0026.pdf

Federal Aviation Administration (FAA). (2013). *Integration of Civil Unmanned Aircraft Systems (UAS) in the National Airspace System (NAS) Roadmap*. Retrieved from https://www.faa.gov/about/plans_reports/modernization/media/Sec.332(a).pdf

Federal Aviation Administration (FAA). (2020a). *FAA UAS Data Exchange (LAANC).* Retrieved from https://www.faa.gov/uas/programs_partnerships/data_exchange/

Federal Aviation Administration (FAA). (2020b). *FAA DroneZone.* Retrieved from https://faadronezone.faa.gov/#/

Federal Aviation Administration (FAA). (2020c). *FAA Urban Air Mobility and Advanced Air Mobility.* Retrieved from https://www.faa.gov/uas/advanced_operations/urban_air_mobility/

Federal Aviation Administration (FAA). (2016, August 24). *Pilot's Handbook of Aeronautical Knowledge,* Chapter 15. Retrieved from https://www.faa.gov/regulations_policies/handbooks_manuals/aviation/phak/media/17_phak_ch15.pdf

Federal Aviation Administration (FAA). (2016). The FAA's New Drone Rules are Effective Today [Press release]. Retrieved from https://www.faa.gov/news/updates/?newsId=86305

Hubbard, S., Pak, A., Gu, Y., & Jin, Y. (2017). UAS to support airport safety and operations: opportunities and challenges. *Journal of unmanned vehicle systems*, *6*(1), 1-17. https://doi.org/10.1139/juvs-2016-0020

Kamienski, J., & Semanek, J. (2015). ATC perspectives of UAS integration in controlled airspace. *Procedia Manufacturing*, *3*, 1046-1051. https://doi.org/10.1016/j.promfg.2015.07.169

Loura, J. (2014). Human factors and workload in air traffic control operations-a review of literature. *International Journal of Management and Social Sciences Research*, *3*(3), 1-5. https://citeseerx.ist.psu.edu/viewdoc/download?doi=10.1.1.428.6877&rep=rep1&type=pdf

Mackie, T., & Lawrence, A. (2019). Integrating unmanned aircraft systems into airport operations: from buy-in to public safety. *Journal of Airport Management*, *13*(4), 380-390. https://woolpert.com/resource/integrating-unmanned-aircraft-systems-into-airport-operations-from-buy-in-to-public-safety/

Ronczka, J. (2018, December). System engineering and logistical UAS corridors: Challenge the present to drive the future. In *2018 26th International Conference on Systems Engineering (ICSEng)* (pp. 1-8). IEEE. https://doi.org/10.1109/ICSENG.2018.8638249

Vu, K. P. L., Strybel, T., Chiappe, D., Morales, G., Battiste, V., & Shively, R. J. (2014, July). Air traffic controller performance and acceptability of multiple UAS in a simulated NAS environment. In *Proceedings of the International Conference on Human-Computer Interaction in Aerospace* (pp. 1-7). https://doi.org/10.1145/2669592.2669693

APPLYING EYE-TRACKING TECHNOLOGY TO EXPLORE THE VISUAL SCANNING PRACTICES OF AIR TRAFFIC CONTROL TOWER CONTROLLERS

Jerry Crutchfield
Federal Aviation Administration
Oklahoma City, Oklahoma
Ziho Kang
Ricardo Palma Fraga
Saptarshi Mandal
University of Oklahoma
Norman, Oklahoma

In the domain of Air Traffic Control (ATC), visual scanning refers to a systematic and continuous effort to acquire all necessary information to build and maintain a complete awareness of activities and situations which may affect the controllers' area of responsibility. Our research team has supported FAA efforts to improve training of the important scanning skill, by conducting research to identify characteristics of successful tower visual scanning behavior. In addition to conducting multidisciplinary working groups and structured one-on-one interviews, we have collected eye-movement data from tower control experts while they controlled high fidelity air traffic simulations of airports at which they are certified. Participants included fifteen air traffic control tower instructors (employed by the FAA Academy) and twelve front line controllers (from Centennial, Denver, Minneapolis, and Orlando airports) each operating Local control for multiple 20-30 minute scenarios. Additionally we ranked instructor performance using time to detect off-nominal scenario events (e.g. smoking aircraft engine, noncompliant vehicles, occurrence of birds, etc.). We subsequently compared number and duration of eye fixations occurring during a scenario across our high and low ranked instructors and found no reliable differences. We also analyzed fixations within and transitions between identified Areas of Interest including Final, Touchdown, Downwind Midfield, Runway Midfield, Runway Intersection, and Departure Corridor. In this presentation, we will discuss what these analyses showed about the usage, by our participants, of scanning best practices identified by our working group (i.e. frequently scanning "hotspots", airfield-out, segmented scanning, and backward scanning).

Since 2014, researchers at the University of Oklahoma (OU) and the Civil Aerospace Medical Institute (CAMI) have been conducting research in support of the Federal Aviation Administration's efforts to improve the training of Air Traffic Control Specialists (ATCS) in the visual scanning skill. In the domain of Air Traffic Control (ATC), scanning refers to a systematic and continuous effort to acquire all necessary information to build and maintain a complete awareness of activities and situations which may affect the controllers' area of responsibility. ATCS must continually scan their environment to gather information that is vital to maintaining the safe and expeditious movement of air traffic. Initial research examined the usefulness of eye-tracking technology for the characterization of ATCS scanning behavior and focused on En

Route ATC (Kang, Mandal, Crutchfield, Millan, & McClung, 2016; Mandal & Kang, 2018). In 2016, however, a working group of FAA human factors specialists and pilot and controller Subject Matter Experts (SME) met to discuss the training of contoller scanning in the airport tower environment specifically.

The working group produced a list of scanning best practices and a recommendation that research be done to examine the possibility of training tower controllers to use a specific scanning pattern in much the same way a pilot is taught to scan on the flight deck. Although Tower ATCS are trained to frequently scan hotspots (locations visible out the tower window where aircraft paths most frequently cross and where errors or off-nominal events can have severe consequences) and often rules of thumb, they are not currently taught to consistently use any specific scanning patterns in the way pilots are taught during their training. Our team subsequently initiated a program of research to determine:

- If eye-tracking technology could be used to characterize Tower controller scanning behavior and identify scanning patterns
- How the rules of thumb taught during training are applied in ATC and whether they can be linked to detection performance
- If individual controllers use particular scanning patterns consistently
- If particular scanning patterns are used across different controllers
- If identified scanning patterns can be linked to performance
- If controllers can be trained to use specific identified scanning patterns

During our research we found eye-tracking technology to be a very useful tool as have others (Kearny & Li, 2018; Pinska, 2006). In addition to structured one-on-one interviews with tower control experts, we recorded eye-movement data from the SMEs while they controlled high fidelity air traffic simulations of airports at which they are certified. We found that by demarcating Areas of Interest (AOI) for analysis around meaningful regions in the tower controller experts' visual field and examining the frequency of eye fixations in and the direction of transitions between the AOI we are able to characterize controller scanning behaviors and identify patterns. A fixation is said to occur when a participant's gaze stayed in one location for at least 100 ms.

In this paper we discuss what we found regarding the usage, by our participants, of scanning best practices that were identified by our working group. A review of the curriculum at the FAA Academy shows that Tower controllers are taught to frequently scan hotspots. Hotspots include both ends of an active runway and places where traffic crosses the runway. A rule of thumb also taught at the Academy, referred to here as Airfield-Out, is to prioritize looking at the airfield over looking further out into the airspace. Another rule of thumb that is taught by some instructors in the field, referred to here as Backward Scanning, is to look at where an aircraft is going and scan backward to where its current location. The last best practice we will discuss, referred to here as Segmented Scanning, is the practice of stopping ones eyes, long enough to take in sufficient visual information, between hotspots when scanning along an active runway. This best practice was initially suggested by human factors subject matter experts. When tower control SMEs have subsequently been surveyed and interviewed during our studies, roughly two thirds of them expressed doubt that they used Segemented Scanning or that it would be useful, relating that a scan should be continuous and that slowing down one's scan would be ill-advised in a dynamic airport environment.

Method

Fifteen retired Tower controllers, employed as instructors by the FAA Academy in Oklahoma City, controlled simulated air traffic during 14 scenarios presented on an Adacel tower simulator. Twelve current front line Tower controllers (4 from Orlando International Airport, 2 from Denver International Airport, 2 from Centennial Airport, and 4 from Minneapolis-St. Paul Airport), controlled simulated air traffic during 4 scenarios presented on Adsync tower simulators located at their respective airports. Scenarios ranged between 22 and 38 minutes in length and represented busy air traffic during daylight hours with high visibility conditions. The instructor participants had between 10 and 42 years of experience as Tower controllers, averaging 26 years. The data from 3 instructors were dropped due to technical simulation and participant non-compliance issues. The current controllers had between 9 and 32 years of experience in Towers with an average of 16 years. Due to differences between Centennial operations and the operations at the other airports, the analyses in this paper does not include the data from the 2 Centennial controllers.

In all cases the training simulators used presented a high fidelity representation of an airport across ten or more 55" or larger screens configured to wrap greater than 180 degrees around trainees. All the simulations also included flight progress strips and simulated BRITE Radar displays. Operations at all airports (apart from Centennial) were configured to use two non-crossing runways. The airport presented to the instructors, although based on a real tower, was greatly modified to meet Academy training needs. Instructors therefore had never worked real traffic at this airport but were highly familiar with the simulation. In all cases the participants fulfilled the role of the Local controller, issued clearances to aircraft using a standard communication headset and wore a Tobii Pro Glasses 2, head-mounted eyetracking system (equipped with prescription lenses as necessary).

Results

We derived ATC performance levels for the instructors by measureing time to detect 6 off-nominal events scripted to occur within the scenarios. Off-nominal events included: a smoking aircraft engine, the appearance of a flock of birds, a non-compliant ground vehicle near the runway, an aircraft attempt to land on an incorrect runway, an aircraft attempt to taxi to the wrong runway and an aborted takeoff. Time to detect was measured using eye-tracking video recordings, starting at the time the participants verbally responded to an event and viewing backward until the initiation of the participant's last fixation on the presentation of a target associated with the event. Then we subtracted the time of event onset from the time of fixation initiation. Average fixation times for events ranged from 14 to 204 seconds. We designated the six instructors with the shortest detection times as the high scoring group and the six with longest detection times in the low scoring group. Given the short length of time the front line controllers were available to participate, we did not collect performance measures at the airports we visited.

Working with SMEs, we identified AOI based on operational significance. We identified AOI associated within the airspace (where aircraft could be seen on final approach, downwind midfield, the departure corridor, and the BRITE Radar), AOI associated with the airfield (where aircraft touched down on runways, runway crossings, runway midfield, the departure end of the runway and ASDE surveillance screens) and other general information sources like the flight progress strips and the Automatic Terminal Information Service display. Figure 1 shows the AOI for Academy Tower.

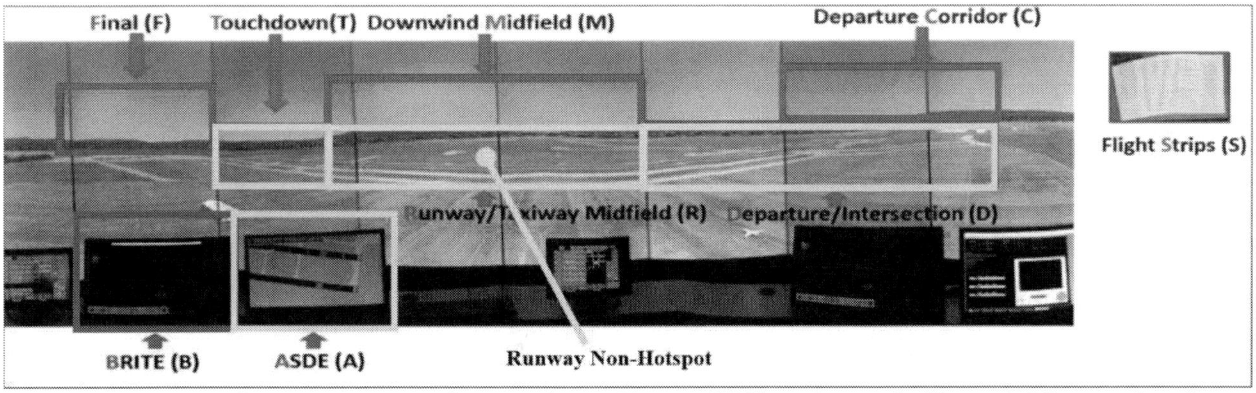

Figure 1. Areas of Interest (AOI) for Academy Tower. Yellow boxes indicate AOI associated with "Airfield" and red boxes indicate AOI associated with airspace or "Out".

Number of fixations at AOI was useful for characterizing participant scanning behavor. Heat maps, that depict relative number of fixations superimposed on pictures of the simulated environment, were a useful way to visualize those numbers. For example, Figure 2 shows how controllers do fixate more often at hotspots specified in the training curriculum, than at other locations on the surface.

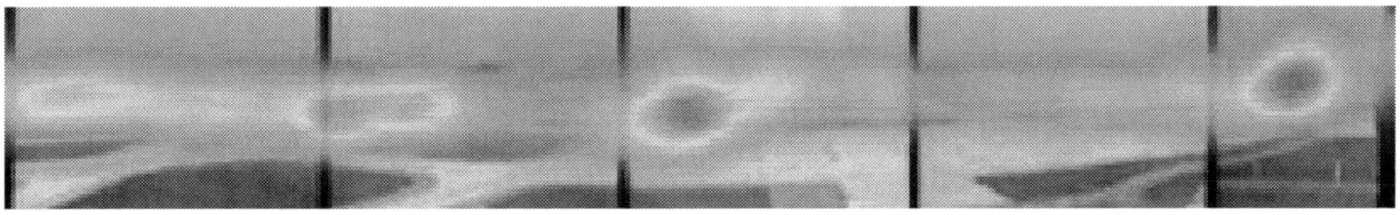

Figure 2. Heat map showing example of hotspots at Minneapolis-St. Paul Airport. Green shading indicates controller fixations with moderate frequency, yellow with higher frequency and red with the highest frequency corresponding to the ends of the runway and at runway crossings.

We compared number of fixations at AOI to answer questions about the usage of the Airfield-Out and Segmented Scanning rules of thumb as well. Both the high and low scoring instructors and the current controllers fixated rougly twice as often at Airfield AOI as they did at AOI associated with Airspace (see Figure 3). There were no significant differences between high and low scoring groups. Usage of Segmented Scanning was assessed by comparing the number of fixations at hotspot AOI (touchdown and runway crossing AOI at Academy Tower, Denver International and Orlando International airports) with an AOI not associated with hotspots that was located on the runway between the two hotspots. Although a greater percentage of fixations were at hotspots, between 7 and 14% of Airfield fixations did occur at the runway location between them (see Figure 4).

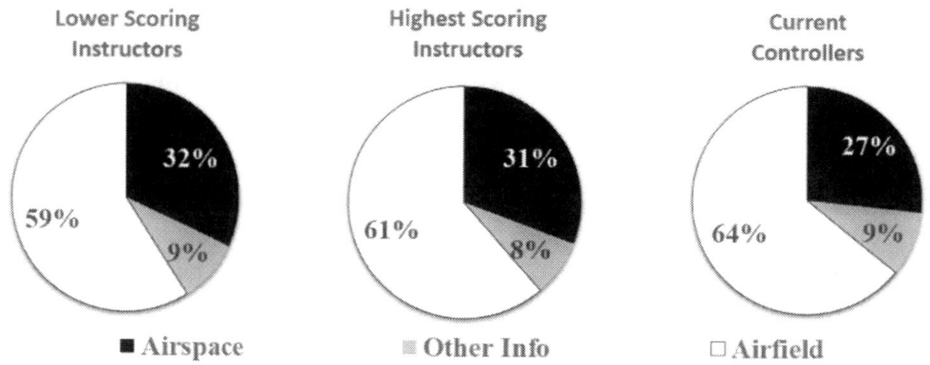

Figure 3. Percentage of fixations associated with Airspace, Airfield and Other Information sources across both instructor groups and the current controller group.

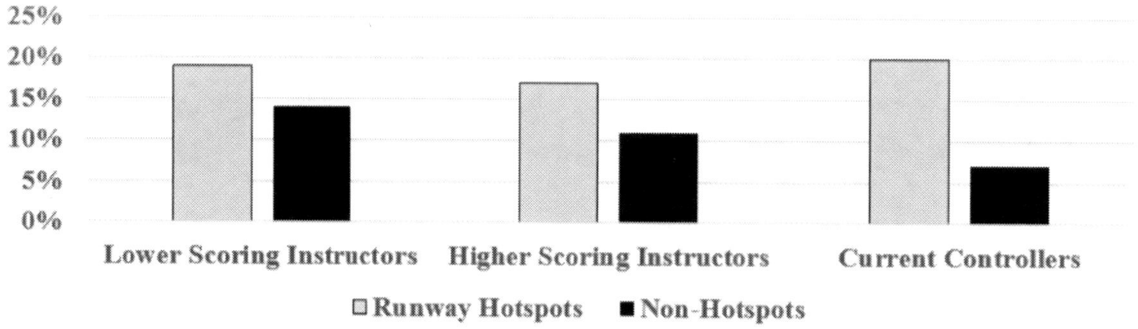

Figure 4. Average percentage of Airfield fixations for Hotspot and Non-Hotspot AOI across both instructor groups and the current controller group.

The usage of Backward Scanning was explored by looking at the eye-movement transitions that occurred after the cleared for take-off clearance. For aircraft arrivals, Backward Scanning is similar to Airfield-Out, with a controller's eyes moving from the runways where an arriving aircraft is headed, back to the airspace where the aircraft is currently located. For departures, Backward scanning means the controller's eyes will move from the departure corridor in the airspace back to the runway surface on the airfield. Therefore we looked at the number of transistions to and from the departure corridor that participants made during a time when they were working departures. We found no evidence of Backward Scanning over and that done in association with Airfield-Out.

Discussion

Analyses of eye-tracking data, recorded from both Tower controller instructors and current Tower controllers as they controlled simulated air traffic, allowed us to characterize the scanning behavior of air traffic controllers fulfilling the Local control function. Number of fixations at AOI and number of transitions between particular AOI support that expert controllers frequently scan hotspots on an active runway and practice the Airfield-Out rule of thumb. Additionally, data supported that controllers will sometimes use the Segmented Scanning rule of thumb although this usage is less prevalent.

We grouped the Instructors into low and high performing groups based upon the speed at which they detected off-nominal events. We compared the usage of rules of thumb across the groups but were unable to find an impact on detection performance related to usage. One consiideration is that these rules of thumb may not impact performance. Otherexplanations for the lack of relationship include that our measurement was not powerful enough to detect the impact, that there weren't enough samples to test adequatly, or that the performance of the experts who participated in the study produced a ceiling effect. Should further studies include novices that make use of the rules of thumb less often, the impact on detection performance may show up as statistically significant. Regardless, the fact that the three groups showed similar usage of the rules of thumb suggests that we could emphasize these in training novices to use standard visual scanning techniques.

Although the research presented in this paper addresses many of the questions we set out to explore in response to the 2016 working group recommendations, several questions still remain. The data collected up to this point have positioned us to address the questions about the use of specific scanning patterns as well. We are currently analyzing eye-movements made by our participants, during these scenarios, that occured near in time to when the participants delivered certain types of clearances. Clearances include "hold short", "line up and wait", "cleared to land" and "cleared for takeoff". In our analyses we will try to find consistent patterns within and among controllers such that we can attempt to teach novices to practice the same patterns before, during and after delivering those clearances.

Acknowledgements

This research was supported and funded by the FAA NextGen Organization's Human Factors Division, ANG-C1.

References

Kang, Z., Mandal, S., Crutchfield, J., Millan, A, and McClung, S. (2016). Design concepts and algorithms in eye tracking research for a multi-element objects tracking-and-controlling task to support human performance analysis. *Computational Intelligence and Neuroscience*, 2016(1), 1-18.

Kearney, P. and Li, W.C. (2018). *Multiple remote tower for Single European Sky: The evolution from initial operational concept to regulatory approved implementation*. Transportation Research Part A: Policy and Practice, 116, 15-30.

Mandal, S. and Kang. Z. (2018). Using eye movement data visualization to enhance training of air traffic controllers: A dynamic network approach. *Journal of Eye Movement Research*, 11(4), 1-20.

Pinska, E. (2006). *An investigation of the head-up time at tower and ground control positions*, EUROCONTROL Experimental Centre, Bretigny.

SPATIAL-TEMPORAL CLUSTER APPROACH TO DISCOVER VISUAL SCANNING BEHAVIORS IN VIRTUAL REALITY

Ziho Kang
University of Oklahoma
Norman, Oklahoma
Jerry Crutchfield
Federal Aviation Administration
Oklahoma City, Oklahoma
Ricardo Palma Fraga
Saptarshi Mandal
University of Oklahoma
Norman, Oklahoma

If we could discover the visual scanning patterns of expert air traffic control operators (ATCOs), we could use those findings to better train novices. One critical issue is that visual scan paths can be complex even for a short period of time, therefore, a systematic approach is required to obtain clear and meaningful visual scanning patterns. We transformed the raw eye movement data of expert ATCOs into visual scanning patterns by using the collapsed eye movement sequences that occurred on important areas of interest, then visualized them based on accumulated time frames. We collected and subsequently analyzed controller eye movements that occurred before and after controllers issued takeoff clearances, in a high-fidelity virtual reality airport tower. We obtained clear visual scanning patterns from our analyses of eye movement data. We plan additional investigation to determine whether tower controllers can be trained to employ effective visual scanning behaviors.

Airport tower controllers perform a cognitively complex and visually demanding task. They must monitor multiple aircraft by visually scanning runways and airspace located outside the tower and the surveillance displays and flight progress strips located inside the tower. They then must integrate information from various sources and issue clearances and advisories to aircraft. Tower control instructors teach trainees what information is needed before clearances can be delivered, and when to look for the information. For example, trainees must learn to continually scan, making visual contact with the aircraft in their area of responsibility. They also must scan the arrival end and crossing points of active runways frequently to identify potential traffic conflicts. Beyond this, however, we currently lack evidence of how these experts visually scan their operational environment.

Our research team seeks to characterize the visual scanning behavior of Air Traffic Control Operators (ATCOs) controlling traffic in the airport tower environment. If we could discover clear visual scanning patterns, especially how the expert ATCOs visually scan important information sources based on their eye movement transitions among those sources, we should be better able to understand their underlying cognitive processes. Training organizations can use the findings to improve the instruction of novices.

Other researchers have previously applied the use of eye-tracking technology to air traffic control at airport towers (Li, Kearney, Braithwaite, and Lin, 2018; Manske and Schier, 2015; Svensson, 2015). These researchers collected eye fixation numbers/durations or pupil sizes at Areas of Interests (AOIs) within the ATCO's field of view. These studies were conducted primarily as a way to evaluate new systems and look for ways to improve controller efficiency. In the past, our research team has used eye-tracking analyses specifically to inform controller training but these studies were conducted in the En Route air traffic control environment and not in airport towers (Kang and Landry, 2014; McClung and Kang, 2016). We analyzed the visual scan paths of expert ATCOs scanning a primary Radar display for aircraft conflicts. In these studies, we were successful at discovering expert visual scanning patterns that might be taught during training and improve conflict detection performance of novices.

We analyzed eye-tracking data recorded from ATCOs while they controlled a high fidelity simulation of airport traffic. The sheer number and apparent randomness of eye fixations and saccades makes it challenging to extract clear visual scanning patterns, suitable for use in training, from complex visual scanning paths. Figure 1 shows an example of an expert's raw visual scan path over a 10 minute interval.

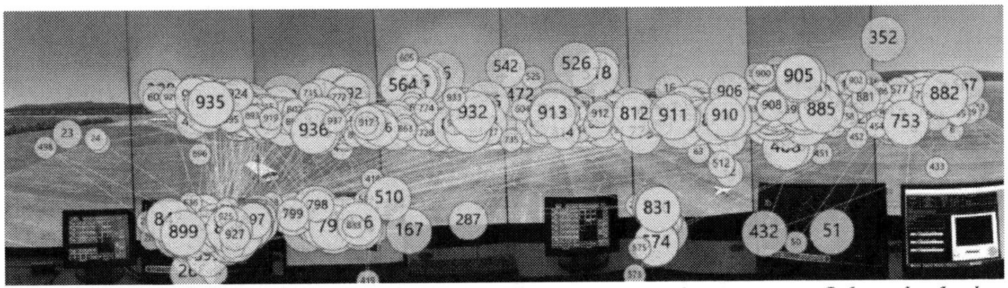

Figure 1. Visual scan path of an expert for a 10 minute period: Center of the circle is the eye fixation location, numbers indicate the time-ordered eye fixation sequences, and the yellow lines indicate the fast transitions among the eye fixations.

The purpose of this study is to investigate whether we could discover consistent and effective visual scanning patterns used by expert ATCOs when at safety-critical moments, such as when issuing a takeoff clearance. Ultimately, we expect that our research will enable us to identify visual scanning best practices that we can teach to new ATCOs to improve their visual scanning skills.

Visual Scanning Data Analysis

We created a systematic data analysis approach by combining the three approaches briefly explained in the Introduction section. Here are the eight steps that we used:

Spatial-temporal clustering approach

Step 1. Define AOIs to cover the full visual field of view.
Step 2. Apply the existing I-VT, I-DT, or similar algorithm to create eye fixations (Komogortsev et, al., 2010).
Step 3. Aggregate consecutive eye fixations that occurred on the same Area of Interest (AOI).
Step 4. Identify important AOIs (where most of the eye fixations occurred).

Step 5: Filter only the visual scan paths that occurred on those important AOIs.
Step 6. Identify the directions of the eye movement transitions from one AOI to another AOI.
Step 7. Visualize the eye movement` transitions based on accumulated time frames before and after an important event occurs.
Step 8. Express the amount of eye movement transitions based on the thickness of the transition lines (i.e. the more the transitions, the thicker the lines).
Step 9. Create aggregated AOIs to further simply the visual scanning patterns (e.g. if some AOIs are physically closers to one another, group them into a larger AOI).

Method

Three retired tower controllers were recruited for the experiment held at the Civil Aeronautical Medical Institute (CAMI) in Oklahoma City. Twelve 55" HD (1080p) monitors were used to simulate the out the window view of the airport. Tobii Pro Glasses II (100 Hz) were used to capture participant eye movements. A MaxSim simulator, developed by Adacel Systems Inc., was used to develop and present the simulated traffic. Scenarios were scripted to run for approximately 22 minutes and to have multiple aircraft take-offs and landings. Only eye movements during take-offs were analyzed for this paper. The layout is shown in Figure 2. The task was to have the local controller (i.e. the participant) communicate with the ground controller and the pseudo pilot (who followed the commands provided by the local controller) and issue commands such as takeoff clearances.

The raw eye movements were processed using our eight step analysis approach provided above. The eye fixation durations were not considered since the ATCOs were vigilantly viewing the field of view and our interest was on the directional eye movement transitions among the important AOI. The AOI depicted in Figure 2 were identified by asking subject matter experts to indicate the location of operationally significant information. The scanning patterns were analyzed for 10 seconds, 20 seconds, and 30 seconds before and after the time at which a clearance was issued.

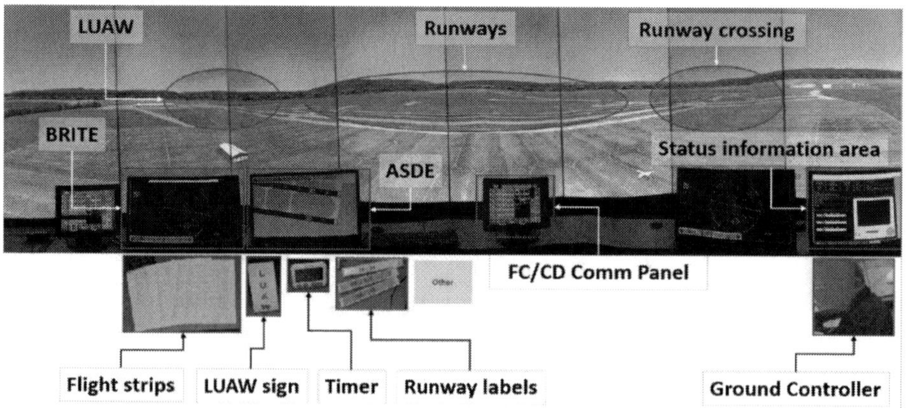

Figure 2. View of the expert local controller (i.e. the participant). LUAW stands for Line Up And Wait, BRITE stands for Bright Radar Indicator Terminal Equipment, and ASDE stands for Airport Surface Detection Equipment.

Results

Figure 3(a) shows the results of the aggregated eye movements of one participant using a 60ms threshold for a 30 second period before the cleared for takeoff clearance was issued. Since the direction of the eye movement transitions are not clearly visible, the eye fixations sequences (represented as numbers 1 through 56 in Figure 3(a)) were replaced with direction indicators (i.e. arrows) and the location of eye fixations were substituted with the AOIs (see Figure 3(b)). We defined the important AOIs to be "LUAW," "runway," "runway crossing," "flight strips," "ASDE," and "runway labels" since most of the eye fixations (in our case, 100% of the eye fixations) occurred on those AOIs.

Figure 4(a) shows how the eye movement transitions were simplified using line thickness. Figure 4(b) shows how we further abstracted the AOIs into simpler expressions and aggregated the AOIs within the "inside view" (closer to one another compared to other AOIs) into a larger burgundy color AOI. When we observe Figure 4(b), a visual scanning trend begins to emerge.

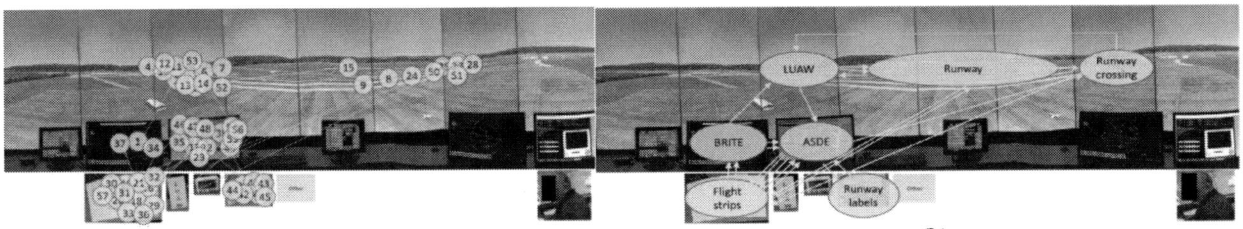

(a) **(b)**

Figure 3. Example outputs for a 30 second duration before the "clear to take off" command was issued.

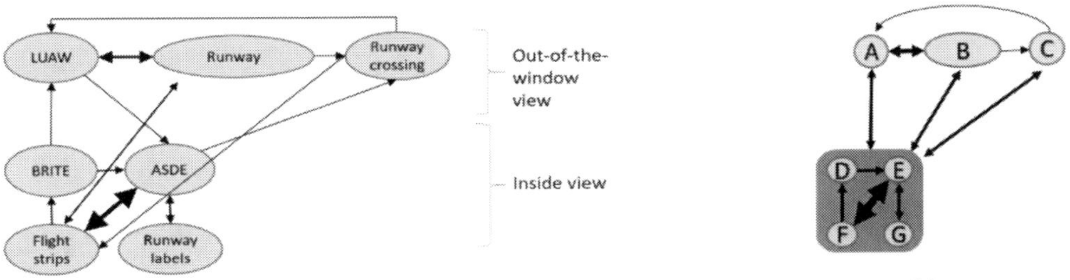

(a) (b)

Figure 4. Effective visualization of the visual scanning behavior

The output of all three expert ATCOs are provided in Figure 5. The dotted lines (in the center) indicate the point in time when the clearance was issued. The eye movements were accumulated from the point in time that the ATCO issued the takeoff clearance. We accumulated eye movements for three intervals before and after the takeoff clearance. The visualized intervals included eye movement data for 10s, 20s, and 30s after the clearance was issued.

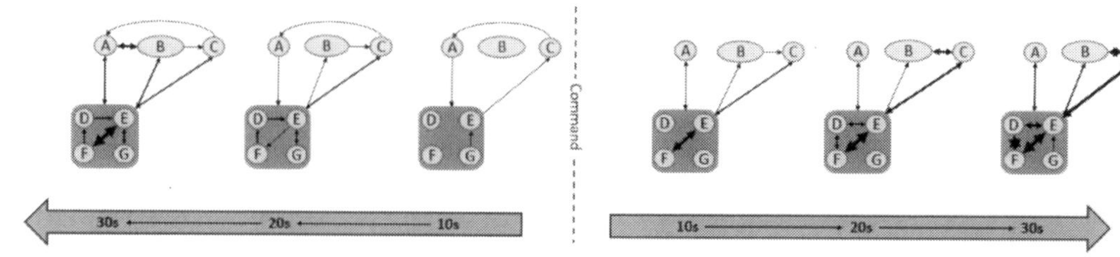

(a) Expert 1

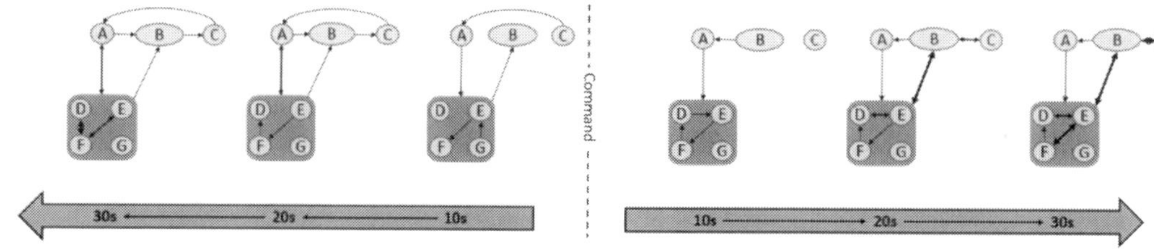

(b) Expert 2

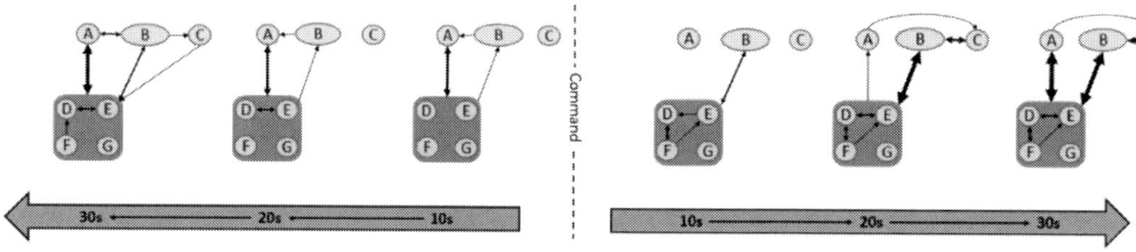

(c) Expert 3

Figure 5. Accumulated visual scanning patterns of three experts.

Discussion

Using our spatial-temporal clustering approach in the analysis of ATCO visual scanning eye tracking data, we were able to extract clearer visual scanning patterns. Our approach provides a foundation to aggregate and visualize patterns from many data sets as well as multiple time frames within each data set. In this study, we were able to express eighteen visualizations in a single figure (Figure 5) to create a holistic view of the visual scanning patterns. This enabled us to discover a visual scanning pattern that could be easily described and taught to new ATCOs.

By analyzing the visualizations, the "vigilant" scanning behavior of the expert ATCOs can be explained as the controllers actively interrogate the information shown on the BRITE/ASDE radar displays and flight strips, then verify the information by looking out the window and observing the aircraft on the LUAW, runway, and runway crossing. The thickness of the lines shows that many eye movement transitions occur between the inside view and out-of-the-window view which coincides with the expert ATCOs' vigilant scanning efforts on knowing what they are looking for and when to look for it. The runway labels were not always observed as intensely as other AOIs, but we believe that the runway labels might be more actively used/observed if the number of aircraft at the airport increases or the runway configuration becomes more complex.

We were able to clearly show the evidence of the experts consistently and vigilantly interrogating the inside view information and out-of-the-window information before and after issuing a command. In addition, the visualization shows that the controllers attend to "hot spots" such as the LUAW and the runway crossing, but also attend the runway to check for any other anomalies. If we could track and analyze their eye movement data, we would be able to provide timely feedback to the novices.

Acknowledgements

This research was based on collaboration with the Aerospace Human Factors Division, FAA Civil Aerospace Medical Institute at the Mike Monroney Aeronautical Center, OKC, OK (Project #: A18-0232). In addition, the concept of the spatial-temporal clustering introduced in this paper is an integral part of the first author's NSF CAREER project titled "Non-text-based smart learning in multi-person virtual reality" (Award #: 1943526).

References

Kang, Z., & Landry, S. J. (2014). Using Scanpaths as a Learning Method for a Conflict Detection Task of Multiple Target Tracking. *Human Factors: The Journal of the Human Factors and Ergonomics Society*, *56*(6), 1150–1162.

Komogortsev, O. V., Gobert, D. V., Jayarathna, S., Koh D. H., and Gowda, S. M. (2010). Standardization of automated analyses of oculomotor fixation and saccadic behaviors, *IEEE Transactions on Biomedical Engineering*, 57(*11*), 2635–2645.

Li, A., & Chen, Z. (2018). Representative Scanpath Identification for Group Viewing Pattern Analysis. *Journal of Eye Movement Research*, *11*(6), 1–17.

Li, W. C., Kearney, P., Braithwaite, G., & Lin, J. J. H. (2018). How much is too much on monitoring tasks? Visual scan patterns of single air traffic controller performing multiple remote tower operations. *International Journal of Industrial Ergonomics*, *67*(May), 135–144.

Manske, P. G., & Schier, S. L. (2015). Visual Scanning in an Air Traffic Control Tower – A Simulation Study. In *6th International Conference on Applied Human Factors and Ergonomics (AHFE 2015) and the Affiliated Conferences, AHFE 2015* (Vol. 3, pp. 3274–3279). Elsevier.

Mcclung, S. N., & Kang, Z. (2016). Characterization of Visual Scanning Patterns in Air Traffic Control. *Computational Intelligence and Neuroscience*, *2016*(1), 1–17.

Svensson, Å. (2015). *Air traffic controllers' work-pattern during air traffic control tower simulations : A eye-tracking study of air traffic controllers' eye-movements during arrivals*. Linköping University Examiner: https://doi.org/10.1108/02686900410549457

Wickens, C. D., Hooey, B., L., Gore, B., F., Sebok, A., & Koenicke, C. S. (2009). Identifying black swans in NextGen: Predicting human performance in off-nominal conditions. *Human Factors, 51*, 638-651.

MEASURING THE RANGE OF ATTENTION TO PREVIEW AND ITS MOMENTARY PERSISTENCE IN SIMULATED DRIVING

Tyler N. Morrison
Department of Mechanical Engineering, Ohio State University
Columbus, Ohio
Richard J. Jagacinski
Department of Psychology, Ohio State University
Columbus, Ohio
Jordan Petrov
City and Regional Planning, Ohio State University
Columbus, Ohio

Participants used a position control system to track the center of a simulated winding roadway with preview that ranged from 0.3 to 1.0 s. Participants' spatial distributions of attention were measured by perturbing the roadway with different frequency sinusoids at different roadway positions and then measuring the degree to which those frequencies were present in their tracking movements. Participants exhibited a continuous range of attention, and it lengthened with the amount of displayed preview. When preview disappeared for 5 s, longer time to regress to feedback control was strongly correlated with the amount of preview that was withdrawn. During preview withdrawal, visual sensory memory of the previewed roadway may be used for a fraction of a second to prolong the period of feedforward control. Attention may be shifted to relevant positions of the sensory memory image to anticipate the roadway curvature. The present methodologies may be useful in aviation contexts.

ATTENTIONAL TRAINING IN MULTITASKING FOR UAS SENSOR OPERATORS

Christopher Wickens, (Colorado State University Department of Psychology)
Angelia Sebok; (TiER1 Performance)
Nathan Herdener, Kimberly Spahr, Benjamin Clegg (Colorado State University)

We developed a simulation training battery for the multi-tasking skills required by the sensor operator of a Navy Unmanned Air System when managing subtasks. Specific attentional skills of scanning, dynamic task prioritization, and interruption management were adaptively trained. Six hours of training was administered followed by a transfer trial, and performance was compared with a control group who performed on the same task without specialized adaptive training of the three attentional skills. Although no benefit was observed by the final transfer test, the ATTICUS training did produce significant benefits during training on the important continuous monitoring and situation awareness tasks, and a cost on one of the discrete tasks.

Introduction

Unmanned Aerial Systems (UAS) in the Navy are commonly controlled and supervised by two aviators on the ground, the pilot and the sensor operator (SO). The focus of the current research is on the demands of the SO which frequently exceed the redline of workload from the array of tasks they must concurrently perform. SOs monitor large number of ships using multiple sensors. These sensors are vulnerable to temporary failures when the UAS enters unfavorable operating conditions, due to temperature, humidity, visibility, or other changes. Under these conditions, their operation must be restored, even as the surveillance is continued.

The current research developed a multi-task battery to train multitasking skills necessary to preserve workload below the redline, and avoid any attentional narrowing or cognitive tunneling, that might compromise the primary task of surveillance monitoring. The research progressed in five phases. First, interviews of subject matter expert (SME) sensor operators at Naval Air Station Patuxent River were used to validate concerns on workload overload (one reported that "sensors fail about 10% of the time), understand their tasks, and design a prototype of our battery for them to approve. Our design addressed two competing goals: (a) Achieve sufficient fidelity to capture the information processing (particularly attentional) demands of the SO and some component of the "look and feel" (i.e., greater realism than alternative platforms such as MATB); (2) Attain sufficiently generic and simplified elements so that participants in our initial validation, with none of the specialized training of the SO, could adequately master the task after a few hours of training. With these factors in mind, our second phase completed the design.

The third phase developed adaptive training strategies to foster necessary time sharing skills through explicit training, and adapt the nature of the task as skill developed. To carry out this phase, an extensive literature review identified well understood attentional components of multi-tasking and documented both their "trainability" and their transfer to environments beyond the training tool. Three such attentional skills were identified: visual scanning (S), task prioritization (P) and Interruption management (I).

Visual scanning (S) encouraged the operator keep his or her eyes moving across the various sources of information. Such skills are taught to aircraft pilots and vehicle drivers, where they learn to monitor the out-the-window view as well as other sources of information (e.g., displays, mirrors). Visual scanning is a skill whose performance can be improved by training (Fisher & Pollatsek, 2007). Different domains require specific visual scanning patterns, driven by the importance of information and the frequency with which it changes (Wickens & McCarley, 2008). Training aimed to allocate visual attention to different areas in proportion to the importance and bandwidth (frequency of change) of information at those areas (Fisher & Pollatsek, 2007).

Prioritization (P) is invoked when multi-tasking, resource allocation must respond to dynamic changes in difficulty or priority of those tasks (Gopher, 2007). Research indicates that these skills can be taught and transferred to a more complex environment (e.g., Gopher, Weil, & Barakeit, 1994; Gopher, 2007).

Interruption management (I) is invoked when periodic interruptions require the operator to divert attention from an ongoing task to deal with an interrupting event, and then return fluently to the ongoing task (Wickens & McCarley, 2008). Fluency represents latency to resume the ongoing task, and accuracy at the point of resumption. Exposure to, and practice with, interruption management can improve performance in multitasking situations. Specific interventions can train operators to make a mental note of their "place" and next step needed in an ongoing task, prior to shifting attention (Trafton & Monk, 2007). This intervention supports prospective memory, and is an effective technique for improving interruption management (Dismukes & Nowinski, 2007; Loukopoulos, Dismukes & Barshi, 2009).

Having identified SPI as a trilogy to be instructed, our fourth phase developed adaptive means of training these via **scaffolding removal.** Each attentional skill was given initial instruction, and then scaffolding techniques (visual highlighting to guide attention) were developed to guide the learner through deployment of the strategy early in training, much like training wheels (Hutchins Wickens & Carolan, 2013). Support is adaptively removed as the S, P or I skill improves. Adaptive training, in other skills, has been found to be a reasonably successful technique (Landsberg et al., 2012), although often challenging to implement.

The fifth phase was to collect experimental data to determine how well and rapidly the skills could be trained, and how they might transfer and be retained. Our adaptive ATTICUS training regime was administered over the course of 14 x 20 minute scenarios. Two different comparison conditions were also run. The **maximum difficulty** condition presented the six tasks together from the very beginning (like the ATTICUS group), but contained neither the specialized SPI instructions and scaffolding nor, (obviously) their adaptive removal (since there was nothing to remove). The **fixed increase** condition again contained no SPI instructions, but incrementally increased the difficulty of all subtasks, at the same rate—schedule—for all participants, until reaching the same final level of difficulty as for the other two groups.

Methods

The study was approved by the Colorado State University Institutional Review Board.

45 paid participants were recruited from posters and on-line advertisements to participate in the 6 ½ hour experiment, carried out over 3 sessions within a one-week span, plus a final half hour retention test approximately 2 weeks later.

The ATTICUS task battery, shown in Figure 1, was displayed on a 23" computer screen, subtending a visual angle of 38 x 22 degrees when the participant was seated approximately 75 cm from the screen. The display hosted information for five tasks, described below.

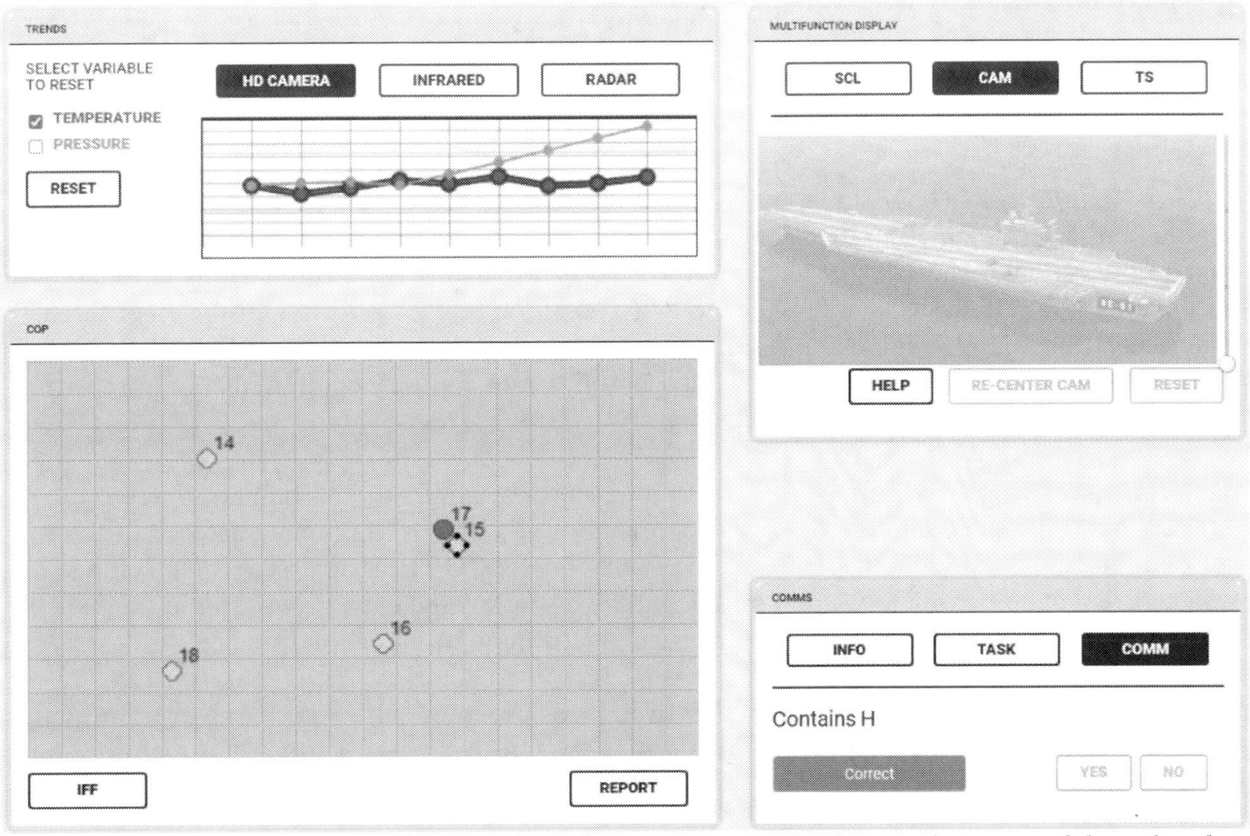

Figure 1: The ATTICUS test bed display (see text for further explanation of the subtasks).

1. ***Common Operating Picture (COP)***. The primary task of the SO is building and maintaining situation awareness of the maritime traffic. Icons representing ships were present in the display on the lower left, updating positions every second. Participants had to identify and flag those engaged in suspicious behaviors: altering course, accelerating (to 50% increased speed), rendezvous with another ship and then separating, pairs of ships moving in parallel (formation), and entering the screen from a pre-designated suspicious direction. These events occurred at random intervals of approximately every minute. There was an average of five ships on the screen.
2. ***Camera task.*** Ships that engaged in suspicious behavior required the operator to seek detailed information on that ship, using three analog controls to control a camera view (upper right window) to locate the hull number, and enter that into the Ship Classification interface.
3. ***Sensor trend monitoring***. Sensor parameters (e.g., temperature, pressure) and trends were displayed on tabs of the window in the upper left. Participants needed to cycle through three

different sensors, each with temperature and pressure indicators, updated every second. If sensor data indicate parameters moving out of range, the operator would intervene to correct the problem by clicking a reset icon. Trend failures were not indicated by any discrete alert.

4. ***Sensor troubleshooting***. A repair sequence required the operator to look up the code associated with an unreliable sensor, diagnose the failure, and then choose the appropriate repair code. This was accomplished through an interactive display that could be called up within the same window as the camera task. In contrast to the trend monitoring task, these major failures were signaled by a red alert. These major troubleshooting events occurred with a mean frequency of one event per min.

5. ***Communicate***. Operators listened to periodic auditory communications, and responded to only designated call signs through entering a corresponding alphanumeric sequence. These also occurred randomly with a frequency of one per minute.

Task scheduling. The COP and the Trend monitoring task characterize what Wickens, Gutzwiller, and Santamaria (2015) characterized as "ongoing tasks," in that they require **continuous** situation awareness (to be performed perfectly) and hence are heavily demanding of visual attention. Of these, as noted, the COP task is of the highest priority. Of the other three tasks, all **discrete** tasks, camera and ship classification are necessary and predictable follow-on's to events within the COP. In contrast, troubleshooting and auditory communications are true **interrupting tasks,** occurring unpredictably.

Experimental Design and Procedures. In an initial two-hour session, participants were introduced to the study and provided approximately five minutes of practice on each of the single task components. Participants performed a pre-test on the COP task, to assess monitoring skill and assure that approximately equal initial skill levels populated each group. Based on their pre-test performance, participants were assigned to one of three training conditions.

1. ***ATTICUS Adaptive Training***. This group was provided with approximately 15 minutes of specific instructions on the three critical attention strategies, as well as the procedures of scaffolding and its removal.

2. ***Maximum Difficulty.*** This group was identical to the Atticus group, except they were given none of the attentional strategies instructions nor received any scaffolding on subsequent trials (and hence adaptive scaffolding removal).

3. ***Difficulty Increase***. In contrast to the other two conditions, this group started dual task training with all five tasks adjusted to very easy levels (e.g., initially only 2 ships on the COP), and as the training progressed these were incrementally increased in difficulty, on a fixed schedule (the same for all participants) to eventually reach a target level identical to the first two groups, one scenario prior to the final one.

Participants then proceeded through 14 twenty-minute training scenarios, scheduled over three days, each within 2-hour sessions. Each scenario was generated with different sequencing of events, so that this was unpredictable by the participant, and for the Atticus and Maximum Difficulty conditions, all scenarios were the same, and of approximately equivalent difficulty. On two thirds of the scenarios, the ship-load of the COP task would ramp up from 5 to 10 for 1 minute, and then decrease back to 5, so that we could provide an explicit period to assess task prioritization in response to those increased demands. On a final transfer trial (Scenario 15), participants in all three groups received the identical scenario. Approximately two weeks later, participants received a final delayed assessment, consisting of a different scenario.

Results.

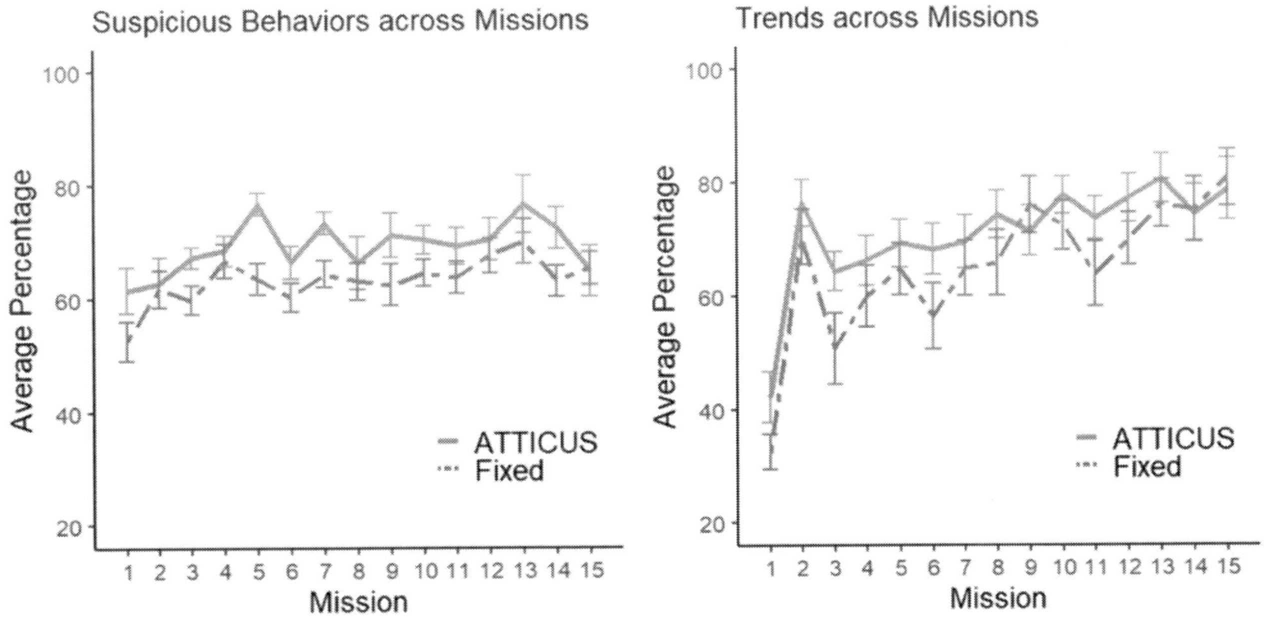

Figure 2: Training data for COP task and Trends task for the Atticus versus Fixed Difficulty groups across 14 training trials plus the final test trial.

Training trial data are only presented for the two groups training at the maximum difficulty, because performance on the increasing difficulty condition throughout most of the experiment is much better (since the tasks were much easier). A multilinear regression model fit through the four curves shown in figure 2 revealed, for the COP, a marginally significant advantage for the ATTICUS group ($F(1,28) = 3.96$, $p= .06$, $\eta^2_p=.12$), and, for the trend monitoring task, a significant ATTICUS advantage ($F(1,28)=7.58$, $p=.01$, $\eta^2_p=.21$). The Trend monitoring task also showed a significant improvement over trials (training effect $F(1, 28) = 10.90$, $p<.01$, $\eta^2_p=.59$), while the COP task did not. Neither the comms task nor the troubleshooting task revealed a difference in accuracy during learning, while the Camera task showed a marginally significant cost for the ATTICUS group ($F(1,28)= 3.89$; $p=.06$, $\eta^2_p=.12$). There were no significant differences between the three training conditions on the transfer trial, on either of the speed and accuracy measures of any of the six subtasks (all p values > 0.10).

Discussion

The results revealed no overall benefit of ATTICUS transfer after 14 sessions of dual task training. At the same time, selective benefits and some costs were interpretable and meaningful. In particular, the two continuous tasks, that can both described as "maintaining situation awareness" revealed an ATTICUS benefit throughout training. This benefit was modest for the most difficult (and important) task of maintaining the common operating picture, and quite large for the other, somewhat easier supervisory task – trend monitoring. The idea that more efficient training can be attained through the use of ATTICUS has important practical implications. ATTICUS training did not produce unmitigated improvement in multi-task efficiency but rather some tendency toward re-allocation of resources in that one of the discrete tasks – camera management followed by ship classification was slightly inhibited. Whether this was from neglect of the task, in the accuracy of reading and remembering the hull numbers, or the accuracy of their entry on the interface is yet to be determined.

The finding that the overall ATTICUS advantage to the continuous tasks, and minor cost to the camera task, is eliminated by the final transfer trial, suggests that the benefits of such training are realized early, and may wash out with extensive training signaling, among other things that such training need not be extensive.

Acknowledgments

This project was supported by the Naval Air Warfare Center Training Systems Division, Contract N68335-18-C-0138. Any opinions, findings, conclusions, or recommendations expressed are those of the authors and do not necessarily reflect the views of the Navy. The authors gratefully acknowledge the guidance and support of Laticia Bowens and Tashara Cooper (NAWCTSD), and our colleagues at CSU and TiER1.

References

Cades, D. M., Boehm-Davis, D., Trafton, J. G, & Monk, C. (2011). Mitigating disruptive effects of interruptions through training. *Journal of Experimental Psych: Applied*, *17*, 97-109.

Dismukes, R. K., & Nowinski, J. (2006). Prospective memory, concurrent task management, and pilot error. In A. Kramer, D. Wiegmann, & A. Kirlik (Eds.), *Attention*. (225-236). New York: Oxford University Press.

Fisher, D. L., & Pollatsek, A. (2007). Novice driver crashes: Failure to divide attention or failure to recognize risks. In A. Kramer, D. Wiegmann, & A. Kirlik (Eds.), *Attention: from theory to practice* (pp. 134–156). Oxford, UK: Oxford University Press.

Gopher, D. (2007). Emphasis change as a training protocol for high demands tasks. In A. Kramer, D. Wiegmann, & A. Kirlik (Eds.), *Attention: From theory to practice* (pp. 30-60). New York: Oxford University Press.

Gopher, D., Weil, M., & Bareket, T. (1994). Transfer of skill from a computer game trainer to flight. *Human Factors, 36,* 387–405.

Loukopoulos, L., Dismukes, K., & Barshi, I. (2009). *The multitasking myth: Handling complexity in real-world operations*. Burlington, VT: Ashgate.

Wickens, C. D., & McCarley, J. M. (2008). *Applied attention theory*. Boca-Raton, FL: CRC Press, Taylor & Francis.

Applying Human Factors Heuristic Evaluation Tools to Improve Aviation Weather Displays: A Mismatch

John Kleber, M.S.

Paige Lawton, B.S.

Beth Blickensderfer, Ph.D.

Embry-Riddle Aeronautical University

Daytona Beach, FL

Background. Weather-related accidents have one of the highest fatality rates among General Aviation (GA) accidents. Pilots obtain graphical and textual weather information from aviation weather displays during both preflight planning and while inflight. Interpretation scores of weather products remain low regardless of pilot certification/rating. Previous research identifies poor usability as one cause of weather displays' low interpretability. Given the frequency of updates to weather displays, a rapid usability assessment technique is needed. Heuristic evaluations are a common method for quickly identifying usability problems. Heuristics evaluations entail expert reviewers evaluating an interface using a validated set of heuristic guidelines. This paper examined using current heuristic tools to evaluate aviation weather displays. **Method.** Human Factors (HF) specialists identified 14 published heuristic sets and subsets. This included heuristics designed to evaluate information visualizations, user interfaces, and online documentation. The research team applied the heuristic tools to evaluate several types of aviation weather displays (e.g., Graphical Forecast for Aviation and Low-Level Significant Weather Chart). **Results.** The evaluation identified numerous characteristics of the tools that yielded them unusable for aviation weather displays. Mismatches include limitations to color usage, error prevention/recognition, and auditory elements. **Discussion.** The inspection of heuristic evaluations found no suitable sets for the use of evaluating aviation weather display. Current heuristic sets often include recommendations that are either not applicable to the aviation weather domain or do not match well with the domain's characteristics. Future research is needed to develop a validated set of heuristics specific to the domain of aviation weather.

ASSESSMENT OF A HORIZONTAL-VERTICAL ANISOTROPY IN UTILIZING AN AIRCRAFT ATTITUDE SYMBOLOGY

George A. Reis[*,**], Michael E. Miller[*], Eric E. Geiselman[**], David W. Dommett[**], Brent T. Langhals[*], Christine M. Schubert Kabban[*], and Julie A. Jackson[*]
[*]Air Force Institute of Technology, Wright-Patterson Air Force Base, USA
[**]711 Human Performance Wing, Wright-Patterson Air Force Base, USA

This study examined how visual field performance asymmetries affect visual processing of complex, meaningful visual stimuli, such as the Arc-Segmented Attitude Reference (ASAR). Participants' performance was collected in their recall and report of attitude information after briefly presented ASAR symbology within the peripheral visual field. Visual processing of the ASAR was assessed when making coordinate and categorical judgments at cardinal display locations for each of several flight contexts (roll left/right, climb, and dive). Primarily in coordinate tasking, performance trends were consistent with the horizontal-vertical anisotropy literature—performance is better on the horizontal meridian over the vertical meridian in the field of view. Such effects should be considered for determining symbology placement, particularly within head-worn displays.

The visual processing of stimuli is not equivalent across the visual field, to include locations that are isoeccentric. These perceptual asymmetries in the visual domain have a long standing research history, showing their manifestation to vary (Karim and Kojima, 2010). Much of the applicable past research methodologically employed simple visual stimuli, e.g., dots, lines, or Gabor patches. The present research examined whether these visual performance field asymmetries extend to real-world, complex stimuli in an attempt to aid interface design for head-worn or near-eye displays.

Reis, Geiselman, and Miller (2019) explored some performance differences utilizing the Arc-Segmented Attitude Reference (ASAR) at cardinal and ordinal locations in the visual field. The ASAR exemplifies a meaningful, real-world, symbology by representing aircraft roll and vertical flight path (VFP) angles relative to a natural horizon (Fuchs and Fischer, 1995). The results from Reis et al. suggested a concordance of a horizontal-vertical anisotropy (HVA) similar to past findings (Carrasco, Talgar, and Cameron, 2001); the visual processing of the ASAR was better in the "west" and "east" locations (left and right from center of display) over the "north" and "south" locations (top and bottom from center).

The current study tested the robustness of the Reis et al. (2019) study by analyzing the ASAR with more participants, a switch of response handedness, a more compact set of ASAR angles, and the inclusion of a Gabor stimulus to validate the experimental methodology. Lastly, we evaluated performance at four Gabor and eight ASAR operational flight contexts: collecting responses to either a categorical or coordinate task to an observed Gabor showing a particular left or right roll angle, an ASAR showing a particular left or right roll angle, and to an ASAR showing a VFP representing a particular climb or dive angle. Responding in a categorical manner entailed responding with an assessment of the ASAR's general status, such as representation of roll direction, i.e., left or right, a representation of VFP direction, i.e., climb or dive, or just roll direction in the case of the Gabor presentation (left or right). Responding to coordinate stimuli entailed assessing the ASAR or Gabor for its extent of roll or VFP change from a straight and level flight representation. Past research has shown that coordinate visual processing is better performed in the left visual field versus the right, whereas categorical visual processing is better performed in the right versus left (Kosslyn et al., 1989).

Method

Participants

Twelve participants (6 males, 6 females; mean age of 39) completed the study. All but one female were right-handed. One male and one female had experience piloting aircraft. All participants had normal or corrected-to-normal vision. The institutional review board from the Air Force Research Laboratory approved the study and participants gave informed consent prior to participation.

Experimental Set up and Stimuli

The experiment was run on a Dell Precision 5820 X-series with a 24.5" Sony PVMA250 Professional OLED Production Monitor (1920 x 1080 pixel resolution, 60 Hz refresh rate). The experiment was administered in the Unity 3D programming environment. Participants' heads were stabilized with a chin rest while they binocularly viewed the display at a distance of 57 cm. Their responses were registered on a ZD-V+ USB wired gaming controller gamepad. The test stimuli consisted of a Gabor patch at various roll (slant) orientations and the ASAR at various representations of an aircraft rolling left, rolling right, diving, and climbing. Gabor patches of dimension 100 horizontal x 100 vertical pixels were generated by multiplying 12 cycles of a sinusoid with a Gaussian function (see Figure 1). This stimulus subtended a visual angle of 3.5 degrees.

Figure 1. Gabor patch oriented at a 45° right roll.

The ASAR includes a fixed 'ownship' symbol that represents the VFP (climb/dive) angle by its relation to a half-circle arc surrounding the symbol. During straight-and-level flight, the upper portion of the circle is not visible and represents the area above the horizon as shown in Figure 2-A. The visible arc represents the area below the horizon. As the climb angle increases, the visible angle area of the arc narrows in proportion to the angle, as shown in Figure 2-B. Conversely, as the dive angle increases, the arc closes towards a circle, as shown in Figure 2-C. During rolling maneuvers, the arc rotates about the ownship symbol as shown in Figure 2-D and 2-E. The ASAR's lines in this study were white, and both the Gabor patch and the ASAR were presented against a gray background. Gabor roll orientations of 3°, 6°, 9°, 12°, 15°, 18°, 21°, 24°, 27°, 30°, 33°, 36°, 39°, 42°, 45°, and 48°, left and right, were presented to the observers. These degree increments were also tested for ASAR roll (left/right) and VFP (climb/dive) depictions.

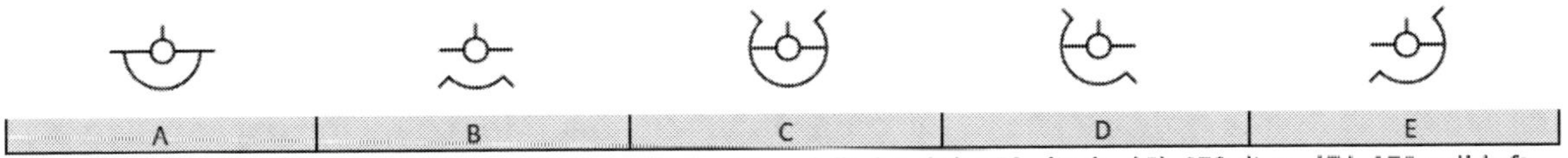

Figure 2. The ASAR representing (A) straight and level flight, (B) 45° climb, (C) 45° dive, (D) 45° roll left, (E) 45° roll right. At a climb/dive of 0° (panels A, D, E), the half-circle subtended 3.5° of visual angle.

Procedure

Participants performed six different 'situation' blocks (see Table 1), each with its own type of trials, repeated three times, totaling 18 blocks. Presentation order of these six types of blocks was counterbalanced across all participants. A ten day window was allowed to complete the totality of the 18 blocks.

Table 1. *The six different types of situation blocks.*

(1) Gabors rolling left or right and responding in a categorical manner.
(2) Gabors rolling left or right and responding in a coordinate manner.
(3) ASAR representations showing rolling left or right and responding in a categorical manner.
(4) ASAR representations showing rolling left or right and responding in a coordinate manner.
(5) ASAR representations showing climbing or diving and responding in a categorical manner.
(6) ASAR representations showing climbing or diving and responding in a coordinate manner.

A training session was administered at the beginning of the study to explain the mechanics of the ASAR. At the beginning of every test session, the participants spent one minute on a simulator, maneuvering an aircraft with the coupled ASAR behavior on the screen. Preceding any block of test trials, the participants performed 20 random trials from that block's situation type for familiarization. Each situational block contained 256 trials (8 x 2 x 16) where the stimulus was presented for 80 ms randomly, without replacement, across: 8 hemimeridian locations on the monitor, 13° of visual angle from center, in the polar coordinate system, with angles of 0°, 45°, 90°, 135°, 180°, 225°, 270°, 315°, 2 flight parameter directions (left or right if block contained roll trials; dive or climb if block contained VFP trials), and 16 angle deviations from straight and level, 3°, 6°, 9°, 12°, 15°, 18°, 21°, 24°, 27°, 30°, 33°, 36°, 39°, 42°, 45°, and 48°. Participants performed three repetitions of every combination of block, hemimeridian, direction (left/right or dive/climb), and angle deviation.

For each trial, participants were instructed to fixate on a crosshair symbol centered on the display. They initiated a trial by pressing a button on the left side of the controller and provided responses with their right thumb on the right joystick of the controller. The ASAR was presented for 80 msec in one of the hemimeridian locations. After presentation, the gray screen was replaced with a mask of static Gaussian noise to reduce visual persistence. If the situational context was roll/categorical, participants were instructed to respond by pushing the joystick in the appropriate direction to match the actual roll direction indicated by the ASAR or Gabor. Likewise, if the situational context was VFP/categorical, the participants responded by pushing the joystick distally, reporting that the symbology represented aircraft dive, or proximally, representing aircraft climb—mapping to the mechanization of aircraft control stick input. In both, roll and VFP categorical trials, response time (RT) and accuracy (reporting representation correctly or incorrectly) were recorded for each trial. After a response, the Gaussian noise disappeared and the crosshairs reappeared to begin the next trial. For all trials, participants were instructed to prioritize accuracy.

If the situational context was roll/coordinate, after the ASAR or Gabor disappeared in the periphery, the stimulus reappeared in a straight and level attitude in the middle of the screen with Gaussian noise as a backdrop. The participants then attempted to match the exact roll angle of the stimulus by moving the joystick left or right. Likewise, if the context was VFP/coordinate, after the ASAR disappeared, the participants changed the straight and level ASAR by moving the joystick proximally or distally to match the climb or dive angle. After the participants obtained the attitude they thought they observed, they pressed a button on the left side of the controller with their left index finger and the crosshairs reappeared. Absolute error (AE) was the dependent variable (DV) of interest: i.e., the absolute value between the actual ASAR/Gabor attitude presented and the participants' attitude responses.

Data Analysis

For this reporting, the data are limited to the 0°, 90°, 180°, and 270° hemimeridian positions. Additionally, to remedy possible attentional lapses and anomalous anticipatory responses, the response data set was curtailed to the median values from the three repetitions at each combination of situational context x hemimeridian x angle deviation x direction. These medians were then averaged across all angle deviations at each situational context x hemimeridian x direction combination. For this analysis, the derived DVs for categorical trials were labeled A-RT and A-PI (percent incorrect for the accuracy data); for coordinate trials, the DV was labeled A-AE. The 'A' signifies that the value represents the aggregate of the attitude angles presented. A one-way repeated-measures analysis of variance (ANOVA) procedure was then applied for each of the 12 operational flight contexts to assess the A-RT and A-AE. Six non-parametric Friedman tests were used to analyze the A-PI data in which the participants were used as blocks for the six different scenarios of Gabor roll left/right, ASAR roll left/right, and ASAR VFP dive/climb.

Results

The analyses revealed, in general, for responding categorically or coordinately to the Gabor stimulus, the effect across the performance measures of A-RT, A-PI, and A-AE was dependent upon the Gabor's hemimeridian location as shown in Table 2. Table 3 shows pairwise comparisons for the select flight contexts that showed significant impact on A-AE and A-RT due to hemimeridian. Figure 3 shows the trends in the data for select flight contexts and response variables. While the trends, as a function of hemimeridian location, for the ASAR were similar to the trends for the Gabor stimulus, the size of the differences between conditions was subdued, as shown in Figure 3, which compares Gabor and ASAR results. In fact, just one categorical flight context for the ASAR was statistically different ($p = .01$) across hemimeridian location, namely, ASAR Roll Right Categorical (the 0° hemimeridian having lower A-RT than 90°, 180°, and 270°). Further inspection into the coordinate flight contexts show larger effects in the VFP coordinate flight contexts over the roll coordinate flight contexts. In categorical flight contexts, only the Gabor data showed evidence that A-PI was different across hemimeridians (the 0° hemimeridian having lower A-PI than 90° and 180°; roll left, $p = .006$, roll right, $p = .001$).

Discussion

The results indicate that, in general, our testing replicated a performance HVA and to some degree the vertical meridian asymmetry (Carrasco et al., 2001) for the Gabor stimulus. This result validated our current experimental methodology. More interestingly, we found evidence that elements of the HVA occur with the ASAR, matching the results of Reis et al. (2019).

However, this finding was primarily evident in the coordinate context. This may be a result of the coordinate task being subjectively more difficult than the categorical task. There was no reliable evidence that performance at left or right of center hemimeridians were any better than the other for categorical or coordinate responses. Arguably, the ASAR stimulus changes in angular deviations may not compare directly to the categorical/coordinate framework of past research where relations were linear in space. In summary, we found replicating evidence of some elements of a performance HVA while using the ASAR and such effects should be considered for determining symbology placement within head-worn displays.

TABLE 2. *Statistics from Analyses.*

CATEGORICAL	*F(df1, df2)*	$\chi^2(df)$	*p*-value	*S*-value	η^2	*W*
Gabor Roll Left (DV: A-RT)	$F(3, 33) = 3.95$		**.016**	6	.26	
Gabor Roll Left (DV: A-PI)		$\chi^2(3) = 12.53$	**.006**	7		.348
Gabor Roll Right (DV: A-RT)	$F(3, 33) = 6.25$		**.002**	9	.36	
Gabor Roll Right (DV: A-PI)		$\chi^2(3) = 19.99$	**.001**	10		.555
ASAR Roll Left (DV: A-RT)	$F(3, 33) = 0.34$		.794	0	.03	
ASAR Roll Left (DV: A-PI)		$\chi^2(3) = 0.32$	.955	0		.009
ASAR Roll Right (DV: A-RT)	$F(3, 33) = 4.44$		**.010**	7	.29	
ASAR Roll Right (DV: A-PI)		$\chi^2(3) = 0.84$	.856	0		.024
ASAR VFP Dive (DV: A-RT)	$F(3, 33) = 0.29$		.835	0	.03	
ASAR VFP Dive (DV: A-PI)		$\chi^2(3) = 3.20$	.318	2		.098
ASAR VFP Climb (DV: A-RT)	$F(1.80, 19.85) = 0.30$		.828	0	.03	
ASAR VFP Climb (DV: A-PI)		$\chi^2(3) = 3.49$	.856	0		.024
COORDINATE	*F(df1, df2)*		*p*-value	*S*-value	η^2	
Gabor Roll Left (DV: A-AE)	$F(1.66,18.24) = 8.19$		**.004**	8	.43	
Gabor Roll Right (DV: A-AE)	$F(1.92,21.11) = 14.98$		**< .001**	13	.58	
ASAR Roll Left (DV: A-AE)	$F(3, 33) = 8.41$		**< .001**	12	.43	
ASAR Roll Right (DV: A-AE)	$F(3, 33) = 2.58$		.070	4	.19	
ASAR VFP Dive (DV: A-AE)	$F(1.71,18.79) = 12.37$		**.001**	11	.53	
ASAR VFP Climb (DV: A-AE)	$F(3, 33) = 3.91$		**.017**	6	.26	

Note: *F* values with non-integer degrees of freedom indicate corrections for violations to sphericity. χ^2 values pertain to the Friedman test. The *S*-value is the binary Shannon information (Rafi and Greenland, 2020). η^2 is the effect size measure. Kendall's *W* conveys observer agreement. *P*-values < .05 are bolded.

TABLE 3. *Pairwise Comparisons of Hemimeridians, within Selected Flight Contexts.*

Level	- Level	p-value	Level	- Level	p-value	Level	- Level	p-value	Level	- Level	p-value
Gabor Roll Left Coordinate			ASAR VFP Dive Coordinate			ASAR Roll Left Coordinate			Gabor Roll Left Categorical		
90	0	**.001**	90	0	**< .001**	90	180	**< .001**	90	0	**.012**
90	180	**.002**	90	180	**< .001**	90	0	**.002**	90	270	.121
270	0	.063	270	0	**.016**	90	270	**.025**	90	180	.121
270	180	.097	270	180	**.019**	270	180	.275	180	0	1
90	270	.177	90	270	.090	270	0	.607	270	0	1
180	0	.758	180	0	.865	0	180	.607	180	270	1
Gabor Roll Right Coordinate			ASAR VFP Climb Coordinate			ASAR Roll Right Categorical			Gabor Roll Right Categorical		
90	180	**< .001**	90	0	**.031**	90	0	**.022**	90	180	**.004**
90	0	**< .001**	270	0	.104	270	0	**.025**	90	0	**.004**
270	180	**.008**	90	180	.131	180	0	**.041**	90	270	**.032**
270	0	**.035**	270	180	.316	90	180	1	270	180	1
90	270	**.035**	180	0	.903	270	180	1	270	0	1
0	180	.511	90	270	.903	90	270	1	0	180	1

Note: p-values are adjusted for family-wise error rate using the Holm-Bonferroni method; *p-values* < .05 are bolded.

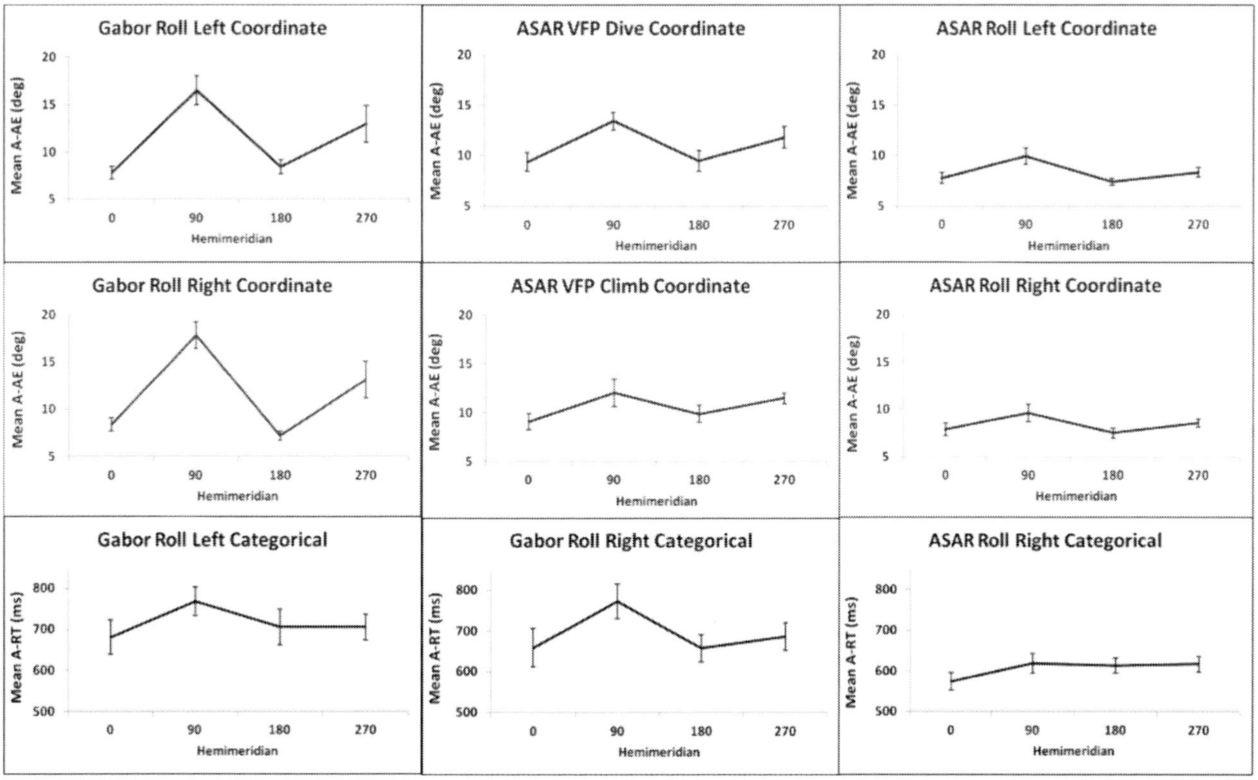

Figure 3. Plots are a subset of the operational flight contexts. In general, the trends show better performance at 0° and 180° hemimeridians over that of 90° and 270° hemimeridians. Error bars represent ± standard error of the mean.

References

Carrasco M, Talgar CP & Cameron EL (2001). Characterizing visual performance fields: effects of transient covert attention, spatial frequency, eccentricity, task and set size. Spatial Vision, Vol. 15, No. 1, pp. 61–75.

Fuchs, W., & Fischer, G. (1995). "Arc Segment Attitude Reference – ASAR," presented at the AGARD Symposium on Flight Simulation: Where are the challenges? AGARD-CP-577, Edinburgh, Scotland, May 1995.

Kosslyn, S. M., Koenig, O., Barrett, A., Cave, C. B., Tang, J., & Gabrieli, J. D. E. (1989). Evidence for Two Types of Spatial Representations: Hemispheric Specialization for Categorical and Coordinate Relations. *15*(4), 723-735.

Rafi, Z., Greenland, S. (2020). Semantic and cognitive tools to aid statistical science: replace confidence and significance by compatibility and surprise. BMC Med Res Methodol 20, 244. https://doi.org/10.1186/s12874-020-01105-9

Reis, G. A., Geiselman, E. E., & Miller, M. E. (2019). Effects of Visual Perceptual Asymmetries on Performance While Using an Aircraft Attitude Symbology. 20th International Symposium on Aviation Psychology, 247-252. https://corescholar.libraries.wright.edu/isap_2019/42

Rezaul, Karim, A. K. M., & Kojima, H. (2010). The what and why of perceptual asymmetries in the visual domain. *Advances in Cognitive Psychology, 6*(6), 103-115. doi:10.2478/v10053-008-0080-6

THE CUSTOMER IS ALWAYS RIGHT… TOWARDS RHINO POINTING AND EYE TRACKING INTERFACES FOR COMBAT AVIATORS

Patrick Highland[a,b], Daniel Harp[b], Thomas Schnell[a], Eric Geiselman[c], Paul Havig[c]
[a]The University of Iowa, Iowa City, Iowa
[b]United States Air Force, Edwards Air Force Base, California
[c]711 Human Performance Wing, Wright-Patterson Air Force Base, Ohio

At warfighter request, research was conducted to determine the military utility of eye tracking (ET) as a human machine interface (HMI) for large area displays (LADs) in a tactical live-flight environment. Flight test determined ET felt effortless until rapidly changing lighting conditions and pupil sizes as well as elevated gravitational load factor induced ET slippage rendered the requested interface unusable. In the spirit of, "the customer is always right," researchers proved the warfighter ultimately wise in her request for a novel LAD HMI by testing a head tracking algorithm, called "Rhino Pointing (RP)", alongside ET. RP was simpler to implement and offered improved accuracy and decreased task completion times. Though not as effortless as ET from a physical workload standpoint, evaluation of the RP concept during flight test demonstrated significant improvements over traditional touchscreen LAD HMIs and offered the warfighter a superior alternative to ET in most measures of performance.

Researchers experimented in a tactical live-flight environment with a fifth-generation fighter operationally representative flight helmet (ORFH) to determine if Eye Tracking (ET) as a combat large area display (LAD) human machine interface (HMI) could meet warfighter needs. A headtracking algorithm called Rhino Pointing (RP) was specifically added to the research because the warfighter requested ET HMI was assessed at a NASA technological readiness level (TRL) of four and was hypothesized to be marginally successful as a combat aviation HMI, at best. Subjects used three HMIs to execute three tasks, as can be seen in Figure 1 a-c. These tasks were executed across five environments onboard an L-29 aircraft (Figure 1d) while seated at the back seat crew station (Figure 1e) while wearing an instrumented OFRH (Figure 1f).

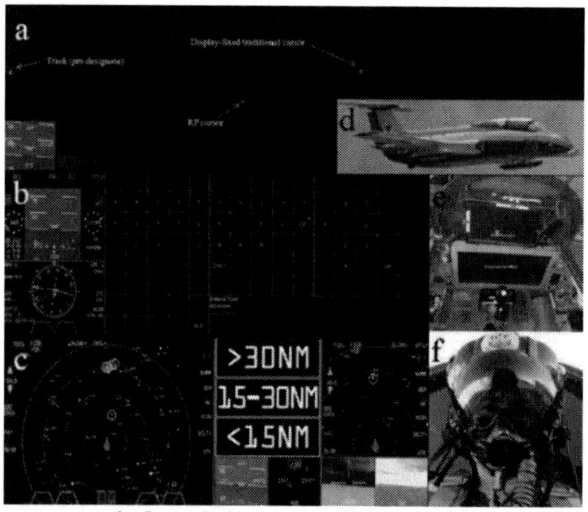

Figure 1. a) Designate a target task *b)* Alphanumeric data entry task *c)* Air to air intercept task *d)* L-29 aircraft *e)* Back seat crew station of L-29 *f)* Helmet with eye and head tracking

Method

Participants and Apparatus

Five expert United States Air Force (USAF) fighter aircrew, called evaluation pilots (EPs), who had at least 1000 hours of fighter time as well as experience in the last five years with air to air intercept were selected for this paper from a larger dataset (Harp, Armstrong, Whiting et al., 2020). All EPs had experience with the control HMI (hands-on-stick and throttle (HOTAS) and touchscreen) which was implemented in the aircraft in similar fashion to fifth-generation fighter aircraft. None of the participants had experience with the two interventions, ET nor RP.

The researchers used an Aero Vodochody L-29 aircraft for this research. The L-29 was operated in the experimental research and development category and was fully aerobatic and capable of performing high dynamic maneuvers up to +7.5/-3.5 Gz at speeds up to Mach 0.7. This L-29 was highly instrumented and used state of the art avionics that incorporated onboard and netcentric air warfare simulation capabilities, weapon models, fire control radar simulation, and helmet mounted display (HMD) capabilities.

The OFRH used in the study was a fifth-generation fighter aircraft representative HMD integrated into the L-29 with a head-tracked graphics processor (Geiselman, 1999). While wearing the HMD the EP experienced realistic instrument meteorological condition flying scenarios while operating the L-29 aircraft from the back seat crew station as if she was in a single seat fifth-generation HMD fighter environment (Schnell, Reichlen, & Reuter, 2017).

Intervention

The ET subsystem consisted of a Dikablis Professional ET integrated into the OFRH. Two eye cameras were mounted directly onto the helmet visor, in front of and slightly below the eyes of the EP. A scene camera was mounted facing outward on the front of the oxygen mask. The eye camera used infrared video to track the pupil of the wearer. The ET algorithm assumed that a vector orthogonal to the surface of the eye at the pupil indicated the instantaneous user gaze location.

The RP subsystem consisted of a Polhemus Scout head tracker integrated into the EP station and OFRH. Three orthogonal alternating current powered electromagnetic coils were mounted in the rear cockpit and a small magnetic receiver was mounted onto the OFRH. The head tracker measured variations in the local electromagnetic field due to helmet motion to determine the six degrees of freedom position and orientation of the OFRH. The system used pre-flight calibration offsets to define a vector known as the "rhino boresight" rigidly affixed to the helmet, which indicated the location of interest of the EP on the display, as if a Rhino horn was protruding from the head of the EP and touching the lower display at a point of interest (Harp, Schnell, Armstrong et al., 2020).

Tasks and Maneuvers

Two tasks across six maneuvers were flown. The first task was to slew a control cursor, ET cursor, or RP cursor to a track as quickly and as accurately as possible as soon as it appeared. Once the cursor was slewed, a HOTAS button was used to designate the track for all HMIs (Figure 1a). The second task was to use touchscreen (in lieu of HOTAS), ET, or RP to enter a predefined alphanumeric text string using an onscreen keyboard (Figure 1b).

The first maneuver was deemed baseline and consisted of the EP flying straight and level while executing the two tasks. The second maneuver was deemed low workload (LWL) and required the EP to maintain 30 degrees angle of bank, level altitude, and a constant airspeed while executing all tasks across all HMIs. The third maneuver was deemed high workload (HWL) and required the EP to maintain a three degree per second turn rate, while climbing or descending at 500 feet per minute, at a constant airspeed while executing all tasks across all HMIs. The next maneuvers were flown by the safety pilot (SP) in the front seat while the EP executed the tasks while not flying the aircraft. The fourth maneuver was a SP flown 2G normal load factor turn flown while the EP executed all tasks across all HMIs. The fifth maneuver was the same except at 3.5Gs.

The final maneuver was deemed the operationally relevant air to air intercept maneuver (Figure 1c). During this maneuver, multiple air to air adversaries appeared on screen. The EP was responsible for using the run HMI to designate each track as quickly and as accurately as possible, then to assign a distance appropriate weapon to that target, all while flying the aircraft safely (no specific altitude or airspeed tolerances assigned).

Dependent measures

There were five dependent measures. The first dependent measure was time to an initial action in response to a track. This was defined as the amount of time from track appearance until a HOTAS, ET, or RP cursor was slewed to the track. The second dependent measure was the amount of time taken to enter a predefined alphanumeric data string using each HMI. The third dependent measure was the track first action distance error which was defined as the distance the cursor was from the track when the EP designated it with the HOTAS, or the distance from the track where the ET or RP cursor was slewed before designation (since designation was always accomplished with a HOTAS action).

Electrocardiogram (ECG) data was obtained from the EPs throughout the flight and used to calculate objective real-time cognitive workload as a fourth measure. The ECG time series data was transformed into phase space using the cognitive assessment toolset (CATS) software tool (Schnell, 2021). This step established the ergodicity transition matrix (ETM) that represented the dynamics of the ECG signal in phase space for the different conditions in the flight (Schnell & Engler, 2014). The transitions within the ETMs were summarized into a single metric, termed the Transition Probability Variance (TPV). TPV calculated the variance of the probabilities of transition from one cell to another different cell of the course-grained ETM. The TPV therefore captured the variability in the dynamics of the ECG signal as the EP underwent different levels of cognitive loading (Martin, Calhoun, Schnell et al., 2019). TPV varied inversely to the degree of workload with higher TPV numbers seen under low workload conditions and low TPV numbers seen under high workload conditions. TPV-based workload was calculated continuously for the EP throughout each flight.

Finally, as a fifth measure, the EPs filled out post-flight questionnaires using a six-point subjective rating scale (six being very satisfactory) to evaluate user satisfaction with the HMIs. All results, other than subjective satisfaction, were presented using boxplots where large dots were means, solid horizontal lines were medians, vertical lines were quartile boundaries, and small dots were outer quartile data. Test statistics were produced by Wilcoxon tests.

Results and Discussion

As can be seen in Figure 2, the control HMI suffered from very long times to slew the cursor across the LAD, which is what drove this research into ET and RP. While ET and RP were always better than the control, they only differed from each other in the 3.5G maneuver where RP far outperformed ET due to G-force-induced eye tracker slippage.

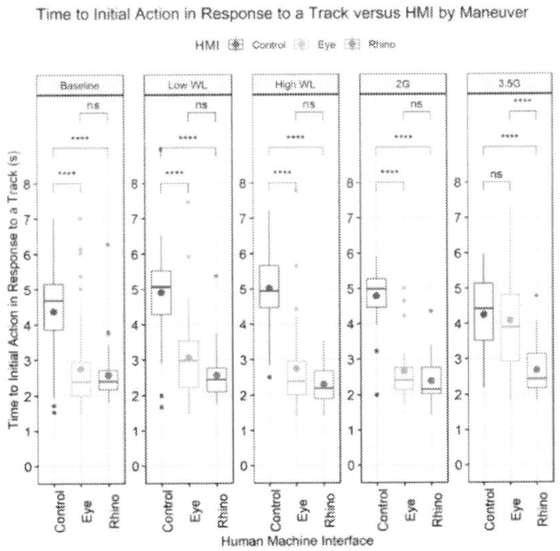

Note: Large dots are means. Solid horizontal lines are medians. * P<=.05, ** P<=.01, *** P<=.001, **** P<=.0001.

Figure 2. Time to initial action in response to a track versus HMI by maneuver

As can be seen in Figure 3, the control outperformed both ET and RP in time to enter alphanumeric data, due to the use of touchscreen over HOTAS. Of note, RP was always better than ET, to statistically significant amounts in three of five maneuvers.

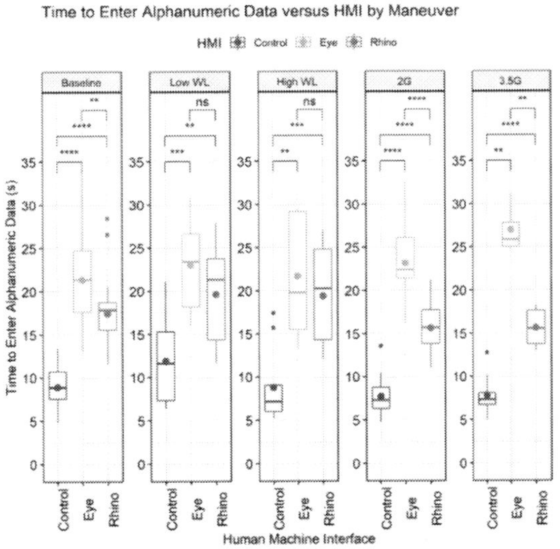

Note: Large dots are means. Solid horizontal lines are medians. * P<=.05, ** P<=.01, *** P<=.001, **** P<=.0001.

Figure 3. Time to enter alphanumeric data versus HMI by maneuver

As can be seen in Figure 4, the control was always the most accurate, however, RP matched control accuracy in the baseline maneuver and was typically only about one centimeter further from the target than the control. In contrast, ET suffered unsatisfactory accuracies across all maneuvers. At elevated G forces, ET was unusable for selecting a specific track.

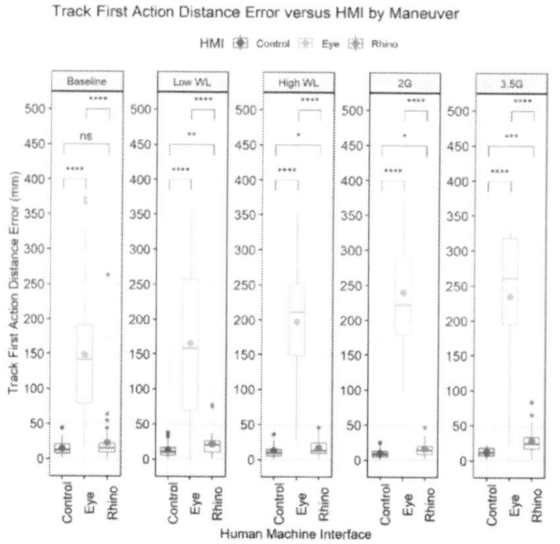

Note: Large dots are means. Solid horizontal lines are medians. * P<=.05, ** P<=.01, *** P<=.001, **** P<=.0001.

Figure 4. Track first action distance error versus HMI by maneuver

As can be seen in Figure 5, objective mental workload was always increased for ET over the control across all three workload scored maneuvers. Of note, RP did not increase workload over the control, except for the tactical maneuver. ET always increased workload over RP. Of note, workload without a reference to performance has little utility (Schnell, Reichlen et al., 2017). Flight technical performance was calculated, however, no statistically significant differences were noted between HMIs, and so were omitted here for brevity.

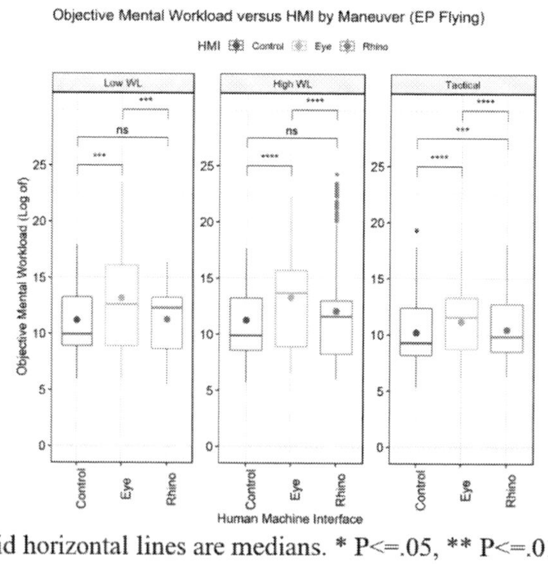

Note: Large dots are means. Solid horizontal lines are medians. * P<=.05, ** P<=.01, *** P<=.001, **** P<=.0001.

Figure 5. Objective mental workload versus HMI by maneuver (Evaluation Pilot flying)

For subjective HMI satisfaction, comfort using RP was found to be lower than the control, but not statistically worse than ET. An investigation into subjective feedback found that EPs were experiencing neck strain (an increase in physical workload) to use RP at the edge of the display. Finally, overall usability of ET and RP were found to be lower than the control, however, RP was found to be statistically superior to ET.

Conclusion

While the control performed better when it came to accuracy and some measures of satisfaction, the control showed the limitations of HOTAS use with a LAD. The warfighter desired to overcome these limitations with ET. Results also showed that while ET felt effortless at times from a physiological and mental workload standpoint, ET performance degraded with time and failed under operationally representative G loadings. While RP did not offer the low physical workload in some environments that ET sometimes offered, RP did consistently offer better fighter aircrew performance across all environments.

This research showed that ET suffered from unacceptable accuracy and mental workload increases (with no performance changes). The warfighter customer was correct in that the LAD HMI problem needed to be solved. RP, while not as accurate as the control, offered almost all the desired ET performance while only incurring a small physical workload penalty. Given that RP was less costly and simpler to implement than ET, it was an ideal candidate for implementation into fifth-generation tactical aircraft until ET matured to a higher TRL. Without the "customer is always right" mentality, this research would have found ET abandoned as a 2021 combat aviation HMI. Instead, researcher foresight and the desire to prove the combat aviator ultimately correct ensured the warfighter found a satisfactory alternative LAD HMI in Rhino Pointing.

References

Geiselman, E. (1999, 01 September 1999). *Practical considerations for fixed wing helmet-mounted display symbology design.* Paper presented at the Proceedings of the Human Factors and Ergonomics Society Annual Meeting, Los Angeles, CA.

Harp, D., Armstrong, R., Whiting, H., Purser, Z., Dold, S., & Nevala, S. (2020). *Comparison of Eye-Tracking, Rhino-Pointing, and Traditional Human System Interface (Project Have Rhino)* (AD1113400). Retrieved from USAF Test Pilot School: https://www.dtic.mil

Harp, D., Schnell, T., Armstrong, R., Dold, S., Nevala, S., Purser, Z., . . . Highland, P. (2020, 24 September 2020). *Report to the Aerospace Profession, Towards an Improved Interface for Large-Format Displays: A Comparison between Eye Tracking, Rhino Pointing, and Traditional Tactile Input.* Paper presented at the The Society of Experimental Test Pilots 2020, Lancaster, CA.

Martin, P., Calhoun, P., Schnell, T., & Thompson, C. (2019). *Objective Measures of Pilot Workload* (AD1088654). Retrieved from https://www.dtic.mil/

Schnell, T. (2021). *Cognitive Assessment Tool Set (CATS) User Manual.* University of Iowa. Iowa City, IA. Retrieved from https://iti.uiowa.edu/our-research/centers/operator-performance-laboratory

Schnell, T., & Engler, J. (2014). Entropic skill assessment of unmanned aerial systems (UAS) operators. *Journal of Unmanned Vehicle Systems, 2*(02), 53-68.

Schnell, T., Reichlen, C., & Reuter, C. (2017). *Spatial Disorientation Threat Characterization for F-35 Representative Helmet-Mounted Display Use in the Flight Environment* (AD1038022). Retrieved from https://www.dtic.mil

SURVEYING COLLEGIATE STUDENT PILOTS FOR THEIR PERSPECTIVES ON THEIR COLLEGIATE FLIGHT TRAINING EXPERIENCE

Nicoletta Fala
Oklahoma State University
Stillwater, OK

Denise K. Whitford
Purdue University
West Lafayette, IN

Women make up roughly half of the population, but continue to be an underrepresented group in aviation. They constitute nearly 14% of student pilots, but only 8% of Federal Aviation Administration (FAA) pilot certificates, leading to a lack of understanding why a substantial number of women do not complete their training. Past diversity initiatives have not resulted in a sufficient increase in the numbers. This paper discusses a study in progress that is intended to gain perceptions of women's experiences and obstacles in collegiate flight programs using a longitudinal survey of collegiate pilots. The aim of the study is to generate guidelines on how flight programs can impact change to welcome and retain their female students. We provide a preview of the survey structure based on a thematic analysis of the literature, present metrics we will track throughout the study from the themes uncovered, and discuss survey dissemination plans.

Historically, women have been underrepresented in aviation disciplines. Among pilots, in particular, women held only 7.94% of active FAA pilot certificates in 2019. While that number has been increasing in recent years, change is slow—from 6.73% in 2010 to 7.94% in 2019. By comparison, women hold a larger portion of the active student pilot certificates—12.40% in 2010 and 13.79% in 2019 (FAA, 2020). Based on the data, a large proportion of women pilots do not complete their private pilot license. While several organizations have attempted to improve diversity in aviation in the United States overall (for example, programs such as Girls in Aviation Day), we have not investigated what makes women leave aviation, particularly at the student pilot level and at a faster rate than men. Being severely underrepresented in a male-dominated field has been shown to create challenges for women pilots, impacting their performance (Matthews, Ender, Laurence, & Rohall, 2009). Correcting the pilot shortage while improving aviation safety therefore hinges on our industry providing appropriate training and support to all stakeholders. However, the factors that impact diversity within aviation have not been sufficiently researched in recent years (Sobieralski & Hubbard, 2019).

Past research and incentives have not sufficiently corrected the disparities among women and men in the aviation industry. The literature consistently recommends we continue efforts to recruit women and other underrepresented students, but also that we monitor the status of women in those efforts (Ison, Herron, & Weiland, 2016). Through our research, we aim to identify causes of these disparities and develop guidelines which will help collegiate programs, aviation organizations, and private flight programs address them. We will use a nation-wide longitudinal survey, distributed among collegiate pilots over four years, to identify (1) any obstacles they face that make them question their fit in the program or otherwise impact their experience, (2) perspectives on the challenges they face or they expect to face within aviation, both in their current program and in their future career, and (3) changes to individual perspectives as they navigate their collegiate program. In this paper, we discuss the literature on diversity improvement efforts and perspectives on discrimination in aviation and related industries, and identify themes and measures to further investigate. We then develop Likert-scale items to score participants on the identified measures and discuss survey design and dissemination plans.

Literature Review

Historically, research has attempted to identify obstacles that women face in their journey towards becoming pilots, but also in their flying careers. Such studies have identified learning style, gender stereotyping, the male-dominated industry, work-life balance, and representation pressure, among others, as reasons why women either do not begin their training or do not complete their studies.

Academic aviation curriculum design was identified as a potential roadblock for diversity improvements in the industry in the early 2000s, when the Department of Education and the Alfred P. Sloan Foundation both funded a three-year study to maximize retention of female students (Karp, et al., 2001). The study investigated *learning style considerations* and how learning style impacts retention, and recommended that educators present their curriculum using all learning style environments, rather than relying on the traditional auditory environment. The same project resulted in a national survey of 390 pilots at nine institutions to identify the factors that influence women in collegiate aviation and found that women in the early stages of their training responded differently to the provided statements than women later on in the curriculum, while the responses of men remained unchanged (Turney, et al., 2002). Additionally, the responses of experienced women matched those of men, bringing into question the nature versus nurture debate. If we expect that women will adapt into more men-like attitudes and thought processes to survive their training, or that training will change women to think more like men, what happens to those women who are not able to adapt to the perceived mold of what a pilot resembles? Unfortunately, two decades after this important finding, there is no evidence of our academic environment changing to address different learning styles or adding a diversity component to crew resource management training.

Semi-structured interviews querying women airline pilots, based in Turkey, on their experiences in the flight deck resulted in *gender stereotyping and prejudice, male-dominated industry and industry prejudice against female pilots, discrimination from male cockpit crew, pressure of showing masculine behavior and controlling attitude, and difficulty of balancing family life* as themes that were present in their interviews (Yanıkoğlu, Kılıç, & Küçükönal, 2020). A recent phenomenological study investigating the factors that impact success among minorities in aviation identified three categories that women have talked about: *open communication, friendship and community*, and *positive faculty support* (Kim & Albelo, 2020).

In the literature, we also observed some flawed research methodologies that we need to avoid, both in the research described here, but also as a discipline. For example, while students are already aware that fewer women participate in collegiate aviation, as well as the airline industry (Casebolt & Khojasteh, 2020), the disproportionate nature of the aviation population can introduce bias in direct questioning. Casebolt and Khojasteh (2020) concluded that "collegiate aviation students do not perceive that women have negative experiences at their collegiate aviation institutions because of their gender," since 75% of students disagreed with the statement "Gender biases for female aviation students exist at my collegiate aviation institution" and 83% of students disagreed with "Gender barriers for female aviation students exist at my collegiate aviation institution." However, the researchers also cited a sample with 96% male participation—out of 124 responses—which means that if the five women surveyed had a difference in opinion, their opinions would not be heard. Other attempts at measuring the perceptions of bias and discrimination among women did not attempt to make correlations based on demographics or compare the results to male pilots (Depperschmidt & Bliss, 2009).

Survey Design

In this research, we are designing a survey to study perceptions of collegiate pilots of all genders as they progress through their flight training. This longitudinal study will identify if (1) pilots with varying demographics (starting by comparing populations based on gender) perceive their personal progress, struggles, and learning differently, and (2) if the students' perceptions change over the years (i.e., if going through their program changes the way they respond to our questions). The aim of this study is to identify if there are things that flight programs, both in the collegiate and private sectors, are doing to either deter a section of the population from applying to the program or initiating their training, or discourage (or not motivate) a student, while they are in the program.

The described type of survey can introduce biases into the data which we need to be aware of. The two most relevant biases are due to the population, and due to the type of questioning. In the first case, we will minimize the potential for bias by comparing groups of people separately based on the traits we are investigating. For example, we will need to treat responses by women, men, and nonbinary pilots as different groups. We therefore do not aim to collect a dataset from a representative sample, which would result in less than 10% responses from women, but rather focus on collecting a large dataset which will allow us to break the sample into the desired groups. To address the second bias, we are identifying themes in the literature, converting them to measures, and designing our survey to measure them through indirect questioning. As a simplistic example, we would not be asking students to rate their own math skills, because they are likely unaware of the existence of concepts they do not know about, but also because they are likely to provide the answer they think we want from them or is desirable. Instead, we could ask them to rate their knowledge on specific math topics, such as algebra, differential equations, etc., or we could give them short math problems to solve so that we could measure their skills more objectively. Similarly, we cannot simply ask respondents if they experience discrimination in their program. The themes identified in the literature resulted in a list of *measures* that we want to track. Each measure is mapped to a number of *items* that will collectively assess it. For example, one of the measures identified is the feeling of *safety*. To track the safety measure, we will ask participants to indicate their agreement with statements such as *"I know how to report discrimination if needed."* and *"I would feel comfortable with reporting discrimination."* among others. Table 1 highlights some additional examples of measures and representative items.

The survey instrument will consist of a web-based questionnaire which will be disseminated to collegiate pilots yearly for four years, to capture any changes to perceptions and responses as students progress along their program from their freshman year to graduation, and any "generational" changes, albeit small, as new students start their programs.

Survey structure

The questionnaire will consist of six sections. The respondent will (1) be introduced to the research and its importance and be provided with instructions, (2) provide the most necessary demographics (gender and school year classification), (3) respond to Likert-scale statements, and (4) provide the remaining demographic information we request. Moving most of the demographic questions to the end maximizes the amount of useful information received if the respondent does not fully complete the survey. If they respond to some or all of the Likert statements presented to them, but do not complete the demographics section, we can include their datapoints in any generalized assessment that does not make correlations based on demographic data. Additionally, the respondents will only be shown approximately half of the items for each measure, to shorten the total survey. The list of statements they see will be randomly generated. The two remaining sections will ask the respondents to volunteer more of their time if they would like to. They will first be asked (5) if they want to be considered for future focus groups or survey re-runs, or if they want to receive updates once we compile the results, in which case a separate instrument will be used to collect their contact information. Lastly, they will be asked (6) if they have a few more minutes and want to respond to more statements, in which case they will be shown the items we previously withheld.

Survey dissemination

Although the response rate of web-based surveys is typically low, the number of people we will be able to reach will likely outweigh any recall and self-report biases. We will disseminate the survey through collegiate flight programs, by compiling a list of contact information for the chief flight instructor and/or academic advisor at each of the 39 (currently) AABI-accredited schools. All students, regardless of gender, will be eligible to participate. In the case of local schools, we will ask to visit the campus and introduce the survey to the students through their classes, but for more distant schools we will rely on administration to propagate how important participation is and the impact it may have. We also plan to make use of student organizations to promote survey participation, especially through organizations such as Women in Aviation student chapters and campus-based flying clubs and flight teams.

Theme Identification

The aim of the survey is to investigate and report on differences in perspectives surrounding biases and obstacles that may impact women's careers in aviation at the collegiate level. The research reviewed has not adequately identified what actions will allow women to join or stay in aviation programs. The same longitudinal study will have to be administered consistently over a period of four years to track changes as pilots advance from their freshman year to graduation, making validity in the metrics and survey design important.

To identify metrics that we should be tracking, we reviewed the literature and identified themes that the work brought up. We generated a list of 37 papers ranging from 1979 to 2020 by (1) searching online databases for papers that included the terms "*gender* AND *aviation*" and excluding papers that had to do purely with accident investigation, and (2) including relevant papers from the original set's lists of references that the databases had not identified. For example, a study of pilots' perspectives towards gender in Thai aviation (Thatchatham & Peetawan, 2020) brought up differences in *how flight instructors treat female students*, *women pilots having to prove their skills*, *perspectives on gender bias*, and *women adopting male behaviors*. The identified themes were then arranged in 13 metrics: *perceptions*, *belonging*, *safety*, *support*, *learning*, *preparation*, *planning*, *finances*, *community*, *communication*, *prejudices*, *expectations*, and *identity*. Examples of metrics and items are shown in Table 1.

Table 1.
Each measure listed is described by multiple items, in the form of Likert-scale statements. The number in parenthesis refers to the number of items that currently make up the measure.

Metric (number of items)	Representative examples of items
Perceptions (15)	Gender barriers exist for women in my program. My program is sufficiently diverse.
Belonging (17)	I sometimes feel isolated in my program. I have to prove myself to be accepted.
Safety (7)	I know how to report discrimination if needed. I would feel comfortable with reporting discrimination.
Support (9)	It is easy to find information and resources in my program. My family is excited about my career choices.
Learning (9)	It often takes me longer than others to build a skill. I spend a lot of time studying.
Preparation (4)	I prepare diligently for each flight lesson. I was sure of my decision to pursue a flying career before enrolling into my program.
Planning (6)	I am concerned about my ability to have a family as a pilot. I am confident that I will have valuable employment upon completion of my program.
Finances (4)	I have to fund my training myself. I always apply for scholarships that come my way.
Prejudices (3)	By nature, men make better pilots than women. Some jobs are better suited to people of a particular gender.
Communication (4)	I participate in class discussions. I prefer to ask questions in office hours than during class.
Community (4)	I rely on my friends for help with decisions. Most of my friends are pilots.
Identity (5)	My gender is important to my sense of identity. I think of myself as a confident person.

Conclusion and Future Work

In this paper, we discussed the gender inequity among licensed pilots in the United States and recent literature on perspectives and reasoning for the imbalance. Through the literature, we identified common themes for the biases and barriers that women have mentioned in the past, and converted them to measures that we want to track in the future. We designed a survey around these measures that consists of Likert-scale statements and demographic information. The survey will be distributed to all collegiate pilots, regardless of gender, in a longitudinal study that aims to investigate differences in perceptions among genders, but also any potential differences in perceptions over time, as these pilots progress through their programs.

We expect to distribute the survey to collegiate programs in late August 2021 as students are starting their academic year by contacting academic advisors and chief flight instructors, as well as campus-based flying clubs and aviation organizations which interact with our population of interest. We hope to get approximately 300 responses from the first run of the survey. This year's survey will help us establish a baseline understanding of perceptions of our student body so that we can start forming guidelines for diversity policies and initiatives. We will run the same survey yearly to track students' perceptions as they advance in their schooling and training and report our results in future publications.

References

Casebolt, M. K., & Khojasteh, J. (2020). Collegiate aviation students perceptions of female representation in collegiate aviation and the U.S. aviation industry. *Journal of Aviation/Aerospace Education & Research, 29*(2), 17-37.

Depperschmidt, C. L., & Bliss, T. J. (2009). Female Flight Students: Perceptions of Barriers and Gender Biases within Collegiate Flight Programs. *Collegiate Aviation Review International, 27*(2), 18-27.

FAA. (2020, April 18). *U.S. civil airmen statistics.* Retrieved December 17, 2020, from https://www.faa.gov/ data_research/aviation_data_statistics/civil_airmen_statistics/

Fala, N. (2019). Data-driven safety feedback as part of debrief for general aviation pilots. West Lafayette, IN: Purdue University.

Ison, D. C., Herron, R., & Weiland, L. (2016). Two decades of progress for minorities in aviation. *Journal of Aviation Technology and Engineering, 6*(1), 25-33.

Karp, M. R., Turney, M. A., Niemczyk, M., Green, M., Sitler, R. L., & Bishop, J. (2001). Retaining women in collegiate aviation by improving learning style considerations. *Collegiate Aviation Review International, 19*(1), 92-101.

Kim, E., & Albelo, J. L. (2020). Women in aviation: A phenomenological study exploring the needs and wants necessary for graduation. *Collegiate Aviation Review International, 38*(2), 67-81.

Matthews, M., Ender, M., Laurence, J., & Rohall, D. (2009). Role of group affiliation and gender on attitudes toward women in the military. *Military Psychology, 21*(2), 241-251.

Sobieralski, J., & Hubbard, S. (2019). Women in Aviation: Identifying Trends in Industry and Higher Education. *Navigating Careers in the Academy: Gender, Race, and Class, 2*(2), 37-51.

Thatchatham, C., & Peetawan, W. (2020). An Exploration on Pilots' and Student Pilots' Perspectives towards the Gender Bias in Thai Aviation Industry. *Proceedings, 39*(24).

Turney, M. A., Bishop, J. C., Karp, M. R., Niemczyk, M., Sitler, R. L., & Green, M. F. (2002). National survey results: Retention of women in collegiate aviation. *Journal of Air Transportation, 7*(1), 69-92.

Yanıkoğlu, Ö., Kılıç, S., & Küçükönal, H. (2020). Gender in the cockpit: Challenges faced by female airline pilots. *Journal of Air Transportation Management, 86.*

HOW TO TEACH COLLEGE AVIATION STUDENTS ABOUT SITUATION AWARENESS IN A VIRTUAL CLASSROOM SETTING

Andrew R. Dattel
Hui Wang
Nathaniel Booker
Andrei Matveev
Syaza R. M. Haris
Hanzi Xie
Zihao Zhang

Embry-Riddle Aeronautical University
Daytona Beach, FL

Creating a group class project to demonstrate situation awareness (SA) can be an effective pedagogical approach. By engaging students in such projects, they can understand the meaning of SA — comprehension of relevant information in a dynamic environment (Durso et al., 2007). In an SA class taught at an aeronautical university during the past several years, students engaged in observational and interactive projects. However, COVID-19 has prohibited in-person activities. In lieu of such in-person activities, students enrolled in a virtual SA class during summer 2020 utilized online road cameras found at traffic intersections in Madrid, New York, and Tokyo. This alternative approach allowed students to observe apparent SA of pedestrians across different cultures. Distinct differences were found. Pedestrians in New York seemed to have better SA than pedestrians in the other cities. Pedestrians in Tokyo seemed more compliant; whereas pedestrians in Madrid were less compliant and appeared to have poor SA. This virtual group project satisfied a viable approach to collecting SA data, as well as an effective pedagogical lesson for teaching SA.

A typical graduate college course includes lectures, discussion, assigned papers, and individual and group projects. Group projects can be an effective pedagogical tool because it permits students to participate in active learning. Active learning engages students to be creative in a meaningful learning process through relevant activities (Robertson, 2018). As part of the course objective of a "Situation Awareness and Performance" class taught at an aeronautical university, students engaged in collaborative learning such as group projects. By engaging students in such projects, they can develop a deeper appreciation of theories and applications related to situation awareness (SA). In past years, class projects included observations and surveillance at EPCOT (Dattel et al., 2017), a baseball game, a gambling casino night cruise, and at a paintball range. These group outings to achieve the goals of the project were conducted in person. However, due to COVID-19 the class decided to conduct a real-time virtual observation that would be an effective alternative to the previous classes' SA projects to develop SA skills. This paper discusses two objectives: a) benefits and techniques of learning SA in a virtual group project and b) the results found from the research design.

SA is an important aspect of safe operation of an aircraft. Endsley (1988) defined SA as "the perception of elements in the environment within a volume of time and space; comprehension of their meaning; and projection of their status in the future" (p.97). Durso, Rawson, & Girotto (2007) stated that SA is the comprehension of relevant information in a rapidly changing environment. The Federal Aviation Administration (1991) considers SA as a key element to safe decision making. The "Situation Awareness and Performance in Aviation/Aerospace" class taught during the last several years included students with a variety of interests and occupation, including instructor pilots, airport managers, aviation technicians, and air traffic controllers. Student course evaluations invariably included two main themes: how little knowledge of SA students realized they had before enrolling in the course (especially students such as instructor pilots who thought they were quite knowledgeable about SA prior to enrolling in the course); and how much better their task performance was by successfully applying SA to their current jobs.

In addition to aviation, SA is also utilized in other fields such as healthcare. Chang et al. (2017) conducted a study that revealed the advantages of simulation-based training of SA over lecture-based training. Chang et al. found that simulation-based training of SA yields better SA at the perception level compared to lecture-based training. Unfortunately, there is almost no research available on teaching SA in the aviation domain. Lectures and scenario-based training are targeted at improving decision-making ability and the application of learning. However, it would be beneficial to develop training that is primarily aimed at improving SA of student pilots, as well as instructor pilots.

Virtual learning (VL) has been a mode of delivering education for several years. The current COVID-19 pandemic situation has prompted online learning to become more practical and probable in many educational institutions. Several universities have transitioned to teaching online applications using internet platforms (e.g., Zoom, Microsoft Teams). The number of virtual schools in North America has increased significantly since 1995 (Beck & LaFrance, 2017). VL is a process conducted through a digital platform that commonly takes place in an online environment. There appears to be no specific benefit or detriment to online learning other than convenience. If designed properly, online student satisfaction and academic performance are equal to traditional in-person class instruction (Hara & Kling, 1999; Nguyen, 2015). Petrides (2002) argued that students felt it was more convenient to collaborate with classmates in an online course because the arrangement of in-person schedule would not be required. Online learning can be a good substitute for the traditional classroom learning, especially in circumstances like a pandemic situation.

The importance of online learning needs to be acknowledged for two reasons. First, the current unprecedented pandemic situation can impose difficulties to host traditional in-person class meetings; thus, virtual education needs to be incorporated to manage the situation. Second, online learning should be maintained even after the pandemic is over if the benefits of this education mode can be extended. Our experience of online learning currently has prepared us to adapt this learning mode as an effect pedagogical approach to future education. Muthurprasad, Aiswarya, Aditya, and Jha (2021) analyzed the benefits of online learning including the time flexibility and convenience where students could study at their own pace at their own convenient time and locations. Conversely, conflicting factors such as technological constraint, lack of direct interactions and engagement, and students' inefficacy toward learning were impacting the effectiveness of online education. VL expands education accessibility which primarily benefits students in rural areas, assuming they have a stable internet connection. The asynchronous courses allow students to participate according to their own time suitability, which can improve time management skills. Accessibility of internet during a learning process can offer opportunities for students to explore broad information related to the curriculum. VL encourages the development of student's digital skills and independence.

Due to COVID-19, many educational institutions have been forced to restructure in-person teaching and VL has been imposed at nearly all levels of education. Because the pandemic has shifted the way of teaching, revised pedagogy and andragogy approaches are constantly being modified that tests the effectiveness of education quality in such unprecedented times. This paper provides an example of a modified approach to learn about SA by conducting a group project in a virtual environment.

Method

Thirteen graduate students were registered for the Situation Awareness and Performance in Aviation/Aerospace class offered in the summer of 2020. Due to COVID-19 restrictions, the class was conducted synchronously online. Students considered several SA projects that could be executed remotely and chose to study pedestrian behaviors via webcam observations at busy crosswalk intersections in Tokyo, Madrid, and New York City. Students agreed to observe the behaviors during a weekend evening local time, when it was expected that residents and tourists would be active. The students also considered a variety of behaviors to observe, ultimately deciding on a set of five behaviors that could be observed at each chosen intersection.

Participants

A total of 7,058 pedestrians were observed at three locations: 5,214 at the Shibuya intersection in Tokyo, 1,548 at the Puerta Del Sol intersection in Madrid, and 296 (small number due to rain) at the 45th Street and 7th Ave. intersections in Times Square in New York City. Pedestrians in Tokyo were observed from 10:00pm – 11:00pm local time on Friday, July 10, 2020. Pedestrian in Madrid were observed from 10:30pm – 11:30pm local time on Friday, July 10, 2020. Pedestrians in New York were observed from 9:50pm – 10:50pm local time on Saturday, July 11, 2020.

Materials and Procedure

Each student agreed to observe two of the three intersections identified; thus, each intersection was observed by eight or nine students. During the hour-long observation, two students were assigned to observe each behavior using the frequency observation method and to record the behaviors using event sampling. The students viewed webcams located at each location in real time. For scoring purposes, the students recorded behaviors through Zoom. The specific pedestrian behaviors observed are indicated in Table 1.

After the initial observations and data collection, six students volunteered to continue work on the project. Each video (recording of a city intersection) was reviewed by two student researchers for accuracy. Video recording playbacks were conducted for every light change cycle until full agreement of pedestrian behaviors between researchers was established.

Table 1. *Specific Pedestrian Behaviors Observed in Each City*

Questions
How many people were outside the marked lines of the crosswalk when they crossed?
How many pedestrians started to cross before the crosswalk light turned green? (False Start)
How many people hesitated to start crossing once the crosswalk light turned green?
How many people were still in the crosswalk with the crosswalk light turned red?
How many people were trying to overtake others in the crosswalk?

Results

Five types of pedestrians' abnormal behaviors were observed when 7058 pedestrians crossed the roads in three cities. A chi-square test for independence was run. There was a significant difference between the number of pedestrians outside of crosswalks and cities, χ^2 (2) = 1048.06, p <.001. A significant difference was between cities for the number of pedestrians who conducted false starts, χ^2 (2) = 2175.02, p <.001. There was a significant difference between cities for the number of pedestrians who hesitated when crossing, χ^2 (2) = 76.043, p <.001. The chi-square test also showed a significant difference between cities for the number of pedestrians remaining in the crosswalks when the signal turned red, χ^2 (2) = 307.96, p <.001. However, no difference was found between the number of pedestrians who overtook others when crossing in the crosswalks, χ^2 (2) = 3.169, p = .205. Observed frequencies and percentages of pedestrian numbers for each type of crossing behavior are presented in Table 2. The data indicate that pedestrians in different cities are associated with different crossing behaviors. Figure 2 illustrates the percentages of each type of participants' behavior in different cities.

Table 2. *Frequencies for Participants' Behaviors in Different Countries (N = 7058)*

Behaviors	Tokyo		New York		Madrid	
	n	%	*n*	%	*n*	%
Outside of Crosswalks	1640	30.45	14	4.73	1141	73.71
False Start	64	1.23	20	6.76	659	42.57
Hesitation	332	6.37	13	4.39	13	.84
In Crosswalks when Red	1339	25.68	9	3.04	111	7.17
Overtaking Others	103	1.98	9	3.04	24	1.55

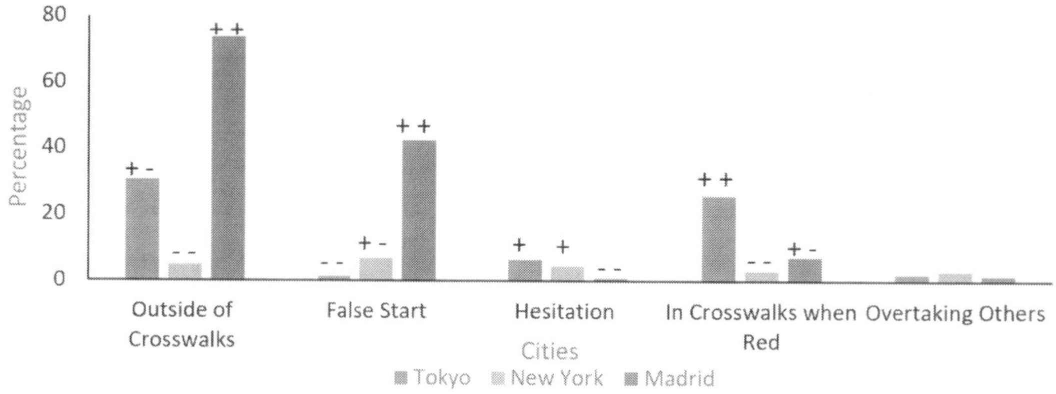

Figure 2. Percentage of Participants' Behaviors in Different Cities. A positive (+) symbol indicates a significant greater number of behaviors than a corresponding negative (-) symbol. For example, Madrid had more behaviors outside of the crosswalk than both Tokyo and New York, and Tokyo had more behaviors outside the crosswalk the New York.

Discussion

The Situation Awareness and Performance in Aviation/Aerospace graduate course has been offered for 5 years. Typically, the course is taught in a seminar format where students and instructor sit at a conference table and share ideas about theories and applications of assigned articles, job–related experiences, and personal insights and intuitions. There were challenges for the most recent class due to the move to a synchronous online environment. Nonetheless, two themes emerged from the course evaluation comments. First, many students felt that the course was quite successful despite it being offered online. Second (as can be seen in Table 3), it is hard to recognize that the course was not offered as an in-person traditional classroom setting.

Table 3. *Students' Comments after Taking the SA Course in a Synchronous On-line Format*

Elements That Most Helped You Learn	Elements That Least Helped You Learn
"I can learn about Situation Awareness through various important articles, and actual examples of aviation. This class has been very helpful for me to understand the subject clearly." "Guided discussion session during class." "Doing it online was different, but I think we achieved the same learning outcome as if we were to do it in person. The professor was engaging and available to answer our questions." "The open class discussions are fantastic. It supports for open-mindedness and a co-creative environment. Really enjoyed the feedback on out presentations." "The readings and class discussions were outstanding! I have learned so much about SA that I didn't know was possible"	"I am very much satisfied with this lecture so I do not have any opinions about this question." "Nothing it was ***** 5 star!! Outstanding."

The group project brought the class "together" in a virtual environment. The class discussed some pedestrian behaviors of SA, even pointing out how several vehicles in Madrid had to stop because pedestrians were in the crosswalk when the vehicles had the right-of-way. Interesting class discussion about cultural differences and possible relationships to SA were proposed. The synchronous online Situation Awareness and Performance class proved to be just as successful as previous in-person class, in addition to being able to complete the class' favorably mentioned and renown group project.

References

Beck, D. & LaFrance, J. (2017). Online schooling in the United States: A response to Saultz and Fusarelli. *Journal of School Choice, 11*(1), 42-59. https://doi:10.1080/15582159.2016.1272937

Chang, A. L., Dym, A. A., Venegas-Borsellino, C., Bangar, M., Kazzi, M., Lisenenkov, D., Qadir, N., Keene, A., & Eisen, L. A. (2017). Comparison between simulation-based training and lecture-based education in teaching situation awareness. A randomized controlled study. *Annals of the American Thoracic Society, 14*(4), 529-535. https://doi.org/10.1513/AnnalsATS.201612-950OC

Dattel, A. R., Babin, A., Li, T., Dong, Z, Fussell, S. G., & Yang, Q. (2017). A pedagogical approach to teach aviation students how to conduct situation awareness research. Proceedings paper published in the *19th International Symposium on Aviation Psychology*, 443-448, Dayton, OH.

Durso, F. T., Rawson, K.A., & Girotto, S. (2007). Comprehension and situation awareness. In F.T. Durso, R. S. Nicerson, S. T. Dumais, S. Lewandowsy, & T. J. Perfect (Eds.). *Handbook of Applied Cognition* (2nd ed.) (pp. 163-193). Wiley & Sons.

Endsley, M. R. (1988). design and evaluation for situation awareness enhancement. *Proceedings of the Human Factors and Ergonomics Society Annual Meeting, 32*(2), 97-101. https://doi.org/10.1177/154193128803200221

Federal Aviation Administration. (1991). *Aeronautical decision making* (AC No. 60-22). U.S. Department of Transportation. https://www.faa.gov/documentLibrary/media/Advisory_Circular/AC_60-22.pdf

Muthuprasad, T., Aiswarya, S., Aditya, K. S., & Jha, G. K. (2021). Students' perception and preference for online education in India during COVID -19 pandemic. *Social Sciences & Humanities Open, 3*(1), 100101-100101. https://doi.org/10.1016/j.ssaho.2020.100101

Nguyen, T. (2015). The effectiveness of online learning: Beyond no significant difference and future horizons. *MERLOT Journal of Online Learning and Teaching*, *11*(2), 309-319.

Hara, N., & Kling, R. (1999). A case study of students' frustrations with a web-based distance education course. *First Monday, 4*(12). https://doi.org/10.5210/fm.v4i12.710

Petrides, L. A. (2002). Web-based technologies for distributed (or distance) learning: Creating learning-centered educational experiences in the higher education classroom *International journal of instructional media*, *29*(1), 69.

Robertson, L. (2018). Toward an epistemology of active learning in higher education and its promise. In A. Misseyanni, M. D. Lytras, P. Papadopoulou, & C. Marouli (Eds). *Active Learning Strategies in Higher Education: Teaching for Leadership, Innovation, and Creativity*. Emerald Publishing, Ltd.

COVID-19 IMPACTS ON COLLEGIATE AVIATION TRAINING

Wu, Xiaoyu O.
Griffith University

Jacob J. Shila
Bowling Green State University

Abstract

In 2020, the COVID-19 pandemic has had a significant impact on the aviation industry. The regular day-to-day flight training routines have been altered for several reasons, including physical distancing requirements, disrupted training schedule, and the increased level of concern. Aviation educators will likely need to adapt their programs to optimize the learning experience, maintain effective safety delivery, and ensure competent graduates. The impact of the month-long break in flight training on the airmanship skills is also unaccounted for. This study surveyed collegiate aviation students to identify the COVID-19 shut down impacts on both their flight skill developments and the overall collegiate degree progress. Specifically, the study investigates the significance of improving communication, being innovative, requiring collaboration and flexibility, and better planning within aviation education programs. The study also provides practical tips on how the aviation education programs can adapt to the the continuously evolving pandemic landscape.

In early 2020, the COVID-19 pandemic shocked the world; this ultimately led to the significant slowing down of the then steady growth of the aviation industry. Various policy measures have since been implemented across the various countries to contain the spread and impact of the pandemic (Debata, Patnaik, & Mishra, 2020). These measures, including border closures and lockdown measures, have consequently led to a rapid decrease of aviation activities. The International Civil Aviation Organization (ICAO) confirmed that passenger traffic suffered a dramatic 60 per cent drop over 2020, bringing air travel totals back to 2003 levels (International Civil Aviation Organization, 2021). In higher education institutions, students are among those impacted by these measures such as the social restrictions – including social distancing and self-isolation measures(Al-Taweel et al., 2020). For the aviation students, that has slowed the hiring process for the recent graduates as well as the hands-on training such flight instructions(Ley, 2020). Due to the lockdown and social distancing measures, flight training activities have been slowed down to enable the colleges accommodate all the measures while conducting the trainings (FlightLogger, 2020). As a result, flight training activities were down to approximately 60 percent of the level before

Covid-19 during the lock-down periods (FlightLogger, 2020). These challenges have raised concerns among the students on how to best move forward; while the students are most passionate about the future career options, they are also concerned on how to accommodate the delays and resulting financial penalties(Ley, 2020; Plane and Pilot, 2020)

This paper attempts to shed light on the impact of COVID-19 pandemic on college aviation students. This study describes and quantifies the causal effects of the COVID-19 pandemic on students' flight training experiences. In particular, the study analyzes flight training progress, financial situation, career expectation, online learning experiences, and program supporting. For this purpose, we surveyed about 200 student pilots who enrolled in a college aviation program, in late February 2021. To that end, this study aimed to address the following research questions:

1. What is the perception of the collegiate aviation students regarding the impacts of the covid-19 pandemic on their financial situations?

2. What is the perception of the collegiate aviation students regarding the impacts of the covid-19 pandemic on their flight skills?

3. What is the perception of the collegiate aviation students on their academic progress given the changing nature of the course offering (such as online, hybrid, etc.) due to the pandemic?

4. What are the career challenges that the collegiate aviation students are facing due to the onset of the covid-19 pandemic?

Method

This study aims to shed light on the impacts of COVID-19 pandemic on college aviation students and their flight training experiences. The study describes and quantifies the causal effects of the COVID-19 pandemic on students' flight training experiences. In particular, the study analyzes flight training progress, financial situation, career expectation, online learning experiences, and program supporting.

About 200 pilot students, all enrolled in a Part 141 collegiate aviation program, were contacted to participate in a survey in early 2021. The survey was delivered to the participants through an email correspondence. These student pilots were asked to participate in an online survey about their experiences in light of the COVID-19 pandemic. The survey was delivered in February 2021. The survey is structured as self-reporting 5-point Likert-scale survey one means disagree, five means agree, and three represents a neutral attitude.

The survey was programmed using a Google Form. Additional participants' information that was collected in the survey included students' demographics such as GPA, flight type rating, school tenure, gender, age, etc. The survey also collected the flight training location for future comparison study. A total of 209 respondents completed the survey. Three respondents were deemed ineligible for the study (such as enrolled in graduate degree programs or diploma programs) and were dropped from the sample. The final sample size of the survey participants is 206; the breakdown of the survey participants is provided in Table 1.

Table 1
Sample Information

Gender	Number	GPA	Age	Total Flight Hours
Male	162	3.6	21.3	151.1
Female	39	3.7	20.3	124.9
Not disclosure	4	3.73	19.3	146.7

Overall, the sample of this study is a reasonable representation of students at many college flight programs. Usually, these programs have predominantly large number of male students. The average flight hours may also reflect the on-going training situation. It is important to acknowledge that college aviation programs may have additional resources to address the need for a global pandemic than general flight schools.

The study used the Principal Component Analysis (PCA) as well as descriptive statistic to answer the research question. PCA is a statistical analysis method used to reduce dimensions of data by clustering data into multiple factors. Factors that are revealed by PCA shed a light on students' concerns during the pandemic. One advantage of using PCA is that the study could obtain stable estimates even if there are violations of certain assumptions. On the disadvantage side, the quality of the results depends on the quality of the sample. The descriptive statistic helps the study in showing the impact of COVID-19 pandemic in an amicable way.

Results

One of the significant findings is about the delay. Many students expressed that the COVID-19 pandemic has made them delay their flight training as well as graduation. The study finds that 80.3% students have experienced delay in their flight training, and 54.8% students have delayed their graduation.

PCA Factors

The PCA yields 5 main factors (See Figure 1). They are *COVID-19 Problems*, *Job & Career*,*Financial*,*Program Supports*, and *Online Learning*. Each of the factors is a snap shot of student pilots' situation as of February 2021.

COVID-19 problems. The first heavy loaded factor is about the problems/concerns that are caused by problems. This factor includes questions regarding about the concern of their flight skill degradation, worrying the flight training progress, danger of COVID-19 infection during the flight training, and concerns regarding the potential job markets.

Generally, students believe that their flight skill has degraded during the pandemic, especially after the lock-down. As shown in Figure 2, more than 61% of students reported that the flight performance has been different after the lock down. In addition, more than 69% of students indicated that their light skill had degraded degradation due to the lockdown measures.

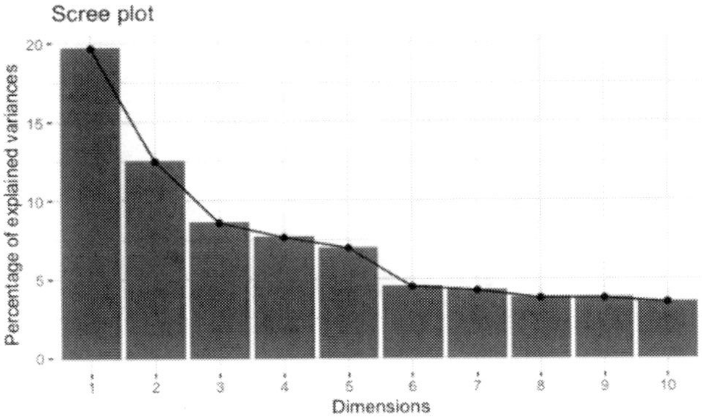

Figure 1. PCA Dimension Reduction results

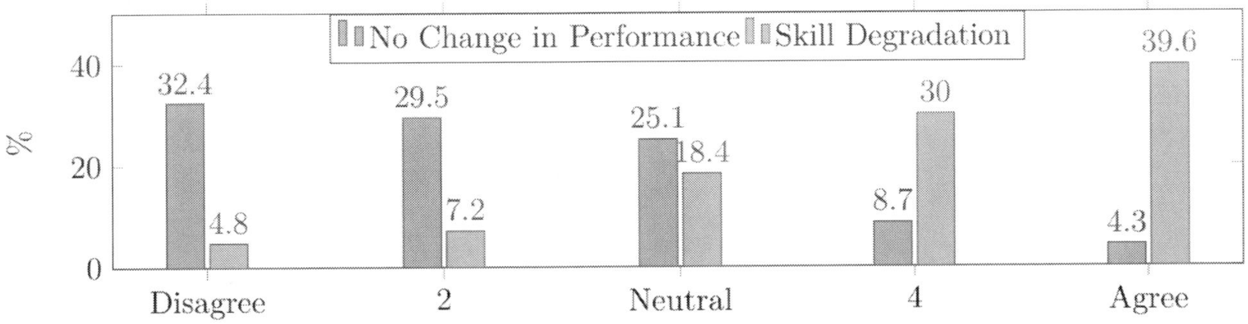

Figure 2. Stduents' Perception regarding flight performance and flight skill

Online Learning. The attitude towards online learning is negative. Students indicated that the quality of online learning is not as good as in-person teaching; they indicated not gaining the same knowledge with online learning as in-person learning, nor their experience is positive. However, the only positive feedback for online teaching is regarding the flexibility of making their own learning schedule as well as universities have maintained the quality of teaching.

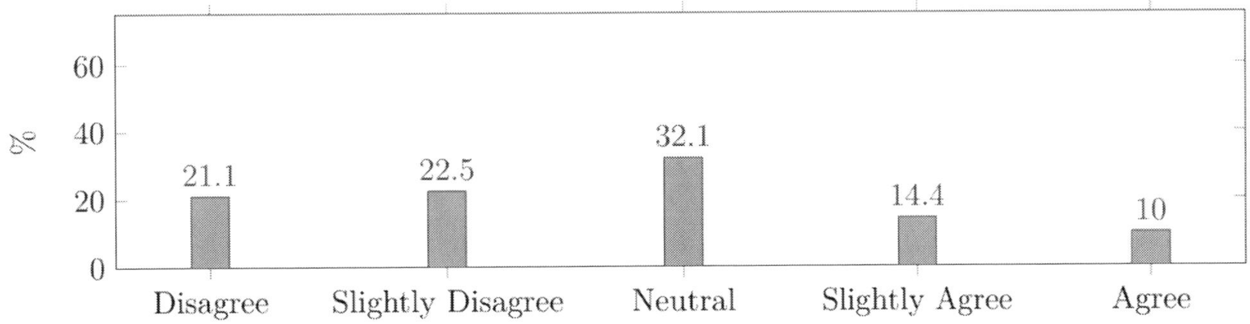

Figure 3. Perception of statement of having a positive online learning experience

Figure 3 shows that more than 40% of students hold a negative view regarding the online learning experiences. Only 10% of participants state that they had a positive online learning experiences during this pandemic.

Job & Career. The attitude toward the aviation job market is negative. Many students believe that they will have a difficult time in finding a job in the aviation industry.

This was expected by the researchers given the layoff and furloughs that have been happening in the aviation industry especially the airline sector. More than 79% of students hold a very negative view regarding the job market. Due to the current uncertainty in the job market, 42% of students are considering continuing their education (graduate school, etc.,) due to the gloomy job market.

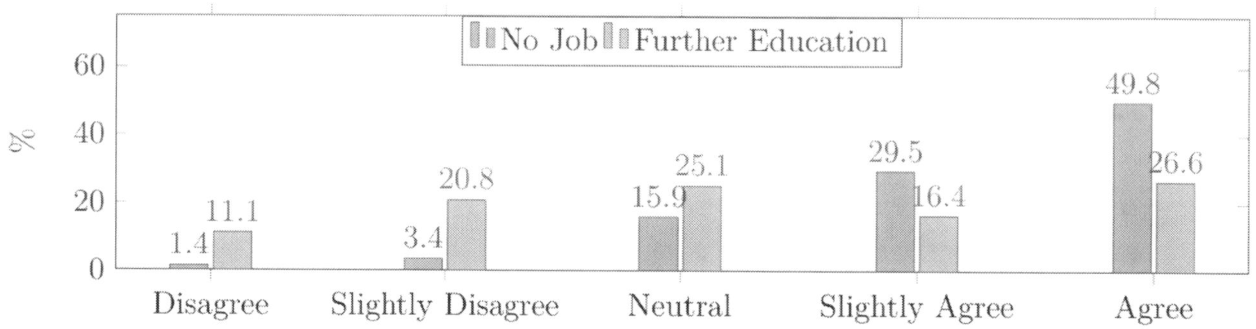

Figure 4. Students' Attitude to Job Market and Possibility of Further Education

Financial. The COVID-19 pandemic has impacted students financial budget for flight training. They also feel the flight training gets more expensive. Meanwhile, more than 40% of students think it is very hard to access financal resources for flight training.

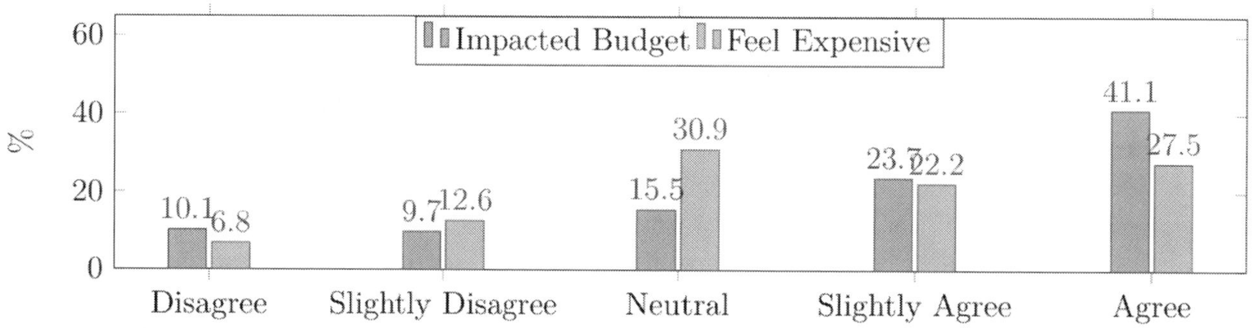

Figure 5. Students' Flight Training Financial Situation

Program support. Only 26% of participants agree that their program has offered some help in searching for internships and job positions. More than half of the participants believes that the program should offer more information regarding financial resources.

Conclusion

The COVID-19 pandemic impacts the student pilots in many ways. More than half of them actually are need financial support for flight training during the pandemic. Most of the students feel that their flight skill has degraded due to the lock-down and lack of practice during the pandemic. Students' online learning experiences are not positive at all. They prefer face-to-face learning and teaching. Students also lack of confidence in the job market and the recovery of the aviation industry.

The impacts of the COVID-19 pandemic brought up challenges which if not addressed on a timely manner, they might also impact the aviation workforce in the near future. The distractions brought by the pandemic restrictions might also impact the skills of the future pilots which if left to go unchecked might compromise the safety quality of the aviation industry in the future. In addition, there is also a chance that some students might consider changing career plans due to the uncertainty that surrounds the aviation industry. This goes against the efforts that various governments are implementing to ensure there will be sufficient workforce for both aircraft pilots and aircraft technicians in the next two decades.

As shown from the results, the COVID-19 pandemic has impacted the student pilots in various ways. The COVID-19 pandemic is actually increasing the student pilots mental stress due to the lock-down and cannot progress in training. At the same time, according to Xiong et al. (2020) the COVID-19 pandemic is associated with psychological distress significantly. The lock-down and stop of flight training delay students' training progress as well as increasing their financial burdens. The metal distress for student pilots should be emphasized in the aviation program and create some mitigation methods to help students coping with the situation.

The negative attitude towards online study is not surprising either. The sudden change of learning environment makes faculty have less time to adapt the situation. According to this study, faculties have maintain the quality of teaching, but students' learning experiences are still negative. The face-to-face interaction may be an important part in learning experiences. Especially for flight training, it is impossible to learn to fly based on a computer.

The last, but not the least, students need more support from the program. They need support in locating financial resources, in finding internship opportunities, job positions, and confidence building. It is important for the program to consider students' special needs during the pandemic. For example, in many collegiate aviation program, it is very difficult to find financial supporting information regarding flight training, as well as mental health support during the pandemic.

References

Al-Taweel, D., Al-Haqan, A., Bajis, D., Al-Bader, J., Al-Taweel, A. M., Al-Awadhi, A., & Al-Awadhi, F. (2020). Multidisciplinary academic perspectives during the covid-19 pandemic. *The International Journal of Health Planning and Management*, *35*(6), 1295–1301.

Debata, B., Patnaik, P., & Mishra, A. (2020). Covid-19 pandemic! it's impact on people, economy, and environment. *Journal of Public Affairs*, *20*(4), e2372.

FlightLogger. (2020, Nov.). *New statistics show the impact of covid-19 on pilot training.* Online.

International Civil Aviation Organization. (2021, Jan). *2020 passenger totals drop 60 percent as covid-19 assault on international mobility continues.* Online. Retrieved from `https://www.icao.int/Newsroom/Pages/2020-passenger-totals-drop-60-percent-as-COVID19-assault-on-international-mobility-continues.aspx`

Ley, M. (2020, Spet). *Aviation program students, faculty stay optimistic as industry struggles.* Online.

Plane and Pilot. (2020, March). *Covid-19 disrupts pilot training industry.* Online.

Xiong, J., Lipsitz, O., Nasri, F., Lui, L. M., Gill, H., Phan, L., ... others (2020). Impact of covid-19 pandemic on mental health in the general population: A systematic review. *Journal of affective disorders.*

EXPLORING METHODS TO COLLECT AND ANALYZE DATA ON HUMAN CONTRIBUTIONS TO AVIATION SAFETY

Jon Holbrook
National Aeronautics and Space Administration
Hampton, VA

Focusing on undesired operator behaviors is pervasive in system design and safety management cultures in aviation. This focus limits the data that are collected, the questions that are asked during data analysis, and therefore our understanding of what operators do in everyday work. Human performance represents a significant source of aviation safety data that includes both desired and undesired actions. When safety is characterized only in terms of errors and failures, the vast majority of human impacts on system safety and performance are ignored. The outcomes of safety data analyses dictate what is learned from those data, which in turn informs safety policies and safety-related decision making. When learning opportunities are systematically restricted by focusing only on rare failure events, not only do we learn less (and less often), but we can draw misleading conclusions by relying on a non-representative sample of human performance data. Changes in how we define and think about safety can highlight new opportunities for collection and analysis of safety-relevant data. Developing an integrated safety picture to better inform safety-related decision making and policies depends upon identifying, collecting, and interpreting safety-producing behaviors in addition to safety-reducing behaviors. Opportunities and challenges in collecting and analyzing the largely unexploited data on desired, safety-producing operator behaviors are discussed.

Focusing on undesired operator behaviors is pervasive in system design and safety management cultures in aviation. This is evidenced by the extent and range of resources put into eliminating, reducing the likelihood, reducing the consequences, and conducting investigations of adverse states or events. This focus, however, limits the data that are collected, the questions that are asked during data analysis, and therefore our understanding of what operators do in everyday work. Humans play an integral role in aviation safety. Therefore, human performance represents a significant source of aviation safety data. Human performance includes both desired and undesired actions. Most of the time, those actions *promote* safety, but sometimes those actions can *reduce* safety. Commercial aviation hull-loss accidents today are significantly below 1 per million flights, and have been steadily decreasing since the advent of commercial jet operations in 1958. Since as early as 1967, no year has had more than 4 hull losses per million flights (Airbus, 2020). While it is difficult, if not impossible, to tease out the individual contributions of the significant advances that have been made in hardware, software, and human factors to the steady reduction in these accidents over the years, the fact that hull loss accidents were exceedingly rare even in the decades preceding these advances suggests that humans have been making significant contributions to aviation safety throughout its history.When our safety thinking systematically restricts the data we collect and analyze, however, this restricts our opportunities to learn from human performance. Importantly, when this restriction is systematic, it can bias what we learn, which can, in turn, affect our safety policies and decision making.

Most of our learning about human performance in aviation comes from studying relatively rare errors and failures. The magnitude of the discrepancy between human performance that is representative of operations and human performance that is actually analyzed can be difficult to grasp without data to provide some base rate context. For example, it is widely reported that human error has been implicated in 70%-80% of aviation accidents (e.g., Wiegmann & Shappell, 2003). Additionally, analysis of Line Operational Safety Audit (LOSA) data indicates that pilots intervene to manage aircraft malfunctions on 20% of normal flights (PARC/CAST, 2013). If those percentages are examined within the context of 10 years of world-wide jet data (Boeing, 2017), a contingency table can be constructed, depicting Outcome (not accident or accident) by whether human intervention was identified as being associated with that outcome (see Figure 1). When only data from errors and failures are analyzed, the vast majority of human impacts on system performance are largely ignored. Not only does this indicate significant missed opportunities for learning, but it implies that learning is focused on a very small sample of non-representative data, which can have significant impacts on what is learned and how those insights inform safety policies and safety-related decision making.

Outcome

Attributed to Human Intervention	Not Accident	Accident	
No	195,199,690	78	195,199,768
Yes	(48,799,922)	(310)	48,800,232
	243,999,612	388	244,000,000

Figure 1. Human contributions to safety successes (solid oval) far outweigh their contributions to failures (dashed oval), but are relatively unstudied and poorly understood. Note: Accident is defined as hull loss and/or at least one fatality.

While making generalizations from a sample of population data is common practice, responsible researchers work to ensure that their sample is representative of the population of interest. Failure to do so can result in sampling bias, in which what is learned from the sample is erroneously attributed to the larger population. When a sample is systematically non-representative, generalizations from the sample data are suspect. An assertion, for example, that human error contributes to accidents, therefore removing humans will reduce accidents, ignores that humans are also a significant source of successful system performance, and in fact contribute to safety far more than they reduce safety. Indeed, extrapolating from the data in Figure 1 suggests that pilots intervene to keep flights safe over 157,000 times for every time that pilot error contributes to an accident.

Changing Our Safety Thinking

Changes in how we define and think about safety can highlight new opportunities for collection and analysis of safety-relevant data. Hollnagel (2016) has proposed that we update our definition of safety to include not only minimizing opportunities for undesired states, but also maximizing opportunities for desired states. This approach is better aligned to understanding human performance, which contributes to both. Furthermore, to maximize opportunities to learn from human performance, data should be collected and analyzed on routine performance, not just exceptional performance. Learning only from rare events means that learning only occurs rarely. While learning from frequent successes has the advantages of increasing sample rate, sensitivity, and timeliness of safety learning, it raises important issues about determining exactly what data to capture, how to analyze and manage this potentially massive expansion of safety data, and translating learned insights into policy and design decisions.

Collecting and Analyzing Data on Safety-Producing Behaviors

Most aviation organizations already collect data on operator performance from various sources, including operator-, observer-, and system-generated data. As we expand our understanding of what constitutes a safety-relevant occurrence to include both desired as well as undesired behaviors, this raises questions about the collection of data on operators' safety-producing behaviors. Can we leverage data that are already being collected, and are there new opportunities for data collection based on our expanded safety thinking? Much of the data we collect are only analyzed for "safety exceedances". While we are actually collecting significant data on "what goes right," even in our safety reporting systems that focus on "what goes wrong," our analysis processes do not often consider these data. Similarly, flight data recorders capture "what happens," including both desired and undesired actions, but our analyses typically focus on the undesired. Thus, there are potential opportunities for data collection (i.e., systematically collecting data on what goes well), as well as data analysis (i.e., analyzing data we have already collected, but may not have previously considered relevant).

Operator-Generated Data

Operator-generated data include interviews, questionnaires, and event self-reports about an operator's own lived experience. These data represent perhaps the best source of insights into what the operators may have been thinking about, including their motivations, intentions, goals, and pressures, and how they believe those considerations may have influenced their decisions and actions. Although some modes of thought are not open to introspection (see Kahneman, 2011), insights can still be gained from how operators think about their thinking. Similarly, operator-generated data affords an opportunity to learn from how operators talk about their own safety-producing performance. We have well-established shared terminology for describing risks, hazards, and errors (e.g., Wiegmann & Shappell, 2003), but do not yet have such a vocabulary for safety-producing behaviors. Can we identify the parlance already in use for describing safety-producing behaviors, and can we use that to bootstrap development of a new safety language that can address both desired and undesired behaviors? This question is explored elsewhere in these Proceedings by Feldman et al. (2021), using an existing event report collection: NASA's Aviation Safety Reporting System (ASRS). Although ASRS reports are collected for the purpose of describing something that went wrong, they may represent a valuable source of data on things going well, particularly related to noticing, tracking, responding to, and learning from the described problem event (e.g., Holbrook et al., 2020).

Observer-Generated Data

Observer-generated data include data from observations of line operations as well as training and simulated events. These data represent an excellent source of insight into overt behaviors, particularly those that may not be salient to operators. This is particularly relevant for safety-producing behaviors that operators may see as "routine" or "just part of the job" and therefore may be less likely to self-report those behaviors. In rich and complex environments like aircraft flight decks, not every behavior is realistically observable – there is simply too much going on, and not every action is meaningful to capture. Knowledge frameworks are often used to train and prepare observers for what to focus on. These knowledge frameworks can be thought of as one way of embodying safety thinking. There are some well-established knowledge frameworks, such as Threat and Error Management (TEM), which is the basis for LOSA (Klinect, Wilhelm, & Helmreich, 1999). While TEM uses undesired behaviors and states as the primary triggers for data collection, this framework still affords opportunities to collect data on how pilots safely managed threats and errors. American Airlines has developed a "Learning and Improvement Team" framework for flight line observations that is explicitly designed for the collection of flight crew resilient performance (American Airlines, 2020). Exploration of how the knowledge frameworks of observers affect the insights they derive from an observation is discussed elsewhere in these Proceedings (Mumaw, Billman, & Holbrook, 2021).

System-Generated Data

System-generated data include flight data records as well as documentation of flight regulations and procedures. Automation enables collecting a large volume of system data on what is happening, via flight data recordings, with less overhead at the time of collection than operator- and observer-generated data. Here, the focus on risks and hazards manifests in terms of which data we choose to analyze – that is, a failure state or adverse event triggers analysis that leaves the vast majority of collected data unconsidered. Indeed, commercial airlines with Flight Operations Quality Assurance (FOQA) programs use data from flight data recorders to monitor daily operations, but often only look at the data from flights with known adverse events (i.e., flights that violate some pre-determined "safety exceedance" criterion). "Non-event" flight data may be analyzed to establish a baseline for comparison, but not as a valuable source of learning, themselves. These "non-event" flights, however, can afford opportunities for insights into safety-producing behaviors, such as actions taken by flight crews to mitigate or prevent adverse events from manifesting (e.g., Holbrook et al., 2019). That is, the occurrence of the adverse event does not have to be a pre-requisite for learning. The amount of flight data collected opens up application of "big data" approaches to analysis, and flight data represent an excellent source of data on quantitative performance parameters, such as timing or frequency. But while flight data can provide many quantitative details about operator and vehicle performance, they cannot provide information about the knowledge state, motivation, or broader context for the event. This information could be obtained through observer- or operator-generated data to supplement system data and provide a more complete understanding of the event.

Human-in-the-Loop (HITL) Flight Simulation

HITL flight simulations represent an additional approach to collecting data on human performance. While data from real-world operations offer the most veridical glimpse into everyday work, HITLs provide opportunities to collect multiple sources of data from the same event, as well as the capability to more efficiently test new approaches to data collection. One of the challenges in designing HITL simulations from a safety mindset focused primarily on errors and failures is that, to have something to measure, scenarios must be designed to induce those errors and failures, which can be difficult when studying high-performing workers. Ironically, in this situation, the resilient, safety-producing performance of the test participants is seen as an impediment to data collection, rather than important data to be collected – an obstacle that can sometimes require experimenters to create scenarios that are far-removed from representative flight operations in order to induce the errors and failures "required" for performance measurement. Scenarios can be designed, however, to include events and perturbations that might not otherwise involve enough risk or hazard to trigger data collection or analysis in real-world operations, and thus afford opportunities for observation and measurement of more "routine" performance. While it is always a concern that participants in any simulation will not perform in the same manner that they do in actual operations, perhaps this effect may be somewhat mitigated by designing simulation scenarios that are more representative of real-world operations. Exploration of HITL simulation as an approach to learning about pilots' safety-producing behaviors is discussed elsewhere in these Proceedings (Stephens et al., 2021).

Implications for system design

Changing the way we think about safety is not just relevant to system operators but also to system designers. Our safety thinking affects our design assumptions, which are influenced by our understanding of human performance. We are just beginning, however, to build an understanding of safety-producing behavior. A focus on failure alone can lead to design assumptions about improving safety by minimizing human roles and the need to protect the system from error-prone humans. While we certainly should acknowledge human limitations and the consequences of human error, system designs should also leverage the capabilities of humans to create and sustain safe operations. If these capabilities are poorly understood, what assumptions are going into the design of the increasingly autonomous machine systems to which these functions may be relegated? Challenges and opportunities for system design that leverage human capabilities and new ways of thinking about safety are expanded upon elsewhere in these Proceedings (Lachter, Hobbs, & Holbrook, 2021; Nemeth & Holbrook, 2021).

Conclusions

While we should continue to learn from what goes wrong, we should also try to learn from what goes right. Learning from what goes right can enable us to make data-informed adjustments to operations and policies without having to wait for something to go wrong. Changing how we define safety – expanding our understanding of what constitutes a safety-relevant issue – can inform this learning, and is relevant to both operations and system design. This expansion in thinking brings with it a need to expand methods of data collection and analysis, representing an important opportunity for human factors and human performance communities.

Acknowledgments

This work was funded by NASA's System-Wide Safety Project, part of the Aeronautics Research Mission Directorate's Aviation Operations and Safety Program.

References

Airbus. (2020). A statistical analysis of commercial aviation accidents 1958-2019. Reference X00D17008863 Issue 4. France: Art & Caractere.

American Airlines' Department of Flight Safety (2020). Trailblazers into Safety-II: American Airlines' Learning and Improvement Team, A White Paper Outlining AA's Beginnings of a Safety-II Journey.

Boeing. (2017). Statistical Summary of Commercial Jet Airplane Accidents: Worldwide Operations 1959 – 2016. Seattle, WA: Aviation Safety, Boeing Commercial Airplanes.

Feldman, J., Barshi, I., Smith, B., & Matthews, B. (2021). Reports of resilient performance: Investigating operators' descriptions of safety-producing behaviors in the ASRS. In Proceedings of the 2021 International Symposium on Aviation Psychology.

Holbrook, J., Prinzel, L. J., Stewart, M. J., & Kiggins, D. (2020). How do pilots and controllers manage routine contingencies during RNAV arrivals? In Proceedings of the 11th International Conference on Applied Human Factors & Ergonomics.

Holbrook J. B., Stewart, M. J., Smith, B. E., Prinzel, L, J., Matthews, B. L., Avrekh, I., Cardoza, C. T., Ammann, O. C., Adduru, V., & Null, C. H. (2019). Human performance contributions to safety in commercial aviation. NASA/TM-2019-220417.

Hollnagel, E. (2016). Resilience Engineering. erikhollnagel.com/ideas/resilienceengineering.html. Accessed March 1, 2021.

Kahneman, D. (2011). *Thinking, fast and slow.* New York: Farrar, Straus, & Giroux.

Klinect, J.R., Wilhelm, J.A., Helmreich, R. L. (1999). Threat and Error Management: Data from Line Operations Safety Audits. In Proceedings of the Tenth International Symposium on Aviation Psychology. Columbus, OH: The Ohio State University.

Lachter, J., Hobbs, A., & Holbrook, J. (2021). Thinking outside the box: The human role in increasingly automated aviation systems. In Proceedings of the 2021 International Symposium on Aviation Psychology.

Mumaw, R. J., Billman, D., & Holbrook, J. (2021). How do different knowledge frameworks help us learn from aviation line operations? In Proceedings of the 2021 International Symposium on Aviation Psychology.

Nemeth, C., & Holbrook, J. (2021). Resilience engineering's potential for Advanced Air Mobility. In Proceedings of the 2021 International Symposium on Aviation Psychology.

PARC/CAST Flight Deck Automation Working Group. (2013). Operational use of flight path management systems. Final Report of the Performance-based operations Aviation Rulemaking Committee/Commercial Aviation Safety Team Flight Deck Automation Working Group. Washington, DC: Federal Aviation Administration.

Stephens, C., Prinzel, L. J., Kiggins, D., Ballard, K., & Holbrook, J. (2021). Evaluating the use of high-fidelity simulator research methods to study airline flight crew resilience. In Proceedings of the 2021 International Symposium on Aviation Psychology.

Wiegmann, D. A., & Shappell, S. A. (2003). *A human error approach to aviation accident analysis: The Human Factors Analysis and Classification System*. London, UK: Ashgate.

HOW DO DIFFERENT KNOWLEDGE FRAMEWORKS HELP US LEARN FROM AVIATION LINE OBSERVATIONS?

Randall J. Mumaw
San Jose State University Foundation
San Jose, CA

Dorrit Billman
NASA Ames Research Center
Moffett Field, CA

Jon Holbrook
NASA Langley Research Center
Hampton, VA

Human performance includes actions that increase safety, as well as actions that can reduce safety. Ensuring safety in complex dynamic operations like commercial aviation depends on the ability to institute appropriate responses based on what is learned from flightcrew performance and the contexts in which it occurs. To do this systemically at the organization level requires collecting data on flightcrew performance, developing effective approaches to analyzing those data, and understanding how to translate what has been learned into policies, procedures, and practice. Systematic observation of front-line operators is a vital source of human performance data. Much has been learned from such observations, including methodological principles. Most observations have been based on a framework focused on managing safety challenges and the ensuing unsafe events. A complementary perspective focuses on flexibility and actions that promote continued safe and effective operation. We consider lessons learned about observational methods from an established framework focused on undesired actions and how these might be extended for a framework focused on desired actions.

Holbrook (2021) makes a case to expand our approach to aviation safety to include the ideas introduced through the concept of Safety II (Hollnagel, 2014), sometimes characterized as "what goes right." Safety II is focused on understanding and documenting how operators (for aviation, the flightcrew) successfully manage the high workload, variable operating conditions, goal trade-offs, limitations on procedures, unexpected threats and disruptions, and other normally occurring challenges to achieving mission objectives safely and efficiently.

The more traditional approaches to safety—that focus on the rare things that go wrong—have not focused on the extensive knowledge and skills that pilots bring to the job to achieve safe, effective operations despite these challenges. In aviation, the term "airmanship" is often a reference to these hallmarks of professionalism and skill. Unfortunately, few have articulated the elements of airmanship, which makes it more difficult to provide to less-experienced pilots through training. One pathway to capturing how flightcrews create safe and efficient flights is through structured observations in a real or realistic setting.

Structured observations can reveal a range of performance dimensions within the current operational context. Further, pilot training, standard operating procedures (SOP), and equipment design need to be held up to the actual demands of line operations; what is sometimes called "work as done" (as opposed to "work as imagined"; see Hollnagel, 2017). Specifically, observation of practitioners in a real or realistic operational setting can provide a deeper understanding of:

- the operational demands, such as pace, time-pressures, forced trade-offs, pop-up tasks from the cabin crew, time zone shifts and their effects on sleep, and workload.
- the demands from operational variability, such as noise in the communications, unfamiliar accents in ATC English, changing weather and visibility, shifts in safety margins, and changing runway conditions and wind direction.
- the threats to efficient and safe operations, such as slam-dunk approaches, thunderstorms, and strong tailwinds.
- the limits of the engineered solutions, such as procedures that may need to be adapted, system alerts that confuse more than clarify an equipment issue, and unexpected autoflight mode changes and consequences.
- the types of knowledge and skill that pilots call on to operate safely and effectively, such as detailed airport knowledge regarding how it typically operates, and flight path management methods that create room for downstream flexibility.
- how flightcrews manage potentially conflicting operational goals, including which goals are prioritized and how safety is maintained when meeting efficiency goals.

Two Frameworks for Guiding Flightcrew Observation

Currently, opportunities to observe flightcrews exist: Airlines routinely put check airmen in the flight deck (real or simulated) as assessors of SOP and operational policies, and airlines observe flightcrew performance through LOSA (Line Operations Safety Audit, see ICAO, 2002), which uses the TEM (Threat and Error Management) perspective to capture work as done. These evaluations focus on the threats present in the operational environment, any wrong decisions and erroneous actions taken by pilots and others, and if/how those get linked to an undesired outcome. For example, a pitot tube becomes clogged by ice while in flight, it generates a set of alerts, the airspeed indication decreases rapidly from the air data system failure, and the pilot, not aware that airspeed indications are invalid, pitches the nose down to regain airspeed. Certainly, it is important to understand how the airplane presents the pitot failure and the pilot's poor response. More broadly, when these types of events occur, it is important to understand pilot response. This is the traditional approach to safety: **How are undesired outcomes produced and how do we prevent or manage those situations?**

The other framework for observation is a "Safety II"-oriented assessment (e.g., American Airlines, 2020). Safety II, as a complement to Safety I, asks instead: **How do flightcrews/pilots manage the variability, disruptions, and threats of normal operations to ensure that mission objectives are achieved?** Applications of Safety II attempt to understand and characterize flightcrew/pilot performance that creates capacity, anticipates problems that may be developing, adapts to the needs of the moment, and learns from operational experience. The focus is not on undesired outcomes but on how flightcrew performance manages the normal (and unexpected) variability in the operational environment.

Lessons about Observational Methods

The current observational protocols, shaped in the Safety I framework, have helped establish basic principles that make observational data more reliable and generalizable:

- A high level of domain expertise – The observer is an interpreter of flightcrew actions and verbalizations. The observer needs to understand the operational situation to see what the flightcrew is faced with and how they should respond, for example, to know that a tailwind can threaten their ability to make an altitude restriction. Observers, therefore, need to be highly skilled practitioners.
- An agreement about which aspects of performance are relevant – Observers are limited in what they can capture about flightcrew performance, especially given the need to capture the larger context and the appropriateness of the actions in that context. The observer has to focus on selected decisions and actions that are relevant to the observation goals, for example, for TEM, the flightcrew's ability to identify and manage an operational threat.
- A set of performance standards – For TEM, the observer is working from an understanding of what the typical flightcrew can do. Given the threat that is presented, what is expected of the flightcrew and did they perform at a requisite level? For example, given a severe failure or environmental hazard, there may be limits on what the flightcrew is able to do to mitigate or manage it. Ideally, these expectations of flightcrew performance are shared across observers to ensure that the observers score consistently.
- An agreement about how observations will be aggregated and reported – Finally, the observations need to be aggregated and reported to influence policy, procedures, training, or some other component of airline operations. This understanding of how the data are gathered and reported needs to be shared across the set of observers to ensure a consistent approach to data collection.

Translation into Safety II: Considerations in Collecting Observational Data

Safety II expands the set of relevant behaviors that could be observed, and it, therefore, becomes even more important to maintain standards. Observation focuses on how the entire flight is managed; acknowledging the complexity and variability of line operations, Safety II captures how pilots "make it all work." In some sense, Safety II is trying to capture the "extra" things that pilots do beyond what is prescribed in the SOP, which is a large behavior space. Hollnagel (2014) has identified four broad categories for Safety II behaviors: anticipate, monitor, respond, and learn. These four categories also suggest a temporal perspective.

- Anticipate: anticipating what might happen in the near future and preparing for the range of situations that may occur.
- Monitor: gathering and integrating information about the operational situation, as it develops.
- Respond: responding appropriately to an event that just happened or an unexpected shift.
- Learn: using experience to expand skills and knowledge from the present to be useful in future operational situations.

For example, observations may capture the following

- Pilot Flying (PF) saw that the Pilot Monitoring (PM) was overwhelmed with radio communications with dispatch on taxi out and used the relief pilot to complete the Before Takeoff checklist.

- The crew added a briefing at 18,000 feet because runway conditions had changed significantly during the initial descent.

Gathering useful observational data demands both an effective coding scheme and trained observers. A coding scheme for complex situations, implicitly or explicitly, requires three interdependent aspects: a prompt or trigger that a codable event occurred, a classification system for categorizing the identified events of interest, and criteria that segment the behavior stream or mark the beginning and end of the event. Safety II presents several challenges to forming an effective coding scheme.

Coding Schemes for Safety II

Prompts. Consistent coding for Safety II is difficult because relevant behaviors may be both frequent and not marked by a single, observable event or prompt. Rather, pilot activity is often driven by internal goals and intentions, which may be expressed in a variety of specific actions, as appropriate in context. Thus, there are these concerns about prompts:

- Prompts may be less connected to a single event – There is not always an easily identified failure or error or environmental threat to prompt an observation. The prompts for the relevant behaviors may lie in normal operational complexity: clearance changes, shifts in workload, changing weather. This variability or complexity is not infrequent. Indeed, it is incredibly routine, and therefore, there is a much larger set of relevant behaviors.
- Prompts may not be directly observable – In many cases, a pilot responds to some anticipated event, such as anticipated workload, anticipated flight path deviation, or maybe anticipated limits to performance by the autoflight system in its current configuration. In other cases, a pilot is attempting to better understand the operational environment by seeking or sharing knowledge, or even just encouraging knowledge-sharing.

Because of the absence of unambiguous behavioral cues, classification may require inferring and identifying an intent and associated behavioral indicators. The mapping between intent and behavior may be quite varied. While the importance of behavioral indicators is well-established, its application here may be relatively difficult. Indeed, it is hard to determine how much the intent effectively cues a search for behaviors or behaviors cue the intent for the observer. For example, the cues to intent might come from a statement about a plan that occurs considerably prior to the event. Because many categories of interest are grounded in intentions, we suggest that useful prompts or triggers may be organized around anticipation and intent.

Classification. Classification may rely on a mapping between intentions and the most commonly used mechanisms for satisfying those intentions. The intention categories may be quite general, though more specific than Hollnagel's categories of anticipation, monitoring, responding, and learning. For example, the PF may intend to distribute workload more evenly prior to entering a busy period by changing who makes entries into the flight management system (FMS). The observer should capture the mechanism for distributing workload (the re-assignment of FMS duties) and the intention of that mechanism (redistributing workload to prevent overload at a later time). The intention captures an important Safety II skill (referred to as a proficiency), and there is value in understanding the various ways that pilots can satisfy that intention. That is, other duties could have been re-distributed to accomplish a workload reduction. While it is easier for the observer to latch onto the mechanism, one might argue that the pilot's intent is, at least, equally important to capture. It shows that the pilot recognized the need to manage workload. However, this approach puts the observer in the position of "seeing" intention, and it suggests a data collection form that is intention-oriented.

Segmentation. Setting the event start and stop boundaries may also pose additional challenges. Effective action is based on both stability of intent across changes that require behavioral adjustment and flexibility to change intent with changing conditions. An intent might be formulated some time before any action is taken and might or might not be expressed when formed. An event might take several actions to complete. Thus, there may be ambiguity and tradeoffs in what forms the unit of activity, and if these are very small, how relations among small events to accomplish an intention should be represented.

Observer Requirements

Capturing both the intention and the action places a heavy burden on the observer. The observer now has to "see" the intention. Notably, the pilot may not have been able to articulate the intention in the moment; it may be too well embodied in the actions. However, in order to capture pilot proficiencies, it is critical to provide the intentional context for the actions.

To achieve this as a scoring scheme, observers will need to be facile with the set of intentions (which should be kept relatively small for this reason). In addition, observers will need strong familiarity with the range of mechanisms (behaviors) that can be applied to fulfil the intention. The data collection tool could make explicit many of these links as an aid. The observers might first see behaviors to recognize intent, or see intent to recognize behaviors, or rely on the interplay. In any case, following training, observers must be able to apply the scheme with acceptable reliability.

Uses of Observer Data

The objective of collecting Safety II observations is to influence operations going forward. Capture and analysis of observations might impact operations through changes in training, procedures, or interface.

To Influence Training

The observational data, ideally, allows one to create a mapping between Safety II-type intentions and the specific behaviors that pilots are using during line operations to enact those intentions. An intent can be realized in many ways, and it may be possible to train at a more-generalized level than individual behaviors. Gathering positive (but not proceduralized) behavior might make "higher order" behavior more visible and thus amenable to training; this might include management of multiple goals such as building pilot experience and immediate operational efficiency or prioritizing different briefing topics. For airlines, this articulation adds another layer of description for skilled performance—on top of "technical" skills, which are focused on flight maneuvers, and Crew Resource Management (Helmreich et al., 1999) skills, such as leadership. Finally, observations of line operations will make trainers better aware of operational environment complexities and variability, which can lead to the development of more realistic training scenarios, as well as better feedback and debriefing.

To Influence Procedures and Interface Design

Initially, observational data may point to areas where flightcrews need to adapt existing written procedures due to operational factors. Eventually, these data may identify ways that the procedures should signal intent and meaning of the individual procedure steps. Alternatively, the data may point to a need to rewrite procedures to be better aligned with how system failures present. Similarly, there is the potential for feedback into interface design. While airplane operators do not determine design directly, they have the opportunity to point out to manufacturers that operational realities make it hard to perform necessary tasks effectively. Perhaps the interface requires too many actions to meet performance time limits. The larger point is that airplane operators have to deal with design deficiencies that are revealed by the operational environment.

Summary and Conclusion

The observational methods that have been refined in the context of Safety I provide a valuable foundation for Safety II methods and observer requirements. Important adaptations for a Safety II framework may derive from the importance of anticipation and intent within Safety II.

Acknowledgments

This work was funded by NASA's System-Wide Safety Project, part of the Aeronautics Research Mission Directorate's Aviation Operations and Safety Program.

References

American Airlines' Department of Flight Safety (2020). Trailblazers into Safety-II: American Airlines' Learning and Improvement Team, A White Paper Outlining AA's Beginnings of a Safety-II Journey.

Helmreich, R.L., Merritt, A.C., & Wilhelm, J.A. (1999). The evolution of crew resource management training in commercial aviation. *The International Journal of Aviation Psychology, 9*(1), pp.19-32.

Holbrook, J. (2021). Exploring methods to collect and analyze data om human contributions to aviation safety, Proceedings of the International Symposium on Aviation Psychology

Hollnagel, E. (2014). *Safety-I and Safety-II: The past and future of safety management.* Farnham, UK: Ashgate.

Hollnagel, E. (2017). Why is work-as-imagined different from work-as-done?, Resilient Health Care, Volume 2: The Resilience of Everyday Clinical Work. https://doi.org/ 10.1201/9781315605739.

International Civil Aviation Organization (2002). *Line operations safety audit (LOSA)*, Doc 9803, AN/761, Montreal: ICAO.

REPORTS OF RESILIENT PERFORMANCE: INVESTIGATING OPERATORS' DESCRIPTIONS OF SAFETY-PRODUCING BEHAVIORS IN THE AVIATION SAFETY REPORTING SYSTEM

Jolene Feldman
San Jose State University Research Foundation
San Jose, CA

Immanuel Barshi
NASA Ames Research Center
Moffett Field, CA

Brian Smith
NASA Ames Research Center,
Moffett Field, CA

Bryan Matthews
KBR Wyle Services
Moffett Field, CA

While many existing taxonomies and frameworks provide a common vocabulary for describing how human operators fail in the context of sociotechnical systems, at present, there is no common vocabulary to describe how humans succeed. Such a framework would facilitate systematically collecting and analyzing data on how human performance can produce safety, not just how it can reduce safety. One potentially rich source of currently available information for exploring desired performance is the reports submitted to NASA's Aviation Safety Reporting System (ASRS). These de-identified, confidential, and voluntary narrative reports are submitted by pilots, controllers, ground operators, and others within aviation operations. While these reports are primarily submitted to describe safety risks, incidents, and problems, they also often describe how those risks were mitigated, and provide a window into aspects of everyday work in aviation. This paper describes an analysis of ASRS narratives to understand how operators talk about their own resilient behaviors during adverse safety conditions and events. Guided by Erik Hollnagel's Resilience Assessment Grid framework (i.e., anticipate, monitor, respond, learn), we illustrate our approach and methodology with examples from reports. We also highlight some of the challenges and how further research is needed in developing a taxonomy of operators' descriptions of resilient performance.

Safety has largely been viewed and defined by examination of threats, errors, and undesirable behaviors. Although most of the time things go right, there are no common methods to collect information from operations when they do, and there is no common vocabulary that exists to discuss these successful behaviors nor taxonomy to organize them. However, one potentially rich source of currently available information to learn about resilient performance is reports submitted to the ASRS. These de-identified, confidential, and voluntary narrative reports are submitted by pilots, controllers, ground operators, and others within aviation operations. These reports present an opportunity to look at "small" (incidents not accidents) but frequent events in everyday work.

Operators submit reports to ASRS typically after something went wrong. Pilots' behaviors that result in successfully handling complex situations may often not be reported, as they are likely viewed as part of their job. Our premise is that nearly every ASRS report demonstrates resilient potential(s) and is an example of positive outcomes, since operators survived the event and were able to write the report. Terms such as "resilience" or being "resilient" are not currently part of the language of how operators talk about their own performance in ASRS reports, despite descriptions that clearly demonstrate resilience.

Resilience, described in a general way, is an adaptive capacity that contributes to successful outcomes. Looking at resilience can increase our understanding of how safety is produced and where to focus safety training efforts. Detecting resilient performance in these reports was guided by Erik Hollnagel's Resilience Assessment Grid framework (Hollnagel, 2015). Hollnagel identifies four potentials that are necessary for resilient performance. A system's resilience is described by its potential to: Anticipate (knowing what to expect, anticipating disruptions, new or changing situations), Monitor (knowing what to look for in one's performance as well as in the environment), Respond (knowing what to do, respond to changes and adjust), and Learn (knowing what happened, learning from the experience, and learning the right lessons). In ASRS reports, these potentials manifest themselves in behaviors. These descriptions of behaviors can be analyzed and identified (or labeled) as resilience potentials.

Method

The ASRS database can be searched in a variety of ways. The database is vast, with currently over a million and a half reports in its repository, and growing by the hundreds on a daily basis. Searches can result in thousands of reports to review which can be labor intensive and time consuming. Descriptions of resilient performance can also include more than key words and phrases. Key word searches can be informative, but can fail to detect resilient behaviors that are not specifically named. Refined investigations can include using combinations of selected keywords, phrases, filters, codes (e.g., ASRS coding taxonomy), and/or focusing on specific targeted situations or events (e.g., Chandra et al., 2020). With the quantity of data contained in narratives, software tools may facilitate and enhance searches and analysis. For example, some software programs can be utilized as discovery tools (e.g., based upon frequency of terms in documents) or as query tools (e.g., based upon searching for similar documents that have a shared context). Ongoing research efforts continue to explore the possibilities and limitations of various tools for analyzing data from ASRS reports. Using such tools for ASRS data analysis has been conducted (Paradis, et al., 2021) but further investigations are necessary.

Using Hollnagel's framework, Holbrook et al. (2019) identified resilient performance strategies, employed in routine aviation contexts, as a way to organize identified behaviors, for each of the four potentials. For example, for the Learn potential, Holbrook et al. identifed four strategies: 1) Leverage experience and learning to modify or deviate from plan., 2) Understand formal expectations., 3) Facilitate others' learning., and 4) Conduct after-action debriefing.

These strategies can be used to identify and code operators' statements. To illustrate our approach and to address some of its challenges, specific statements from selected reports are used as examples below, with accompanying notations and explanations. Examples #1-4 come from ASRS report #1516535.

Example #1

"Assigned CHSLY3 RNAV Star into CLT, FO was PF, I was PM... FO had checked ATIS and noted that Runway 23 was in use. He also mentioned at that time that since Runway 23 is on the ATIS we would most likely get that runway from our direction. So FMC was planned with Runway 23 during cruise..."

Based on the proposed coding scheme, the First Officer (FO) is both Anticipating and Learning through conducting a pre-action briefing, leveraging his experience, and facilitating others' learning (FO to Captain [CA]). He discusses the planned action and identifies the variable that affects the plan. The CA describes how the FO anticipated what to expect, and compared his experience to the current situation to develop real-time assessment and to modify the plan. Since neither crew member used the word Anticipate or Learn in the narrative, this report would not be returned using those resilient terms in a keyword search (although a search using the keyword "Respond" would return this report, because it appears later in the narrative). Another issue is that the action of the FO can be identified as both Anticipate and Learn. The coding scheme does not account for strategies that fit into more than one potential. One strategy may be described by both potentials, or even be associated with or linked to another potential (e.g, Learning leads to Anticipating). Hollnagel describes this as "the interdependence of the potentials". Similarly, Kiernan et al. (2020) describes these enablers of resilient performance as "exhaustive but not mutually exclusive". The same words can represent different strategies depending on context, and different words can represent the same strategy. This makes such coding challenging, particularly when developing a taxonomy to group data in a meaningful and systematic way.

Example #2

"At about that time the FO asked what should I do? I said start slowing."

In this situation, the Learning opportunity occurs between the FO and the CA. The CA is facilitating the FO's learning. This strategy is described as, "sharing information with others to increase their immediate understanding and long-term learning" (Holbrook et al., 2019). Individuals bring with them their past experience and knowledge, which affects their actions, so determining the source of the Learning may require further investigation in this work. This is evident in the next example.

Example #3

"I have found out the hard way that this aircraft will not come down and slow down simultaneously. Phone call with OPD guy was enlightening as he says this is a huge ongoing issue..."

Though he doesn't go into detail about "the hard way", the CA is describing his Learning through experience, and his Learning is further facilitated through his conversation with a colleague.

Example #4

"Approach saw our problem and said nicely, I'm going to spin you around and get you back on a heading to intercept localizer for 23."

The opportunity to Anticipate, Monitor, and Respond is attributed to the Approach air traffic controller (ATC). ATC anticipates and notices the problem, and adjusts the current plan. ATC monitored the crew's ability to make the descent, with cues signaling a change from normal or expected operations. This statement depicts the role of ATC and describes how different actors contribute to resilience in the system. Another method of evaluating resilience in ASRS reports can focus the analysis on the turning points and triggers of an event. In this situation, it was ATC that changed the course of the flight.

Example #5 (ASRS #1741671)

"...I called for a Go Around... the Go Around was done to look at a problem...Captain decided to keep Flaps 1 .. to avoid a potential emergency gear extension procedure...Crew Established and Communicated a plan that entailed to [advise] with ATC to have CFR in place and RWY priority in [if] *necessary. We communicated our intentions with flight attendants and advised passengers and Dispatch. We requested delay vectors to run the required QRH Decent Checklist...We followed the QRH procedures, and [advised ATC], requested fire rescue to be ready...Captain..decided to stop on runway to have CFR look over outside of aircraft... I decided that taxi was too difficult...and requested a tow...Crew debriefed for clarification, any issues and ways procedure/crew actions could be improved.*

An FO complied with a CA's call for a go-around while handling a loss of hydraulic fluid. The Captain was Monitoring and Responding to a rapidly changing situation, and called for a go-around. The operators describe several strategies during and after the incident that indicate resilient performance. They also include some explanations of their understanding, communication process, assumptions, and motivations. Narrative self-reports can provide this kind of information.

Actions are often judged based on the subsequent outcome. However, the decision to call for a go-around may be appropriate in one situation, but not in the next. Had the outcome been poor, a go-around would not be considered "desired" performance, but the capabilities to anticipate the need, monitor for relevant cues, and adjust the flight by executing a go-around all contribute to the potential for resilient performance. Furthermore, some operators are very explicit in describing what they were Anticipating, or were Monitoring for, etc., but other times these descriptions are implicit. Thus, reading for a deeper understanding of the meaning in context and the subtle nuances can increase insight.

Example #6 (ASRS #1759282)

[Leg 1] *"I read the clearance back and wrote down "HFR" at the bottom of my clearance..."*

[Leg 2] *"I put the checklist between the throttles to remind me and verbally repeated "Hold for Release" on every radio call on CTAF frequency."*

By leveraging his experience and managing available resources, pilot Learning and Responding occurred between two flight legs in this narrative. The flight crew received a Hold for Release (HFR) due to a maintenance delay, so the pilot wrote this down as a reminder. Bag loading was completed, and the flight crew started the engines, taxied, and took off. However, they took off without receiving ATC clearance. The pilot recognized the risk of forgetting the clearance from the first leg and so, to prevent this from happening again on the next leg where they again receive an HFR, devised additional memory cues by putting the checklist between the throttles, and by repeating the phrase to keep the information active in working memory. These reports can provide Learning opportunities for both individuals and organizations. The examples above provide an opportunity to learn how operators describe their own resilient performance, and illustrate the complexity of labeling and coding these behaviors.

Discussion

There is no common vocabulary to describe how humans succeed, and a method to systematically capture success does not yet exist. Taxonomies can be created in many different ways and for many different purposes (e.g., Operator's Guide to Human Factors in Aviation, 2010; International Civil Aviation Organization Accident/Incident Data Reporting System Taxonomy, 2013). Our analysis highlights the challenges in determining how to capture and group expressions of resilient performance meaningfully into a taxonomy that could be useful for both machine analysis of the data and support human analysis.

In creating a framework that increases understanding and where to focus safety training efforts, it may be helpful to determine both who and what it was that created the opportunity for resilience or directly contributed to resilient performance: Was it an individual, a flight crew team, or a procedure created by the organization? And what did it or they do to contribute to resilience? This may help in learning what kind of support is needed to promote resilient performance as operators do their jobs and within the organization. It is worth considering Vesel's (2020) descriptions of the challenges and biases present in attributing causality in an event, particularly in written investigation reports. Vesel addresses the biases, linguistic framing and shortcuts that occur which can hinder safety promoting efforts. Attempting to "fit human action into preset, limited categories," such as in a taxonomy is an example. Vesel proposes ways to increase the opportunities for learning by looking at the context of an event and the interrelations between factors.

Conclusion

Operators describe their resilient performance in ASRS reports in many ways. Operators' narratives provide descriptions of resilient performance, even for reports intended to describe something that went wrong. In the above examples, the resilient potentials of Anticipate,

Monitor, Respond, and Learn are present in operators' descriptions. Hollnagel's framework created opportunities for identifying resilient performance, yet presented new challenges. Machine learning (ML) and natural language processing (NLP) applications may also be helpful in the analysis of large datasets, but the assumptions and implications of using search and query tools for the ASRS database must be understood. An interesting avenue currently being explored is "sentiment analysis." It utilizes NLP techniques to detect emotions and tones in written text, which may be useful in helping to identify expressions of positive, resilient human behaviors. Addressing the present challenges in future research can enable further learning from operators' safety producing behaviors and resilient performance.

Acknowledgements

This work was funded by NASA's System-Wide Safety Project, part of the Aeronautics Research Mission Directorate's Aviation Operations and Safety Program.

References

Aviation Safety Reporting System: ASRS Database Online. (2020). https://asrs.arc.nasa.gov/search/database.html. Accessed November 16, 2020.

Chandra, D., Sparko, A., Kendra, A., & Kochan, J. (2020). *Operational complexity in performance-based navigation (PBN) arrival and approach instrument flight procedures (IFPs).* https://rosap.ntl.bts.gov/view/dot/43835

Flight Safety Foundation European Advisory Committee. (2010, October). Using "Positive" human factors data. Operator's Guide to Human Factors in Aviation. https://www.skybrary.aero/index.php/Using_"Positive"_Human_Factors_Data_(OGHFA_BN)

Holbrook, J., Stewart, M. J., Smith, B. E., Prinzel, L. J., Matthews, B. L., Avrekh, I., Cardoza, C. T., Ammann, O. C., Adduru, V. & Null, C. H. (2019). *Human performance contributions to safety in commercial aviation* (NASA/TM-2019-220417). https://human-factors.arc.nasa.gov/publications/NASA-TM-2019-220417-web.pdf

Hollnagel, E. (2015). RAG – Resilience Analysis Grid. https://erikhollnagel.com/onewebmedia/RAG%20Outline%20V2.pdf

International Civil Aviation Organization. (2013). *Explanatory factors.* Accident/Incident Data Reporting System (ADREP) Taxonomy. https://www.icao.int/safety/airnavigation/AIG/Pages/ADREP-Taxonomies.aspx

Kiernan, K., Cross, D. C., Scharf, M. (2020). *Developing a taxonomy for success in commercial pilot behaviors.* Collegiate Aviation Review International, 38(1). https://doi.org/10.22488/okstate.20.100203

Paradis, C., Kazman, R., Davies, M. D., & Hooey, B. L. (2021). Augmenting topic finding in the NASA Aviation Safety Reporting System using topic modeling. AIAA SciTech Forum. https://arc.aiaa.org/doi/abs/10.2514/6.2021-1981

Vesel, C. (2020). Agentive language in accident investigation: Why language matters in learning from events. *ACS Chemical Health & Safety, 27*, 34-39.

RESILIENCE ENGINEERING'S POTENTIAL FOR ADVANCED AIR MOBILITY (AAM)

Christopher Nemeth
Applied Research Associates, Inc.
Albuquerque, NM

Jon Holbrook
National Aeronautics and Space Administration
Hampton, VA

The national airspace (NAS) will rapidly evolve in the next ten to twenty years. Plans for Advanced Air Mobility (AAM) during that period envision highly automated airspace management systems and electrically powered vehicles. AAM concepts also anticipate limited human roles. The goal of limiting the human role is to minimize the potential for misadventures, yet how the human role is limited needs to be carefully considered in order to also preserve the potential for human successes. The field of resilience engineering (RE) focuses on how systems can change in order to seize an opportunity or withstand an unforeseen challenge. RE methods rely on the use of empirical data to optimize the ability of any system to adapt. RE studies have shown how individual and team initiatives ensure resilient system performance by creating safety through flexibility. Benefits of the RE approach include improved awareness of operational circumstances and how system elements depend on each other, and the ability to allocate limited resources and prepare for surprise. RE offers the ability to account for and incorporate the human role as an essential element in order to ensure NAS systems' resilient performance. Data on the human contribution to safe and resilient system performance, which is termed "work as done," are available but are not being considered as the NAS evolves. We present an approach that describes how use of RE can enable the evolving NAS to adapt, and perform, in a resilient manner.

Incremental advancements in computer software, sensors, energy storage, and electric propulsion are fueling the development of new air vehicles that promise to change the way that cargo and people are moved. Simplified electric vehicles capable of lower noise levels and vertical flight, with lower operating and maintenance costs than today's vehicles, could lead to a vast expansion of opportunities for tasks to be accomplished by flight that are currently accomplished using ground-based systems. Urban Air Mobility (UAM) represents one such opportunity, focused on moving people and goods within and around densely populated urban centers, with the eventual goal of providing the public with airborne personal transportation and cargo services. Services may be scheduled, on demand, or part of an intermodal transportation link, connecting passengers or goods to ground-based networks of road or rail systems. UAM vehicles with electrically powered vertical takeoff and landing capabilities will range from small drones that deliver packages to passenger-carrying aircraft that operate in and around metropolitan areas.

Opportunities to leverage these emerging aerial technologies also exist in non-urban areas, or in support of other missions including longer range regional transport of people and goods using electric vehicles that operate out of more conventional airstrips; air-ambulance and medical transfer services; search-and-rescue or disaster relief operations; power-line inspection or other visual surveillance operations, et cetera. This larger ecosystem, known as Advanced Air Mobility (AAM), would represent a highly complex system of individual vehicles, local fleet operations, and regional networks that must all work cooperatively, not only within the AAM ecosystem, but with connected ground-based systems and any adjacent conventional aviation airspace. As a result, AAM comprises a broad range of stakeholders, living and operating in a wide range of locations with different geographic features, using different classes and sizes of airborne vehicles to accomplish a diverse set of missions.

AAM Challenges and Barriers

Despite ongoing technological advances, the potential benefits of AAM do not come without challenges (National Academies of Sciences, Engineering, & Medicine, 2020). Implementing a versatile advanced aerial mobility system with multiple applications and users is a complex, multidisciplinary challenge. No entity within the US, however, has the mandate to promote the development, adoption, and commercialization of new aviation technologies and applications. Nor does any single entity currently have sufficient oversight or responsibility to effectively make advanced aerial mobility a reality, while maximizing the potential societal benefits. Commercialization of AAM will require clarity from regulators and a timely regulatory progress, to support new flight operation types or applications. Without regulatory certainty, advanced aerial mobility systems may develop in an ad-hoc manner, with private point-to-point systems instead of open many-to-many systems. Closing the AAM business case means lowering "cost-per-mile" to the point that perceived benefits to the consumer make the cost acceptable. Highly trained expert human operators represent a significant cost in today's air operations – a cost that will have to be addressed in AAM. Expanding use cases for aviation across the economy and increasing the scale of airspace activity by orders of magnitude are key components of the AAM vision. More vehicles operating in more densely packed airspace, will require increased data needs to schedule, track, and separate those vehicles. The pilot recruitment and training pipeline is not expected to be able to keep up with the anticipated massive expansion of vehicles and operations. Insufficient pilot supply and high pilot cost are driving demand for increased levels of automation and increased demands on automation capability and reliability.

The ultimate success of AAM will depend on providing benefits not only cost-effectively but also safely. The increased levels of automation in AAM systems, however, create challenges for traditional safety assurance methods. Testing and simulation alone will not be adequate to ensure safety in these complex software-intensive systems, which can fail very differently than the more hardware-based systems of the past for which legacy hazard analysis tools were developed. The demand for automated systems that can learn and adapt will require new methods of certification – methods that address automation capabilities that can change how they perform over time. Despite supply and cost demands to reduce human involvement in the control of AAM systems, to date, humans remain the most capable source of information ingestion, situation understanding, and real-time decision adaptation.

In today's systems, human operators participate directly in the control and safety management of the system. Human roles in AAM will share vehicle control and contingency management responsibilities with automation and will be expected to perform with less training. Although well-understood vehicle control tasks may be simplified for UAM vehicles, contingency responses will still be required when vehicle or infrastructure systems fail, environmental conditions are hazardous, passengers are disruptive, et cetera. Given the dramatic anticipated increase in the number of operations in AAM compared to today, the overall frequency of contingency operations is also likely to increase. One barrier to effective and timely safety management of a complex dynamic system on this scale is identifying, collecting, and analyzing the key system configuration, health status, and performance data from all of the entities that operate in or support the ecosystem. Most plans for enabling automated contingency management involve coding "well-established" contingency management procedures into the automation. However, many of these "well-established" procedures depend upon significant interpretation and adaptation by human operators to be successful. A second barrier is that these "well-established" procedures may not be as "well understood" as some may believe. A third, and largely unrecognized, barrier that applies to both data needs and to contingency management is a barrier that results from how we think about safety. Contingency management is not limited to responding and recovering from anomalies, but also routinely preparing for and preventing them from happening in the first place. Our safety thinking can limit the performance data we choose to collect and analyze (Holbrook, 2021).

Resilience Engineering

Safety and risk management thinking has often led to the assumption that human error was the cause for adverse outcomes, that counting "errors" is a way to limit adverse results, and removal of humans would mitigate this risk. AAM, just like the NAS, is a socio-technical system. In order to be effective, socio-technical systems must reflect intense attention to behavior of operators, users, and maintainers who work as participants in what can be considered a joint cognitive system (Woods and Hollnagel, 2006). Resilience Engineering (RE) has evolved from safety studies over the past 10 years to enable systems in high stakes sectors to anticipate and sustain operation when confronted by unforeseen threats (Hollnagel, Woods, & Leveson, 2006). Hollnagel (*in press*) more recently defined resilient performance as "the ability to succeed under varying conditions, so that the number of intended and acceptable outcomes (in other words, everyday activities) is as high as possible" both in the face of adversity as well as during normal conditions. RE studies collect empirical data on work as it is actually done (rather than as it is imagined), and what goes right, and why, at the system level. Results demonstrate how operators ensure resilient performance, making adaptation possible in the face of complexity. They reveal barriers to cognitive work operators perform and show actual (rather than assumed) system performance and interdependencies among system elements. Methods such as the Resilience Analysis Grid (RAG) (Hollnagel, 2010) can be used to identify opportunities to anticipate, monitor, respond, and learn during routine and exceptional conditions. The RE approach can help to understand the emergent, interdependent, irregular nuances and complexities that can be expected in AAM, because those traits and operator performance data already exist in the manned NAS. AAM requires a valid grasp of how operators create resilient performance. RE makes it possible to develop an understanding of what goes well in the NAS, and how to capitalize on that understanding as an asset for AAM.

Human Performance and AAM

Efforts to accomplish goals in high hazard domains such as AAM include individual behavior as well as macrocognitive activities such as contingency planning that was mentioned previously. Cacciabue and Hollnagel (1995) describe macrocognitive activities as "the cognitive functions that are performed in natural (rather than artificial laboratory) decision-making settings." The macrocognitive view (Table 1) can be used to develop descriptive models of these activities, in order to identify and understand how they occur in the NAS and the AAM.

Table 1. *Macrocognitive Activities* (adapted from: Crandall, Klein, & Hoffman, 2006)

Activity	Description
Naturalistic decision making	Reliance on experience to identify a plausible course of action and use of mental simulation to evaluate it.
Sensemaking/situation assessment	Diagnosis of how a current state came about, anticipation of how it will develop.
Planning	Changing action in order to transform a current state into a desired state.
Adaptation/re-planning	Modification, adjustment, or replacement of implemented plan.
Problem detection	Ability to notice potential problems at an early stage.
Coordination	How team members sequence actions to perform a task.
Developing mental models	Mental imagery and event comprehension, based on abstract knowledge and domain concepts and principles.
Mental simulation and storyboarding	Use of mental models to consider the future, enact a series of events, and ponder them as they lead to possible futures.
Maintaining common ground	Ongoing maintenance and repair of a calibrated understanding among team members.
Managing uncertainty and risk	Coping with a state or feeling in which something is unknown or not understood.
Turning leverage points into courses of action	Ability to identify opportunities, turn into courses of action.
Managing attention	Use of perceptual filters to determine the information a person will seek and notice.

While automated systems can follow rules, human intervention is routinely required at the knowledge level (Rasmussen, 1983). Humans ensure resilient system performance in multiple high hazard settings, including the NAS, as data on manned NAS human performance demonstrate (Weick and Sutcliffe, 2007; Holbrook et al, 2019). Automated system inability to function beyond the rules level has shown how understanding human performance is essential to understand and manage risk in the NAS. Organizations must now consider the interplay of different types of risk. More automation reduces the risk of human errors, most of the time, as shown by aviation's excellent and improving safety record. But automation also leads to the subtle erosion of cognitive abilities that may only manifest themselves in extreme and unusual situations (Oliver, Calvard and Potocnik, 2017). Data that describe desired performance already exist. The research and development literature (e.g., job design, work procedures, standards) already describes aviation performance as it is intended and is routinely accomplished. Even reports of adverse outcomes included in the Aviation Safety Reporting System (ASRS) (Billings et al., 1976) include data on resilient pilot performance.

How RE Can Incorporate Human Performance into AAM

While increased automation is a key tenet of AAM, RE offers some direction in thinking about what kinds of functions we might want to automate. For example, how could automation be used to enhance a system's capability to anticipate unforeseen events? How could automation be used to enhance a system's capability to monitor the environment and its own performance? By recognizing that humans are a source of flexibility and resilience, and not just a source of errors and hazards, we can focus design on how the automation can support human performance, rather than on trying to replace the human or protect the system from the human. Or, in the event that we must replace the human, we can better recognize and understand the range of capabilities that we are attempting to replace. We can start by investing in how we think about safety, which informs the safety data we choose to collect and analyze. However, because data collection and analysis are typically triggered by failure outcomes, we rarely study how failure preparation, response, and recovery lead to successful outcomes. We typically wait for something to go wrong before we start to learn from "what happened." We diligently learn from our mistakes, but do we systematically learn from our successes? The answer, far too often, is "no."

Rethink Safety Policy. When we only analyze data from errors and failures, we are ignoring that vast majority of human impacts on system performance. Without understanding how safety is produced, claims about the predicted safety of autonomous machine capabilities that cannot account for this are inherently suspect. Plans to minimize or even remove the only demonstrated, reliable source of safety-producing behavior, without first understanding the capability being minimized or removed, introduces unknown, potentially unaccounted-for risk.

Collect and Analyze Human Performance Data. Fortunately, there are opportunities to address these risks. Data already exist or could be collected with minimal effort on successful preparation for, response to, and recovery from failure. While we don't have the many decades of experience and infrastructure for doing this like we do for human error, emerging approaches to safety and risk management, such as Resilience Engineering, offer a useful place to begin.

Use RE to build new performance models. Start by broadening the data sources on both desired and undesired human performance. Use those data to distill criteria that will define resilient performance. Validate those criteria through the collection of empirical data using rigorous methods such as analysis of simulator runs. Develop requirements and use cases on human roles in AAM. Build means to learn from experience, and to grow the field of practice.

Summary

The UAM/AAM environment will evolve over the next 10+ years, and prior risk/safety models may not serve this new domain well. Organizations can systematically drive change, which can begin by using tools already at their disposal (e.g., policies, procedures, training, equipment) to effectively translate insights into action. Resilience Engineering offers new means to develop the AAM environment through deep insight into how humans ensure resilient performance. Data exist that RE methods can use to create an effective AAM.

Acknowledgments

This work was funded by NASA's System-Wide Safety Project, part of the Aeronautics Research Mission Directorate's Aviation Operations and Safety Program.

References

Billings et al. (1976, September). *NASA Aviation Safety Reporting System Quarterly Report* Number 76-1, April 15, 1976 through July 14, 1976. NASA TM X-3445. Ames Research Center, NASA, Moffett Field, California 94035. Available at: https://ntrs.nasa.gov/archive/nasa/casi.ntrs.nasa.gov/19760026757.pdf

Cacciabue, P.C. & Hollnagel, E. (1995). *Simulation of cognition: Applications.* Expertise and technology: Cognition and human-computer cooperation. In J.M. Hoc, P.C. Cacciabue, & E. Hollnagel (Eds.). Mahwah, NJ: Lawrence Erlbaum Associates. 55-73.

Crandall, B., Klein, G. & Hoffman, R. (2006). *Working Minds: A Practitioner's Guide to Cognitive Task Analysis.* Cambridge, MA: MIT Press.

Hollnagel, E. (2010). Epilogue: RAG. The Resilience Analysis Grid. In Hollnagel, E., Paries, J., Woods, D. & Wreathall, J. (Eds). *Resilience Engineering in Practice: A Guidebook.* Ashgate Publishing. Farnham, UK.

Holbrook, J. (2021). Exploring methods to collect and analyze data on human contributions to aviation safety. In Proceedings of the 2021 International Symposium on Aviation Psychology.

Holbrook, J. B., Stewart, M. J., Smith, B. E., Prinzel, L, J., Matthews, B. L., Avrekh, I., Cardoza, C. T., Ammann, O. C., Adduru, V., & Null, C. H. (2019). Human Performance Contributions to Safety in Commercial Aviation. NASA Technical Memorandum 2019-220417. Available at: https://ntrs.nasa.gov/api/citations/20190033462/downloads/20190033462.pdf

Hollnagel, E. (in press). From Resilience Engineering to resilient performance. In C. Nemeth and E. Hollnagel (Editors). *Advancing Resilient Performance.* New York: Springer Nature.

Hollnagel, E., Woods, D. D. & Leveson, N. (Eds.) (2006). *Resilience Engineering: Concepts and Precepts.* Aldershot, UK: Ashgate Publishing.

Klein, G., Ross, K. G., Moon, B. M., Klein, D. E., Hoffman, R. R., & Hollnagel, E. (2003). Macrocognition. IEEE Intelligent Systems, 18(3), 81-85.

National Academies of Sciences, Engineering, and Medicine 2020. *Advancing Aerial Mobility: A National Blueprint.* Washington, DC: The National Academies Press. https://doi.org/10.17226/25646.

Oliver, N., Calvard, T. & Potočnik, K. (2017, Sept 5) The Tragic Crash of Flight AF447 Shows the Unlikely but Catastrophic Consequences of Automation. *Harvard Business Review.* H03W8I-PDF-ENG.

Rasmussen, J. (1983) Skills, Rules and Knowledge: Signals, Signs and Symbols, and Other Distinctions in Human Performance Models. *IEEE Transactions on Systems, Man and Cybernetics.* Vol. SMC-13. No.3. May. 257-66.

Weick, K and Sutcliffe, K. (2007). *Managing the Unexpected: Resilient Performance in an Age of Uncertainty.* New York: John Wiley and Sons.

Woods, D. D. & Hollnagel, E. (2006). *Joint Cognitive Systems: Patterns in Cognitive Systems Engineering.* Boca Raton, FL: CRC Press.

THINKING OUTSIDE THE BOX: THE HUMAN ROLE IN INCREASINGLY AUTOMATED AVIATION SYSTEMS

Joel Lachter
NASA Ames Research Center
Moffett Field, California

Alan Hobbs
San Jose State University Research Foundation
Moffett Field, California

Jon Holbrook
NASA Langley Research Center
Hampton, Virginia

Rapid advances in automation are enabling transport systems to operate in an increasingly autonomous manner. From time to time, these systems encounter operational conditions that fall outside a "competency box" of scenarios and environments for which the system was designed. Human operators add resilience because they can see and act outside the automation's competency box. Advanced aviation concepts envision fleets of highly automated air vehicles providing on-demand transport for people and goods. We examine one such concept, Urban Air Moblity (UAM) and explore how humans can best be incorporated to maintain resilience. A human-autonomy teaming approach is suggested.

Advances in automation are changing many aspects of everyday life, including the way goods and people are moved from place to place. Remotely-operated trains, robotic warehouse delivery systems, and "self-driving" cars are showing us what a future transport industry might look like. The urban air mobility (UAM) concept is such a case. Several companies are proposing UAM systems in which electric-powered vertical takeoff and landing aircraft (eVTOL) would routinely transport people and products.

In recent years, some proponents of automation have envisioned future transport systems that will operate with limited or no oversight from a human operator. Proponents of UAM note that this final state reduces cost as well as eliminating pilot error, identified as a contributing factor in many aircraft accidents (e.g., Uber Elevate, 2016). This viewpoint ignores the possibility that human operators add resilience because they can perceive and act outside the "competency box" of the automation. We use the term "competency box" to refer to the scenarios and environments within which the automated system has earned trust that it can operate safely without the need for human intervention. This is similar to the "competence envelope" discussed by Hoffman and Hancock (2017) and the system boundaries discussed by Woods (2015). During the design process, the intended competency box may be expressed explicitly in terms of performance specifications, but some aspects of the intended competency box may also remain unstated. As operational experience accumulates, the actual competency box will sometimes turn out to be smaller than intended, as the system fails to deal with scenarios and environments, including some anticipated by the designers. In other cases, the system might fail to deal with scenarios and environments that had not been anticipated. A safety critical system possesses resilience when it is able to adjust its functioning to maintain safety in the face of expected and unexpected conditions (Hollnagel, 2015). We propose that achieving a resilient UAM system must involve the complementary capabilities of automation and humans working together.

An automated system's capabilities can be expanded over time with modifications to software, sensors, and other components. A characteristic of machine learning is that automated systems have the potential to expand their capabilities as experience is accumulated. However, even the most capable automated systems have limits, and it is unclear at what point, if ever, the competency box becomes large enough to safely eliminate the role of the human operator. The designers of UAM systems face the challenge of how to make the best use of intelligent automation, while also leaving room for the resilient performance potential of humans.

The UAM Concept

The FAA UAM Concept of Operations (ConOps; FAA, 2020) covers operations occurring in dedicated corridors in urban environments. This ConOps envisions an initial stage of UAM operations in which aircraft operated by an on-board pilot fly within the current air traffic management (ATM) system. Pilots would exert direct "within the loop" control of the automated systems, much as they do today. The next stage, referred to as ConOps 1.0, would involve aircraft flying in UAM corridors that are not under direct control of air traffic controllers (ATC). ATC would, however, have the authority to open and close these corridors. An on-board pilot would monitor systems and would have the ability to take control when required or desired, under a "human-on-the-loop" supervisory control model. The necessity of carrying an on-board pilot clearly reduces the carrying capacity of the vehicle and would probably make this stage economically unviable for high tempo operations (Uber Elevate, 2016, p38). With the corridors in place, the ConOps 1.0 stage is envisioned as one where automation can mature, operational tempo can increase, and use cases can evolve. The FAA ConOps envisions a mature stage with remote pilots who "passively monitor" aircraft and are prompted to take action in situations outside the automation's competency box ("human-over-the-loop" operations). Below we discuss how a human-automation teaming approach can aid in defining the competency box of the automation and help to reveal the extent to which automated systems can safely conduct flights under the full range of operational and environmental conditions, including gradually changing conditions and sudden threats.

A Human-Automation Teaming (HAT) Approach

The field of Human Computer Interaction (HCI) has often viewed task decomposition in terms of assigning some tasks to automation and others to human operators; often with a background assumption that Machines Are Better At some things while Humans Are Better At others (MABA-HABA; Fitts 1951). The HAT philosophy is to break down those roles in much the same way as Crew Resource Management (CRM) sought to break down the strict hierarchy of mid-twentieth century flight decks (Shively et al., 2018), allowing both "partners" to contribute to the performance of any given task. The term "teaming" is aspirational; it indicates a desired objective, but not necessarily the current state of affairs.

In a well-functioning human team, all members of the team share an understanding of the goal. In contrast, we expect that for the foreseeable future, in human-automation teams, only the human will understand the context of the high-level goals. Nevertheless, current automation can work jointly on tasks with human operators, each monitoring the other's performance and negotiating task assignments, resulting in more efficient and resilient performance than if either were to perform the task alone. Shively et al. (2017) proposed three tenets of Human Automation Teaming (HAT): Bi-Directional Communication, Transparency, and Dynamic Delegation (which they call Operator Directed Interface). The first two of these tenets serve to enable joint task performance, while the third serves to maintain the operator's situation awareness and ensure that the automation is assigned tasks within its competency box.

Bi-directional Communication

Bi-directional communication is central to the concept of HAT. For there to be joint task performance, the automation and the human operator must be able to share information, goals, and strategy. As noted above, the automation does not have the same deep understanding of the system goals that the operator does (e.g., why are we going to Gilroy?); however, it can have information about what is necessary to achieve the goals (strategy) and the states that it must achieve on the way (sub-goals). It can display this information, along with reasons one strategy is preferable to another, give feedback on operator proposed strategies, and take into account information that the operator has that it may not be able to independently sense. Perhaps most importantly, the automation must be able to inform the human when it has encountered conditions that fall outside (or may be approaching) the limits of its competency box. Such a "call for help" from the automation triggers a non-normal state for the human operator.

Transparency

In some cases, we delegate a function to automation (e.g., an electronic engine controller) and leave the machine to perform its function, only informing the human if the system fails. However, if we expect humans and automation to work interactively, the human needs to be able to perceive *what* the automation is doing, and *why* the automation is doing it. With this understanding, the operator can judge if the automation is missing information or insight, or, conversely, whether the automation has information that the operator was unaware of. While this may seem obvious, automation is not always particularly transparent to the human operator. This lack of transparency can be the result of interface design choices, but it can also be the result of machine learning algorithms that obscure the cues used by the automation in making decisions.

Dynamic Delegation

In contrast to a static division of tasks between the human and automation, dynamic delegation involves a more flexible allocation of work, taking into account factors such as workload and time pressure. While, traditionally, automation operates with a particular level of human oversight (e.g., Sheridan & Verplank, 1978), a feature of dynamic delegation is that working agreements (Gutzwiller et al., 2017) allow the level of human oversight to vary according to the conditions. In particular, automation can be restricted to acting autonomously only under conditions that are clearly within its competency box. For example, automation might be trusted to land an aircraft autonomously on an unoccupied landing pad but require operator verification before landing on a pad with concurrent operations. Importantly, as automation earns increasing trust, the range of conditions within the competency box increase and the need for operator oversight decreases. Shively et al. (2017) propose that dynamic delegation can also involve predetermined sets of actions that are grouped together so that they can be quickly implemented, referred to as "plays." In aviation they act much like a checklist or quick reference handbook, in that they define the tasks needed for a particular situation; however, they also contain working agreements governing the level of automation expected for each task.

Applying HAT to UAM

The success of a mature UAM system will depend on its ability to demonstrate resilience in the face of anticipated and unanticipated conditions. We maintain that when highly automated systems operate in complex environments, human operators contribute to system resilience via their ability to see and act outside the competency box of the automation. HAT principles can help humans perform this role. The FAA UAM ConOps outlines several functions and roles where human operators' ability to see and act outside the competency box of the automation, may protect the system while allowing automation to earn trust where appropriate.

Pilot in Command (PIC)

The PIC is the only role called out by the FAA's UAM ConOps that is clearly assigned to a human. The Code of Federal Regulations states that "the pilot in command of an aircraft is directly responsible for, and is the final authority as to, the operation of that aircraft" (14 CFR 91.3a). The evolution of UAM envisioned by the ConOps is largely the evolution of this position from an onboard pilot who is in the loop on flying the aircraft, in much the same way as helicopter pilots are today, to a remote pilot, operating "over the loop" (HOVTL), managing contingencies for multiple aircraft. While the responsibility of the remote pilot remains unchanged, their ability to exert authority when operating HOVTL relies on appropriate bi-directional communication between human and automation. Ideally, an automated system would have the capability to alert the pilot when it is about to encounter a condition that falls outside its competency box. Transparent automation enables the human operator to understand how the automation will respond. In contingency conditions, dynamic delegation can ensure that responses are assigned to the entity most able to appropriately respond in the time available. In order to achieve HOVTL operations, further increases in automation will be necessary with most routine operations becoming fully automated. That automation will need to be trusted, and that trust will need to be earned. The HAT paradigm discussed above gives us an incremental path for verifying this automation in an operational environment.

Air Traffic Control (ATC)

The FAA ConOps envisions a significant change in the role of ATC between the initial operations stage, which, similar to current helicopter route operations, require the PIC to interact with ATC, and the ConOps 1.0 stage, where UAM vehicles operate within corridors with minimal ATC interaction. The FAA ConOps specifies that ATC will "respond to UAM off-nominal operations as needed" (FAA, 2020). This is a potentially difficult task (particularly if there were to be any large-scale system failure) and it is to be added to ATC's normal workload managing aircraft. Controllers could be aided in performing this task through appropriate system transparency that allows them to gain situation awareness as rapidly as possible, and a play structure that allows them to quickly organize and delegate the tasks necessary to mitigate contingencies safely. Further, ConOps 1.0 could be implemented in an incremental way by the gradual delegation of tasks from ATC to automation. For example, corridors could start off as default flight paths, and assignment of aircraft to corridors could initially be manual with automated recommendations or mixed initiative depending on working agreements. This incremental development would serve several purposes: Testing the automation while controllers are still in (or at least on) the loop, developing and calibrating trust in the automation, and creating a hierarchy of plays that operators can fall back on under off-nominal conditions.

Operator

UAM operators are commercial entities that are responsible for regulatory compliance and all aspects of UAM operation execution. Prior to a flight, the UAM operator obtains information such as weather conditions and aerodrome availability, plans the flight, and provides the information necessary to operate in a UAM corridor. It is envisioned that for larger "airline" operations, the operator would also perform a role coordinating individual aircraft operations, akin to modern day dispatch. Setting the high-level goals and policy are intrinsically human roles; however, these roles are increasingly informed by automated interactive computer modeling. The dispatch-like roles are likely to be highly automated, even in initial operations, however, these roles will involve mitigating contingencies, and thus tasks that are likely to be at or beyond the borders of the automation's competency box. To maintain safety and efficiency we expect human operators will typically delegate dispatch-like tasks to automation; although under dynamic delegation, a human will need to be more closely involved in these operations at times.

Provider of Services for UAM (PSU) & UAS Service Supplier (USS)

As defined in the FAA UAM ConOps, PSUs and USSs are the information integration and dissemination backbone envisioned for UAM. They collect intent information for aircraft, availability information for air corridors and landing sites, weather information, and other operationally relevant information, provide this information to operators and assist with scheduling flights. PSUs and USSs may be automated, although this is not specified in the FAA's ConOps. Presumably someone will have to manage PSUs in off-nominal situations. Building PSUs and USSs to be transparent and creating plays that allow human operators to take control without handling all network traffic would seem to improve system resilience.

Aerodrome Managers

The FAA UAM ConOps specifies that UAM aircraft takeoff and land at "aerodromes"; although others refer to these as "vertiports." While the ConOps does not specify human interaction at these aerodromes, some human interface is likely required, as aerodrome operations will periodically present conditions that fall outside the competency box of automated systems. Loading and unloading aircraft is somewhat unpredictable, which, in turn, adds unpredictability to the availability of the aerodrome. Unlike buses or subways where a car can simply wait when arriving at an occupied station, battery capacity on UAM aircraft is unlikely to allow for extended hovering. Dynamic delegation will be critical, as aerodrome managers will likely be required to recognize the state of the aerodrome, assist in smoothing traffic flow, and interface with the PIC and PSU about availability windows.

Conclusion

As progress towards a future UAM system continues, designers must not overlook the positive contribution made by human operators to system resilience. A challenge facing designers of UAM systems is to integrate the characteristics of humans and automation to produce an effective human-automation team. Rather than assigning tasks in a static manner to either automated systems or humans, future UAM systems are likely to involve a flexible approach to task delegation. This will require operational personnel to possess an appropriate awareness of the functioning of automation and be equipped to monitor performance, anticipate conditions that will fall outside the automation's competency box, and respond as necessary. A HAT framework may be useful in achieving a resilient UAM system that involves the complementary capabilities of automation and humans working together.

Acknowledgements

This work was funded by NASA's System-Wide Safety Project, part of the Aeronautics Research Mission Directorate's Aviation Operations and Safety Program.

References

FAA (2020). Urban Air Mobility (UAM) Concept of Operations. Available at https://nari.arc.nasa.gov/sites/default/files/attachments/UAM_ConOps_v1.0.pdf

Fitts P.M. (1951) Human engineering for an effective air navigation and traffic control system. National Research Council, Washington, DC.

Gutzwiller, R. S., Espinosa, S. H., Kenny, C., & Lange, D. S. (2017, July). A design pattern for working agreements in human-autonomy teaming. In International Conference on Applied Human Factors and Ergonomics (pp. 12-24). Springer, Cham.

Hoffman, R. R., & Hancock, P. A. (2017). Measuring resilience. *Human factors*, *59*(4), 564-581.

Hollnagel, E. (2015). RAG – Resilience Analysis Grid. https://erikhollnagel.com/onewebmedia/RAG%20Outline%20V2.pdf.

Shively, R. J., Lachter, J., Brandt, S. L., Matessa, M., Battiste, V., & Johnson, W. W. (2017) Why Human-Autonomy Teaming? To be presented at the 8th International Conference on Applied Human Factors and Ergonomics (AHFE 2017), Los Angeles, CA.

Shively, R. J., Lachter, J., Koteskey, R., & Brandt, S. L. (2018, July). Crew resource management for automated teammates (CRM-A). In International Conference on Engineering Psychology and Cognitive Ergonomics (pp. 215-229). Springer, Cham.

Uber Elevate: Fast-Forwarding to a Future of On-Demand Urban Air Transportation, Oct. 27, 2016. [Online]. Available: https://www.uber.com/elevate.pdf. [Accessed Nov. 20, 2019].

Woods, D. D. (2015). Four concepts for resilience and the implications for the future of resilience engineering. *Reliability Engineering & System Safety*, *141*, 5-9.

Evaluating the Use of High-Fidelity Simulator Research Methods to Study Airline Flight Crew Resilience

Chad L. Stephens[1], Lawrence J. Prinzel III[1], Daniel Kiggins[2], Kathryn Ballard[1], Jon Holbrook[1]

NASA Langley Research Center[1]
Hampton, VA

San Jose State Research Foundation[2]
San Jose, CA

As it evolves, aviation will continue to require integration of a wide range of safety systems and practices, some of which are already in place and others that are yet to be developed. New concepts in system safety thinking have emerged to consider not only what may go wrong, but also what can be learned when things go right during commercial flight operations. Taken together, these complementary perspectives form a more comprehensive approach to system safety thinking that can help to recognize and preserve the resilient performance capabilities currently provided by humans. A need exists, however, for research methods to enable better understanding of the human contributions to aviation safety. NASA's System-Wide Safety Project supports research on using flight simulation methods to study operator resilience and safety-producing behaviors. Building on prior NASA efforts investigating procedural non-adherences during area navigation standard terminal route arrivals, a high-fidelity commercial aviation line operational simulation (LOS) experiment has been designed to study how flight crews anticipate, monitor for, respond to, and learn from expected and unexpected disturbances during these operations. A diverse set of LOS scenarios were developed to simulate highly realistic, complex, but routinely encountered operational situations. Each scenario provided multiple opportunities to collect data on how flight crews manage threats and errors, as well as novel opportunities to observe resilient and safety-producing behaviors. The experimental design, implications for the study of safety-producing behaviors using simulation, and considerations for airline pilot training will be discussed.

New innovative technologies and operational concepts will be required to meet the ever-increasing global demands on air transportation. The NASA System-Wide Safety (SWS) Project is focused on how future aviation advances can meet demand needs while maintaining today's ultra-safe system safety levels. Aviation safety as it evolves shall require new ways of thinking about safety, integrating a wide-range of existing and new safety systems and practices, creating and enhancing tools and technologies, leveraging the access to system-wide data and data fusion, improving data analysis capabilities, and developing new methods for in-time risk monitoring and detection, hazard prioritization and mitigation, safety assurance decision-support, and in-time integrated system analytics (Ellis, Krois, Davies, & Koelling, 2020). To meet these needs, the SWS project has developed research priorities including In-time System-wide Safety Assurance (ISSA) and development of an In-time Aviation Safety Management System (IASMS; Ellis et al. 2020). As part of this effort, the concepts of "resilience" and "productive safety" are being studied (e.g., Hollnagel, 2016). Traditional approaches to aviation safety have focused on what can go wrong and how to prevent it. Another approach to thinking about system safety should reflect not only "avoiding things that go wrong" (protective safety) but also "ensuring that things go right" (productive safety) that together enable a system to exhibit resilient performance.

The SWS project has focused on development of domain-specific safety monitoring and alerting tools, integrated predictive technologies, and adaptive in-time safety threat management (Ellis et al., 2019). Significant research challenges include how to identify data sources and indicators for in-time safety critical risks, how to analyze those data to detect and prioritize risks, and how to optimize safety awareness and safety action decision support. One focus area within the SWS Project is understanding how to evaluate and measure resilient performance and productive safety and application for in-time safety assurance and safety management systems. The research outcomes are intended to both significantly expand the knowledge base of resilience engineering through empirical data collection and analysis, and also help to inform ISSA and IASMS for traditional and emerging operational concepts, such as Advanced Aerial Mobility (AAM; e.g., National Academy of Sciences, 2020).

The challenges associated with ISSA and development of IASMS are significant even for existing air transportation system operations where work-as-imagined and work-as-done can actually be compared. These challenges include collecting productive safety in-time data, granularity of data types and measurement, need for new analytical methods, issues of identifying in-time productive safety metrics and indicators, and potential approaches toward quantification of resilient performance indices. On-going SWS research is focused on application of these concepts for ISSA and design of IASMS, and a test case for this effort concerns non-adherence of area navigation standard terminal arrival route (RNAV STAR) procedures used at major airports. Through initial focused research efforts (i.e., understanding productive safety through test case of non-adherences of RNAV arrivals), the benefits shall provide for a more comprehensive system-wide safety research approach.

Alternative and Complementary Approach for Risk and Safety Management

Stewart, Matthews, Janakiraman, and Avrekh (2018) analyzed aircraft flight track data for more than 10 million flights into 32 domestic airports, which revealed that only 12.4% of flights fully complied with the vertical and lateral profiles on published arrivals. Building on that NASA research, Holbrook et al. (2020) further examined safety producing behaviors during RNAV STAR by collecting data from pilots, from mainline and regional airlines, and terminal radar approach control (TRACON) air traffic controllers. Interviews were conducted to understand how they "anticipate", "monitor" for, "respond" to, and "learn" (Hollnagel, 2014) from routine disturbances during RNAV arrivals into Charlotte Douglas International Airport (KCLT). As reported by Holbrook et al. (2020), different data sources resulted in different estimates of the frequency of RNAV STAR non-adherences at KCLT, ranging from 30% (TRACON controller estimate) to 43% (pilot estimate) to 84% (estimate based on flight track data specific to KCLT arrivals collected for Stewart et al., 2018).

These previous findings highlight how published procedures can be misaligned with routine and safe operations. The reasons for the misalignment and interpretation of these findings cannot be addressed with traditional approaches to risk and safety management. An alternative and complementary approach for risk and safety management is necessary to explore whether non-adherences may reflect desired safe behaviors. Examples from aviation operations such as those reported by Stewart et al. (2018) and Holbrook et al. (2020) indicate that the definition of safety should reflect not only "avoiding things that go wrong", but also "ensuring that things go right". Global demands on air transportation drive increasingly complex operations, and to maintain safety, humans in the system continuously adjust their work to match their operating conditions (Hollnagel, 2014).

Proposed Human-in-the-Loop (HITL) Flight Simulation Study

Currently, the resilience engineering literature and very limited empirical published research have focused almost entirely on the conceptual aspects of productive safety. There remains a need to systematically collect empirical data to explore the practical application of these concepts toward improving aviation safety. The current research study is intended to meet that need with identification and collection of data sources and indicators for in-time safety critical risks, examination of how to analyze those data to detect and prioritize risks, and specifications for how to optimize safety awareness and safety action decision support toward development of ISSA and IASMS, for both traditional and future aviation concepts of operations.

Research Questions

The central research question is, "how do commercial airline pilots manage routine contingencies and productive safety during RNAV arrivals?" The present study was designed to identify and capture real-world operational behavior through replication of known actual line operational events that have occurred at KCLT in which observable resilient behavior had been described. A "structured observation" methodology (Gray, 2013) was chosen for this study to allow for careful observations of specific behaviors in a setting that is more structured than the settings used in naturalistic and participant observation. A structured observation approach, when combined with other methodologies (e.g., interviews) has significant benefits, particularly in comparison to naturalistic observation, by providing for costs, time, access, safety, and validity controls needed to meet the experimental objectives.

A challenge in studying such events in line operations is the limited data that can be collected and coded for the purposes of productive safety research for ISSA and ISAMS, or that is unavailable for collection or analysis for various logistical, procedural, or regulatory reasons. The proposed research study's primary objective is to obtain a comprehensive data set of identified candidate measures, in order to facilitate anticipated data science efforts and to help better understand the phenomena of interest.

Experimental Design Considerations

The proposed study will investigate how pilots respond to expected and unexpected disturbances during RNAV arrivals. Boeing 737NG rated professional commercial pilots from a major airline will be recruited to perform multiple RNAV STAR arrivals and data will be collected with regard to how they anticipate, monitor for, respond to, and learn from routine disturbances during RNAV arrivals into KCLT. The purpose of the present paper is to describe the methodology for this experimental study of productive safety that requires high-fidelity simulation of commercial aircraft line operations and scenario constructions that enable collection of these data types. A validation of the scenarios was performed with carefully screened participants in an off-site high-fidelity flight simulation facility that enacted substantial

COVID-19 participation protection protocols. The objective was to assess the efficacy of the scenarios for the study in the NASA high-fidelity simulators with airline participants. The remainder of the paper describes the testing methodology that shall be utilized in the planned NASA study.

Study Test Participant Considerations

Twelve (12) 737-800 Part 121 commercial airline flight crews (24 pilots) shall be the test participants for the study. The pilots shall be recruited to serve as a flight crew in respective roles (Captain, First Officer) and have familiarity with KCLT RNAV arrivals. The selection of 737-800 pilots is to ensure high familiarity with the NASA Langley Research Center Cockpit Motion Facility (CMF) 737-800 high-fidelity simulator (Figure 1).

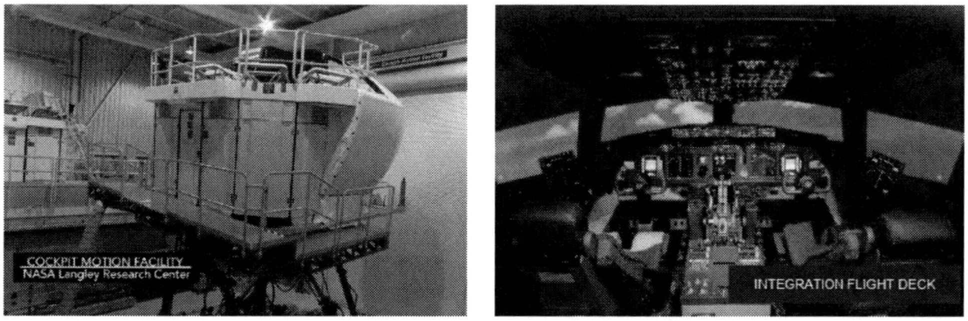

Figure 1. NASA 737-800 High-Fidelity Full-Motion Simulator

Flight Simulation Scenario Considerations

Seven scenarios were constructed based on current KCLT RNAV arrivals. The scenarios were designed to simulate anticipatable and un-anticipatable "routine" disturbances that are well-documented for occurrence in Aviation Safety Reporting System (ASRS) and Aviation Safety Action Program (ASAP) reports, airline crew reports, and known high frequency events that occur during KCLT operations (e.g., weather, traffic). Each of the experimental scenarios included documented events in which flight crews have exhibited resilient performance in response to the disturbance(s) encountered. The scenarios each have two events that were designed to present an "opportunity" for the flight crew to evince "anticipate", "monitor", "respond", and/or "learn" behaviors.

An important component of each of the scenarios is the ecological validity of the emulation of the scenario to replicate real-world line operations, including dispatch releases, weather reports (e.g., TAF, METAR, WSI), required and unanticipated cabin calls, dispatch communications (e.g., ability to contact dispatch including data communications and ACARS and text-to-speech capability), live air traffic control (TRACON, APPROACH), and other aspects, often neglected in research studies but critical to replicating actual operating conditions. Essentially, the scenarios and simulation environment are designed to provide high fidelity re-creation of commercial line operation arrivals into KCLT based on the following event categories: (a) energy management; (b) traffic compression and high flows; (c) convective weather; (d) unanticipated tailwind; (e) autoflight issues; (f) icing conditions and ice crystal icing; (g) system caution-level events; (h) wake encounter during arrival descent; (i) ATC and/or pilot clearance errors (e.g., hearback/readback error); and (j) high workload.

Each scenario begins after top-of-descent (TOD) and the aircraft is positioned on the RNAV arrival track with operationally appropriate attitude and airspeed. Pilots shall be provided with a detailed synopsis of the scenario before scenario start (including time to conduct a detailed arrival briefing and any FMS entries, etc., nominally completed prior to TOD in line operations). Dispatch paperwork has been created for the simulated RNAV arrivals. Since all airlines have different paperwork, our dispatch paperwork is a conglomeration of different airline formats with all the required information included. The dispatch release, Notices to Airmen (NOTAMS), Terminal Area Forecasts (TAF), Meteorological Terminal Aviation Routine (METAR) report, terminal weather, Automatic Terminal Information Service (ATIS), and other typically available flight/weather information shall be provided.

The pilots will perform as a flight crew in respective roles at the airline. All airline company hardcopy (plastic) checklists and other normally available documents shall be provided. All pilots shall also utilized their company supplied tablet which contains all necessary plates, aircraft reference manuals, Quick Reference Handbook (QRH), etc. All required company arrival briefings, standard calls, all standard operating procedures, etc., that would nominally be conducted in actual line operations shall be conducted in the study.

For this study, the key research questions for characterizing and measuring productive safety for in-time system-wide safety assurance include: (a) What data can and should we collect and analyze to understand existing productive safety capabilities?; (b) How can we measure the productive safety capability of a system?; and (c) How can productive safety support safety assurance of emerging systems?

Scenario Validation

The research study was designed based on a "structured observation" methodological approach combined with dialog, interview, and observer-based rater data and analyses. One goal of the research is to develop a system-level framework/taxonomy to understand operator's resilient performance, and develop organization-level strategies that promote recognition and reporting of resilient performance. Therefore, key to this work is understanding what those data are, how to collect them, and how to utilize them for in-time system-wide safety assurance and emergent risk prediction. The study attempts to contribute to the development of new metrics based on quantification and measurement of behaviors that support resilient performance through, in part, conducing high-fidelity simulation of "work-as-done" in traditional commercial airline operations.

The preliminary check-out validation of the scenarios was conducted at the Boeing Miami B737-800 flight simulator facility with active commercial airline pilots highly experienced with RNAV arrivals into KCLT. Using COVID-19 screening and safety protocols, four commercial airline flight crews were required to don personal protective equipment that limited the realism in simulating actual line operations in addition to the inability to fully implement all the scenario aspects that are critical to the study of productive safety. Despite these limitations, the validation flight crews confirmed the capability of these scenarios to achieve the experimental objectives. Based on SME preliminary assessment, the flight crews that were highly responsive early in the scenarios and exhibited resilient behaviors (i.e., monitor, anticipate, respond, learn) were better able to address potential threats well before they emerged to become significant hazards and additional data analyses are ongoing to confirm this finding. Review by subject matter experts and post-hoc discussions with the pilots confirmed the efficacy of the scenarios to enable simulation study of both productive safety and more traditional threat and error management (e.g., line operations safety audit type observable behaviors).

Conclusions

A challenge intended to be addressed in this research study (and project, more generally) is that data from system, observer, and operator sources are rarely (if ever) all available for the same set of events. Thus, there is little opportunity to explore the integration of factors that contribute to operators' resilient performance. Additionally, existing methods for measuring resilient performance are immature. Safety monitoring, prediction, and mitigation technologies based only on hazards and risks will address an incomplete picture of safety. Furthermore, the low frequency of undesired outcomes may impact the temporal sensitivity of safety assessments, challenging the notion of "in-time" mitigation. Building a more thorough, data-rich, and representative understanding of safety is needed to achieve NASA's vision of in-time system-wide safety which includes developing methods to enable the systematic study of productive safety. The high-fidelity simulation research study described here is an important step forward toward this goal for aviation safety, and the NASA team is prepared to begin data collection impacted by the unprecedented situation the pandemic has presented and has impacted so many involved in scientific study involving human participants.

Acknowledgments and Notes

This work was funded by NASA's System-Wide Safety Project, under the Aeronautics Research Mission Directorate's Aviation Operations and Safety Program. Dr. Lawrence (Lance) Prinzel is the corresponding author, but has co-primary authorship with Mr. Chad Stephens who contributed significantly to the paper.

References

Ellis, K., Krois, P., Davies, M., Koelling, J. (2019). In-Time System-Wide Safety Assurance (ISSA) Concept of Operations. ARC-E-DAA-TN74516-1. Moffett Field, CA: NASA Ames Research Center.

Gray, D. E. (2013). Ethnography and participant observation (Chapter 17). Doing research in the real world. London: Sage.

Holbrook, J., Prinzel, L. J., Stewart, M. J., & Kiggins, D. (2020). How do pilots and controllers manage routine contingencies during RNAV arrivals? In Proceedings of the 11th International Conference on Applied Human Factors & Ergonomics.

Hollnagel, E. (2014). Safety-I and Safety-II: The Past and Future of Safety Management. Farnham, UK: Ashgate.

Hollnagel, E. (2016). Resilience Engineering. erikhollnagel.com/ideas/ resilienceengineering.html. Accessed March 1, 2021.

National Academies of Sciences, Engineering, and Medicine 2020. Advancing Aerial Mobility: A National Blueprint. Washington, DC: The National Academies Press. https://doi.org/10.17226/25646

Stewart, M., Matthews, B., Avrekh, I., Janakiraman, V. (2018) Variables influencing RNAV STAR adherence. IEEE/AIAA. Proceedings of the 37th Digital Avionics Systems Conference (DASC). London, UK.

DISPLAY DESIGN TO AVOID AND MITIGATE LIMIT CYCLE OSCILLATIONS ON THE F-16C

Captain David Feibus
Air Force Institute of Technology
Wright Patterson Air Force Base, OH

The U.S. Air Force F-16C's flight envelope is defined by its external weapon stores configuration, and the employment of some munitions at certain speeds can cause the F-16 to enter a flutter-like state called Limit Cycle Oscillations (LCO). In LCO, the pilot experiences turbulent vibrations reducing their fine motor control. The current research attempted to help pilots anticipate an LCO-susceptible configuration by projecting the consequences of employing certain munitions. It was hypothesized that the new displays would result in fewer flight envelope violations, fewer LCO occurrences, and increased situation awareness. The results show that there are situation awareness benefits if the pilot is not task overloaded, but the performance results were inconclusive. Further design maturation is necessary to understand the implications of the new display.

The flight envelope of the F-16, also known as the operating bounds in terms of airspeed, altitude, and wing loading, is dynamic. The addition or deployment of underwing-stored missiles, bombs, fuel tanks, sensors, or other devices changes the aerodynamic properties of the vehicle as well as the recommended flight envelope. Violation of the flight envelope incurs maintenance and sustainment impacts but does not constitute immediate structural failure.

A particularly problematic state occurs when wing tip missiles, typically air to air missiles, are employed. In the F-16, mass located at the wingtip damps oscillation of the wings. Thus, deploying stores from the wingtip can lead to turbulent oscillation of the wings and aircraft, known as Limit Cycle Oscillations (LCO). LCO is non-catastrophic and is closely linked to classical flutter but does not diverge (Bunton & Denegri, 2000). LCO causes loss of a pilot's fine motor control and reduces visual acuity like low frequency vibration due to flutter (Tung et al., 2014). To recover from LCO, a pilot must reduce airspeed and wing G-loading. As this is not a problem in today's mostly Air-to-Ground wars where the United States and enjoys air superiority, it presents a potential issue should the F-16 be employed in Air-to-Air battles. The flight envelope depends on the mass and drag properties of the underwing-stored articles, which defines the current and future flight envelopes, where a particular future envelope, referred to as the downloaded envelope, takes effect immediately after expending or jettisoning a store.

A pilot can be within their current flight envelope but have an airspeed that is beyond the downloaded envelope's limits, and when a missile is deployed, the F-16 will be operating outside its new designed flight envelope and will experience heavy LCO without warning. This research seeks to develop display symbology which improves the pilot's situation awareness (SA) by aiding their understanding of current circumstances, anticipating the impact of deploying a wing-stored article, and permitting them to project future actions (Endsley, 1995). As the pilot may intentionally violate the downloaded limits this research will additionally explore the use of status or command displays to aid recovery from LCO or other flight envelope limits. In the experimental evaluation of the design alternatives developed in this research, it was hypothesized that showing pilots their current and future envelope limits will enhance their SA and reduce both number and duration of flight envelope violations. It was also hypothesized that a status display would be more effective than a command display for recovering from LCO by enhancing the pilot's SA, allowing the pilot to make more educated decisions about this secondary task and avoid task overload (Weinstein & Wickens, 1992).

Method

The research method consisted of three phases. First, extensive interviews of Subject Matter Experts (SMEs) from a Flight Test Squadron were conducted using the Cognitive Work Analysis method. Second, display prototypes were developed and reviewed by SMEs. Last, an experiment was designed and conducted to test the effect of the displays on SA and performance.

Design Method

From the Cognitive Work Analysis, the pilots' needs, operational constraints, and environment resulted in the main design goal: to take the limits out of the pilot's head and put them into the world, as Norman suggests (Norman, 1988). Pilots expressed a strong interest in quickly cross referencing the current and future flight limits, and desired a display that operates in the background and not consume unnecessary real estate on the already-crowded displays.

The display consisted of two elements: Predictive Feedback (PF), and LCO Recovery (LR), when combined create the LCO Support System. PF used common SA design criteria, including supporting knowledge of both the current and future envelope limits to help pilots understand the consequences of their actions on their flight envelope. The PF display showed the current munition selected, if they were at risk for future LCO based on the current munition, and the current and future envelope limits in Mach and airspeed. The yellow wingtip missile in Figure 1 below shows that if that missile is selected, they will still be beyond their download limits and continue in LCO.

The LR display provided feedback to show the pilot that they were out of their flight envelope and a change in their flight conditions was necessary to recover. The LR display consisted of a colored banner with either a status or a command message telling the pilot they are OOB, and if LCO was present. A yellow banner meant the pilot was past the download limits based on the currently selected munition, and a red banner meant they were in LCO or OOB. These two elements were combined to create the LCO Support System for the experiment, as shown below in Figure 1.

Figure 1: The left side shows the download envelope (DWN), and the right side shows the current envelope, both based on the current missile selected (shown in white). The LR banner appears when the participant was either: currently experiencing LCO, past their download limits, or out of bounds of their current flight envelope.

Ideally an indicator of LCO would be available in the Heads-Up Display (HUD), similar to all other important flight information. However, the Center Display Unity (CDU) was the target display due to its easily modifiable architecture. The top of the CDU's screen was ideal for two reasons: First, it was determined that with a helmet and oxygen mask on, the bottom half of the CDU was obscured, requiring the pilot to move their head to crosscheck the display. No other major alert on the F-16 requires head movement which might be difficult under high G maneuvers. Second, the turbulent motion induced by LCO disturbs the vestibular system making displays farther from the resting visual angle of the HUD difficult to read.

Experimental Design and Procedure

The experimental design was a three-by-two-by-two, mixed-subjects, experimental design including the within-subjects variables of LCO Support (off or active), and the Scenario (1, 2, or 3), and the between-subjects variable of the Display Type (Status or Command). The Display Type was randomly assigned. The participants were tasked to fly a flight simulator and achieve air superiority against several Sukhoi Su-27 aggressors. Their secondary task was to stay within the flight envelope. The simulated scenarios were derived from real training missions.

The simulated F-16 was loaded with 4 long range and 2 short range Air-to-Air missiles. To ensure an LCO-susceptible configuration would be entered, participants were told their wingtip missiles were "superior" long range missiles and should be used first. Participants were invincible but instructed that 50 points were subtracted if they were hit, and 100 points were awarded for each kill. Their overall scores were collected to understand their performance.

Before the experiment began, each participant was first familiarized with employing air-to-air missiles using the F-16 interfaces and displays. Next, they were briefed on the envelope restrictions for the loadout for the experiment. Participants then flew a training scenario in which they had to face multiple enemies, from a starting distance of 45 nautical miles (nm), but without LCO Support and were encouraged to redo the training scenario until they felt comfortable.

Participants flew three different scenarios, each time flying the control case first with LCO Support inactive, and then flying the scenario with it active. The starting distance from the enemies decreased with each scenario, starting at 45 nm, to 30 nm, to 15 nm. The SAGAT method was used to measure SA by pausing the scenario during high points of stress: when focused on trying to fire a missile, evade an enemy missile, or for the missile's radar to go active. Once the scenario was paused, participants were asked to answer questions about their flight envelope, the state of the scenario, and their aircraft. Lastly, an end survey asked for feedback on the CDU design, as well as HUD designs.

Participants

Nine males with flight experience volunteered from among the military and civilian workforce on Wright-Patterson Air Force Base. Participants included two active and three retired F-16C pilots with an average of 1950 flight hours. Additionally, two active F-15 pilots with an average of 1160 flight hours, and two participants with general flight experience participated.

Apparatus

Digital Combat Simulator (DCS) 2.5.6 was used for the flight simulator, using an F-16C Block 50 airframe. Unity was used to create the LCO Support System and CDU. A Thrustmaster Hands on Throttle and Stick (HOTAS) Cougar was used. An LG 65" TV was used to display DCS, and the CDU was displayed using a ViewSonic VG2455-2k 24" Monitor. The participant was seated 29 inches away from the CDU, mimicking the viewing angle present in the F-16C, without needing to recline the chair. Participants were not allowed to use the unreliable Missile Step button on the HOTAS to switch between missiles of the same type to avoid a possible disparity between DCS and Unity. The Fire Control Radar (FCR) cursor was augmented to include the option to either use the left index finger or thumb, accounting for the HOTAS differences in the F-16 and the F-15.

Metrics

All variables were subjected to a mixed factor ANOVA. Score within each scenario as well as envelope violations and durations were collected for the LCO and OOB conditions, respectively. SAGAT scores were scaled by multiplying incorrect answers by -1 and correct answers by 1, and a confidence rating was collected for each SAGAT response on a 1 to 5 scale.

Results

It was expected that participant familiarity with the F-16 would lead to superior performance, but there were no statistically significant results on the sample population. For analysis, the population was divided into two groups: 5 High Performers (HP) and 4 Low Performers (LP). This was based on the sum of their total scenario scores being above or below the mean. The current F-16 and F-15 pilots were classified in the HP group while the general aviation participants fell within the LP group. Surprisingly, two of the retired F-16 pilots were classified in the LP group.

Performance

LCO support had a significant effect on the mean duration of LCO envelope violation for the HP group (F (1, 3) = 18.071, MSE = 26.84, p = 0.024, $\eta_p^2 = 0.858$). As shown in Figure 2, the mean duration of LCO envelope violation decreased for the HPs by 8.21 s (SD = 1.93), in accordance with the hypothesis. LCO support also had a significant effect on the mean duration of LCO envelope violation for the LP group (F (1, 2) = 38.682, MSE = 45.68, p = 0.025, η_p^2 = 0.951). However, for the LPs it significantly increased by 17.16 s (SD = 2.77), counter to the hypothesis.

The ANOVA indicated a significant increase in the number of LCO violations for the LP group (F (1, 2) = 81, MSE = 0.01, p = 0.012, $\eta_p^2 = 0.976$) as shown in Figure 3. The number of LCO violations was not significant for the HP group (F (1, 3) = 0.008, MSE = 0.73, p = 0.93, $\eta_p^2 = 0.003$). There was also no effect of LCO support on the total number of envelope violations (LCO or OOB) for either the HP group (F (1, 3) = 0.692, MSE = 1.81., p = 0.466, $\eta_p^2 = 0.187$) or the LP group (F (1, 3) = 5.31, MSE = 1.89, p = 0.148, $\eta_p^2 = 0.726$).

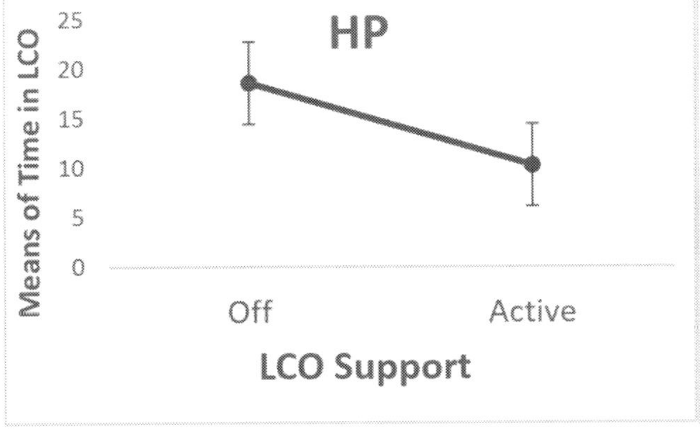

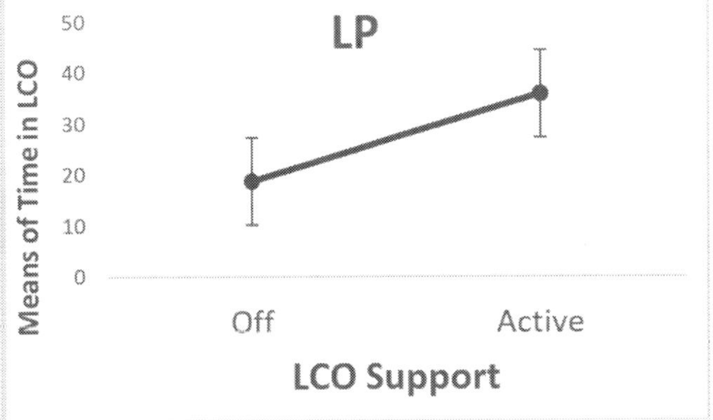

Figure 2: LCO Support versus Mean Time spent in LCO, where lower values are desired. Error bars represent the standard error.

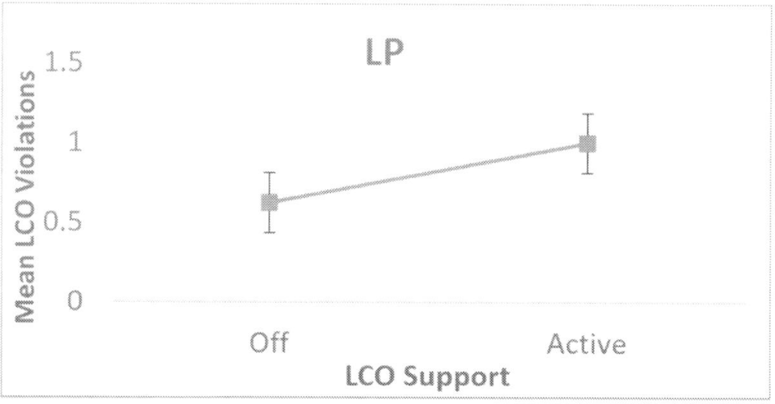

Figure 3: LCO Support versus Mean LCO Violations for the LP group, where lower values are desired. Error bars represent the standard error.

Situation Awareness and Display Aids

It was expected that LCO Support would yield large SA benefits, but due to the small sample size there were no significant main effects of LCO Support for either group on the SAGAT questions. The HP group's mean number of correct answers and confidence levels increased with active LCO Support, while the LP group's both decreased. From the interaction of LCO Support and Display Type, there were no significant main effects on either group. While it was expected that the group using the Status display would have greater SA and performance, participants consistently could not recall the content of the LCO Recovery banner indicating that the type of display did not have an effect. Most participants reported they did not have time to read the LCO Recovery banner during the intensive scenarios but relied on the color of the banner to determine their status in the flight envelope.

Discussion

The current research explored the application of a predictive feedback display to project a participant's future state after employing munitions, permitting pilots to make changes to their tactics to avoid inducing LCO. It was theorized that this system would change the task structure for the pilot, replacing the working memory required to recall the dynamic envelope limits with a simple item added to a pilot's normal instrument crosscheck. On average, pilots in a high performing group reduced the time they violated the flight envelope while participants in the low performing group tended to violate the flight envelope limits more and for longer durations when LCO Support was active. The research provided evidence that, if not task saturated, the participant's SA will benefit from the new display, as was true for the HP group.

The experiment did not simulate the negative LCO consequences, such as display shaking or flight control issues, so participants had no true incentive to avoid LCO. With LCO consequences, a participant may have been more willing to adjust their tactics and an improvement in performance might have occurred. Real feedback would update the participant's mental model of their aircraft and permit them to experience the weight of their decisions. The F-

16's FCR in DCS also displayed unreliable performance, randomly losing radar lock on enemies at critical moments in the kill chain, forcing a closer engagement and exacerbating an already stressful and difficult situation, leading to poorer performance.

It is possible the participants needed more exposure and training on the new display before being subjected to difficult scenarios. Participants in this experiment had to contend with their envelope limits throughout the entire scenario, which is not operationally representative. The desired outcome of the integration of this display is to augment training and influence the use of weapons deployment strategies to reduce the likelihood of LCO.

While participants liked the design and said that without LCO Support it required too much mental energy to stay within the limits and be tactically effective, participants said the design still required some maturation and fine tuning before fielding. The flashing color acted as truth data, updating the mental models of the participants, especially those who were unaware of their position in the envelope. All participants expressed an affinity for LCO Support indicators in the HUD, and felt the design was incomplete otherwise.

Due to experimental limitations, the performance results are inconclusive which demonstrates a need for a higher fidelity experiment with actual LCO consequences. Such a system may provide additional performance and SA advantages to the pilot, but further design maturation is required. Additionally, the research suggests that without indicators in the HUD, the high workload which occurs prior to LCO onset may prevent one from receiving information from the CDU. As a result, further research is required using a higher fidelity apparatus to determine the impact of LCO Support on a pilot's performance and SA.

Acknowledgements

The author would like to thank the F-16 System Programs Office for motivating this research, as well as the 40th Flight Test Squadron for their many hours of SME time for their assistance in creating the experimental designs and research method. The views in this article are those of the authors and do not necessarily reflect the official policy or position of the Department of the Air Force, Department of Defense, nor the U.S. Government.

References

Bunton, R. W., & Denegri, C. M. (2000). Limit cycle oscillation characteristics of fighter aircraft. *Journal of Aircraft*, *37*(5), 916–918. https://doi.org/10.2514/2.2690

Endsley, M. R. (1995). Toward a theory of situation awareness in dynamic systems. *The Journal of the Human Factors and Ergonomics Society*, *37*(1), 32–64. https://doi.org/10.1518/001872095779049543

Norman, D. (1988). *The Design of Everyday Things: Revised and Expanded Edition - Kindle edition by Norman, Don. Health, Fitness & Dieting Kindle eBooks @ Amazon.com*. Basic Books.

Tung, K., Miller, M., Colombi, J., Uribe, D., & Smith, S. (2014). Effect of vibration on eye, head and helmet movements while wearing a helmet-mounted display. *Journal of the Society for Information Display*, *22*(10), 535–544. https://doi.org/10.1002/jsid.286

Weinstein, L. F., & Wickens, C. D. (1992). Use of Nontraditional Flight Displays for the Reduction of Central Visual Overload in the Cockpit. *The International Journal of Aviation Psychology*, *2*(2), 121–142. https://doi.org/10.1207/s15327108ijap0202_4

OBSERVING ON-DEMAND AIRCREW TRANSITIONING FROM PAPER TO ELECTRONIC FLIGHT BAGS: THE IMPACT ON WORKLOAD

Charla L. Long, Ph.D.
Federal Aviation Administration
Oklahoma City, OK

Kevin M. Gildea, Ph.D.
Federal Aviation Administration
Oklahoma City, OK

The introduction of electronic flight bags (EFBs) for flight crew use has reduced the overall workload, except in some situations if not designed properly or employed effectively. Researchers from the Civil Aerospace Medical Institute (CAMI) undertook an observational study combined with crew interviews to assess overall flight crew operations including flight demands, procedures, and the methods the crews used to integrate EFBs into all aspects of their flights from preflight planning to postflight debrief. The researchers also examined the EFB applications (apps) themselves for general usability and developed some recommendations for ways EFB use in operations could be improved. General recommendations and specific recommendations for each phase of flight are provided and include:1) adopting EFBs that are intuitive, 2) standardizing procedures for EFB usage, and 3) incorporating EFB usage best practices into training.

Electronic flight applications and the use of such applications began with the invention of the Electronic Flight Bag (EFB) meant to supplement and/or replace the conventional flight bag in the late 1990s. In 2002, the Federal Aviation Administration (FAA) approved the use of EFBs by publishing an Advisory Circular, (AC) 120-76. The AC defined EFB as an electronic display system intended for cockpit/flight deck use; displaying a variety of aviation data as well as calculations for basic performance in the aircraft. The EFB was originally designed to replace some of the paper products and tools used for flying, such as manuals, as well as serve as a supplemental device to paper documentation (FAA, 2020a). The FAA identifies an EFB as any portable electronic device (PED) a consumer could purchase off-the-shelf with functionality that replaces "conventional paper products and tools, traditionally carried in the pilot's flight bag" (AC) 120-76D.

The FAA created a requirement for Parts 91K, 121, 125, and 135 mandating the use of the approved EFBs in lieu of privately-owned devices to limit distraction from other apps (such as personal email, etc.) during the operation of aircraft (Federal Aviation Administration, 2020b). Part 91 operators (apart from 91K) are allowed to use privately-owned devices. Recent ACs (including the current AC 120-76D) published by the FAA, further defined the EFB as any device actively displaying dynamic and interactive applications such as weather, aircraft parts manuals, chart supplements, daily flight logs (DFLs), crew member qualification logs, weight and balance calculations, performance calculations, electronic checklists, and flight planning. This new designation means iPads and similar electronic devices can be considered replacement devices for flight bags, not merely as a supplement to paper-based flight bag documents. However, the AC requires the failure of an EFB to be a minor hazard when flying with minimal effect on safety if a failure of the EFB occurs. Thus, flight crews must be able to perform normal duties without a fully functioning EFB. Thus, EFBs are used in different ways during all phases of the flight. Reviewing airport configurations, approach plates, and other information pertaining to the flight occurs during preflight with crews at cruising altitude using the EFB to monitor flight progress, review future flights, and access information pertaining to the next airport.

Previous Research on Effects of EFB Use on Flight Crew Performance and Workload

There is some disparity within the literature as to whether EFBs increase or decrease workload. Some researchers state that EFBs increase workload, distractions, and head-down time (Chase & Hiltunen, 2014; Sweet, et al., 2017) while other studies suggest the EFBs decreases workload, increases situational awareness, and improves performance (Haddock & Beckman, 2015; Solgård & Oppheim, 2019). Additional studies suggest there are no statistically significant differences in workload between operations using an EFB and those using paper charts and materials (Suppiah, 2019). A couple of possible explanations for this disparity include variability in the relation between EFBs and workload due to organizational differences (Solgård & Oppheim, 2019), software application or system differences (Sweet et al., 2017), or Human Factors (HF) design differences (Laursen & Ludvigsson, 2017). Other possibilities include differences in study methodology and/or sample type. Many of the studies show a decrease in workload with EFBs when student pilots are participants in the research (Haddock & Beckman, 2015; Suppiah, 2019).

Previous research has shown EFBs increase risk of errors if designed poorly (Chase & Hiltunen, 2014). Risk increases if the EFB's interface is complicated, presents data in a way that is difficult to read, or is ambiguous. Identifying these issues to find the most appropriate tool(s), providing training on the tool(s), and limiting the total number of applications available can decrease errors using the EFB (Haddock & Beckman, 2015).

Providing flight crews with too much information or information that is difficult to access, requiring numerous applications can lead to HF issues such as information overload, increased distractions during flight operations, and the necessity to flip back and forth between applications during critical flying maneuvers (Sweet, et. al., 2017). Flight crews have noted the need to be more "heads down" during flight. However, with the increase in electronic information many aircrew members stated the tendency to fixate inside the aircraft can lead to more safety mishaps. Training must emphasize attention control strategies including proper scans of the instrument panel, EFB, and outside environment (Haddock & Beckman, 2015; Lylte, 2015).

Researchers have noted transferring skills from paper-based flight bags to EFBs for pilots well-versed in the features and functions available show significant increases in pilot performance (Sweet et al., 2017). Tasks such as monitoring other traffic (traffic alerting), accessing paper charts, following checklists, receiving weather updates, standard manuals, and performance calculations became easier when using the EFB (Haddock & Beckman, 2015).

Current Research

Most research on EFBs has been conducted in simulators, where scenarios were presented to pilots and EFB usage was documented (Haddock & Beckman, 2015; Sweet et al., 2017). Research in simulators is common because access to pilots in flight is difficult. However, for this study, Civil Aerospace Medical Institute (CAMI) researchers had access to flight crews, being able to conduct an observational study of flight operations during week-long trips. Prior to the observations, researchers conducted interviews and observed flight crews from planning to flight completion, including the flight debrief, in order to gain a better understanding of the impact of the EFBs on workload, operations, and safety. Researchers recorded real-time workload information, usage of EFBs during flights, and crew interaction to provide recommendations regarding EFB usage, training requirements, and safety challenges when using EFBs.

The overarching scientific question examined was the impact of the EFB on flight crew workload when switching from paper to electronic flight bag materials. More specific questions included: 1) What information on the EFB is important during flight? 2) What is difference between the information presented on EFB vs presented on paper? 3) Is there a difference in workload between EFB and paper-based materials? 4) Does the workload-related risk increase at any specific time during the flight, and if so, how do EFBs change that risk? 5) In what ways are distractions increased or decreased when using EFBs (compared to paper) during flight?

Methods

Participants included pilots and crewmembers from various locations. Data were de-identified and demographic information was separated and not stored with participant responses in an electronic or physical data file. In addition to Informed Consent and Demographics, flight crews were asked a set of prepared interview questions. The study was reviewed and approved by the CAMI Institutional Review Board (IRB) for the protection of research participants.

A sample of 30 pilots and 11 additional crewmembers participated in the study. Flight crews consisted of two pilots and an additional crewmember for each itinerary. The pilots were either the Pilot in Command (PIC) or Second in Command (SIC) during each flight. The PIC was responsible for the flight, while the SIC was often the flying pilot. Most crews switched responsibilities (PIC or SIC) every other day of the itinerary. However, trainees (regardless of years of experience as a pilot) were not eligible to be certified as the PIC, thus they were primarily the flying pilot for the week-ling trip. The additional crewmember collected data during the flight.

Demographic information was collected from crewmembers including position, location, and total flight time to ensure a representative sample of the target population was reached. Flight hours ranged from 5,200 to 27,000. Primarily, pilots fell into one of three categories: 1) 5,000 – 9,999, 2) 10,000 – 19,999, or 3) 20,000 hours. Seven pilots are in category 1, three pilots are in category 2, and one pilot is in category 3. Supervisor pilots' years of flying ranged from 24 to 35 years with the average being 29 years while the flight hours ranged from 7,500 to 14,500. The additional crew members did not provide flight time, ranging from 2 to 30 years in their current position.

Procedures

The procedures consisted of researcher observation of crew interactions with EFBs during flights and in-person interviews at field locations. The typical itinerary consists of a planning day and four flying days, with two flight periods each day. The CAMI researchers took the role of non-participant observer, seated in a jump seat behind the pilots and observed the crew for an entire workweek (40+ hours). The observations included preflight through postflight operations and additional informal meetings in the evenings.

Interview protocol. In-depth, semi-structured interviews allow researchers to explore and reconstruct meanings from events without personally experiencing them (Rubin & Rubin, 1995). Interviews are extensions of ordinary conversations and can be described as conversations with a purpose. The primary difference between interviews and conversations is the intentional listening for verbal and nonverbal cues to better understand the phenomenon (Long, 2006). The interview questions were created following a topical approach with probing questions added for clarification. Steering probes were employed to keep the participants on target and restrict the information being collected (Rubin & Rubin 1995). Interviews were conducted for 30-60 minutes per individual. Interviews were conducted with 41 participants.

Interviews conducted with flight crews covered topics including flight complexities, workload, and use of tools (EFBs, paper, etc.). Questions included, 1) what available information on the EFB is important during flight, 2) what is different between the information presented on EFB vs presented on paper, 3) are distractions increased or decreased when using EFBs (compared to paper) during flight? Questions were not provided to the flight crews prior to conducting the interviews. This allowed the researchers to take notes during the interviews and allowed for more reliable data collection to occur. Individual comments and statements were noted, but no identifying information was included in the note-taking process.

Observation Protocol. Non-participant observations were conducted at each facility after the interviews to provide a nuanced and dynamic approach to situations not easily captured through other methods (Lui, & Maitlis, 2010). In addition, observations were conducted to document any site-specific nuances vs consistent and widespread organizational issues, e.g., workload differences based on region (Lui, & Maitlis, 2010).

Researchers used a three-stage technique described by Lui & Maitlis (2010) in order to observe and document preparation and planning at each facility and during flight in the aircraft. The first stage involved descriptive observation, broadly defining the setting of the observation. The second stage, focused observation, helps researchers narrow the focus by observing the activities directly related to research questions. Finally, selected observation, allows the researcher to investigate the relationship and make connections to ensure a comprehensive understanding of the phenomenon documented.

Observations began with the planning phase of the flight where the researcher took notes by hand or computer, focusing on the use of the EFB vs paper to plan and brief the flight requirements. The researcher observed at least one flight planning session, prior to joining the flight crew. During flight operations, the researcher observed crew interactions and radio communications. Each flight consisted of 2 legs (approximately 3.5 hours in the airplane) breaking for lunch and fuel. After landing, observation continued both formally (end-of-day debriefing) and informally (after 8 hour workday).

Results

Based on observations, interview responses, and discussions with the flight crews, workload using the EFB was noted as moderate to high during all phases of the flight, including planning, preflight, and flying. Minimal impact on workload was noted postflight. Data were analyzed using qualitative techniques discussed below.

Flight Planning

Flight planning was noted by 80% of the crews as a moderate workload phase. Pilots noted the EFB was a positive tool decreasing workload during this stage due to the access to weather briefings, NOTAMs, updated airport information, and the ability to file flight plans quickly and efficiently. The crews stated the ability to import information from numerous sources and share among crewmembers, as well, the capability of overlaying information from one application onto the next, made the flight planning easier. However, 83% of aircrew members stated EFB sources did not always match, requiring further research prior to finalization, impacting workload at times.

Preflight

Flight crews go through a series of procedures to ensure the aircraft and flight plans are still achievable prior to each flight. Preflight checklists were often accessed through the EFB. The PIC reviews the itinerary and informs the crew of changes to the schedule completed each time a full stop occurs. The preflight procedures are considered a moderate workload phase by 64% of the crews if no major obstacles (e.g., weather, maintenance, late clearance) occur. Flight crews find these procedures routine and predictable. The challenge to the crews is being diligent during this stage to assess and record changes. When the routine becomes mundane, errors can occur. Catching errors prior to flying is critical to the success of the flight and organization.

Flying

Approach and landing are considered the highest workload phases by 100% of the crews. While most pilots fly from point A to B with general predictability, the observed pilots work on-demand operations, requiring flights to infrequently-visited or unfamiliar airports. Researchers observed 95% of the crews using the EFB as the primary tool used in flight. Aircrew members reviewed procedures, charts, NOTAMs, monitored weather conditions, and completed checklists using the EFB. The EFB increased situational awareness for the crews during flight when managed properly. However, when system changes (e.g. FMS) occurred during flight, the EFB became more difficult as multiple touches were used to update the procedure, charts, etc. according to 73% of the crews.

Depending on the complexity and requirements of the procedure, the pilot who is flying (in the left seat) is likely to be head-up, remaining vigilant to look for traffic. The PIC who is in the right seat may be head-down or head-up depending on the point in the flight. Since the PIC is responsible for planning and performing the runs, they will look at the approach plates, procedures, policies, and/or notes during the flight using the EFB. Researchers observed this methodology occurring approximately 98% of the time. However, observations were made where both the PIC and SIC were head-down reviewing a procedure, policy, or chart on the EFB during in-flight maneuvers, affecting situational awareness. This occurred more with pilots who were in training as on-demand aircrew member than those with more years in the current position. The impact of the EFB was noted as challenging by 73% of the pilots who were less familiar with the tools and applications used for this job, requiring more touches and head-down time to find information. Workload increased for the PIC when trainees were on board as the trainee was the flying pilot who was focused on flying and less on planning or procedures.

Discussion

Many factors affect the workload of flight crews including the need to complete itinerary planning, and conduct various preflight and in-flight procedures. Adding the complexity of an EFB with various avionics which perform differently can prove challenging. However, the crews stated, and the researchers observed, the addition of an EFB actually decreased workload and made missions safer. However, when trainees were part of the week-long missions, the EFB became more difficult and cumbersome at times. Thus, the following recommendations should be considered to help with workload.

The highest workload occurs during approach and landing (Haddock & Beckman, 2015). Depending on the tasks performed, additional traffic, the airspace, the communication required, and outside forces, workload complexity increases. Additionally, the capability of the tools available and the total number of touches required to access information can impact workload. Standardized training should be created using micro-learning videos for required and supported Avionics applications to minimize the impact of the EFB on workload. The majority of the crews stated learning the functionality in most applications occurred through "trial" and "error", not through training increasing workload and safety risk during flight. Training should be developed on how to limit the total number of touches, reducing the need to pan/zoom in-out during flight, identify appropriate information required for each mission using a HF approach.

Evaluation and refinement of resource and cockpit organization can further reduce workload. Crews identified the need to find more efficient ways to collect and organize information (paper, EFB applications open prior to run, etc.). Highly experienced crews generally have tailored methods for cockpit organization. However, there may be gains made by reevaluating those methods, suggesting SOPs, and passing on lessons learned to new hires.

References

Chase, S. G., & Hiltunen, D. (2014). *An examination of safety reports involving electronic flight bags and portable electronic devices* (No. DOT-VNTSC-FAA-14-12). United States. Federal Aviation Administration.

Federal Aviation Administration. (2020a). *AC-120-76A - Guidelines for the certification, airworthiness, and operational approval of Electronic Flight Bag computing devices*. Retrieved from https://www.faa.gov/regulations_policies/advisory_circulars/index.cfm/go/document.information/documentid/23222

Federal Aviation Administration. (2020b). *AC-120-76D – Authorization for use of Electronic Flight Bags.* Retrieved from https://www.faa.gov/regulations_policies/advisory_circulars/index.cfm/go/document.information/documentid/1032166

Haddock, K., & Beckman, W. (2015). The effect of electronic flight bag use on pilot performance during an instrument approach. *Collegiate Aviation Review International, 33,* 1013. doi: 10.22488/okstate.18.100460.

Liu, Feng and Maitlis, S. (2010) Non-participant observation. In Mills, Albert J. and Durepos, Gabrielle and Wiebe, Elden. *Encyclopedia of case study research.* (Vol. 2). Los Angeles, CA; SAGE.

Long, C.L. (2006). Understanding personal theory building for elementary school teachers implementing curricular change (Unpublished doctoral dissertation. Oklahoma State University, Stillwater, OK.

Rubin, H.J., & Rubin, I.S. (1995). *Qualitative Interviewing.* Thousand Oaks, CA: Sage.

Solgård, R. C., & Oppheim, J. B. (2019). Flight Safety Implications associated with the use of Electronic Flight Bags by Norwegian Operators (Master's thesis, Nord universitet).

Suppiah, S. (2019). Impact of Electronic Flight Bag on Pilot Workload.

Sweet, J., Vu, K.L., Battiste, V., & Strybel, T.Z. (2017). The comparative benefits and hazards of EFBS and paper documents in the cockpit. *Paper presented at the Human Factors and Ergonomics Society 2017 Annual Meeting*. Newbury Park, CA: Sage. Retrieved from: https://journals.sagepub.com/doi/10.1177/1541931213601490

FACTORS INFLUENCING PILOT TRUST IN ELECTRONIC FLIGHT BAG INFORMATION

Warren Pittorie
Summer Rebensky
Leonia Hunt
Andrew Narkushian
Maria Chaparro
Meredith Carroll
Florida Institute of Technology
Melbourne, FL

Electronic Flight Bags (EFBs) are widely used by pilots in the commercial aviation industry. EFBs serve as replacements for some traditional sources of information, such as paper charts, manuals, and checklists, augmentation for flight-related information previously unavailable through older cockpit systems such as temporary flight restriction locations, and supplemental information such as a secondary display of traffic). By having access to this information, pilots are able to make more effective decisions in various situations. Related literature has shown that decision makers in situations of uncertainty are influenced by a range of factors such as experience, the level of risk in a situation, and criticality of information. The purpose of this study was to analyze factors that impact pilot trust in information provided by an EFB. Pilot survey and interview data from a simulation study was analyzed and results indicated that an increase in a pilot's total flight hours, experience with specific EFB applications, and the criticality of the information presented on the EFB increased a pilot's trust in information presented by the EFB. Conversely, the more often a pilot used an EFB and the length of time the pilot's company had utilized EFBs on the flight deck, the less trust a pilot had in the information presented by the EFB. The implications of these findings and areas of future research will be discussed.

The U.S. airline industry has been incorporating newer technology into the commercial flight deck, such as Electronic Flight Bags (EFBs). EFBs are capable of providing pilots with supplementary information that was either previously inaccessible or only available through systems built into the aircraft's cockpit. For example, a pilot must be aware of location of their aircraft in proximity to certain restricted airspace and other aircraft to avoid penetrating restricted airspace or colliding with another aircraft in mid-air. EFB applications can support pilots in this task by displaying the location of the pilot's aircraft (ownship) in proximity to restricted airspace and other aircraft on an interactive moving map. EFBs are commonly incorporated onto the flight deck as a tablet computer, but can also be a permanent device installed into a cockpit instrument panel. Certain EFB functionalities are not currently approved by the FAA (2017) for use on the commercial flight deck as an EFB is not intended to replace certified instruments and systems within the cockpit, but to supplement these systems.

Airlines must ensure that pilot trust in their EFBs is properly calibrated and that pilots are able to properly and safely utilize EFBs in conjunction with existing technologies on the flight deck. Trust, as defined by Mayer et al. (1995) is a pilot's willingness to be vulnerable to the functions of a system based on an expectation that the system will perform its intended action in potentially risky situations. There are a range of different factors which could impact pilot trust in information provided by their EFB (See, for example, Carroll and Sanchez, 2020). The purpose of this paper is to present an analysis of factors that impact a pilot's trust in their EFB during decision making on the flight deck. This research effort analyzed data from a previously published experiment (see Carroll et al., 2020) conducted to study the effects of conflicting information between an EFB and a certifid system in the panel, on a pilot's decision making.

General Experience

Previous research has shown that expertise influences system trust. Mosier et al. (1998) examined the relationship between pilot experience and pilot response to an automated command. A positive correlation was found between numerous experience indicators, including total flight hours and years of experience, with the frequency of omission errors (neglecting to complete a needed action that was not recommended by the automation), suggesting potentially higher levels of trust in the automation. Riley (1994) also examined the relationship between overall aviation experience and trust in automation and found that student pilots are less likely to rely on automation than experienced pilots, even after failure occurs within the automation. As such it was hypothesized that pilot experience would influence pilot trust in the EFB.

Experience with System

Pilot experience with a particular system, such as an EFB, also has an impact on trust in the information. Lee & See (2004) proposed three factors that influence trust: performance, purpose, and process. Performance refers to a pilot's direct observation of system behavior. If an EFB operates as expected over time, the pilot will have greater trust compared to a pilot who experiences unreliable EFB information on a regular basis. Purpose refers to the intended use of the system, and this can be related to whether the EFB is utilized as the users believes it is intended (e.g., for supplemental navigation information). The most relevant factor that is related to the functionality of the EFB is process, and this is defined as a user's understanding of the underlying mechanisms for functionality of the system. A pilot who is familiar with the source and accuracy of the information presented on the EFB will have higher levels of trust in their EFB than someone who is not. Pilots who use their EFB more often are likely to be more familiar with its various functionalities in a variety of situations. Therefore, it is hypothesized that pilot experience with an EFB will impact trust in the EFB's information on the flight deck.

Information Criticality

During the course of a single flight, a pilot can face several situations that vary in the amount of time and effort it takes to reach a decision. Each of these situations can cause a pilot's workload to vary throughout the flight. Campbell and Alexander (2016) found that sources of information that can provide a pilot with the "big picture" of various situations, such as graphical displays of a flight route and traffic, can help a pilot during high workload situations by providing critical information in a quick and dynamic format. The EFB is capable of displaying critical information that can increase situational awareness in time-critical or hazardous situations; potentially resulting in the information being deemed critical to the decision-making process. The results of Lee (1991) support this concept of novel information affecting a pilot's decision making if the information is considered critical. A between-subjects design compared traditional methods of conveying hazardous weather information to a flight crew against a new display of in-flight weather information. Pilots with the new display were more likely to divert around the critical weather and this decision was reached in a more timely manner, even though this display was novel and had not been used by any of the participants. The display of critical weather information, much like an EFB displaying critical traffic or airspace information, presented the hazard in an easy-to-understand manner that clearly informed the flight crew of the risk to the safety of their flight. As such, it is hypothesized that pilots who deem the EFB information more critical will have higher levels of trust than those who indicate it is less critical.

Methods

Participants

Data associated with 25 commercial airline pilots was analyzed in this study. All participants were type-rated in the B737 as either a Captain (16) or a First Officer (9). Total flight time ranged from 8,000 hours or less (5), 8,001 – 12,000 (4), 12001 – 20,000 (10), or greater than 20,000 (6).

Experimental Design and Procedure

This study analyzed a subset of data from a simulation study that utilized a repeated measures design in which participants performed a series of scenarios in a B737 desktop simulator (See Carroll et al., 2021 for full details of the study). These scenarios introduced situations with the risk of penetrating restricted airspace or traffic collisions to assess how various factors influence a pilot's decision making when presented with conflicting information between the ForeFlight EFB and an approved information source, such as a certified system in the aircraft panel. Before data collection, each participant was given a demographic survey to collect information on aeronautical and EFB experience. Next, each participant watched a short video tutorial on the function of the B737 simulator and EFB. After the video tutorial was completed, each participant was given three practice scenarios in order to familiarize themselves with the flight simulator testbed and EFB. Next, each participant was assigned to one of two groups: airspace or traffic and performed scenarios per their condition with half of the scenarios containing information conflicts between the EFB and an approved source of information on the flight deck. After each participant completed each scenario, they completed a post-trial survey that contained self-report measures rating the criticality of the EFB to the pilot's decision-making and how much the pilot trusted the EFB information for each scenario. For the purpose of this study, a set of 10 predictor variables were selected for use in examining how they related to pilot trust in an EFB, including (1) total flight hours, (2) how often a pilot uses an EFB, (3) how long a pilot's company has used EFB's, (4) how familiar a pilot is with ForeFlight, (5) if a pilot has used ForeFlight before, (6) if a pilot has used EFB apps other than ForeFlight, (7) information type (airspace or traffic), (8) cockpit configuration (whether EFB was mounted external to the panel or integrated), (9) if the data was conflicted between sources, and (10) the self-reported criticality of the EFB on decision making.

Measures

Demographics. A demographic survey collected information on participant age, rank (Captain or First Officer), total flight hours, and flight hours in a B737.

EFB Experience. A pre-study survey collected data associated with experience with an EFB. These measures included how often a pilot uses an EFB (daily, weekly, monthly, yearly, or never), how long a pilot's company has approved the use of EFBs in their operations (less than 6 months, 6 months – 1 year, 1 – 3 years, or over 3 years), how familiar a pilot is with ForeFlight (not at all familiar, slightly familiar, somewhat familiar, moderately familiar, or extremely familiar), if a pilot has used ForeFlight on an EFB before (yes or no), and if the pilot has used any other EFB applications other than ForeFlight (yes or no).

Information Criticality. Participants self-reported how critical the EFB information was to their decision-making after each scenario. EFB criticality was measured on a scale of 1 to 5 (1 = "not critical to decision-making", 5 = "very critical to decision-making."

Trust in the EFB. The criterion variable for this study was pilot trust in the EFB, and this was measured after each scenario using one self-report item of how much a pilot trusted the information displayed on the EFB, ranging from 1 to 5 (1 = "did not trust", 5 = "complete trust".

Analysis and Results

A multiple regression analysis was used to predict a pilot's trust in their EFB from the following predictor variables: (1) total flight hours, (2) how often a pilot uses an EFB, (3) how long a pilot's company has used EFBs, (4) how familiar a pilot is with ForeFlight, (5) if a pilot has used ForeFlight before, (6) if a pilot has used EFB apps other than ForeFlight, (7) information type (airspace or traffic), (8) cockpit configuration (whether EFB was mounted external to the panel or integrated), (9) if the data was conflicted between sources, and (10) the self-reported criticality of the EFB on decision making. The analysis began with data points from 150 scenarios, representing each participant's individual scenario ratings. Sixty-one cases were excluded during a preliminary data screening due to missing or incomplete data in the predictor or criterion variables (e.g., participants who reported that they did not use the EFB in that scenario). Next, a multivariate outlier analysis identified eight cases as "extreme cases" and these excluded from the final analysis. After the outlier analysis, the remaining cases were checked for their compliance with the assumptions of multiple linear regression. The following predictor variables failed the assumption of correct specification of the predictors by having leverage values greater than 0.05: how familiar a participant is with ForeFlight, if a participant has used EFB apps other than ForeFlight, information type, cockpit configuration, and if the data was conflicted between sources. The predictor variables that remained within the final analysis were (a) total flight hours, (b) how often a participant uses their EFB, (c) how long a participant's company has used ForeFlight, (d) if a participant has used ForeFlight before, and (e) how critical the EFB was to a participant's decision making. A simultaneous regression analysis was run using JMP (SAS, 2018), and the model was found to be statistically significant, $F(5, 75) = 2.28, p < .01, R^2 = 0.60$. The results indicated that an increase in a pilot's total flight hours, experience with the specific EFB application, Foreflight, and the criticality of the information presented on the EFB increased a pilot's trust in information presented by the EFB. The results also indicated that the more often a pilot used an EFB and the length of time the pilot's company had utilized EFBs on the flight deck, the less trust a pilot had in the information presented by the EFB. See Table 1 for Regression coefficients and standard errors.

Table 1
Multiple regression results for EFB Trust

EFB Trust	*B*	95% CI for *B*		*SE B*	β	R^2
		LL	*UL*			
Model						0.60*
Constant	1.92**	1.08	2.75	0.42	0	
Flight Hours	$2.33e^{-5}$*	$2.57e^{-6}$	$4.40e^{-5}$	$1.04e^{-5}$	0.20	
How Often EFB	-0.59**	-0.96	-0.21	0.19	-0.27	
Company EFB	-0.27	-0.58	0.04	0.16	-0.14	
If a pilot has used ForeFlight before	0.45**	0.19	0.71	0.13	0.27	
How critical EFB was to DM	0.51**	0.35	0.66	0.08	0.5	

$^*p < 0.01$, $^{**}p < 0.0001$

Discussion

Several factors appear to influence the trust a pilot has in their EFB. A pilot's overall aviation experience, measured in total flight hours increased with trust in an EFB. This is consistent with Mosier et al. (1998) and Riley (1994) who found a positive relationship between experience and trust. Trust also increased the more often a pilot used their EFB. Pilots who have used the EFB software, ForeFlight, trusted this application more than pilots who were new to ForeFlight. This is consistent with Lee and See's (2004) process factor's relationship with trust. If a pilot is more familiar with the functionality of an EFB or the accuracy of the information presented, they will have greater trust in this system. Lastly, the more critical a pilot believes the information provided on the EFB is to their decision making, the more the pilot will trust the EFB. This is in line with Lee and See's (2004) trust factor of performance. The more experience the pilot has with the system performing as expected and providing effective information, the more critical that information will be to a pilot and the more they will trust it.

These results should be interpreted with caution given the following limitations: (a) small sample size, (b) participants all flew the same make/model aircraft, (c) use of just one EFB software, (d) and the use of a simulator in place of a real aircraft or related systems. However, these preliminary results may provide insight for the airlines regarding factors that influence pilot's developing appropriate levels of trust in EFBs. Airlines can expect pilot trust in EFBs to increase as pilots gain more experience with the EFB application, and in general. Airlines could potentially encourage pilots to use their EFBs as often as possible to gain this experience with the EFB performance and reliability. Airlines could also consider introducing pilots to the EFB software they will use in their operations as early as possible, as trust appears to increase with familiarity with this specific application. It may also be beneficial for airlines to incorporate the usage of EFBs into pilot scenario-based training that involve high-risk situations. This study has revealed that pilots are less likely to trust a source if they do not have experience using it. If pilots gain experience with the EFB through training, they may develop a better understanding of how their EFB works, resulting in more appropirate levels of trust. These results provide preliminary evidence of the factors that influence pilot trust in the EFB; however, more reseearch is needed to fully understand these factors.

References

Campbell, S., & Alexander, A. (2016). The effect of cockpit-based convective weather notifications on pilot decision making. In *16th AIAA Aviation Technology, Integration, and Operations Conference* (pp. 1-13), AIAA AVIATION Forum.
Federal Aviation Administration. (2017, October 27). *Department of Transportation* (United States of America). Retrieved 2020, from faa.gov.

Carroll, M., Rebensky, S., Wilt, D. F., Pittorie, W., Hunt, L., Chaparro, M., & Sanchez, P. (2021). Integrating uncertified information from the electronic flight bag into the aircraft panel: impacts on pilot response. International Journal of Human Computer Interaction, 37: 7, 630-641. https://doi.org/10.1080/10447318.2020.1854001

Carroll, M. & Sanchez, P. (2020). Decision making with conflicting information: Influencing factors and best practice guidelines. Theoretical Issues in Ergonomic Science. doi.org/10.1080/1463922X.2020.1764660

Federal Aviation Administration, Department of Transportation. (2017). Authorization for Use of Electronic Flight Bags (AC 120-76D).

ForeFlight, a Boeing Company. (2021). ForeFlight Mobile [Computer software]. Retrieved from https://foreflight.com/

Lee, A.T. (1991). Aircrew decision-making behavior in hazardous weather avoidance. Aviation, Space, and Environmental Medicine, February, 158-161.

Lee, J. D., & See, K. A. (2004). Trust in Automation: Designing for Appropriate Reliance. Human Factors, 46(1), 50-80.
Mayer, R. C., Davis. H., Cummings, L. L., & Chervany, N. L. (1998). Initial trust formation in new organizational relationships. *Academy of Management Review, 23,* 473-490.

Mosier, K. L., Skitka, L. J., Heers, S., & Burdick, M. (1998). Automation bias: Decision making and performance in high-tech cockpits. *The International Journal of Aviation Psychology, 8*(1), 47-63.
Riley, V. 1994. "A Theory of Operator Reliance on Automation." In Human Performance in Automated Systems: Current Research and Trends, edited by Mouloua M., and Parasuraman R (pp. 8–14). Hillsdale, NJ: Erlbaum.

SAS Institute Inc. (2018). JMP (Version 14.2.0). Retrieved from https://www.jmp.com/en_us/home.html

Skitka, L. J. (1999). Automation: Decision aid or decision maker? (No. NASA/CR-1998-207767.). Retrieved from NASA Technical Reports Server: https://ntrs.nasa.gov/

THE IMPACT OF THE COVID 19 PANDEMIC ON AVIATION WORKERS AND THE AVIATION SYSTEM

Joan Cahill
Trinity College Dublin (TCD)
Dublin, Ireland
Paul Cullen
Trinity College Dublin (TCD)
Dublin, Ireland
Sohaib Anwer
Trinity College Dublin (TCD)
Keith Gaynor
University College Dublin (UCD)
Dublin, Ireland

This paper presents the findings of an anonymous web-based survey addresing the impact of the COVID 19 pandemic on aviation workers and the aviation system. The analysis indicates that aviation workers are experiencing high level of depression and anxiety.Coping mechanisms are being used by many aviation workers. Considerable barriers still remain in relation to reporting mental health (MH) issues at work. Overall, the response from organisations in terms of helping employees cope with the stress arising from the COVID 19 pandemic and changes to wellbeing is weak. The vast majority of aviation workers indicate that wellbeing is not a priotiy for their organisation. Wellbeing supports are required for aviation workers who are currently working and those who have lost their jobs. The introduction of tools at indivdiual and organisaitonal levels to support fitness for work assessment and wellbeing montioring/support would benefit aviation workers and flight safety.

Introduction

Work is part of our wellbeing and a key driver of a person's health. Worker wellness and mental health is hugely important in safety critical systems such as aviation. Aviation workers need to be fit for duty and aware of risk that compromises their health/wellbeing. Work has the potential to negatively impact on mental health particularly in the form of stress. The World Health Organisation have proposed a model of the healthy workplace in which both physical and psychosocial risks are managed. ISO 45003 will be published in early 2021 and will address worker psychological health.

Prior to the COVID pandemic, there was ample evidence of wellbeing and mental health (MH) issues amongst pilots. For example, studies at Harvard (Wu et al, 2016) and Trinity College Dublin reporting on prevalence of depression (Cullen et al, 2017, Cahill et al, 2021). However, there has been little attention to assessing and supporting wellbeing and mental health for other aviation workers.

The pandemic has put increased stress on aviation workers and the aviation industry. The industry has experienced a decrease in capacity. Many workers working on reduced salary, furloughed or lost their jobs. This has had a detrimental impact on their sense of purpose and financial security. Those who are still working are working in very different environments with additional stressors.

People vary in relation to their ability to cope successfully with stress (including WRS). The practice of healthy behaviours strengthens the person's resistance to stress . The substitution of maladaptive coping with more adaptive coping is an important component of therapeutic interventions for work related stress (WRS) Common stress coping strategies include exercise, the practice of relaxation techniques and seeking social support and/or social participation. It is likely that some aviation workers may develop psychological issues during the period of being off work. Social isolation and confinement may lead some people to develop maladaptive coping strategies. If off work, some of the occupational barriers to maladaptive coping are not there (i.e. intoxicant testing by employer). Further, access to social support - a key enabler of adaptive coping - is less available. Currently, support from within a person's social network, peer support group and/or support groups within the community is limited and accessiable online (i.e. non in person). As such, the current Covid-19 pandemic poses a huge occupational health and safety threat. The Flight Safety Foundation has identified three operational scenarios to be managed during the COVID-19 crisis and beyond. This includes (1) being at work during the COVID outbreak, (2) being off work and (3) returning to work (Nelson et al, 2020). Prior research by the authors had identified the requirements for digital tools to support self care and fitness for work assessment for pilots, and other aviation workers (Cahill et al, 2020).

Methodology

The objective of the survey was to address the impact of the COVID 19 pandemic on (1) job and employment, (2) wellbeing and morale, (3) performance and safety behaviour, and (4) safety oversight. The survey also investigated reporting culture, coping strategies, fitness to work assessment, and the supports provided by aviation companies to workers during the pandemic.

The was a cross-sectional descriptive study. An anonymous web-based questionnaire was developed which elicits feedback pertaining to the topics indicated above. The survey incorporated several standardised instruments to measure levels of common mental health issues. These are these Patient Health Questionnaire -9 (PHQ-9) (Kroenke, Spitzer & Williams, 2001), and the GAD 7 (Spitzer, Kroenke, Williams & Löwe, 2006). Ethics approval was granted by the School of Psychology, Trinity College Dublin (TCD). The survey was administered over three weeks in July and August 2020. Using social media chanels, respondents were invited to participate in an anonymous online survey at a time of their choice. Advertising information informed participants that the survey elicits information of a sensitive nature and included a weblink to the survey. Prior to answering survey questions, respondents received background information about the study and completed the electronic consent. Following this, respondents completed questions across the seven survey sections. The survey concluded with a debriefing which included contacts information for relevant support groups. The survey was powered by Qualtrics and did not collect any identifying information about the person. It was assumed that each participant was an aviation worker and only completed one survey. Descriptive statistics were used to describe the respondents and their responses on various survey items. We evaluated depressive symptoms via the Patient Health Questionnaire (PHQ-9) depression module, and axiety symptoms via the GAD 7. Tests for statistically significant group differences were undertaken.

Results

Summary of Respondents

The survey was completed by **2,050** aviation workers. **2,050** respondents participated in the survey, with **1,523** completing it fully (74 % rate). The respondent breakdown was as follows: 38% Pilots (729), 19% Cabin Crew (376), 11% Air Traffic Control (210), 8% Maintenance/Engineering (152), with the remaining 29% spanning other aviation workers. 1796 respondents completed the PHQ-9 (87.9%), while 1796 also completed the GAD 7 (87.9%). Overall, the respondents can be described as male (70% - 1361) and working full time (86% - 1643). The repondents can be split into the following age brackets; <25 (5% - 94), 25-35 (28%-552), 36-45 (30% - 584), 46-55 (23%- 458) and 56-65 (12% - 242). Respondents had worked in aviation related roles for the following lengths of time; <2 years (3% - 67), 2-5 years (15% - 297), 6-10 years (18.5% - 361), 11-15 years (14% - 268), 16-20 years (12% - 227), 21-25 years (12%- 244), 26-30 years (8%- 152) and >30 years (17% - 339).

Impact on Wellbeing

77% (1,383) of respondents rated their physical health as good/very good, while approximately 56% (1,005) rated their mental health as good/very good. The majority of participapnts perceived their MH as worsening since COVID (63% strongly agree or agree that MH had worsened since COVID). As indicated in Figure 1 below, Cabin Crew appear to be the group whose self reported MH is most negatively impacted by COVID.

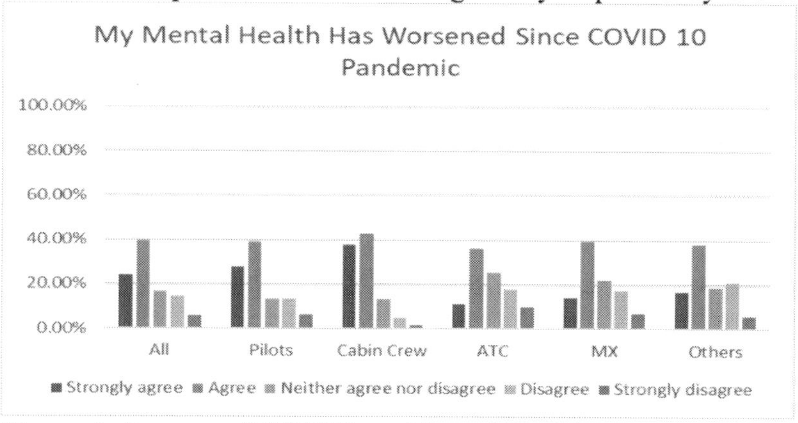

Figure 1: Impact of COVID on Mental Health

34.5% (619) of all aviation workers reported none or minimal depression. A high number met the threshold for mild depression (36%, 647), moderate depression (17.7%, 317), moderately severe depression (7.4%, 134), and severe depression (4.5%, 80). Cabin Crew appear to be most affected. Only 11% (39) Cabin Crew reported no depression symptoms.

Figure 2: Levels of Depression (Role Breakdown)

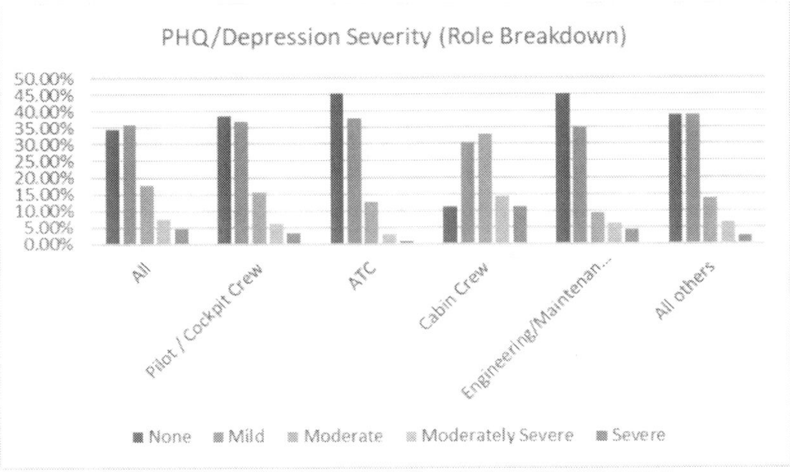

11.69% (all workers) indicate suicidal ideation, with the breakdown as follows: 10% pilots (68), 20% Cabin Crew (71), 7% ATC (15), 15% Engineering/Maintenance (21), 9% all others (35).

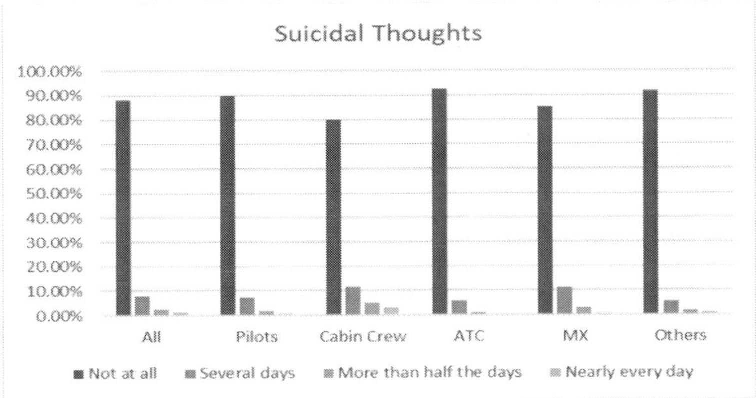

Figure 3: Suicidal Thoughts (Role Breakdown)

Overall, aviation workers reported high levels of anxiety, with 36% (646) meeting the threshold for mild anxiety, 12.8% (230) moderate anxiety, and 11.3% (203), severe anxiety. Cabin Crew are most affected, with only 13% (45) reporting feeling no anxiety.

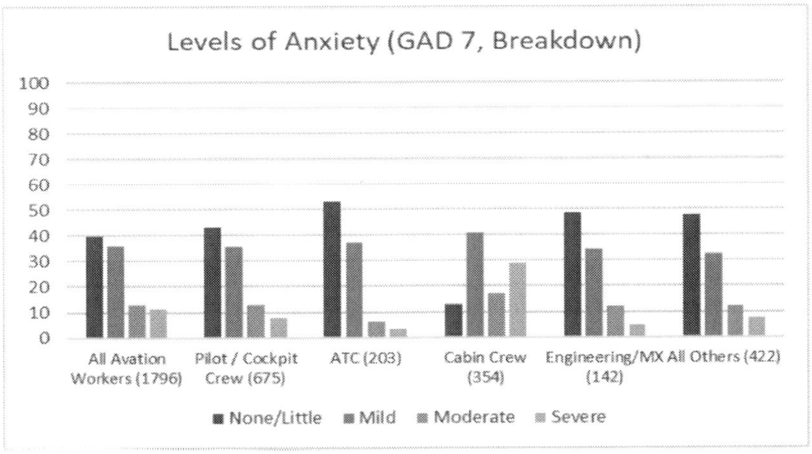

Figure 4: Levels of Anxiety

47.16% (847) indicate that over several days in the last 2 weeks they have felt down, depressed, or hopeless. 29.68% (533) indicate that over several days in the last 2 weeks they have felt bad about yourself, that you are a failure or have let yourself or your family down. 57% strongly agree or agree that wellbeing of family has been negatively affected by changes in their work situation.

Impact on Employment & Job Security

50.95 % (485) of respondents have lost jobs, with 41.41% (200) indicating that this is permanent. Of the 50.95% who have lost jobs, 81.37% (393) not secured another job. 95.07% (444) of those still employed working reduced salary and 93.36% (436) working reduced hours. Of those whose job loss is permanent, 88.94% (370) intend to return to work after pandemic, while 65.84% (239) are actively seeking reemployment within aviation. 56.70% obtaining financial support from government or another agency. Large number (68%) worrying about meeting financial obligations. Only 20% confident about future employment within aviation. A small number of aviation workers (22%) feel that the future of their company looks bright.

Impact on Performance & Flight Safety

69% of aviation workers either agree or strongly agree that changes in morale are negatively impacting on aviation worker engagement. 47% indicate that job motivation has either deteriorated or greatly deteriorated since the COVID 19 Pandemic. Overall, the majority (86%) feel they will be fit to return to work, post the COVID-19 pandemic. 63.44% indicate no change in competence and ability to do the job safely and to the required standard now, as compared to before the COVID-19 pandemic, while 25% of respondents feel their competence to do their job safely has deteriorated. 53.35% indicate that there has been no change to company safety practices since COVID 19, while 14% agree that safey pratice has greatly improved or improved. 56.63% indicate no change to company safety oversight, since COVID 19 pandemic, while 59.29% indicate no change to safety oversight from the regulator.

Coping & Seeking Help

58.27% of respondents indicated that they are using coping strategies/self-care to deal with work related stress (WRS) and wellbeing challenges since COVID. 86% feel they will be fit to return to work, post the COVID-19 pandemic. Overall, survey feedback indicates a strong willingness to seek help if had MH issue (68%), to use org supports if provided (60.14%), and to approach peer support service if provided (68.92%).

Attitudes to MH and talking about MH

Discussion of MH amongst colleagues is low – 33.86% indicated less than once per month, while 31.48% indicated never. 46.32% had previously talked to somebody (other than an employer or colleague) about a mental health issue they are experiencing/have experienced. 67% of respondents either agreed or strongly agreed that there are low levels of speaking out/reporting MH problems amongst colleagues. 78% indicate a lack of willingness to disclose MH issues to employer. Aviation workers are more likely to disclose to spouse (23%) or medical professional (22%) – low figures for Peer Support Service (2.55%) and EAP (1.52%). 59% answered trust in employer has either deteriorated or deteriorated since COVID 19 Pandemic.

Company Supports & Wellbeing Culture

23% indicate that their companies are providing supports for employees to manage wellbeing issues since COVID, but the use of these supports is very low (24.27%).

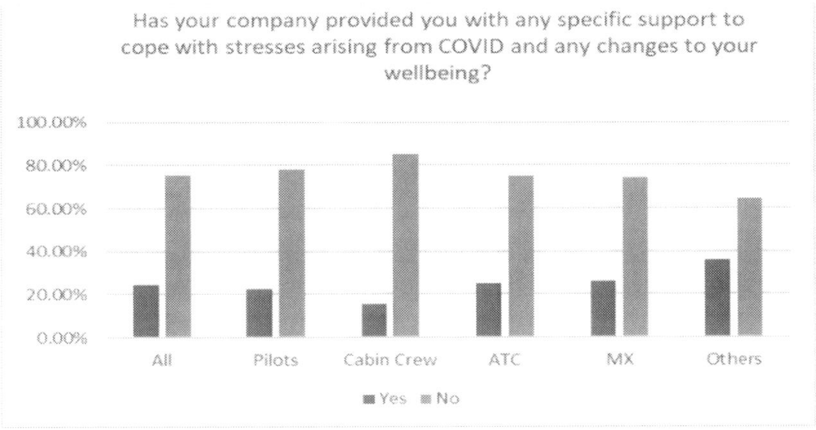

Figure 5: Company Support

A small number of respondents (24.27%) had used existing supports provided by their company to cope with the stresses arising from COVID and any changes to their wellbeing. Only 19.83% had accessed supports outside the company. A very low number of participants (19.83%) agreed or strongly agreed that their company care for employee wellbeing. 80% feel that wellbeing is not a priority for their organisations. Low number reporting supporting and maintaining positive mental health for aviation 'Safety-Critical Workers' during the COVID-19 pandemic is a priority for their company (32% strongly agree or agree). Many companies providing peers support service (69.62% aware of service in company). Almost zero access to Peer Support Programmes provided to Maintenance Engineers.

COVID Experience & Impact on Work

40% report positive impact in terms of productivity. Nearly half suggest resulted in increase in workload (47%). Just under half indicate that remote work makes it harder to achieve a work life balance (46%).

Requirements for Wellbeing Supports

94% indicate need for wellbeing supports for those currently in work, while 92% indicate that these are required for those off work.

Requirements for Fitness to Work Evaluation

61% (1003 aviation workers) indicate need for fitness to work evaluation for all people returning to work. 64% (1060 aviation workers) indicate need for fitness for work assessment for safety critical workers returning to work.

Discussion & Conclusion

In terms of prior studies – which focus on Pilots – survey feedback indicates an increase in depression prevalence. A higher number is meeting threshold for moderate depression as compared with the findings of an equivalent survey in 2018/2019 (Cahill et al, 2019, 2021), and the 2016 Harvard survey (Wu et al, 2016). There is a notable icrease in numbers at the higher end of scale. That is a small number with significant levels of depression (this applies to Pilots and other aviation workers).

Those people who have lost their jobs and/or are experiencing mental health difficulties require immediate support. The roles and responsibilities of different stakeholders in relation to managing wellbeing require rethinking and clarification. Aviation organisations need to rethink their objectives and approach to providing wellbeing supports for those currently in and off work. Organisations and workers need to manage specific sources of stress and anxiety – arising from the job and the specific impact of COVID 19 on aviation workers. Aviation workers are practising self-care – this should be encouraged. A preventative approach is required to ensure that all aviation workers are fit for duty when they return to work. Potentially, the existing supports provided to aviation workers are not fit for purpose. New tools to support wellbeing management on the part of pilots and other aviaiton workers have been proposed and might be consdiered (Cahill et al, 2020). There is a real need for aviation organisations to actively promote and enable a wellbeing culture – supporting healthy behaviour, promoting awareness of mental health and enabling workers to talk about their mental health. The regulator needs to address the timeline for new regulation in relation to the management of wellbeing and mental health for safety critical workers The results of this study should be interpreted with potential limitations in mind. Next steps will involve detailed analysis of survey data. Particpatory co-design activities will also be undertaken with different stakeholders to address wellbeing interventions at different levels.

Acknowledgements

The authors would like to thank the aviation workers for their participation in this study. Further, we would like to thank EASA, RaeS and FSF, who supported this study. The views expressed in this study do not represent the views of the authors's employers.

References

Cahill, J., Cullen P, Anwer S, Wilson, S & Gaynor K. Pilot Work Related Stress (WRS), Effects on Wellbeing (Including Mental Health) & Coping Methods. (2021). The International Journal of Aerospace Psychology, DOI: 10.1080/24721840.2020.1858714, https://www.tandfonline.com/doi/full/10.1080/24721840.2020.1858714?scroll=top&needAccess=true. Published online: 14 Jan 2021

Cahill, J.; Cullen, P.; Anwer, S.; Gaynor, K.; Wilson, S. (2020). The Requirements for New Tools for Use by Pilots and the Aviation Industry to Manage Risks Pertaining to Work-Related Stress (WRS) and Wellbeing, and the Ensuing Impact on Performance and Safety. Technologies (2020), 8, 40.

Cahill, J., Cullen P & Gaynor R. Interventions to support the Management of WRS & Wellbeing Issues for Commercial Pilots. (2019) Cognition, Technology & Work. https://doi.org/10.1007/s10111-019-00586-z

Cullen, P., Cahill, J. & Gaynor, K. (2020). Turbulent Times – How Dow We Fly Straight & Level During Turbulent Times? https://www.tcd.ie/cihs/assets/pdf/TurbulentTimes.pdf

Kroenke, K., Spitzer, R. L., & Williams, J. B. W. (2001). The PHQ-9: Validity of a Brief Depression Severity Measure. Journal of General Internal Medicine, 16(9), 606–613.

Nelson, H., Kremer, P., Anthony, T., Atherton, M., Cahill, J., Cullen, P., Drozdowski, S., Franklin, J., Gillen, M., Talacchia, M., Steinhardt, G., Vernay, A & Whittingham, D. (2020). An Aviation Professional's Guide to Wellbeing. Flight Safety Foundation. https://flightsafety.org/wp-content/uploads/2020/04/Guide-to- Wellbeing.pdf

Spitzer RL, Kroenke K, Williams JB, Löwe B. A brief measure for assessing generalized anxiety disorder: the GAD-7. (2006). Arch Intern Med. 166(10)

Wu AC, Donnelly-McLay D, Weisskopf MG, McNeely E, Betancourt TS, Allen JG. (2016). Airplane pilot mental health and suicidal thoughts: a cross-sectional descriptive study via anonymous web-based survey. Environ Health (2016) 15(1):1–12. doi: 10.1186/s12940-016-0200-6 9.

FLIGHT ALLOCATION IN SHARED HUMAN-AUTOMATION EN-ROUTE AIR TRAFFIC CONTROL

G. de Rooij, C. Borst, M.M. van Paassen, M. Mulder
Delft University of Technology
Delft, the Netherlands

Allocation is a challenge for higher levels of automation in air traffic control, where flights can be dynamically assigned to either a human or an automated agent. Through an exploratory experiment with six professional air traffic controllers, insight was gained into the possibilities and challenges of human-automation teamwork in an en-route environment. Participants showed high levels of automation trust, but mostly ignored automation-suggested allocations, preferring a highly automated sector instead. Most flights were delegated to automation, after they were given a direct and conflict-free path. Flights handled manually were those requiring level changes or non-standard routing. Future research should focus on establishing specifically which flights can be automated.

Air traffic controllers (ATCOs) work in a challenging and demanding environment. The continuous quest for more efficient and safer air travel, drives the development of more advanced automation. Both Europe and the United States aim for higher levels of automation in the coming decades with a more supervisory/strategic role for humans (Prevot, Homola, Martin, Mercer, and Cabrall (2012); SESAR Joint Undertaking (2019)). In such an environment, less people can handle more traffic in larger sectors. Despite high levels of automation, humans are expected to play an important role in supervising these future systems and to intervene when automation falls short (Metzger and Parasuraman (2005)); people will ultimately remain responsible.

To be able to intervene, it is essential that ATCOs maintain vigilance, situation awareness and a sufficient skill level to perform tasks manually (Bainbridge (1983)). This could be achieved by not making the human a supervising bystander, but have him/her work side-by-side with automation in a team, both able to perform and share tasks. This sparks the question of what such co-operation should look like, and what impact it will have on human-automation performance.

Currently, airspace is divided into sectors, each under the responsibility of a different ATCO. This requires considerable coordination between adjacent sectors and may lead to an imbalance in traffic load (and thus workload). To mitigate these issues, Birkmeier, Tittel, and Korn (2016), among others, considered so-called flight centric or sectorless operations. Instead of coupling controllers to geographic areas, a single controller would be assigned to several flights, from departure to arrival, reducing the number of handoffs and possibly providing a better workload balance. This, however, also introduces new challenges. Consider, for example, when two flights under control by different ATCOs are in conflict. Who should then solve the conflict?

What if that other controller is not another human, but an automated system? How are flights then assigned to either a controller or automation? Should all aircraft involved in a conflict be controlled by either the ATCO or the automation, so as to mitigate additional workload related to coordination? If not, who solves a conflict? In addition, with an automated agent, it becomes possible to share (sub)tasks dynamically, back and forth, between human and automation. This could establish true teamwork, but only if above-mentioned questions have been answered first.

This paper discusses an exploratory experiment on the allocation of flights in a shared human-automation en-route airspace. Control over which flights were automated was given to the ATCOs themselves, although initial automation-based suggestions were given for each flight.

Method

Participants and Apparatus

Six professional ATCOs (age M = 38.3, SD = 10.0, years of experience M = 14.8, SD = 8.7), from Maastricht Upper Area Control (MUAC) participated in a real-time simulator experiment. A TU Delft-built Java-based simulator (Fig. 1) was designed to mimic the MUAC interface, to ensure that participants could focus on working with the experimental automation. A 1920 x 1920 pixels 27" display was used with a standard computer mouse for control inputs.

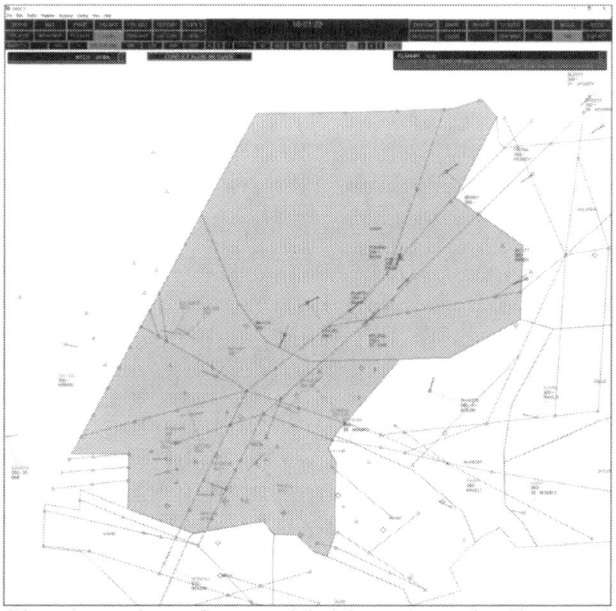

Figure 1. Simulator interface, with blue aircraft allocated to automation and green aircraft to the human ATCO. Background colors have been inverted here for clarity.

Figure 2. Callsign menu, as shown when clicking the callsign in an aircraft label. The ATCO could delegate a flight to automation by pressing "ASSUME TO AUTO". Once the flight was assumed, a "TRANSFER" button was added to the menu.

Airspace and Traffic Scenario

Participants were responsible for traffic above FL245 in the combined DELTA and JEVER sectors, above the Netherlands and part of Germany. Each ATCO experienced the same traffic scenario, resembling an average day in February 2020 (prior to the COVID-19 pandemic). There were between 15 and 30 flights in the sector at any time (M = 21, SD = 4). Flights followed standard routing or directs to their designated exit points. Besides overflying traffic, arrivals and departures to several airports, within or close to the sector, were included. There was no wind.

Automation

During the exercise, the ATCOs were accompanied by an automated "colleague". When flights entered their sector, the ATCOs had to decide whether to manually assume the flight or delegate it to automation (Fig. 2). This allocation remained flexible, such that they could re-assume manual control or delegate flights to automation at any time, anywhere in the sector. All flights had to be manually transferred to the next sector, including those delegated to automation. Automation was capable of performing the following tasks:

- Ensure sufficient separation between automated aircraft (5 NM, 1000 ft),
- Deliver aircraft at their exit point and transfer level, descending as late as possible, and
- Descend arrivals to FL260 to be transferred to lower area control.

When two automated aircraft encountered a conflict, it was always solved in the vertical plane. Automation would never issue any heading commands or direct-to's. In case of a human-automation conflict, it was up to the ATCO to solve it, under the presumption that automation would not know the ATCO's intents. Apart from showing the clearances in the aircraft labels, automation did not provide any feedback on its intentions.

Procedure

After signing a consent form, each participant received a ten-minute training, during which the automation was introduced and participants familiarized themselves with the interface. Both a human-automation and automation-automation conflict were shown to demonstrate how automation would handle both situations. The training was concluded with a short questionnaire.

Next, the measurement run started with a five-minute take-over period, during which no commands could be issued, followed by 90 minutes of real-time simulation. Each ATCO was subjected to one allocation suggestion scheme (Table 1), based on flight type or entry sector. The suggestions were shown by the label color upon sector entry (green = manual, blue = automated), but the ATCOs were not told which scheme was applied to them. In all cases, they could ignore the suggestions and re-allocate each flight at any time, even after delegating it to automation.

Throughout the run, an observer asked the ATCOs to explain their actions and what they were taking into consideration. Every three minutes, the ATCOs rated their instantaneous workload by clicking on an on-screen 0-100 scale. After the experiment, they completed an extensive questionnaire, followed by a radar replay allowing specific situations to be reviewed.

Table 1. *Suggested human-automation flight allocation strategies.*

ATCO	Basic traffic	Complex traffic	DELTA	JEVER
1	Human	Automation	-	-
2	Automation	Human	-	-
3	-	-	Human	Automation
4	-	-	Automation	Human
5	Human	Human	Human	Human
6	Automation	Automation	Automation	Automation

Note. Basic traffic has to descend/climb 2000ft or less in the sector. All other traffic is labelled as complex.

Results and Discussion

Allocation Strategies

All ATCOs delegated at least 50% and up to 100% of traffic to automation, regardless of the suggested allocation (Fig. 3). Whereas most ATCOs largely ignored the suggested allocation, ATCO-3 tried to follow it when he realized that one of the sectors was completely handled by automation. He even delegated/assumed flights as they crossed the border between the two sectors, commenting that solitary manual flights in a predominantly automated area were difficult to handle. The big drop in automation observed for ATCO-4 around 50 minutes, was caused by him purposely re-directing flights manually to "test automation" with a more complex scenario. He stated that he would have been okay with purely monitoring a completely automated scenario.

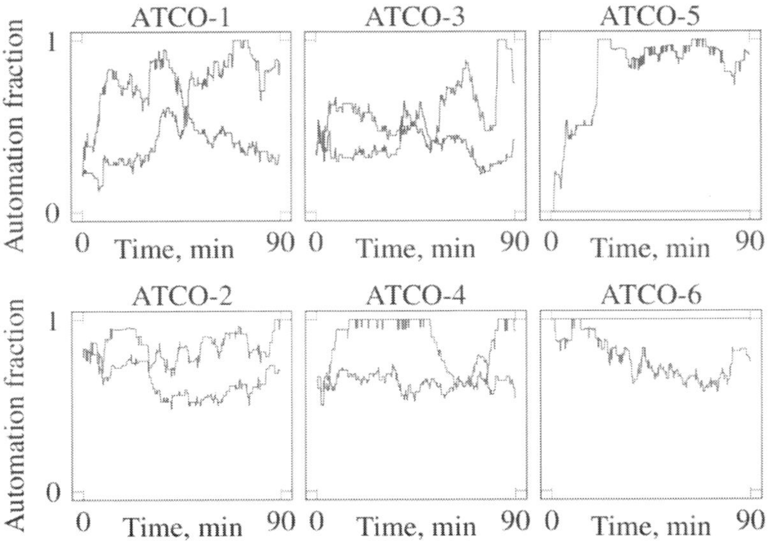

Figure 3. Time traces of the fraction of flights allocated to automation (red). The blue line shows the fraction, if the ATCOs would have followed the suggested allocation on airspace entry (see Table 1).

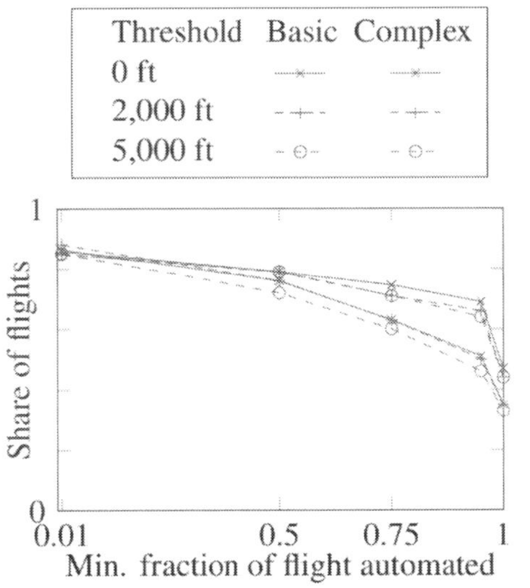

Figure 4. Cumulative share of flights that was delegated to automation for a minimum fraction of their duration, as a function of level change threshold.

When asked about all allocation strategies from Table 1, the ATCOs unanimously agreed that complex flights, here defined as requiring more than 2,000 ft level change, need to be handled manually (potentially with support tools). They indicated a strong preference for delegating basic flights to automation, which is for most ATCOs also reflected in the time that they delegate such flights to automation (Fig. 4). Although some ATCOs said 5,000 ft would have been a more appropriate choice of level change threshold, at which traffic was divided in basic and complex, this is not directly reflected in the figure. All traffic that had to change levels has evoked more manual control than overflights with zero level change and could thus be considered "complex".

Apart from this division into basic and complex traffic, the questionnaire provided more insight into how ATCOs determined whether flights should be delegated or not (Fig. 5). Traffic directly around the flight was especially important when there were many manual flights and delegating a single flight to automation would have added (too) much uncertainty. The suggested allocation was given low priority, or ignored by most ATCOs (except ATCO-3), as confirmed by Fig. 3. In general, flights were assumed manually, sent on a direct to their exit point and only delegated to automation when clear of conflicts, irrespective of the suggested allocation. If automation would have been capable of giving directs, the ATCOs would have delegated more flights.

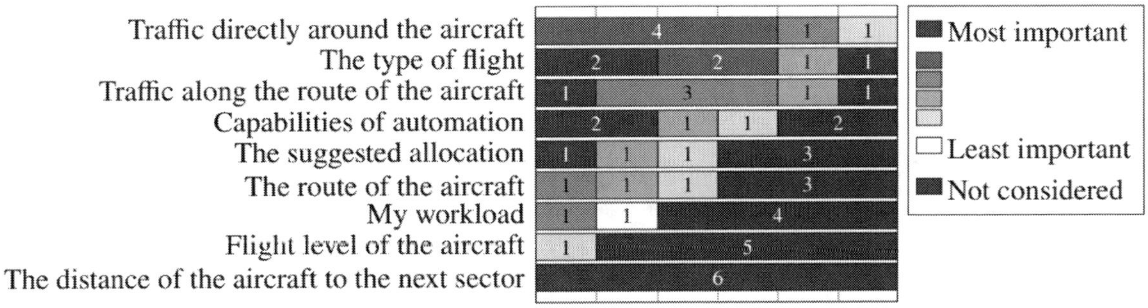

Figure 5. Driving factors that made ATCOs decide to delegate flights to automation, or not.

Trust in Automation

At the start of the experiment, all ATCOs reported to have a high level of trust in automation in general (Fig. 6). Nonetheless, they were suspicious of the experimental automation after the (short) training. Throughout the 90-minute run their trust increased considerably, according to the ATCOs mainly due to seeing the automation perform well. The rule-based form of automation (programmed to be "perfect"), clear separation of responsibilities and absence of uncertainties, such as wind and pilot behaviour, further contributed to this. ATCOs did, however, not like the lack of feedback, a common pitfall in automation design hindering the establishment of human-automation teamwork (Norman (1990)). As automation did not indicate where or when it would descend aircraft, ATCOs sometimes assumed aircraft manually, solely to prevent them from descending unexpectedly. All ATCOs would have liked automation to at least show its intentions about where on the trajectory it would start and end a climb or descent.

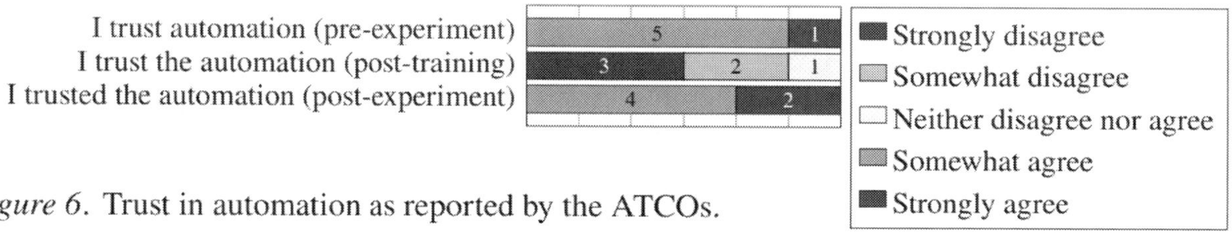

Figure 6. Trust in automation as reported by the ATCOs.

Task Allocation

While this experiment focused on aircraft allocation, a human-automation team may also be created by sharing tasks. Four out of six ATCOs included the capabilities of automation in their allocation strategy (Fig. 5). We replicated part of the study from Prevot et al. (2012), to see what kind of tasks the ATCOs would like to do themselves, share with automation or completely delegate to automation. In line with the findings of Prevet et al., the ATCOs indicated that a considerable number of tasks can be either shared with or completely delegated to automation (Fig. 7). Transfer of control can be automated as a first step towards more automation, but ATCOs should be able to reject auto-transfers as well as to initiate early transfers. The ATCOs prefer to keep short-term, tactical actions manual, while more strategic long-term planning and routine tasks can be (partially) delegated to automation. Presumably this is because automation can introduce too much uncertainty in critical short-term situations.

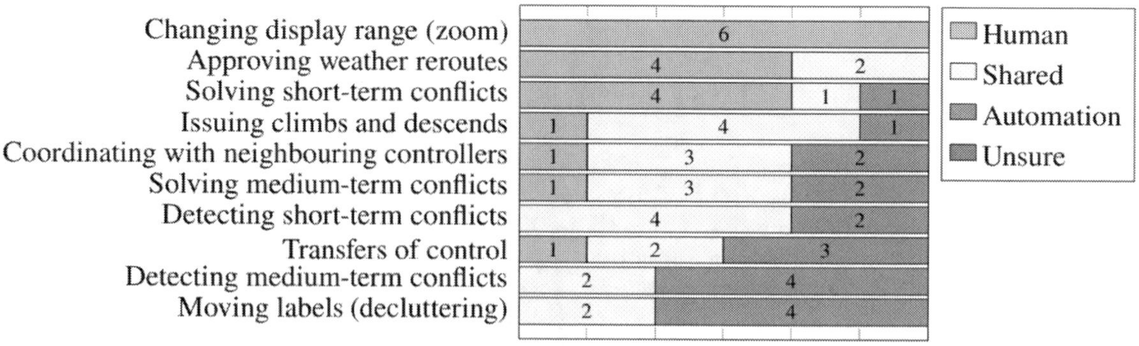

Figure 7. Allocation of tasks between human and automation as desired by the ATCOs.

Situation Awareness

All ATCOs classified their situation awareness as "okay", the middle score on a five-point Likert scale from "poor" to "very good". Several mentioned that they paid less attention to the blue automated aircraft, akin to transferred flights, even though they were still responsible for these flights. At the only (not explicitly programmed) occurrence of a human-automation conflict in the experiment, the involved ATCO was surprised by the short-term collision alert and explained that he had not seen the automated aircraft as it was emerging from, in his words, "a sea of blue aircraft". Future experiments with eye trackers could give insight in changing scanning patterns when aircraft are delegated.

Conclusion

This exploratory study gained useful insights into human-automation teaming in a realistic ATC setting. We showed that professional en-route ATCOs are not averse to sharing their work in a sector with automation. In a simplified situation, lacking uncertainties by wind, emergencies and pilot requests, a high level of delegation to automation was reached, under the condition that flights were on direct routes and free of conflicts. ATCOs generally ignored the suggested allocation, suggesting a need for a different allocation scheme that may be more accepted.

Future research should take a closer look at determining specifically which flights should be considered "basic" or "complex", such that a fitting allocation scheme can be applied. Additionally, the influence of environmental uncertainty (e.g., wind and pilot delays) and automation capabilities should be researched. Together with empirical studies on the various forms of task sharing and distribution, this can help establish human-automation teamwork in a shared ATC environment.

Acknowledgements

The authors would like to express their gratitude to all participating ATCOs, as well as to MUAC for facilitating an experiment in these challenging times of COVID-19.

References

Bainbridge, L. (1983). Ironies of Automation. *Automatica, 19*(6), 775–779. doi: 10.1016/0005-1098(83)90046-8

Birkmeier, B., Tittel, S., & Korn, B. (2016). Controller Team Possibilities for Sectorless Air Traffic Management. In *ICNS 2016: Securing an Integrated CNS System to Meet Future Challenges* (pp. 3–1). Herndon, VA, USA: IEEE. doi: 10.1109/ICNSURV.2016.7486362

Metzger, U., & Parasuraman, R. (2005). Automation in Future Air Traffic Management: Effects of Decision Aid Reliability on Controller Performance and Mental Workload. *Human Factors, 47*(1), 35–49. doi: 10.1518/0018720053653802

Norman, D. A. (1990). The 'Problem' with Automation: Inappropriate Feedback and Interaction, not 'Over-Automation'. *Philosophical transactions of the Royal Society of London. Series B, Biological sciences, 327*(1241), 585–593. doi: 10.1098/rstb.1990.0101

Prevot, T., Homola, J. R., Martin, L. H., Mercer, J. S., & Cabrall, C. D. (2012, 2). Toward Automated Air Traffic Control-Investigating a Fundamental Paradigm Shift in Human/Systems Interaction. *International Journal of Human-Computer Interaction, 28*(2), 77–98. doi: 10.1080/10447318.2012.634756

SESAR Joint Undertaking. (2019). *European ATM Master Plan - Digitalising Europe's aviation Infrastructure.* doi: 10.2829/10044

THE IMPACT OF A SYNTHETIC TEAMMATE ON SITUATIONAL AWARENESS

Aaron Teo
Arizona State University
Mesa, AZ
Kenneth Fieldhouse
Arizona State University
Mesa, AZ
Mary Niemczyk
Arizona State University
Mesa, AZ

Current commercial airplane cockpits generally require two pilots; a captain and First Officer. The autopilot has become an essential aspect of commercial flying as it can reduce the pilots' workload and increase accuracy. Since present autopilot implementations work independently from the pilot, the pilot must adopt a monitoring role. This design is not without its limitations, however, especially concerning the pilot's situational awareness. One possible relief for these limitations is the synthetic teammate, a specialized type of artificial intelligence agent intended to mimic a human operator. It integrates with current autopilot systems and provides pilots with context for action instead of merely recommendations or direct control inputs. The haptic shared control, where both the pilots and the synthetic teammate continuously maintain contact with the control surfaces, is an example of the synthetic teammate implementation. This paper aims to explore a synthetic teammate system's effectiveness in assisting a pilot's situational awareness.

Literature Review

The current commercial flight deck generally includes a captain and a First Officer (FO). Each pilot's role is unique, depending on the phase of flight. While taxiing, the captain usually controls the airplane from the gate to the runway and vice versa. The FO, on the other hand, handles the radio communications, monitors ground navigation through charts or airport diagrams, and provides an additional set of eyes for obstruction clearance. When in the air, the roles of the captain and first officer switch between either Pilot Flying (PF) or Pilot Not Flying (PNF), also known as Pilot Monitoring (PM). Generally, PF and PM roles alternate between each leg of a trip, with the captain at the controls for the first leg of the trip and the FO at the controls during the second, or vice versa. (Jentsch, Barnett, Bowers, & Salas, 1999).

Some tasks, regardless of the phase of flight, remain fixed to each ranking crew member. No matter the pilots' roles for that particular flight, the captain, the higher ranking crew member, has legal responsibility for the aircraft and its actions. Emergencies, tactical, and strategic decisions also remain with the captain, but in some situations, the captain may delegate the actual decision-making process to the FO (Jentsch et al. 1999). Pilot Flying (PF) and Pilot Monitoring (PM) have their specific roles as well. The PF manipulates the trajectory of the aircraft with the controls or via the autopilot control panel. The PM assists with radio communications, setting up instruments, preparing charts, and monitoring the PF's performance (Jentsch et al. 1999). The main reason for separating these duties is to maintain Situational Awareness (SA) throughout a flight. SA is a broad concept that includes gathering every possible bit of information about an event, analyzing it, and deciding how to react to that event. SA is the result of a human's view and awareness of the surrounding environment, which is then processed to understand the received information (Billings, 1995). To gather information and maintain SA, the pilot(s) must attend to both the outside environment and the flight deck instrumentation. While each pilot in a crew is responsible for monitoring and analyzing all of the information, communicating it to each other is critical to team-SA or Crew Resource Management (CRM) (Jentsch et al., 1999). According to Endsley (1994), to attain true situational awareness, pilots cannot simply be provided with multiple pieces of information. Instead, a higher level of understanding and prediction of the situation must be used to formulate and expect a desired result.

Why an autopilot is essential. The autopilot can help improve a pilot's SA, but it can also degrade it. The PF and PM must have proper knowledge of the autopilot's condition and must also be involved in its operation (Billings, 1995); otherwise, SA could be lost because neither human pilot is now in physical control of the aircraft. Pilot and autopilot interaction has become somewhat of a problem which can be described as lack of mode-awareness, mode-confusion, or automation surprises (Degani & Heymann, 2000). Accident and incident reports have indicated that when automation is a causal factor, there either was not enough information presented to the pilots on the status of the automation, or the pilot simply did not understand what the automation was doing (Degani & Heymann, 2000).

Simply put, an autopilot is intended to reduce crew workload so the crew can focus on other flying-related tasks. However, an autopilot's actions are always assumed by the system to be correct, but resulting actions may not be the pilot's desired result. Some pilots may try to reprogram the automation to correct the aircraft's trajectory, but that may make matters worse. Sometimes it is best to simply eliminate automation from the equation, returning to the primal state of manual flying, then bringing automation online when allowable (Curry, 1985). Faulty interactions with automation generally lead to two kinds of reactions, self-blame or blame on the system, the latter of which usually leads to a maintenance write-up claiming a malfunction (Degani & Heymann, 2000). However, Curry (1985) concludes that pilots should simply be trained to turn off the automation instead of programming their way out of an undesirable situation.

AI in the cockpit. As mentioned previously, artificial intelligence (AI) is a massive advancement that integrates into the aviation industry. That being said, it comes in the form of a technology that still requires a lot of learning, correction, and understanding. Within the cockpit, AI can be used in various ways. Button (2019) highlights that a preliminary test for researchers dealing with AI is to examine the reaction time of a pilot in control of the aircraft in contrast with an AI system's reaction time. Instances like these presumably ensure that AI has the capability to outwit even the brightest pilots. Though, that is not always the case. According to Button (2019), edge cases are a critical component to the downfall of artificial intelligence systems. They are described as things that are complex, unique, and seemingly unpredictable. It is not surprising that AI software systems have their critics. Looking back at research from Billings (1995), the claim that AI systems can't be 100% relied upon is validated. A pilot's situational awareness can be altered either positively or negatively through the use of automation. Regardless of how much good comes from the automation systems, the negatives are still there and create a potential hazard to safety.

The synthetic teammate. Autopilot systems used today typically work independently from the human operator; while the autopilot system controls the aircraft, the pilot's role becomes primarily to monitor the aircraft systems. This could result in problems when there is a misunderstanding between both the pilot and the automation. This approach could cause the pilot to lose SA. To better integrate automation into the cockpit, some have proposed implementing the synthetic teammate. Synthetic teammates are a type of artificial intelligence agent intended to replace a human operator in some cases (Doherty, 2003). The author elaborates that the idea of synthetic teammates is to provide the pilots with context for action in addition to merely providing them with recommendations or even direct control inputs. Doherty (2003) continued by listing the synthetic teammate implementation guidelines using a human-centered design approach. Firstly, the pilot must be given the final authority of the flight. Secondly, the pilot has to be informed continually of the synthetic teammate's intentions. Thirdly, the synthetic teammate's actions must be predictable. Finally, the synthetic teammate's goal must support the pilot's goal. A synthetic teammate can be implemented by using the haptic shared control method. Shared haptic control could meet the above objectives. Under the shared haptic control objectives, both the pilots and the synthetic teammate (the autopilot, for example) continuously maintain contact and control with the control surfaces such as the yoke or foot pedals. An example of this approach is to have the system create guiding push and pull forces to guide the operator away from unsafe regions such as outside of the stipulated operating boundaries of the aircraft, for example (Abbink, Mulder, & Boaer, 2011; Rosenberg, 1993).

The pilot must have the final authority of the flight. This approach places the human operator (the pilot) in a position of unchallenged authority (Abbink et al., 2011; Doherty, 2003). This concept allows the synthetic teammate to determine if the operator is about to operate their aircraft outside of its safety boundaries and intervene if necessary. In the shared haptic feedback approach, this can be done through the use of control forces (Rosenberg, 1993), such as friction or stiffness on the control surface (Abbink et al., 2011). However, Doherty (2003) stated that the system must ultimately yield full authority to the pilot by allowing him/her to override its inputs at any time.

The pilot has to be informed continually of the synthetic teammate's intentions. Keeping the pilot informed of the synthetic teammate's intentions is essential to prevent automation surprise. This approach also helps to maintain situational awareness by allowing the pilot to intervene or to react accordingly based on the action of the synthetic teammate (Doherty, 2003). For example, if a plane banked abruptly to the right as soon as the pilot disengaged the autopilot and took manual control, it is evident that both pilots were not aware of the plane's right turning tendencies and that the autopilot intentionally maintained level flight by using the ailerons and spoilerons. With haptic shared control, the pilots would be aware of the synthetic teammate's intentions as the pilot will maintain physical contact with the control surfaces and would notice strong left aileron pressure to compensate for the aircraft wanting to bank to the right.

Automation must be predictable as the pilot must anticipate the future intentions or actions of the synthetic teammate as it aids in his/her decision-making (Doherty, 2003). This predictability gives the pilot sufficient time to coordinate a response if needed. This rule can be implemented with shared haptic control by using visual or audio cues. An example could be using a digital screen that displays the synthetic teammates' future intentions or actions.

Automation must play a supporting role for the pilot. Doherty (2003) states that the ultimate goal of the synthetic teammate must be to provide support in decision-making, problem-solving, information collection, and analysis. The use of shared haptic control can accomplish this by following the rules stipulated above, such as through providing haptic guidance to the pilot if the pilot is straying away from the operational boundaries, while also yielding final decision-making to the pilot.

AI and pilot responsibilities. The introduction of AI to the cockpit, which may mimic human intelligence and reasoning, can present a conflict of duties between who is supposed to be in control of the aircraft. How do these responsibilities change, if at all? Holford (2020) begins by defining the Pilot-in-Command's (PIC) responsibility and authority. He quotes three different aviation authorities for these definitions. The International Civil Aeronautics Organization (ICAO) (2005), the Code of Federal Regulations (CFR) (2021), and the International Air Transport Association (IATA) (2020) all reference the PIC, or Commander in some cases, as the final authority as to the operation of the aircraft. IATA (2020) goes a bit further to say that the Commander "may delegate duties to qualified personnel but remains always responsible" (Clause 3.1.1).

When an airline pilot becomes a Captain, they accept enormous responsibility every time an aircraft moves. This responsibility can be divided into four separate categories, all of which are intertwined. They are causal responsibility, whether directly or indirectly; legal responsibility; moral obligation; and role responsibility (Dalcher, 2007; Holford, 2020). Due to the relation of each of these, Dalcher (2007) provides an example of moral responsibility in relation to causal responsibility. Professional pilots can always be held morally responsible if they do not act appropriately in any given situation. The PIC, quite plainly, must always have the obligation and ability to control the aircraft if danger is imminent. In order to attribute the responsibility defined by the CFR (2021), ICAO (2005), and IATA (2020), Holford (2020) mentions for this technology to be implemented safely, the operator must be able to override the system, therefore meeting all responsibilities at the onus of the pilot.

Gaps in Existing Research. Research gaps can be seen in Unmanned Aerial Vehicle (UAV) flight operations. When adding an AI component to these vehicles, it makes for a very complex piece of machinery. Ackerman (2020) further explains the correlation they have seen between humans and their AI software counterparts. Pilots with manual control of the aircraft still seem to have better agility than automation when reacting to real-world environmental changes (Ackerman, 2020). Ackerman (2020) also highlights the fact that simulator vs. real-world operations can be drastically different. For example, a pilot in training may feel no fear in taking risks (higher tolerance) due to the lack of consequences for their actions. However, a pilot maneuvering equipment in a real-world scenario is likely to take fewer risks (lower tolerance) due to the potential consequences of their actions.

Another missing piece to the puzzle is a lack of accidents strictly due to AI systems. Looking at most of the incidents or accidents that occur, it can be claimed that the pilots were not familiar with precisely how the system worked. When an autopilot system is engaged, the pilots need to be well aware of all aspects of it, including tasks that the autopilot will not do.

Conclusion

There are many use cases for artificial intelligence in the aviation industry. An autopilot is an essential tool in flying; however, it has its fair share of limitations inferring from the literature review above. The goal of the synthetic teammate is to attempt to provide the pilot with additional useful information, such as the context for action instead of just the action, which depends on artificial intelligence. Shared haptic control is a promising approach to implementing the synthetic teammate, as it objectively meets the synthetic teammate implementational recommendations laid out by Doherty (2003). Artificial intelligence systems have proven their effectiveness. However, there is still a lot to be implemented for these programs to be relied upon entirely. Already, it is evident just how big an impact automation can play on the commercial and even up-and-coming drone operations.

Future Research

Some use cases for the haptic shared control exist in the automotive industry, such as the Lane Keep Assist technology found on some modern vehicles. However, this technology has not been implemented in the aviation industry on a scale that allows its efficacy to be researched. Additional research is therefore required in this area.

References

Abbink, D. A., Mulder, M., & Boer, E. R. (2012). Haptic shared control: smoothly shifting control authority? *Cognition, Technology & Work, 14*(1), 19-28.

Ackerman, E. (2020). *AI-Powered Drone Learns Extreme Acrobatics.* IEEE Spectrum. https://spectrum.ieee.org/automaton/robotics/drones/ai-powered-drone-extreme-acrobatics.

Billings, C. E. (1995, November). Situation awareness measurement and analysis: A commentary. In *Proceedings of the International Conference on Experimental Analysis and Measurement of Situation Awareness* (Vol. 1). Daytona Beach, FL: Embry-Riddle Aeronautical University Press.

Button, K. (2019). A.I. in the cockpit. Aerospace America. Retrieved from https://aerospaceamerica.aiaa.org/features/a-i-in-the-cockpit/.

Curry, R. E. (1985). The introduction of new cockpit technology: A human factors study. National Aeronautics and Space Administration. Retrieved from https://ntrs. nasa.gov/api/citations/19850019217/downloads/19850019217.pdf.

Dalcher, D. (2007). Why the pilot cannot be blamed: A cautionary note about excessive reliance on technology. *International Journal of Risk Assessment and Management*, 7(3), 350–366.

Degani, A., & Heymann, M. (2000). Pilot-autopilot interaction: A formal perspective. *Eighth International Conference on Human-Computer Interaction in Aeronautics*.

Doherty, S. M. (2003, May). Human-centered design in synthetic teammates for aviation: The challenge for artificial intelligence. *In FLAIRS Conference,* 54-56.

eCFR. (2021). Clause 91.3 Responsibility and authority of pilot in command. Electronic Code of Federal Regulations. Retrieved from https://www.ecfr.gov/cgi-bin/text-idx?rgn=div5&node=14:2.0.1.3.10#se14.2.91_13

Endsley, M. (1994). Situational awareness in dynamic human decision making: Theory. *Situational awareness in complex systems*, 27-58.

Holford, W. D. (2020). An ethical inquiry of the effect of cockpit automation on the responsibilities of airline pilots: Dissonance or meaningful control? *Journal of Business Ethics*, 1-17.

IATA. (2020). Guidance material and best practices. International Air Transport Association. Retrieved from https://www.iata.org/contentassets/c0f61fc821dc4f62bb6441d7abedb076/guidance-material-and-best-practices-for-command-training.pdf.

ICAO. (2005). Rules of the air. International Civil Aviation Organisation. Retrieved from https://www.icao.int/Meetings/anconf12/Document%20Archive/an02_cons%5B1%5D.pdf

Jentsch, F., Barnett, J., Bowers, C. A., & Salas, E. (1999). Who is flying this plane anyway? What mishaps tell us about crew member role assignment and air crew situation awareness. *Human Factors*, 41(1), 1-14.

Rosenberg, L. B. (1993, September). Virtual fixtures: Perceptual tools for telerobotic manipulation. *In Proceedings of IEEE virtual reality annual international symposium*. 76-82. IEEE.

VIEWING AIR BATTLE MANAGEMENT THROUGH THE LENS OF INTERDEPENDENCE

Jordan E. Peterson[1], John M. McGuirl[1], Michael E. Miller[1], Danielle Bales[2], Brianna Sternitzky[3]
[1]Air Force Institute of Technology, Wright Patterson AFB, OH, USA
[2]United States Air Force Academy, Colorado Springs, CO, USA
[3]337th Air Control Squadron, Tyndall AFB, FL, USA

Recent work has shown the importance of understanding and supporting interdependence relationships among agents engaging in complex, joint activities. Building on the Coactive Design Method of Johnson, the goal of this research was to determine the impact of providing operators with real-time information of team interdependencies. It was hypothesized that allowing operators to focus on maximizing the opportunities for team synergy would result in better planning in a dynamic environment. Operators in the Air Battle Management field used a decision aid that provided information on team interdependence during three combat scenarios. Effectiveness of the decision aid was measured by expert assessment of the operator's decisions. The results of this study could help to inform future training aids and interface design for command and control systems.

Literature Review

Understanding the capabilities of a team requires an understanding of the interdependence relationships that may exist between the team members (Johnson et al., 2014). Interdependence relationships are often not obvious because they depend on the nature of the joint activities the team is conducting, which are often complex and subject to rapid change. A joint activity requires the support of interdependence relationships which "describes the set of complementary relationships that two or more parties rely on to manage [coordinate] required (hard) or opportunistic (soft) dependencies in joint activity" (Johnson et al., 2014 p.56).

These interdependence relationships occur anytime that team members must coordinate their activities to fulfill a common goal. The activity of coordination results in overhead costs including costs to diagnose and select coordination activities, communicate coordination activities, replan coordination activities and time waiting for other entities to complete prerequisite tasks (Klein et al., 2005). To relieve the individual actors of this overhead, many of these tasks are delegated to command and control (C2) structures. An example of how this plays out in a military setting is in Air Battle Management, which involves six core functions: 1) orienting shooters, 2) pairing shooters, 3) solving dynamic problems, 4) expediting decisions, 5) bringing order and 6) developing and disseminating assessments to operational command (Powers, 2018). The individuals responsible for performing these tasks are Air Battle Managers (ABMs), who must have the ability to maintain good situation awareness, perform resource allocation, and mission plan under extreme time pressure and uncertainty (Klein, 1998; Klinger and Gomes, 1993). However, this skill requires time to develop and can be difficult to master.

As part of the Coactive Design Approach for human-robot teams, a method termed Interdependence Analysis (IA) was developed to construct systems that can support the interdependent relationships that exist between human and robotic teammates (Johnson, 2014). This process uses an IA Table (IAT) that consists of a traditional hierarchical task analysis decomposition that identifies the tasks to be performed. Multiple teammates having capacities required for completion each task/subtask, including situation awareness information, knowledge, skills, and abilities; are assigned to each task. The table further provides an enumeration of viable team role alternatives along with an assessment of the member's capacity to perform and capacity to support the associated taskwork. The table employs a color code that helps identify potential interdependence relationships among a primary performer and supporting agents as shown in Table 1.

Table 1. **Interdependence Color Scheme, adapted from (Johnson et al., 2014).**

Team Member Role Alternatives	
Performer	Supporting Team Members
I can do it all	My assistance could improve efficiency
I can do it all but my reliability is < 100%	My assistance could improve reliability
I can contribute but need assistance	My assistance is required
I cannot do it	I cannot provide assistance

While the IA method has proven to be an effective tool for design engineers when developing human-robot teams, this research seeks to extend this work and investigate the utility of an IAT as a decision aid, capable of supporting operator awareness and management of team interdependencies as they evolve in real time. Specifically, this work seeks to apply the interdependence analysis concept and an interdependence table-like representation to represent the interdependencies among aircraft within air battle management scenarios. The utility of this tool is then assessed by having newly trained ABMs perform the air battle management task both with and without the representation.

Methodology

Participants

Eight recent graduates of the Undergraduate ABM training course participated in the study. They had an average of five months experience post Undergraduate training as ABMs, but no experience with operational missions. Half of the participants were randomly assigned to either the control or experimental group.

Scenarios

Three mission scenarios were developed in collaboration with a subject matter expert (SME). Each scenario presented the operator with unique challenges based on the nature of the task.

The first scenario was an offensive mission with a defended, stationary target. It was defined as a time critical target (TCT) with a limited window of opportunity to be destroyed due to the nature of the threat. Updates regarding the nature and number of defensive units were a major complicating factor as they could alter which aircraft was best suited to conduct the strike. Mechanical issues to certain assets also complicated the asset-target decision process. The second scenario was an offensive mission requiring a precision strike on a defended, moving target as it transitioned through areas of varying risk of collateral damage. Depending on the location, the number of strike options would vary. This scenario was also designed to trigger a call to abort the mission as a result of the last update. The third scenario was a defensive mission that focused on protecting a high value asset (airfield) against an unknown number of airborne adversaries. The evolving weather in the area had the potential to interfere with air operations and adversely impact sensor capabilities.

Apparatus

All participants were provided with all of the information that is normally available during a mission to make decisions on assigning assets to mission tasks, such as the mission objectives, physical map of the area of operations indicating objectives, position of friendly and known adversaries, the fuel and weapons status and current assignment of each asset. In addition to this, the experimental group also received the IAT decision aid as shown in Figure 1.

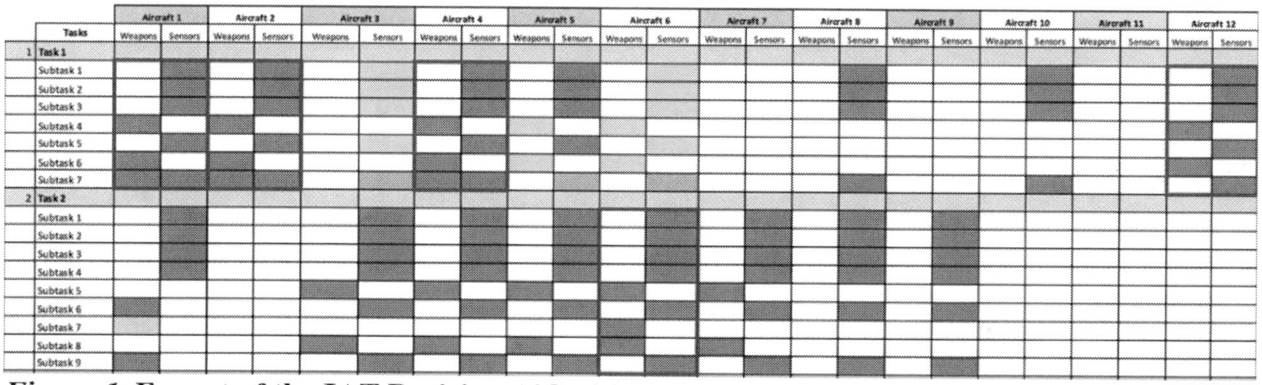

Figure 1. **Except of the IAT Decision Aid with generic entries for tasks and aircraft.**

The IAT was designed as a decision aid to support the operator by highlighting team interdependencies in real-time, specifically those for resource allocation and planning purposes. It was developed from the use of several Excel macros. There were five main parts to developing the decision aid: 1) dissecting the mission objectives into subtasks, 2) identify assets and their capabilities, 3) color-coding the IAT based on the most recent mission update, 4) restricting the capabilities of assets based on the mission timeline and 5) recommending the most capable asset to the operator. The color of a cell mapped the ability of the current asset weapons load out and sensor status to the selected task. To ensure an operator could not assign an asset to two mission objectives occurring at the same time, the macro would grey out the other mission objective rows if the asset was assigned to a task. This feature helped the operator keep track of their resource allocation. Lastly, the IAT made recommendations to the operator by outlining the most capable assets to fulfill a mission objective in a dark blue. The goal of this feature was to help the operator save time during assignment of resources to address a time critical target.

Procedure

The experiment was conducted through Microsoft Teams and took approximately 90 minutes, including a 15 minute briefing, a 60 minute simulation, and 15 minute debriefing. The experimenter acted as the Air Commander and provided additional information or clarifications as needed. Each scenario contained ten mission updates designed to trigger critical decision points around the status of enemy and friendly forces, weather, and other decision factors. The participants were asked to verbalize their thought process while making any necessary adjustments to aircraft assignments or ordering the mission to be aborted if deemed necessary. A debriefing followed to provide further insight into the decision-making process and situation awareness of the mission scenario.

Results

The performance errors among the results were classified into four categories: 1) *Mission Asset Pairing* in which the ABM assigned a mission objective to an aircraft that was better suited for another aircraft, 2) *Crew Coordination* in which the ABM did not properly utilize the interconnected capabilities of assets 3) *Knowledge Gap* in which they made an inadequate decision due to a knowledge gap of necessary information, and 4) *Assumption Error* in which the ABM assumed inaccurate information. The performance of the experimental and control group were analyzed for common errors and compared against the correct predicted response from a SME ABM. No one participant made more than three errors per scenario. The results revealed that the control group made more errors of all types in total and across each mission.

Figure 2 shows the results from scenario one, which involved an offensive mission with a stationary target. No one in the control group completed the mission. Three of four participants aborted the mission by Update 9. The final participant was unable to successfully select an aircraft to perform combat assessment of the target during the tenth update. In comparison, all four of the participants in the control group successfully completed the mission. These results highlight the utility of the decision aid to help the participant keep track of their assets as the mission evolves.

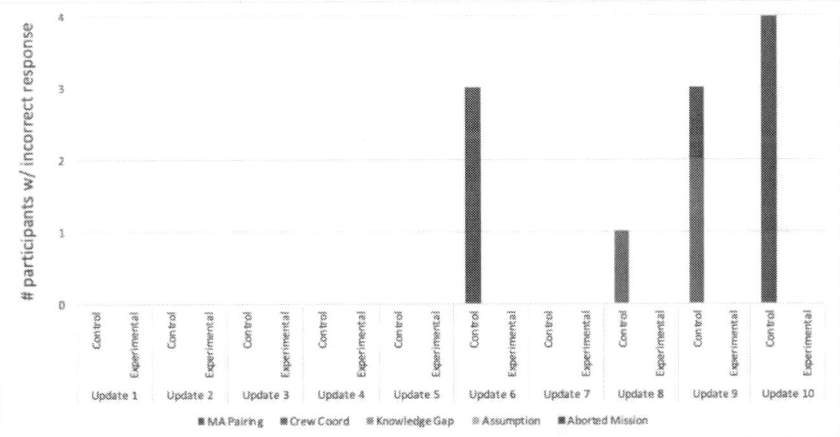

Figure 2. **Participant's Errors during each mission update for Scenario One, color indicates error type.**

During the second scenario, involving the precision strike on a moving target, all participants performed well until the last update. No errors were made by the experimental group. However, three of four participants in the control group made a knowledge gap error on Update 10. This update changed the capabilities of assets due to inclement weather. The Air Commander informed all participants that the target was unable to be detected by any aircraft. Participants in the experimental group were able to use the decision aid to recognize the environmental effects on their asset capabilities. This led to four of four participants making a decision that aligned with the SME's assessment. However, three of four participants in the control group left aircraft hovering over the target in extreme weather conditions due to knowledge gap of aircraft weather capabilities. The responses to this update emphasize how the decision aid can be useful for novice trainees with knowledge gaps from training when making operational decisions.

The third scenario, which focused on defense of an airfield, resulted in the most performance errors. During this scenario, as the updates occurred, the participants were presented with more and more enemy aircraft in the airspace, eventually leading an overwhelming large number of enemy aircraft to be tracked and targeted. The experimental group was able to quickly recognize which assets were able to perform air-to-air defense, while most control group participants were hesitant and made inaccurate assumptions. These results suggest the decision aid was helpful for resource allocation in a task saturated environment.

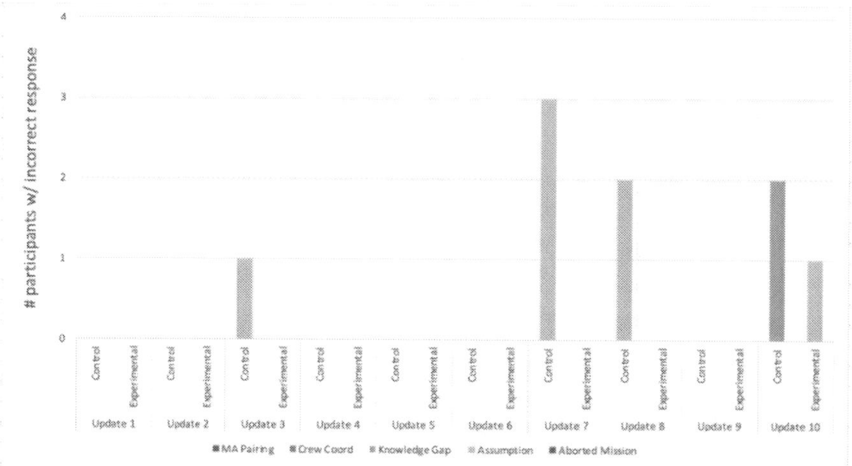

Figure 3. ***Participant Errors during each mission update for Scenario Three, color indicates error type.***

Discussion and Conclusion

As shown in Figures 2 and 3, participants in both groups generally performed well initially in all of the scenarios, however, as the missions continued, more participants in the control group struggled to keep track of their asset capabilities, perform efficient resource allocation, and mission plan. For example, in scenario one, as the number of updates increased so did the number of errors for the control group. It became very difficult for these individuals to keep track of their asset capabilities, which resulted in aborted missions. In scenario two, the Air Commander had more control of assigning aircraft to tasks, which led to fewer errors. However, on the last update, three of the four participants in the control group lacked knowledge of asset weather capabilities and made poor decisions. Lastly, on scenario three, several assumption errors occurred due to the defensive mission type. Students are trained to target enemy aircraft when they reach a particular area of engagement. These participants made inaccurate assumptions about enemy locations and aircraft weapon capabilities. These results highlight how the decision aid was able to support all of these decisions.

Feedback from the participants suggested that having the information on how team interdependencies were changing over time improved their situation awareness, enhanced their resource allocation decisions and ability to plan missions. They also stated that the aid helped them understand how their time critical decisions can have cascading effects on the ability to accomplish competing tasks, ultimately saving time, resources, and increasing resilience.

Future Work

While the results showed some promise for this approach, it was limited to a specific domain and a small subset of AF operators. Future work should focus on increasing the fidelity of the interface, incorporating more complex scenarios, including multiple participants, and potentially artificial agents.

Acknowledgements and Disclosure

The views expressed in this article are those of the authors and do not necessarily reflect the official policy or position of the Department of the U.S. Air Force, U.S. Department of Defense, nor the U.S. Government. The authors gratefully acknowledge the project sponsor Winston Bennett, 711th HPW/RHW, and the study participants for their support of this work.

References

Johnson, M., Bradshaw, J., & Feltovich, P. (2014). *Coactive Design: Designing Support for Interdependence in Joint Activity.* Pensacola: Florida Institute for Human and Machine Cognition (IHMC).

Klein, G. (1998). *Sources of Power: How People Make Decisions.* Leadership and Management in Engineering.

Klein, G., Bradshaw, J.M. (2005). *Common Ground and Coordination in Joint Activity*, in: W.B. Rouse, K.R. Boff (Eds.), Organizational Simulation. Wiley-Interscience, Hoboken, NJ: pp. 139–184.

Klinger, D. W., & Gomes, M. E. (1993). *A Cognitive Systems Engineering Application for Interface Design.* Proceedings of the Human Factors and Ergonomics Society Annual Meeting.

Powers, R. (2018). *Air Force Officer Job: 13BX Air Battle Manager.* Retrieved from The Balance Careers: https://www.thebalancecareers.com/us-air-force-air-battle-manager-job-3356660 (November 15, 2020).

ORGANIZATIONAL SAFETY IN AIRLINE OPERATIONS

Esa Rantanen[1] and Erik-Jan Huijbrechts[2]

[1]Rochester Institute of Technology,
Rochester, NY, USA

[2]Airline Pilot and Independent Aviation Safety Researcher,
Sassenheim, the Netherlands

Organizational accidents is a category of accidents caused by organizational factors. They are rare but have widespread consequences, many defenses and multiple causes, they are associated with judging and deciding, and have a long "history". Organizational accidents are also associated with highly regulated industries, such as aviation. There are several other constructs related to organizational factors undermining safety. Aviation is unfortunately closely associated with the traditional "Safety I" thinking, where adverse outcomes can be found at the end of causal chain and treating, and preferably eliminating, the causes will increase safety by preventing future accidents. An alternative view is "Safety II", where the focus is on what goes right rather than on what goes wrong. Safety II is thus defined as the ability to succeed under expected and unexpected conditions alike. Yet another viewpoint is Work-As-Imagined vs. Work-As-Done. In the former case, the focus is on accident and incident investigation and elimination of non-compliance errors ("name, blame, train"). In contrast, the latter perspective does away with both well-defined functions and malfunctions and accepts that performance is variable as a matter of fact and that the same variability can result in both success and failure. Because performance variability allows for adjustment to changing situations, it is the reason why everyday work is safe and effective.

Commercial aviation is an immensely large and complex system but also one with an unparalleled safety record. One can fairly claim that it is just that complexity that may be credited for the safety of the system. After all, the aviation industry has a relatively short history of just under 120 years (counted from the Wright brothers' first flight) and the rapid development of bigger and faster aircraft has been accompanied with equally rapid development of safety procedures and regulations. Yet, we know that complexity presents its own, inherent, hazards to safety.

Another hazard to safety comes from the distance of aviation managers and regulators from the operations. Through the years the focus of management of aviation companies has migrated from technical/operations oriented to legal/process oriented. In small companies talented workers can still develop management skills and move up to the management level, retaining their good understanding of operational processes. In larger companies managers often originate from management schools or other, sometimes not aviation related, companies. Their focus is to manage the processes within the company with guidance from legal requirements rather than from technical and operational expertise. This development has had a detrimental effect on safety in aviation.

In this paper we examine several concepts related to systemic complexity, poor communications, and management attitude and their impact on aviation safety. Examples are given from operational experience of the second author.

Organizational Factors

Scientific Management is a theory formulated in the late 19th and early 20th century to increase efficiency of work and decrease waste (Taylor, 1914). It introduced empirical methods to study work as it actually took place (i.e., Work-As-Done, WAD), with the intention to prescribe the "one best way" of doing it (i.e., Work As-Imagined, WAI). The principles of scientific management were to analyze tasks to determine most efficient performance, select people to achieve best match between task requirements and capabilities, train people to ensure specified performance, and ensure compliance by economic incentives. The WAI vs. WAD juxtaposition is also evident in behavioral sciences as well. For example, compare the idea of a homo economicus (Mill, 1848) and its assumptions of man completely informed, with infinite sensitivity, and thoroughly rational, and a decision theory that assumes that all options, outcomes and preferences are known and amenable to evaluation (i.e., WAI), to the Naturalistic Decision Making (NDM) properties that include people "muddling through" decisions and satisficing, with a decision theory that recognizes that most situations have incomplete, dynamically changing conditions and competing goal structures (i.e., WAD; Klein, Calderwood, & Clinton-Cirocco, 1985; Klein, 1999). The most recent WAI vs. WAD formulation seems to have appeared first about 10 years ago and it is mentioned by at least 3 authors (Sidney Dekker, David Woods, and Erik Hollnagel) in an edited book (Hollnagel, Woods, & Leveson, 2007). Hollnagel has further developed the WAI and WAD constructs in the healthcare domain (Hollnagel, Braithwaite, & Wears, 2013; Braithwaite, Wears, & Hollnagel, 2015).

Another useful dichotomy, or rather a continuum, is between what are known as "The Sharp End" and "The Blunt End" in an organization. The operators, or people who do the actual work (e.g., pilots), are in The Sharp End, the authorities and regulators in The Blunt End, and the management somewhere in between. In The Blunt End work is being managed by schedules and norms, which describe and prepare work for others, and managing how others do their work (quality controls, productivity standards). Production planning (e.g., "lean" optimization) also takes place here, as does monitoring and managing actions (sampling, level of detail) and investigations and auditing (errors, compliance). People in The Blunt End are regulators and arbiters of right or wrong, but have limited exposure to the actual work.

As organizations operate in time, the gap between The Sharp End and The Blunt End has serious consequences to the information flow between the ends, and the people along the continuum. Two important temporal variables within this model are the time for people to find out what is happening, and the "half-life" of information as it becomes obsolete. A gap between The Sharp End and The Blunt End results in reciprocal (mis)understanding: WAD is what I/we do, whereas WAI is what they (should) do. This creates an "Us" vs. "Them" dynamic in the organization, which is detrimental to safety.

Organizational impact to safety has a relatively short history. The term organizational accidents was coined by Reason (1997) in a book with the same title. Two kinds of accidents may be identified. Individual accidents are frequent, have limited consequences, have few or no defenses, have limited causes, consist of slips, trips and lapses, and have short "history" (i.e., the cause precedes the accident within a very short time frame). Organizational accidents, on the other hand, are rare, have widespread consequences, have many defenses and multiple causes, are associated with judging and deciding, and have a long "history". As such, organizational accidents are much harder to investigate than individual accidents. Consequently, many accidents with organizational "roots" are not investigated as such, but as individual accidents. Unfortunately, such blaming of the flight crew members has a long and fruitless history in aviation safety.

Another type of accidents associated with systems is so-called "normal accidents", also from a book with the same title (Perrow, 1984). These accidents result from multiple and unexpected failures that are built into society's complex and tightly-coupled systems. Tightness of coupling indicates how fast cause and effect propagate through the system. Systems with loose coupling have slack along one or more dimensions such as time or space. Complexity indicates not only how many interactions there are but how hard they are for the operators of the system to see and understand. The tight-complex systems present fast-moving events to operators who may be too overwhelmed by them and react too slowly to rapidly propagating events. Automatic systems installed to react faster may be unreliable or give confusing signals (Whitney, 2003).

A Novel Approach to Safety

Traditional approaches to safety have followed the "causality credo", or a reasoning that adverse outcomes (accidents, incidents, etc.) happen when something goes wrong; adverse outcomes therefore have causes, which can be found; treating or eliminating the causes will increase safety by preventing future accidents. Such reasoning is known as "Safety I" thinking (Hollnagel, 2018). It relies on manifestations, that is, observable adverse outcomes, mechanisms such as assumptions (e.g., the "causality credo" above), and that a root cause for every accident can be determined (e.g., Root Cause Analysis; RCA). Its theoretical foundations lie in the notion that systems are decomposable and accident investigation methods that are based on decomposition of the system to elemental parts.

An alternative approach is to view safety as a "dynamic non-event", where reliability is achieved by constant compensation for changes in system components (Weick, 1987). It is invisible because people often do not know how many mistakes they could have made but did not and have only a vague idea of what produces reliability and how reliable they are. Reliable outcomes are constant, which means there is nothing to pay attention to. Operators see nothing, and seeing nothing, presume that nothing is happening. If nothing is happening and if they continue to act the way they have been, nothing will continue to happen (if it ain't broke, don't fix it). But this diagnosis is deceptive and misleading because it is dynamic inputs that create stable outcomes (Weick, 1987). A possible solution to this problem is to define safety as "a dynamic lack of failures", which is also labeled "Safety-II" (Hollnagel, 2017, 2018). According to this thinking, the focus is on what goes right rather than on what goes wrong. Definition of safety is changed from "to avoid or prevent that something goes wrong" to "to ensure that everything, or as much as possible, goes right". Safety II is thus defined as the ability to succeed under expected and unexpected conditions alike, so that the number of intended and acceptable outcomes (i.e., everyday activities) is as high as possible.

Consider the different perspectives of WAI and WAD to accident and incidents. The "normal" operation, where well-defined work functions, supported by barriers, regulations, procedures, and compliance result in success and acceptable outcomes. Malfunction is viewed as a non-compliance error, that results in accidents, incidents, and unacceptable outcomes. In the WAI case, the focus is on accident and incident investigation and elimination of non-compliance errors ("name, blame, train"). In contrast, the WAD perspective does away with both well-defined functions and malfunctions. Instead, the WAD perspective accepts that performance is variable as a matter of fact and that the same variability can result in both success and failure. Performance variability is inevitable, ubiquitous, and necessary, and because of resource limitations, performance adjustments will always be approximate. Because performance variability allows for adjustment to changing situations, it is the reason why everyday work is safe and effective.

A New Triad

In airline operations, the flying personnel in The Sharp End are not only dealing with airline management in The Blunt End but also aviation regulators and, as is increasingly the case, with aircraft manufacturers, which form a new triad, adding to the complexity of the air transportation system. Certification of aircraft and equipment, and the coupling or aircraft manufacturing and authorities certifying aircraft, was designed to check compliance with legal regulations and to see if possible errors may be overlooked after a manufacturer has developed an aircraft or aircraft system.

However, when certification becomes a goal rather than a check the process becomes counterproductive. If adapting an existing system simplifies the certification process compared to developing a new system, design goals may shift from delivering a safe system to squeezing them into an existing format. For example, Boeing's desire to market the 737 Max aircraft as requiring no additional training for flight crews because its identical handling with the previous generation 737 NG, albeit with the help of sophisticated automation known as Maneuvering Characteristics Augmentation System (MCAS), resulted in two unfortunate accidents (Lion Air Flight 610 on October 29, 2018, and Ethiopian Airlines Flight 302 on March 10, 2019), costly grounding of the aircraft, and redesign of its control laws software (Wendel, 2019).

The blurring of the roles of manufacturers and regulators (certifying agents) further contribute to distancing The Sharp End from The Blunt End. Manufacturers' expertise used for certification introduces a bias to speed up the certification process. Certifying agents often lack expertise and manpower to fulfill their role in the process. Once certification is acquired it may become a barrier to system improvement. Legally-oriented managers may ignore concerns from technical and operational experts so as not to jeopardize the system's certification status. For liability reasons most operators want their crews to follow manufacturers' procedures exactly. This Safety-I thinking. The assumed operational knowledge of a manufacturer is often overestimated. A manufacturer is not an operator and it is the responsibility of an operator to have procedures to be tailored to fit into their operation. Good guidance can be found on how to design flightdeck procedures (Barshi, Mauro, Degani, & Loukopoulou, 2016) and over time changes in common practices, systems, and materials may require adaptation of procedures and regulations. This is Safety-II thinking.

Technical and operational experts, that is, people in The Sharp End, can sometimes foresee operational problems and invest in systems or design procedures that can prevent them. Legally-oriented managers in the Blunt End must be prompted by regulations or operational/safety related incidents to take action. When a company is divided in departments with their own financial responsibility it is even more important that management oversees the consequences such savings in individual departments have on other departments and safety.

Examples

(1) When new aircraft were added to a fleet, a line pilot was consulted about what modifications were needed for making the aircraft operational. This pilot advised the airline management to install a system needed for dispatch in certain environmental conditions. However, the management sought a "second opinion" from another pilot who did not see this need based on his experience in another company abroad. The system was not installed. Only after several flight cancellations the system was installed at higher costs than initially required.

(2) A service bulletin for sealing of electrical connectors was rated as "highly recommended" instead of "mandatory" by the manufacturer of an aircraft. Only after an emergency return of one of the aircraft with a short circuit and smoke in the passenger cabin sealing of electrical connectors was performed.

(3) When remote de-icing started to replace gate de-icing manufacturers' procedures had to be adapted to cover the new situation.

(4) When computer performance calculations replaced "paper" calculations safety margins were diminished resulting in lower safety levels for some take-off procedures. Although the risk is now recognized by authorities, regulations are yet to change to cover the situation (Huijbrechts, Koolstra, & Mulder, 2019). It is difficult to convince management to change procedures just for the benefit of flight safety, even if the cost of change is marginal.

(5) The introduction of Type-4 de-icing fluid (thixotropic) resulted in clogging of drain-holes. Freezing of trapped water on subsequent flight resulted in incidents with jammed flight controls and gears. Extra maintenance procedures had to be introduced to prevent this.

Conclusion

There is a widening gap between the triad of airline management, aircraft manufacturers, and aviation authorities and technical and operational experts, driven by increased role of manufacturers in the certification of aircraft and system. Airline management often side with manufacturers and regulators in enforcing procedures without always adequately understanding the operational processes and consequences of such decisions. Incidents or even accidents have to occur first before procedures and systems are re-evaluated and corrected. Yet, accident investigations do not always reveal the real cause of an accident. Although organizational factors are increasingly recognized in investigations, manufacturers and air- lines usually are part of the investigation team. If a system or procedure fault plays a role in the cause of the accident this can result in liability claims. Investigation teams hence tend to focus on errors made by the individual crew members to avoid such responsibility. To improve flight safety at manufacturer and airline level, a mechanism has to be found to make managers feel a responsibility for the quality and safety of their product that goes beyond just fulfilling legal requirements and following processes correctly. A first step is to improve communication between management and technical and operational experts.

References

Barshi, I., Mauro, R., Degani, A., & Loukopoulou, L. (2016). *Designing flightdeck procedures* (Tech. Rep. No. NASA/TM-2016-219421). NASA.

Braithwaite, J., Wears, R., & Hollnagel, E. (2015). Pathways towards reconciling WAI and WAD: Reconciling Work-as-Imagined and Work-as-Done. In E. Hollnagel, J. Braithwaite, and R. L. Wears (Eds.), *Resilient Health Care*. (p. 171-177). Ashgate Publishing, Ltd.

Hollnagel, E. (2017). *Safety-II in practice: Developing the resilience potentials*. Taylor & Francis.

Hollnagel, E. (2018). *Safety-I and safety-II: the past and future of safety management*. CRC press.

Hollnagel, E., Braithwaite, J., & Wears, R. L. (2013). *Resilient health care*. Ashgate Publishing, Ltd.

Hollnagel, E., Woods, D. D., & Leveson, N. (2007). *Resilience engineering: Concepts and precepts*. Ashgate Publishing, Ltd.

Huijbrechts, E.-J. A. M., Koolstra, H. J., & Mulder, J. A. (2019). Using V_{mcg}-Limited V_1, controllability issues on contaminated runways and in crosswind. *Journal of Aircraft, 56*(4), 1342–1352.

Klein, G. (1999). *Sources of power: How people make decisi*ons. MIT press.

Klein, G., Calderwood, R., & Clinton-Cirocco, A. (1985). *Rapid decision making on the fire ground* (Tech. Rep. No. KA-TR-84-41-7). U.S. Army Research Institute for the Behavioral and Social Sciences.

Mill, J. S. (1848). On the definition of political economy, and on the method of investigation proper to it. *The Collected Works of John Stuart Mill, III–Principles of Political Economy Part II.*

Perrow, C. (1984). *Normal accidents: Living with high risk systems*. Basic Books.

Reason, J. (1997). *Managing the risks of organizational accidents*. Ashgate.

Taylor, F. W. (1914). *The principles of scientific management*. Harper.

Weick, K. E. (1987). Organizational culture as a source of high reliability. *California Management Review, 29*(2), 112–127.

Wendel, W. B. (2019). Technological solutions to human error and how they can kill you: Understanding the Boeing 737 Max products liability litigation. *Journal of Air Law and Commerce, 84*(3), 379–444.

Whitney, D. E. (2003). *Normal Accidents by Charles Perrow* (MIT Working Paper Nos. ESD-WP–Lit-2003-01). MIT.

ASSESSING THE RELATIONSHIPS BETWEEN ORGANIZATIONAL MANAGEMENT FACTORS (4Ps) AND A RESILIENT SAFETY CULTURE IN A COLLEGIATE AVIATION PROGRAM WITH A SAFETY MANAGEMENT SYSTEMS (SMS)

Daniel Kwasi Adjekum, PhD., CSP.
Department of Aviation
University of North Dakota
Grand Forks, North Dakota. U.S.A

Resilient safety culture is a key factor in sustaining safety management systems (SMS) in a U.S. collegiate aviation program. The relationships between four organizational management factors (*Principles*, *Policy*, *Procedures*, *Practices*) and a resilient safety culture model based on Reason's concept was assessed using an online survey instrument. Structural Equation Model (SEM) technique were used to assess measurement models of factors underlying a resilient safety culture. All four management factors had significant predictive relationship with resilient safety culture. *Practices* had the weakest predictive relationship with resilient safety culture and *Policy* had the highest. *Procedures* strongly mediated path between *Policies* and *Practices* and there was no significant predictive relationship between *Principles* and *Practices*. Results suggest a good resilient safety culture within the aviation program even though more focus should be placed on *Practices*. Study adds to the paucity of existing literature on resilient safety culture in U.S. collegiate aviation programs.

A resilient safety culture is a characteristic of an organization that has good safety procedures and practices which enable it to have greater resistance to incidents and accidents, as well as being able to cope better when they occur (Hollnagel, Paries, Woods, & Wreathall, 2011). A safety resilient culture is essential for Safety Management Systems (SMS) which is a formal, top-down, organization-wide approach to managing safety risk and assuring the effectiveness of safety risk controls. It includes systematic procedures, practices, and policies for the management of safety risk (ICAO, 2013). Within the scope of an SMS, a resilient safety culture is known to be reflected in proactive and resilient behaviors of personnel in an organization and also serves as indirect indicator of good organizational management factors (Schwarz, Wolfgang, & Gaisbachgrabner, 2016). Resilient safety culture within the aviation environment has been promoted through extant research (Akselsson, Koorneef, Stewart & Ward, 2009; Reason, 2011; Hollnagel, 2014; Schwarz, Wolfgang, & Gaisbachgrabner, 2016) and the findings of these research advocate for robust and resilient safety systems as the next level in an organization with a fully functional SMS program in place. Some collegiate aviation programs in the U.S. have adopted the SMS voluntary program (SMSVP) facilitated by the Federal Aviation Administration (FAA). The SMSVP provides immense benefits in terms of enhanced proactive safety risk management and a resilient safety culture sustained in their operations (Adjekum, 2014; Adjekum, 2017). Under the SMSVP, a certificate holder that satisfies all the rigorous requirements of SMSVP may be recognized by the FAA and designated an *active conformance* status (FAA, 2015). A key performance metric of a functional and mature SMS is a resilient safety culture under all operational conditions (Paries, Macchi, Valot & Deharvenght, 2018). Measuring resilient safety culture is an essential part of SMS and a path towards continuous monitoring and improvements of organizational safety (Stolzer, Friend, Truong, & Aguiar, 2018). Reason (2011) provides a measurement strategy through a conceptual model of a resilient safety culture engine that drives an organization's safety program based on the Degani and Wiener (1991) model of organizational management factors namely; *Principles*, *Policies*, *Procedures* and *Practices* (4Ps). The aim of this study was to assess hypothesized measurement models showing the strength of relationships between a resilient safety culture and 4Ps management factors in a collegiate aviation program

Research Design & Results

Quantitative survey items modified from Reason's attributes of a proactive safety resilient organization (Reason, 2011) were sent via an anonymous online survey link to all personnel (aviation students, certified flight instructors, faculty, maintenance personnel, dispatch, administrative, and top-management) in a collegiate aviation program located in the mid-Western part of the United States that has an *active conformance* SMS. The collegiate aviation program had a population of 1695 and there were 516 responses (~ 31% response rate) at the end of the survey period which is adequate for most internal online surveys (Tse-Hua & Xitao, 2009). Generally, the response numbers as compared to the non-response suggest minimal response bias. The results show a mean age of about 23 years (M =22.94, SD = 7.944) the modal class being the 20-year old respondents and the highest age being 67 years. There were 396 male respondents (76.7%) as compared to 120 female respondents (23.3%).

Question One

What is the strength of relationships between measurement scale items and their latent management factors (Principles, Policy, Procedures and Practices)?

Principles. According to Reason (2011), *Principles* are a corner stone of policy framework, operational procedures and "sharp-end" practices in aviation organizations. It is determined by an organization's management and becomes a conclusive statement on how operations at the organization is conducted. A resilient safety culture in an organization has an impact on strategic principles, which may not always be clearly stated but will be inferred from procedures, policies and practices (Degani & Wiener, 1991). An item under *Principles* is "Safety is recognized as being everyone's responsibility not just that of the safety management team". A first-order CFA which allows researchers to test hypotheses about a factor structure (Brown, 2006; 2015) was used in assessing the strength of relationship. The goodness-of-fit index for the model was $\chi2$ (5, N =516) = 6.048, p = .302, CMIN/DF = 1.210 NFI=.988, IFI=.998, TLI=.994, CFI=.998, RMSEA = .020 (.000 -.067).

Policy. Reason (2011) suggest that *Policy* (M= 4.39, SD = .443) guides specifications in which management describe how certain operations are to be performed. Management will have policy guidelines that described training, maintenance, line operations and personal conduct etc. They are developed based on the organization's strategic principles but further determined by commercial and operational factors. Example of an item under *Policy* is "Policies ensure that supervisory personnel are present throughout high-risk procedures". The goodness-of-fit index for the model was $\chi2$ (12, N =516) = 21.916, p = .038, CMIN/DF = 1.826, NFI=.948, IFI=.976, TLI=.941, CFI=.975, RMSEA = .040 (.009 -.066).

Procedures. Reason (2011) and Degani and Wiener (1981) suggest that *Procedures* (M= 4.68, SD = .034) should be developed that are in line with an organization's principles and policy framework. *Procedures* should specify the nature of a task, time and sequence for conducting task, actions required, sequence of task and required feedback mechanism. The goodness-of-fit index for the model was $\chi2$ (9, N =516) = 21.473, p = .011, CMIN/DF = 2.386, NFI=.965, IFI=.979, TLI=.951, CFI=.979, RMSEA = .052 (.024 -.080).

Practices. Reason (2011) suggest that *Practices* (M= 3.74, SD = .777) are the actual activities that occur at the 'sharp-end" of any organization and personnel are responsible for ensuring that these are in line with standard operating procedures (SOPs). However, deviations can occur when these actions differ from an organization's procedure. These deviations can be minor or major occurrences and, in some cases, lead to an accident. The goodness-of-fit index for model was $\chi2$ (8, N =516) = 19.623, p = .012, CMIN/DF = 2.452, NFI=.907, IFI=.943, TLI=.893, CFI=.939, RMSEA = .053 (.023 -.083).

Reliability and construct/discriminant validity. The reliability, convergent validity, and discriminant validity of the four management factors that relates resilient safety culture were assessed. A Cronbach's alpha value of 0.7 or higher indicates good reliability of measured items (Nunnally, 1978). In addition, a composite reliability (CR) of 0.7 or higher suggests good reliability and indicating internal consistency exists (Hair, Ringle, & Sarstedt, 2011). Factor loadings and average variance extracted (AVE) methods were used to assess the convergent validity (Fornell and Larcker, 1981; Hair et al., 2011). The Heterotrait-Monotrait ratio of correlations (HTMT) approach with a predefined criterion/ absolute threshold of 0.90 was used for discriminant validity due to its high sensitivity (Henseler, Ringle, & Sarstedt, 2015). Both Cronbach's alpha and composite reliability values for 4Ps indicated acceptable reliability. All the factors had an AVE $\geq$ .50 and above suggesting an acceptable convergent validity for the instrument. The $HTMT_{0.90}$ result suggests evidence of discriminant validity since the inter-construct correlation ratios were less than the criterion of 0.90.

Question Two

What is the strength of relationships between management factors (Policy, Principles, Procedures, and Practices) and the overarching construct resilient safety culture in a collegiate aviation SMS program?

The measured constructs for the four factors were derived by summing the measurement items in each validated CFA model. The new variables were then used to assess the strength of relationships with the over-arching concept of resilient safety culture. The analysis yielded a model with goodness-of-fit index as follows: χ^2 (2, *N*=516) = 5.586; *p* = .061; CMIN/DF = 2.793; NFI = .985; RFI = .925; IFI=.990; TLI =.951; CFI= .990; RMSEA = .029 (.015 - .057). Figure 1 shows the final measurement model of relationship between 4Ps and resilient safety culture.

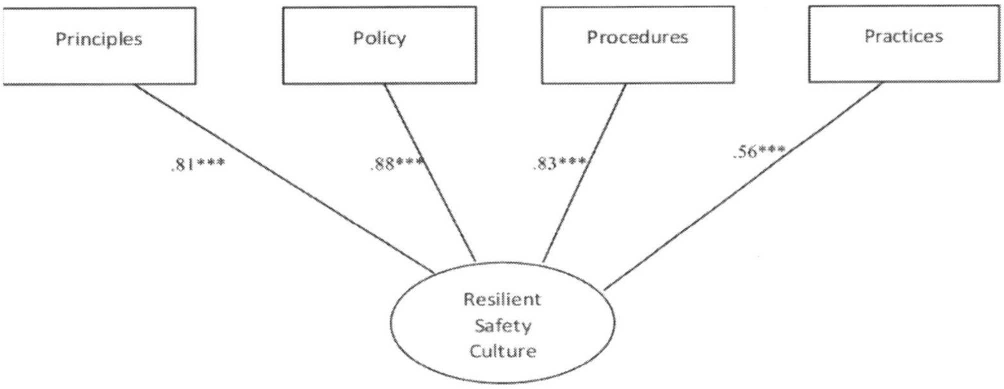

Figure 1. Model showing relationship between 4Ps and resilient safety culture (***$p < .001$)

Question Three

What is the strength of relationships between management factors Policy, Principles, and Practices when mediated by the Procedures in a collegiate aviation SMS program?

The preliminary analysis of a fully mediated 4Ps measurement model failed to produce any acceptable fit. A post-hoc iteration was done on the fully mediated 4P safety resilience model using the modification indices function and a direct path from *Principles* to *Practices* was then removed and a new analysis re-run. The resulting partially mediated model produced a good fit as shown by the fit index: χ^2 (1) = 1.175; p = .278; CMIN/DF = 1.178; NFI = .997; RFI = .968; IFI=.998; TLI =.995; CFI= .998; RMSEA = .019 (.000 -.119). Figure 2 shows the partially-mediated 4Ps measurement model.

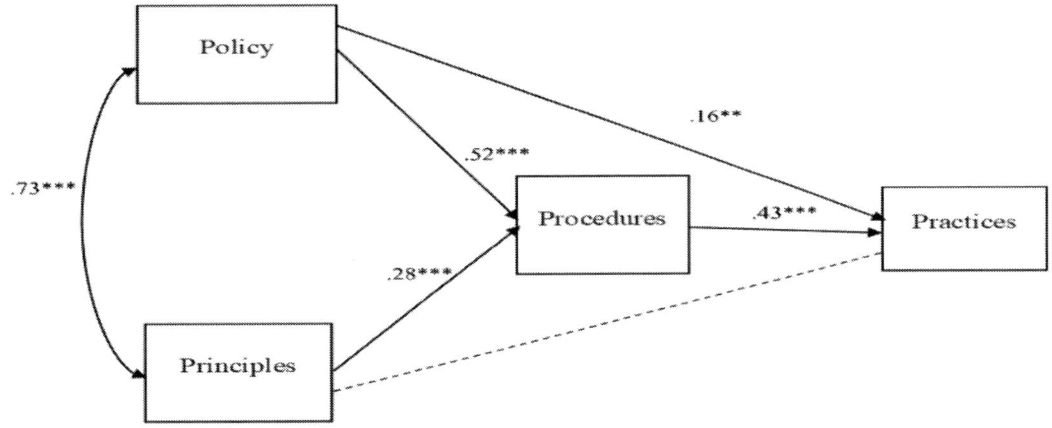

Figure 2. Partially Mediated 4Ps Measurement Model (*** p < .001; **p <. 005)

Discussions and Conclusion

A strategic management implication of this study is that resilient safety culture is strongly influenced by the policies, procedures and principles within an organization and periodic assessments should be conducted to identify gaps and weaknesses related to these factors that can adversely affect a resilient safety culture. This aligns strongly with the findings of Akselsson et al. (2009) on resilient safety culture in air traffic control environment and Reason's conceptual framework on resilient safety culture (Reason, 2011). This finding is also consistent with the need for robust organizational policies and procedures that are primed by a principle that ensure safe and highly resilient flight operations (Degani and Weiner, 1991). The relatively weak relationship between *Practices* and resilient safety culture as compared to the other organizational factors could be attributed to inadequate awareness of resilient safety practices within the collegiate aviation program by some respondents which potentially can affect their perceptions and responses to items related to that factor. An increased focus on resilient safety practices such as safety empowerment may be expedient as part of the SMS promotion activities. This allows for cognizance of safety risk in operational environments and authority to suspend activities when risk exceed tolerable levels required for a task. There was a high correlation between *Policy* and *Principles* that underscores the important role that over-arching principles have in shaping the policy framework of any organization. The results also support literature that suggest that *Policy* framework forges a consistent and pragmatic review of procedures for use by "sharp-end" employees in an organization. (ICAO, 2013; Stolzer, 2018). The strongest predictive relationship was between *Procedures* and *Practices*. However, *Procedures* strongly mediated the path between *Policies* and *Practices*, which suggest that without comprehensive procedures outlining policies, it may be a challenge to sustain resilient practices among "sharp-end" employees. This finding corroborates Hollnagel (2011; 2014) concept of 'work as imagined' and 'work as done' as two contrasting ways of understanding *Practices* at the "sharp-end". 'Work as imagined' is defined by the *Policies* and *Procedures* outlining the way things should work and represents how program leadership and supervisors believe work happens or should happen. 'Work as done', on the other hand, describes the work as carried out by 'front-line' employees at the 'sharp end'; in the case of collegiate aviation, how flight students and instructors practically engage in flight training activities. The finding corroborates suggestion by Schwarz and Wolfgang Kallus (2015) on a need for resilient safety practices in high operational tempo. Variations in resilient safety culture practices could be due to organizational conditions created by those at the 'blunt end' (management); the *Policies* produced, or the way in which standards for *Practices* are perceived. The results substantiate Paries et al. (2018) assertion that excessive attention to SMS formalism (policy, procedures, and traceability) may lead to bureaucratic processes, often at the expense of focusing on desirable resilient factors such as *Practices*. It is therefore essential for collegiate aviation programs to consider seasonal variation within the flight training environment and calibrate resilient safety culture periodically.

Acknowledgements

This work was supported by the Odegard School of Aerospace Sciences, University of North Dakota Seed Grant [# 21267-2205]. Based on original study in Safety Science. Views expressed here are solely that of author and does not reflect that of granting institution.

References

Adjekum, D, K. (2014). "Safety Culture Perceptions in a Collegiate Aviation Program: A Systematic Assessment," *Journal of Aviation Technology and Engineering*: Vol. 3: Issue. 2, Article 8. http://dx.doi.org/10.7771/2159-6670.1086

Adjekum, D. K. (2017). An Evaluation of the relationships between collegiate aviation Safety Management System initiative, self-Efficacy, transformational safety leadership and safety behavior mediated by safety motivation. *International Journal of Aviation, Aeronautics, and Aerospace, 4(2)*. https://doi.org/10.15394/ijaa.2017.1169

Akselsson, R., Koorneef, F., Stewart, S. & Ward, M. (2009). Resilience safety culture in aviation organizations. Retrieved from http://repository.tudelft.nl/view/ir/uuid%3A361fe48d-9390-4a4f-8dba-8a6cf8ec4b64/

Degani, A. and Wiener, E. L. (1991). *Philosophy, policies and procedures: The three P's of flight deck operations.* Columbus, Ohio.

Federal Aviation Administration. (2015, March 9). *Safety Management System Voluntary Program Guide (AFS-900-002-G201)* [PDF]. Washington D.C.: Federal Aviation Administration. Accessed from: University of North Dakota Aviation Safety Department.

Fornell, C. G., & Larcker, D. F. (1981). Evaluating structural equation models with unobservable variables and measurement error. *Journal of Marketing Research, 18* (1), 39–50. https://doi.org/10.1177/002224378101800104

Hair, J. F., Ringle, C. M., & Sarstedt, M. (2011). "PLS-SEM: Indeed, a Silver Bullet." *Journal of Marketing Theory and Practice, 19* (2), 139-51. DOI: 10.2753/MTP1069-6679190202

Henseler, J., Ringle. C. M., Sarstedt. M. (2015). A new criterion for assessing discriminant validity in variance-based structural equation modeling. *Journal of the Academy of Marketing Science*, 43:115–135 DOI 10.1007/s11747-014-0403-8

Hollnagel, E., Paries, J, Woods, DD, Wreathall, J. (2011). Resilience Engineering in Practice — A guidebook, Farnham: Ashgate.

Hollnagel, E. (2014). Safety-I and safety-II. The past and future of safety management. Farnham, UK: Ashgate.

International Civil Aviation Organization. (2013). Safety management manual (Doc. 9859, 3rd ed). Retrieved from https://www.skybrary.aero/bookshelf/books/644.pdf

Nunally, J. (1978). *Psychometric Theory (2 nd ed)*. McGraw-Hill, New York. NY.

Paries, J., Macchi, L., Valot, C., Deharvenght, S. (2018) Comparing HROs and RE in the light of safety management systems, *Safety Science* 117; 501–511, https://doi.org/10.1016/j.ssci.2018.02.026

Reason, J. (2011). The human Contribution – Unsafe Acts, Accidents and Heroic Recoveries. Farnham, UK: Ashgate Publishing Limited.

Schwarz, M., & Wolfgang Kallus, K. (2015). Safety culture and safety-relevant behavior in air traffic management: Validation of the CANSO safety culture development concept. *Aviation Psychology and Applied Human Factors, 5*(1), 3–17. https://doi.org/10.1027/2192-0923/a000068

Schwarz, M., Wolfgang, K., K., & Gaisbachgrabner, K. (2016) Safety Culture, Resilient Behavior, and Stress in Air Traffic Management. *Aviation Psychology and Applied Human Factors*, 6, pp. 12-23. https://doi.org/10.1027/2192-0923/a000091

Stolzer, A. J., Friend, M. A., Truong, D., Tuccio, W. A., Aguiar, M. (2018) Measuring and evaluating safety management system effectiveness using Data Envelopment Analysis, *Safety Science,* 104, 55–69. https://doi.org/10.1016/j.ssci.2017.12.037

Tse-Hua, S., Xitao Fan. (2009) Comparing response rates in e-mail and paper surveys: A meta-analysis, *Educational Research Review*, Volume 4, Issue 1, 26-40. http://dx.doi.org/10.1016/j.edurev.2008.01.003.

SAFETY PROGRAMS, SAFETY MANAGEMENT SYSTEMS AND IMPLICATIONS FOR HUMAN PERFORMANCE[1]

Julia Pounds, Ph.D.[2]
TempoLogics
Mustang, OK

Paul Krois, Ph.D., PMP.[3]
Crown Consulting Inc.
Aurora, CO

A State Safety Program (SSP) and a Safety Management System (SMS) are both part of the Standards and Recommended Practices (SARPs) promoted by the International Civil Aviation Organization (ICAO) to support international safety in air navigation, flight inspections, air accident investigations, training, certification, and related areas. This paper discusses roles and responsibilities of the regulator and service provider focusing on human performance in oversight versus operations. Clarification is intended to help both regulators and operators focus on executing their separate and distinct roles and responsibilities for oversight versus operations. Left unaddressed, role confusion between the organizations can result in safety programs with potentially hazardous gaps, ineffective mitigations or inefficient overlaps. The US SSP is used to illustrate with examples of how these programs can impact human performance.

A major pillar of aviation safety is international cooperation between the national governments that were signatory States to the 1944 Convention on International Civil Aviation (the "Chicago Convention"). As civilian and military air transport grew and the need for standardization increased, the International Civil Aviation Organization (ICAO) - an international framework for cooperation - developed and eventually in 1947 became a United Nations' specialized agency. ICAO works with member States to promote international Standards and Recommended Practices (SARPs). These SARPS include the ICAO State Safety Program (SSP) and with it the Safety Management System (SMS) construct. SMS is a formal framework of processes and practices performed to identify and assess potential hazards. The State is responsible for its SSP but delegates many of its responsibilities to a State regulatory office. The State's delegated regulatory office performs the actual oversight and surveillance of the SARPs and ANSP, e.g., related to air navigation, flight inspections, air accident investigations, training, certification, and related areas.

Our discussion focusses on human performance topics for two groups: regulatory personnel who perform oversight and ANSP personnel perform operations to execute air navigation services. As explained above, these two groups within the SSP act in concert to accomplish safety goals. For the purposes of this paper, we've used examples from the US SSP. FAA Order 8000.369 (2020) helps FAA organizations integrate SMS and/or ICAO SSP guidance into their organizations.

----

[1] The opinions expressed in this paper are those of the authors and do not reflect policies or positions of the FAA or any other entity.
[2] FAA Engineering Research Psychologist and Air Traffic Safety Inspector (Retired).
[3] FAA Manager, NextGen Human Factors Research and Engineering Division (Retired).

Per its international agreements with ICAO, the US government funds its SSP and delegates its aviation SSP responsibilities to the Federal Aviation Administration (FAA). In turn, the FAA Administrator delegates responsibilities for regulation of aviation to Aviation Safety (AVS) an organization within the FAA. Within AVS, the Air Traffic Safety Oversight Service (AOV), is responsible for oversight of the FAA ANSP – the Air Traffic Organization (ATO).

> *"Situated within the Aviation Safety (AVS) organization, the Air Traffic Safety Oversight Service [AOV] establishes safety standards and provides independent oversight of the Air Traffic Organization – the provider of air traffic services in the United States."* (https://www.faa.gov/)

> *"The Air Traffic Organization (ATO) is the operational arm of the FAA. It is responsible for providing safe and efficient air navigation services to 29.4 million square miles of airspace."* (https://www.faa.gov/)

To illustrate the distinction between oversight versus operational responsibilities, the following example is provided: In 2007, AOV conducted an audit of ATO turn to final (TTF) procedures that focused on the methods of separation used while vectoring aircraft to join the final approach. The audit team analyzed data from reports of separation violations (operational errors) from the prior two years. The audit found that the 12 facilities audited were in compliance with then current ATO requirements. The audit also identified that those facilities in the data sample using standardized altitude requirements for all approach configurations had a lower percentage of TTF operational errors than those facilities without standardized requirements. AOV requested that ATO review the report and respond with its determination of the potential impact and national applicability of standardized requirements, and the potential reduction of certain types of operational errors.

The idea that additional clarification of these relationships would be helpful emerged from discussions among AOV personnel about roles, responsibilities, activities and methods for AOV oversight. These discussions were often fraught and complicated, in part, because imprecise foundational guidance allowed for multiple interpretations. For example, see the description posted with FAA Order 8000.369C:

> *"This order establishes the Safety Management System (SMS) policy and requirements for the Federal Aviation Administration (FAA). The requirements contained within this document are intended to help FAA organizations incorporate SMS and/or International Civil Aviation Organization (ICAO) State Safety Program (SSP) requirements into their organizations. FAA organization SMSs work together to form the overall FAA SMS." Retrieved from https://www.faa.gov/regulations_policies/orders_notices/i*

Perhaps, a clearer statement would be: "*FAA organization SMSs work together to form the overall FAA execution of the US SSP framework. The FAA SMS establishes policy for the different FAA organizations in establishing their own individual SMSs for execution.*" For another example of this ambiguity, see Figure 1 retrieved from https://m3rsms.com.au/icao/.

> *"States shall establish a State Safety Program (SSP), in order to achieve an acceptable level of safety (ALoS) in civil aviation"*
>
> Annex 19
>
> The SSP and SMS framework recommended by ICAO are the same.

Figure 1. Example of SSP and SMS ambiguity.

Historically, some things had not been clarified so that both organizations shared understandings and expectations about the meaning of "establish safety standards" and the scope of AOV oversight. While SMS includes formal risk management processes, such as assessing effectiveness of mitigations such as evaluating training success, a lack of clarity might lead to a failure of the regulator or the ANSP to identify an existing risk, e.g., a gap in controller training for a new procedure or that operational personnel had insufficient information about changes in operational procedures.

The relationships between the SSP and SMS programs are illustrated using the line of sight example shown in Figure 2 below. Both oversight of air traffic services and ANSP delivery of those services require that personnel in both organizations have deep knowledge about the operations that support those services. Thus, discussion such as methods for execution of a CE needs to also specify to which organization it pertains; otherwise, misunderstandings are possible. Figure 2 was initially generated to help with explaining these similarities and differences. For example, CE 4 and CE 6 could be interpreted as pertaining to either/both organizations.

SSP CE 4: Ensure personnel are qualified/competent in their functional areas of responsibility.

- ATO ensures that its personnel are hired, trained, and certified to safely perform all activities to support and execute air navigation, including using existing and new equipment systems and accomplishing safe operations according to requirements.
- AOV ensures that its personnel are trained in the role of the regulator and oversight methods, such as how to conduct audits and assessments to determine compliance and how to evaluate ATO safety documents. Since its beginning AOV has conducted audits and assessments to examine many ATO performance requirements, such as, turns to final, fatigue risk management, credentialing of personnel, qualification of personnel on new equipment, and procedures for opposite direction operations.

SSP CE 6: Ensure policies for licensing, certification, authorization and/or approval obligations.

- ATO ensures that its personnel are current in their training and that requirements are met for each to maintain operational and medical certifications.
- AOV established a program to credential its personnel as Air Traffic Safety Inspectors (ATSI) by meeting training requirements. Those maintaining an ATSI credential are qualified to conduct oversight activities.

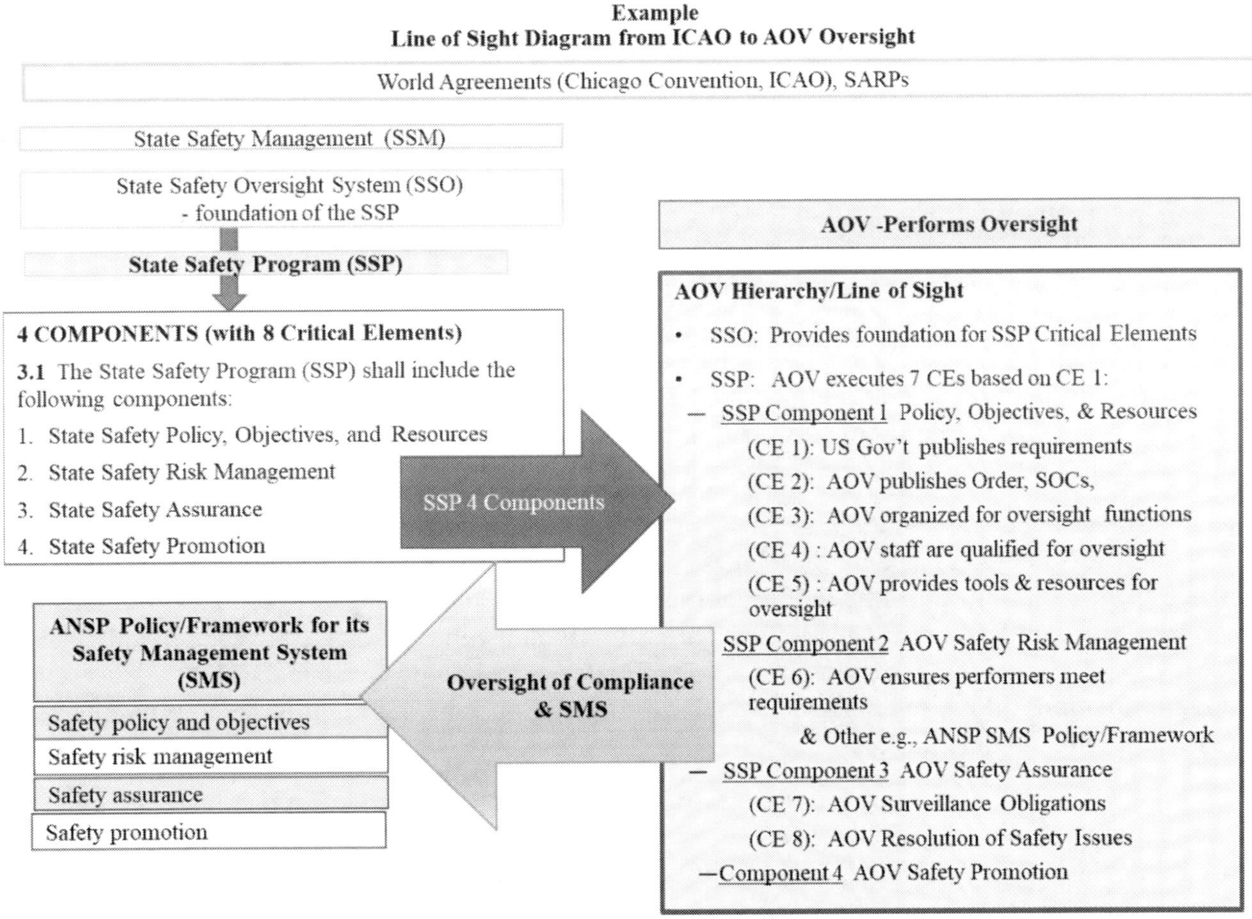

Figure 2. Relationships between US SSP oversight by AOV and the ANSP (ATO).

Because ANSP operations depends on human performance in context, by capturing, measuring and understanding these activities as they are performed in the operational context, they can be better understood in the safety context. Thus the regulator's obligation to focus on actual performance rather than on oversight of the ANSP's SMS formalisms may arguably be relatively more important to safety oversight. Approval of the ANSP SMS may be necessary but not sufficient for oversight and safety assurance. Cambon, Guarnieri, and Groeneweg (2006) recommended that, while a formal SMS corresponds to a formal organizational description, an equally or more important focus would be the influence of the SMS on the operational working environment and practices of people. Reason (2001) speculated how an ANSP would build and operate an effective SMS and how regulators would recognize and evaluate it.

ICAO manual (Doc 10151) discusses how the State can integrate human performance (HP) in its SSP by embedding HP in key oversight responsibilities and activities, such as when evaluating the ANSP's SMS and when conducting surveillance to determine if the ANSP meets these regulatory requirements (ICAO, 2021). CANSO (2019) provided the aviation community with a Standard of Excellence in Human Performance Management that offers information to help ANSPs improve their human performance management.

These oversight activities can take several forms and feedback from the regulator to the ANSP can provide useful human performance information for improving safety that the ANSP may not otherwise have. For example:

- AOV conducted an audit in FY07 to evaluate compliance by ATO personnel with procedures for turns to final arrival configurations for vectoring aircraft to join the final approach course and whether the procedures had an impact on the number of separation violations (operational errors). The audit included the separation procedures used by controllers. The audit found that all facilities sampled were in compliance but that facilities using standardized altitude requirements had a lower percentage of operational errors than those facilities not using standardized requirements. The audit also found that complete records for controllers' turns-to-final refresher training could not be documented. Some facilities did not include turns-to-final procedures in their recurrent training curriculum. AOV requested that ATO determine whether a national standardized requirement would be applicable for potential reduction of operational errors and to update their training records and curriculum maintenance processes.

- AOV conducted an audit in FY17 of the ATO Fatigue Risk Management. Program Office. The audit team noted (1) a lack of delegation of responsibilities for feedback to the program office from facilities after fatigue mitigations were executed. This lack of feedback prevents the program from understanding what efforts are effective for reducing ATC fatigue related errors. (2) a lack of clarity about how the program evaluates or validates the success of its communications and mitigations. This lack of information impedes continuous improvements and (3) out of date documentation.

- AOV conducted an audit in FY15 of ATO compliance with its responsibilities for the Credentialing and Control Tower Operator Certification Program. The audit found that the ATO used inconsistent or deficient management controls to ensure compliance with its credentialing requirements. The ATO conducted activities to come back into compliance with their requirements.

- AOV conducted an audit in FY18 of the management controls for credentialing ATO safety personnel. The program ensures that no person provides direct safety-related air traffic control services or certifications unless the person has the appropriate credentials, ratings, and/or skills evaluations for the safety services provided. Inconsistent management of credentials and ratings introduces the potential for risk into the NAS. The auditors found that the ATO had gaps in execution of management process controls. Inadequate or incomplete verification of compliance with their Credentialing Program requirements creates the potential for personnel providing safety-related services to not be properly credentialed. AOV notified the ATO of planned future surveillance activities to ensure that these gaps were corrected.

- AOV conducted an audit in FY08 to determine ATO compliance with the requirement to provide to controllers with refresher training that maintains and upgrades the knowledge and skills necessary to apply air traffic procedures in a safe and efficient manner. This audit included the delivery of training and evaluation of its effectiveness. AOV found a lack of program guidance for the refresher training program. The lack of program guidance for refresher training leads to inconsistencies in facility-level training when some level of consistency across the NAS might be desirable. The required auditable process for managing refresher training was not observed and training was not being developed or delivered in conformance with directives for administering training. The refresher training records examined did not provide a reliable record for historical accuracy or for making training improvements and no evidence was found that any systematic evaluation of refresher training effectiveness was being done. For example, the focus for evaluation was materials rather than the effects of the training on employees' performance and the responsibilities, partly because ensuring training effectiveness was assigned across different organizations. AOV continued periodic oversight of the program to monitor its improvements.

- AOV conducted an audit in FY15 of the Enhanced Back-up System (EBUS) in the En Route Automation Modernization (ERAM) Program. The audit found the ATO had contradictory training and operational requirements. The contradictory requirements resulted in a variety of responses at the service delivery points as to using EBUS as a mitigation control for a high risk hazard in the national airspace. The ATO addressed these conflicts and was advised that AOV would continue to monitor the ATO management of the ERAM Program.

Discussion

While ICAO promotes aviation safety through SSP standards and recommendations, execution of the SSP is the responsibility of each State and is implemented through the State's regulator. The success of the State's program ultimately relies good communication between the regulator and the ANSP because the goal of both SSP and SMS is to identify and manage safety risks.

Focusing on safety risks introduced by human performance in the system, one could expect each organization to approach it by different means - the regulator through particular oversight and surveillance of influences on human performance with feedback to the ANSP and the ANSP through execution of safety management controls to ensure operational performance of its personnel and successful implementation of safety mitigations. Role clarity supports communication and mitigates confusion so that the regulator and ANSP can collaborate to improve safety in all these areas.

References

Cambon, J., Guarnieri, F., & Groeneweg, J. (2006, November). Towards a new tool for measuring Safety Management Systems performance. In E. Hollnagel (Ed.) *Proceedings of the Second Resilience Engineering Symposium* (pp. 53-62). Antibes-Juan-les-Pins, France.

Civil Air Navigation Services Organisation (2019). *CANSO Standard of Excellence in Human Performance Management*. Retrieved from https://www.icao.int.

Federal Aviation Administration. (2006). Air Traffic Safety Oversight Service (Order 1100.161). Washington, DC.

Federal Aviation Administration. (2020). Safety Management System (Order 8000.369). Washington, DC.

International Civil Aviation Organization (2016). *Annex 19*. Montreal, Quebec, Canada H3C 5H7. https://www.skybrary.aero/index.php/ICAO_Annex_19,_Safety_Management

International Civil Aviation Organization (2021, Draft). *Human Performance Manual for Regulators*. ICAO Doc 10151. Montréal, Quebec, Canada H3C 5H7. Retrieved from https://www.icao.int/safety/OPS/OPS-Section/Pages/Humanfactors.aspx.

Reason J. (2001). Human Factors Aspects of Safety Management Systems. In *Proceedings of the 15th Annual FAA/TC/CAA Maintenance Human Factors Symposium*. London.

EXPLORING THE RELATION BETWEEN DISTRACTOR INHIBITION AND AUDIOVISUAL INTEGRATION

Michelle H. Wang, Wright State University, Dayton, OH

Elizabeth L. Fox, Air Force Research Laboratory, WPAFB, OH

The Stroop paradigm is a great experimental tool to assess the extent that task-irrelevant, but target-related, distractors influence target identification in a variety of contexts. In particular, it has been applied beyond the traditional visual modality (e.g., audio, or audiovisual). However, audiovisual studies using Stroop-like tasks have reported conflicting results. Importantly, these bimodal studies assessed only group-level mean differences and did not investigate whether the degree of bimodal conflict is greater than what is expected of two unimodality distractors that are inhibited in an unlimited capacity, independent, and parallel fashion. In this research, we relied on cognitive-based models of individuals' performance to estimate audiovisual conflict and directly compared the influence that two types of bimodal distractors had on performance: 1) the same conflicting information was presented in both modalities and 2) different conflicting information was presented in each modality. We found unimodal visual, but not auditory, distractors significantly influenced target processing. Most interestingly, we found that despite a lack of unimodal auditory influence some participants performance indicated that bimodal distractors were harder (easier) to inhibit than expected given our model-based predictions, and the direction (limited or super capacity) and degree of deviation from our model prediction depended on cross-modal distractor similarity.

The Stroop Effect (Stroop, 1935) is a phenomenon demonstrating the compelling interference of semantics (i.e., color words) on the identification of the font color of the same word. It also provides insight into one's ability to inhibit such interferences. More recently, interest in the Stroop Effect has expanded into the auditory modality. However, there has been conflicting results in regard to auditory interference: some found evidence of auditory interference in addition to visual interference (Cowan & Baron, 1987), others observed evidence that auditory distractor did not provide additional interference beyond that of visual distractors (Elliott et al., 2014). Perhaps other factors could modulate auditory interference in a Stroop-like paradigm. Francis, McLeod, and Taylor (2014) specifically examined the relationship between interference and distractor similarity: distractors in different modalities both provided color information incongruent to the font color, but the information was either the same across modalities or different. They compared the effect of incongruent distractors to that of the control condition with only uninformative stimuli (i.e., visual control: "xxxx", auditory control: tone). Interference was calculated by subtracting the mean reaction time (RT) of the control condition from the mean RT of each of the incongruent conditions and averaged at the group-level. They found that the presence of a distractor, whether auditory or visual, increased interference as compared to the control regardless of whether the distractors from different modalities were the same or different (agrees with Cowan & Baron, 1987, but disagrees with Elliott et al., 2014). Furthermore, Francis et al. observed that the combined interference in the incongruent same distractors condition were less than that of the sum of the unimodal interferences, whereas the combined interference in the incongruent different condition was approximately equal to that of the sum of the unimodal interferences. This suggests that distractors from both modalities may have integrated when they were the same but may have been processed independently when they were different.

In the currently study, we leverage one measure, the capacity coefficient, from an established mathematical modelling framework, system factorial technology (SFT; see Townsend & Nozawa, 1995 for details), to assess processing capacity to inhibit distractors in a font color judgment where distractors vary in number and modality, at the individual-level. Theoretically, the color judgment in the Stroop task can be made without processing any distractors. If this were the case, using the cumulative reverse hazard function, K, at any time, t, we would expect that processing times to make a color judgment, C, would be the same regardless of the distractors, i. However, the Stroop effect shows that the processing time of the font color may speed-up (when written word semantic and font color are congruent) or slow-down (when semantics and font color are incongruent). This discrepancy can be computed as a capacity measure from SFT called a single-target self-terminating process (ST-ST; Blaha, 2010): $C(t)_{STST} = \frac{K_C}{K_{C(i)}}$. A result equal to 1 indicates no change in color judgement processing capacity (*unlimited capacity*). Otherwise, processing times would speed up (greater than 1; *super capacity*) or slow down (less than 1; *limited capacity*) depending on the distractor information. Analogously, the same ST-ST equation can assess color judgement processing time with auditory distractors by substituting visual distractors with non-informative characters and introduce auditory distractors.

Distractors may also be bimodal, that is, spoken and written color words incongruent with the font color. Therefore, we were also interested in modeling the efficiency to ignore bimodal distractors, and the potential effect of bimodal distractor similarity. One's efficiency to inhibit distractors can be defined as the residual cost of inhibition for each unimodal condition by subtracting the processing time to make the color judgement alone from the total processing time. Using the capacity-AND decision-rule (i.e., both distractors must be inhibited to make a decision; Townsend, 1974), we formed a model prediction of bimodal performance that assumes distractors are processed with Unlimited Capacity, and in an Independent and Parallel fashion (UCIP). The capacity-AND measure is the ratio of the cumulative reverse hazard function of response times of bimodal distractors at a given time, t, for the color of the word, C, the written visual word, V, and the auditory spoken word, A. We do not assume that the written and spoken word need to be processed but define the degree to which their processing occurs depending on the degree to which response times slowed in the single-modality distractor conditions, $K_{C(V)}$ and $K_{C(A)}$. The UCIP model baseline predicts the sum of the processing time to allow the color to influence their judgment, K_C, and the processing time to inhibit written or spoken word interference, $K_V + K_A$, should equal the cumulative reverse hazard function for the combination of the two distractors, $K_{C(AV)}$. The cost of processing time to inhibit distractors was not directly observable. However, we could estimate it by accounting for the processing time of making the color judgment alone with no distractors, compared to in-context of the visual written word to obtain $K_V(t)$. Likewise, for the auditory spoken word $K_A(t)$. Therefore, we can obtain our UCIP prediction to compare to observed performance using a capacity-AND form:

$$C_{AND} = \frac{K_{C(V)}(t) + K_{C(A)}(t) - K_C(t)}{K_{C(AV)}}.$$

Like the ST-ST measure, performance is characterized as *limited* (slower to inhibit bimodal distractors than UCIP model predictions), *unlimited* (as predicted), or *super capacity* (faster).

The current study provided several new contributions to the methodology of the existing literature. We examined the effect of unimodal and bimodal distractors: 1) across the entire distribution of RTs, 2) at an individual-level, 3) using cognitive-model based comparisons, and 4) with approximately 10 times the amount of data compared to previous studies. We predicted a main effect of interference (i.e., distractors would slow RTs), and an interaction between the degree of interference and modality (i.e., visual > auditory). We also predicted performance with bimodal distractors would deviate from UCIP model predictions, and the degree of violation would depend on whether the spoken and written color words were the same or different.

Methods

This study was administered virtually and completed at the subjects' times and locations of choice using the Amazon WorkSpaces, which is a virtual desktop that requires internet connection (experimenters were available virtually). The testing environment and equipment were kept consistent within each subject: the task was generated and administered via PsychoPy3 (Pierce, 2007), the auditory stimuli were delivered via subject's headphones of choice, and subjects used a keyboard to respond. Twelve subjects (reported normal hearing, normal or corrected-to-normal vision, and normal color vision) participated in this experiment. Ten long-term subject panel members (Wright-Patterson Air Force Base) were compensated at an hourly rate, two recruits were compensated at $15/hour, and one was an author of this paper.

Table 1.

Audiovisual Stroop Trial Types

Condition	Stimulus Example		
	Auditory	Visual	Font Color
Control	**<noise>**	**@@@@**	**Blue**
		@@@@	Blue
Congruent			
Visual		blue	Blue
Visual	<noise>	blue	Blue
Auditory	blue	@@@@	Blue
VA	blue	blue	Blue
Incongruent			
Visual		red	Blue
Visual	**<noise>**	**red**	**Blue**
Auditory	**red**	**@@@@**	**Blue**
VA Same	**red**	**red**	**Blue**
VA Different	**red**	**green**	**Blue**
Mixed			
Visual Incongruent	blue	red	Blue
Audio Incongruent	red	blue	Blue

Note. Conditions of interest are bolded.

The visual stimuli included words "red", "green", "blue", and the symbol set "@@@@", which were presented for 650 ms in Arial font, with either red, green, or blue font color at 0.2 normalized letter height (scaled with monitor settings). The auditory stimuli were spoken color words from the same female speaker (i.e., "red", "green", "blue") and white noise delivered at a comfortable listening level. All auditory stimuli were 650 ms in duration, except for "red", which was 601 ms. Subjects were instructed to set the system sound level to the highest setting but were able to adjust the sound settings at their own will. This study was a part of a larger project, which included all trials types shown in Table 1. However, only five trial types (bolded in Table 1) were analyzed in the current study: control, visual, auditory, incongruent same (IS), and incongruent different (ID). The objective was to correctly identify the font color (target) while written and/or spoken words (distractors) were present. Each trial was 4000 ms with the same order of events: a visual fixation cross and an auditory fixation tone (500 ms), randomized inter-stimulus interval (250-500 ms), observation interval (650 ms), response interval (2000-2250 ms; varied duration does not affect response, $RT_{90\text{th percentile}}$ = 1048 ms), inter-trial interval (1000 ms). Subjects each completed 160 blocks (randomized conditions) over 10 experimental session. All experimental session included conditions from the larger project (using the same set of stimuli as the current study) the data of which is not discussed in this paper and should not have altered one's performance in the trials of interests. Fifteen-second breaks were enforced between blocks, but subjects were able to take longer breaks if needed as each block was self-initiated by a key press any time after the enforced break. Prior to each session, subjects completed a 72-trial practice, which included all possible trials present in the experimental session.

Results and Discussion

Performance was highly accurate, 96.7% of the trials were correct. Only correct responses, greater than 100ms in duration, were kept for further analysis. All but one subject showed significantly slowed capacity to process color information in context of visual distractors (written word), $C_{Z\,(V)}$ = [-9.56, -1.50], M = -3.71, successfully replicating the traditional (visual) Stroop effect. But only one subject showed a significant change from unlimited capacity with auditory distractors (spoken words), $C_{Z\,(A)}$ = [-2.24, 1.68], M = -0.37. In general, subjects could more easily inhibit auditory distractor as compared to visual distractor carrying the same information. Some cognitive control seemed necessary to inhibit visual distraction (evidenced by slower RT), which may result from limited capacity processes, an inefficient system structure (i.e., serial processing of each piece of information) or interdependent pooling (i.e., coactive architecture) of the target and visual distractor information (Little, Eidels, Fifić, & Wang, 2018). We found unlimited capacity processing to inhibit auditory distractors, which suggests efficient inhibition of spoken words.

Figure 1 shows individualized estimates of the processing capacity to inhibit both auditory and visual distractors relative to the UCIP model prediction (black solid line, where $C(t)$ = 1). In general, the subjects' processing capacity fall into three groups: unlimited, limited, and similarity dependent. For some (*Fig.* 1a), we found unlimited processing capacity for both distractor similarity types: incongruent-same (IS), $C_{Z\,(\mathrm{IS})}$ = [-1.03, 1.32], $M_{(\mathrm{IS})}$ = -0.16, and incongruent-different (ID), $C_{Z\,(\mathrm{ID})}$ = [-1.54, 0.45], $M_{(\mathrm{ID})}$ = -0.72, which follows from an unlimited capacity, independent, and parallel processing structure. These findings are explained with principles of multisensory integration: multisensory enhancement more often occurs when stimuli from different modalities are presented from the same spatial location, within the same time interval, and/or has similar effectiveness (Holmes & Spences, 2005; Meredith & Stein, 1983). There are several factors in the current study that could have violated these principles and hindered integration: 1) the visual (monitor) and the auditory (headphones) stimuli were not spatially co-located, 2) semantic processing of the distractors were not temporal aligned: the written word was presented instantaneously whereas it takes time to deliver the entire spoken word, and 3) perhaps most importantly, there was a clear modality asymmetry in distractor inhibition: the visual distractor was far more effective (i.e., more difficult to inhibit) than the auditory distractor.

Another group of subjects (*Fig.* 1b) showed significantly slowed processing capacity than the UCIP model prediction (limited) in both IS and ID conditions ($C_{Z\,(\mathrm{IS})}$ = [-3.31, -2.28], $M_{(\mathrm{IS})}$ = -

2.70; $C_{Z\,(ID)}$ = [-2.84, -2.21], $M_{(ID)}$ = -2.50). For this group, the presence of two distractors slowed down processing regardless of whether the distractors were the same or different from one another. Here, limited capacity performance may result from processes that depend on distractors similarity: written and spoken words that were different from one another may result from serial processes (one at a time) and hence slow response times. Alternatively, identical written and spoken distractors may combine to make a stronger composite distractor due to integration or a coactive processing architecture and result in limited capacity performance.

A final group of subjects exhibited processing capacity depended on distractor similarity (similarity-dependent; *Fig.* 1c). Subject 9007 in this group showed unlimited capacity with ID trials ($C_{Z\,(ID)}$ = -1.51) and limited capacity in IS conditions ($C_{Z\,(IS)}$ = -2.15), suggesting potential bimodal integration when the distractors shared the same information. Alternatively, Subject 9001 exhibited the opposite pattern, $C_{Z\,(ID)}$ = -2.49 (limited), $C_{Z\,(IS)}$ = -1.30 (unlimited).

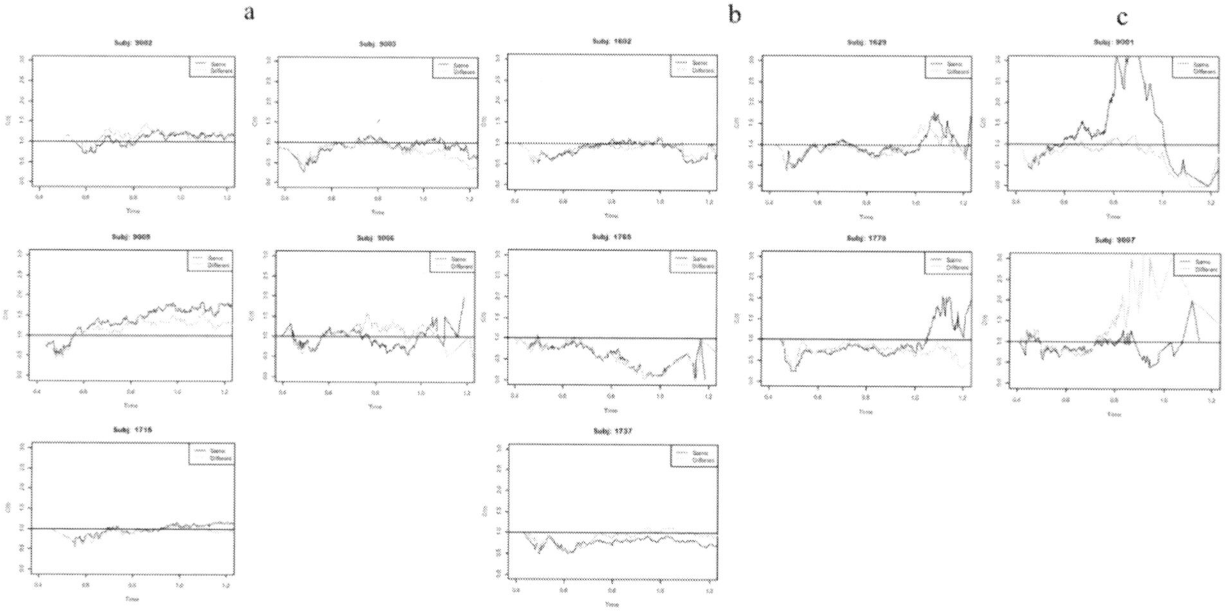

Figure 1. Individual estimates of processing capacity to inhibit same/different bimodal distractors.

Our statistical test assessed capacity across the full function. Further visual inspection of the functions indicated a potential shift in processing capacity between early and late processing times (around 800ms) for some subjects. Specifically, capacity for inhibiting distractors at later processing times increased (i.e., super capacity). Audiovisual integration may have occurred only during later processing times for these subjects when the auditory distractor was fully processed, integrated into a composite and bimodal distractor, and more efficiently inhibited. As discussed previously, the spoken words (and the semantic meaning) took time to convey, but written words were presented instantaneously. We will investigate this in a future study by shifting forward the spoken word onset and by conducting separate statistical tests for early and late processing times.

In conclusion, the current study examined audiovisual distractor inhibition within a Stroop-like paradigm and its interaction with the semantic similarity between two distractors from different modalities. We created a new measure of capacity to examine changes in processing capacity to inhibit bimodal distractors at an individual level. We observed asymmetry in the effect of unimodal distractors, specifically, it was more difficult to inhibit visual distractors than auditory distractors. Also, we found individuals' processing capacity to inhibit bimodal distractors were either: all unlimited capacity, all limited capacity, or similarity dependent. We plan to conduct follow-up studies using another measure of SFT, the survivor interaction contrast, to investigate the processing architecture of the distractors using a factorial manipulation to the processing speed of each distractor type. We will also change the temporal alignment of distractors and present the auditory distractor before the visual distractor to shorten the gap in processing time and facilitate more influence from the auditory distractor on processing times. Indeed, pilot data show incongruent spoken words presented 250ms before the onset of the target significantly slow response times in the color judgment task.

Acknowledgement

This study was supported by the Air Force Research Laboratory and Oakridge Institute of Education and Science. We would like to thank Dr. Brian Simpson for his guidance and support.

Reference

Blaha, L. M. (2011). *A dynamic Hebbian-style model of configural learning* (Doctoral dissertation, Indiana University).

Cowan, N., & Barron, A. (1987). Cross-modal, auditory-visual Stroop interference and possible implications for speech memory. *Perception & Psychophysics, 41*, 393–401.

Elliott, E. M., Morey, C. C., Morey, R. D., Eaves, S. D., Shelton, J. T., & Lutfi-Proctor, D. A. (2014). The role of modality: Auditory and visual distractors in Stroop interference. *Journal of Cognitive Psychology, 26*, 15–26.

Francis, W. S., MacLeod, C. M., & Taylor, R. S. (2017). Joint influence of visual and auditory words in the Stroop task. *Attention, Perception, & Psychophysics, 79*(1), 200-211.

Holmes, N. P., & Spence, C. (2005). Multisensory integration: space, time and superadditivity. *Current Biology, 15*,762-764.

Little, D. R., Eidels, A., Fifić, M., & Wang, T. S. (2018). How do information processing systems deal with conflicting information? Differential predictions for serial, parallel, and coactive models. *Computational Brain & Behavior, 1*, 1-21.

Meredith, M, A, & Stein, B. E. (1983) Interactions among converging sensory inputs in the superior colliculus. *Science, 221*, 389–391

Stroop, J. R. (1935). Studies of interference in serial verbal reactions. *Journal of Experimental Psychology, 18*, 643–662.

Townsend, J. T. (1974). Issues and models concerning the processing of a finite number of inputs. In *B. H. Kantowitz (Ed.), Human information processing: Tutorials in performance and cognition* (p. 133-168). Hillsdale, NJ: Erlbaum Press.

Townsend, J. T., & Nozawa, G. (1995). Spatio-temporal properties of elementary perception: an investigation of parallel, serial, and coactive theories. *Journal of Mathematical Psychology, 39*, 321–359.

SPEED-ACCURACY TRADE-OFFS AND GENERAL SYSTEMS PERFORMANCE THEORY: NOVEL APPLICATION TO FITTS' LAW AND BEYOND

Lawrence R. DiSalvi and George V. Kondraske
University of Texas at Arlington
Arlington, Texas USA

Speed-accuracy trade-offs have long been of interest in human performance. General Systems Performance Theory (GSPT) was motivated by human performance measurement and modeling needs. It has subsequently been applied in those and other areas. In GSPT, all system performance attributes are modeled using a resource construct. Systems are characterized by multi-dimensional performance capacity envelopes (PCEs). The systems of interest here are considered to possess limited speed and accuracy performance resources defining a two-dimensional PCE. When considering human movement, relevance to Fitts' law was conjectured. In multiple Fitts' paradigm tasks, we found a near-perfect correlation between Index of Performance (IP) and PCE area. An almost exact prediction was obtained when scaled by Fitts' index of task difficulty (ID). While the well-known Fitts' law equation does not contain accuracy explicitly, the GSPT-derived expression ($CP_{S\text{-}A}$= ID • Speed • Accuracy) contains both speed (motions/s) and accuracy (hits/attempts). Concepts are applicable beyond human movement; e.g., to visual, auditory, or other information processing types.

The notion of speed-accuracy trade-off has garnered attention wherever human performance is important, including aviation. One definition (Zimmerman, 2011) describes it as "the complex relationship between an individual's willingness to respond slowly and make relatively fewer errors compared to their willingness to respond quickly and make relatively more errors." Motivated by human performance measurement and modeling needs, Kondraske introduced General Systems Performance Theory (GSPT) and has applied it to problems in those areas (e.g., Kondraske, 2011). In GSPT, systems are characterized by multi-dimensional performance capacity envelopes (PCEs) with each dimension representing different performance resource types (e.g., speed, accuracy, etc.). Whereas Zimmerman's definition suggests a behavioral perspective, GSPT provides a performance viewpoint.

In human performance, systems that accomplish human movement represent one type of interest. In this context, despite a plethora of discussions and debates regarding various details, Fitts' law (Fitts, 1954) is widely recognized as that which explains speed-accuracy trade-offs. This is indeed interesting given that it does not explicitly contain speed or accuracy parameters.

For decades, our group has used Fitts' alternating tapping between two targets paradigm in various human performance measurement systems (Kondraske et al., 1984; Kondraske, 1990; Kondraske and Stewart, 2008; Saganis et al., 2020), relying on his *Index of Performance* (IP) with accuracy adjustment (see below) for scoring. When GSPT emerged, it inspired new thinking about many interesting research topics which we have since pursued on a somewhat opportunistic and case-by-case basis as time permits. Among those was an insightful moment regarding composite scores for movement tasks used in our human performance tests. Specifically, GSPT teaches that the mathematical product of speed and accuracy (i.e., the volume of a 2-dimensional PCE) would be a conceptually sound and meaningful composite. Moreover, it was conjectured that a composite so formed might, or should, have some relationship to Fitts' IP. A strong correlation was subsequently confirmed (Kondraske, 1999). While intriguing, other challenges dominated our research agenda until recently when we explored this in greater detail.

Background

Fitts' Law

Fitts's law is an empirical model considered to explain speed-accuracy tradeoffs in human movement with origins in Shannon's information theory and the concept of channel capacity. Fitts' early experiments focused on worker efficiency-related pointing motions in production line and assembly tasks that intrinsically involve speed and accuracy. A key element of Fitts' work is the definition of an *Index of Difficulty* (ID) for such tasks, where ID is a function of motion size or amplitude (A) and the target width (W). While readers are directed elsewhere for details, rationale behind ID, and variations of the initial idea (Fitts, 1954; MacKenzie, 1992), we use a form that has been argued to have desirable characteristics (Sourkoroff and MacKenzie, 2004):

$$\mathrm{ID} = \log_2 (\mathrm{A/W} + 1) \qquad (1)$$

ID has units of bits or, more specifically, bits/motion.

Basic Fitts' Law studies usually employ an alternating upper extremity task (e.g., left-to-right, right-to-left) with multiple motions per trial, measurement of movement time (MT, units = s/motion), and sets of trials that exercise A and W over ranges of interest. To identify where participants work near their capacity limit, trials with too many errors as well as those with none or "near none" are excluded. Generally, researchers follow Fitts' approach, selecting trials with close to a four percent error rate and the corresponding MTs. Assuming a Gaussian distribution of landings at the target region, this corresponds to having target width boundaries at the ±2 standard deviation unit points.

With the above elements in place, Fitts' law is stated in what has been termed its usual form (MacKenzie, 1992) as:

$$\mathrm{MT} = a + b \cdot \mathrm{ID} \qquad (2)$$

where a and b are coefficients determined by linear regression (i.e., best fit to MT and ID data). It states that MT varies linearly with ID. The intercept a is generally a small adjustment. Fitts dubbed the inverse of the slope b the *Index of Performance* (IP) with units of bits/s. When a is zero or not explicitly considered, as was the case in Fitts' original paper:

$$\mathrm{IP} = (1/\mathrm{MT}) \cdot \mathrm{ID} = (1/\mathrm{MT}) \cdot \log_2 (\mathrm{A/W} + 1) \qquad (3)$$

Explicit speed and accuracy terms are not present in either Equation (2) or Equation (3).

Not surprisingly in retrospect, an "adjustment for accuracy" was proposed by Crossman in 1957 in an unpublished report. The method involves the notion of an *effective target width* (W_e). That is, for a repetitive motion dataset with an actual target width (W) where the error rate is not constrained to 4%, determine the target width (W_e) that effectively yields a 4% error rate. While several accounts exist, MacKenzie (1992) provides a concise description of the somewhat cumbersome calculation for the case where only the error rate (*Error*) is known:

$$W_e = W \cdot 2.066/z(1 - Error/2) \text{ for } Error > 0.0049\%, \quad W_e = W \cdot 0.5089 \text{ otherwise} \qquad (4)$$

The term $z(\alpha)$, the inverse cumulative distribution function, returns the z-score where the area under the curve is α%. Considering both bell curve tails, a 4% error rate would require the evaluation of $z(1 - 0.04/2) = z(0.98) = 2.0537$ and $W = 1.00 \cdot W_e$. For *Error* >4%, $W_e > W$.

General Systems Performance Theory

Various aspects of GSPT and their rationale are described elsewhere (e.g., Kondraske, 2011). Briefly, GSPT's objectives are to provide: 1) a conceptual basis to define and measure *all* aspects of *any* system's performance; 2) a conceptual basis to analyze any task and facilitates system-task interface assessments; and 3) identification of the principles that explain success/failure in any given system-task interface. In GSPT, all aspects of a system's performance capacity are modeled with a resource construct. Each *performance resource* represents one dimension of a multidimensional performance space and the goal of system characterization from a performance perspective is to determine its *performance capacity envelope* (PCE). GSPT teaches to expect and how to define a PCE for any system. Another key feature is the *nonlinear, threshold effect* associated with resource economic mathematics at play in system-task interfaces; i.e., performance resource availability must exceed demand ($R_A \geq R_D$) for "success" give rise to new methods of performance prediction in complex tasks.

GSPT and Fitts' Law

As noted, GSPT suggested that speed-accuracy PCEs be considered for human movement systems. For any PCE defined according to GSPT, a single number composite performance measure (i.e., $CP_{S\text{-}A}$) can be obtained as the PCE volume (or, in this case, area):

$$CP_{S\text{-}A} = k \cdot \text{Speed} \cdot \text{Accuracy} \qquad (5)$$

where k is a scaling constant. It was unavoidable but to wonder how this metric would relate to Fitts' IP, which we had been incorporating extensively in the design of instruments to measure aspects of human coordination (e.g., Kondraske, 1990; Saganis et al., 2020).

Methods

Three experiments using de-identified data previously collected during research and development of human performance measurement tools were conducted. Each involves a version of the alternating tapping Fitts' paradigm. Data for Experiments I and II was collected to evaluate a modular human performance measurement system (Kondraske, Potvin, Tourtellotte, and Syndulko 1984; Kondraske, 1990) as part of a center grant. Participants (n = 452; 267 female, 185 male), self-declared healthy, ranged in age from 7 to 83 years (mean = 36.4, sd = 16.6). Many contributed more than one test session. For Experiment III, data was obtained from a dataset created with a web-based tool (RC21X) for cognitive and neuromotor performance measurement (Saganis et al., 2020). Measures from 3rd and 4th self-administered sessions were used for participants (n= 33; 3 female and 30 male) ranging in age from 10 to 74 years (mean = 48.5, sd = 15.8). When asked if healthy, 19 responded "yes" while 14 responded "no".

Experiment I data was collected with a computer-based device with six touch sensor regions (two targets with A = 40.6 cm and W = 1.6 cm, each flanked by two large error regions) during an upper extremity task requiring medial-lateral reciprocal motion. Experiment II employed a similar device (A = 52.0 cm, W = 10.5 cm) in a lower extremity task, while Experiment III data was collected using RC21X in an upper extremity test involving reciprocal tapping between the "A" and "L" keys on a computer keyboard. All set-ups involved the execution of two 10s trials. For Experiments I and II, the better of two trials was retained and available for use and many participants contributed more than one test session. For Experiment III, data from both trials was used.

For each trial (745, 745, and 66 for Experiments I, II, and III respectively), measures of movement speed (i.e., 1/MT) and accuracy (%) were used to compute the accuracy adjusted (Equation(4)) Fitts' IP and the GSPT-based composite $CP_{S\text{-}A}$ using Equation (5) with k = ID for each A/W case (W = *actual width*). Scatter plots were prepared to explore relationships.

Results

Figure 1 facilitates comparison of Fitts' IP and GSPT-based $CP_{S\text{-}A}$. Pearson's r ranged from 0.96 to 0.99. The average of the absolute value of the percent difference (i.e., average of $|100 \cdot (CP_{S\text{-}A} - IP)/IP|$) ranged from 2.6% for Experiment II to 8.1% for Experiment I.

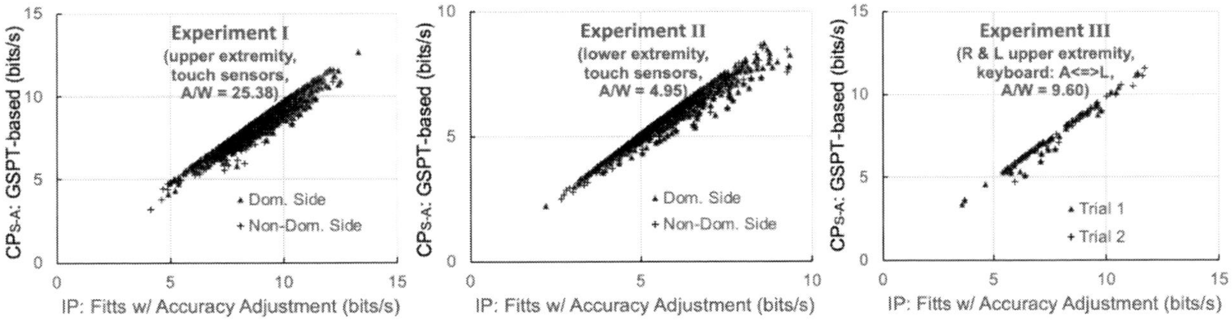

Figure 1. In three contexts and over a wide range of values, strong agreement was found between the simple-to-compute GSPT-based Composite Performance ($CP_{S\text{-}A}$ = ID • Speed • Accuracy) and Fitts' Index of Performance (IP) with accuracy adjustment.

Discussion

In multiple Fitts' repetitive-motion, fixed-target-width pointing tasks with a range of ID and IP values, we found a very strong, near-perfect correlation between Fitts' Index of Performance (IP, bits/s) and GSPT-based Composite Performance ($CP_{S\text{-}A}$), with close agreement of actual values when k in Equation (5) is Fitts' ID computed using the *actual target width*. While we could have used simulated data, we opted to argue by using real experimental data to define the ranges of interest and perhaps communicate this interesting finding in a more direct and powerful way.

The simple and intuitively attractive Equation (5), based on the generalized concept of a PCE and its volume, provides essentially the same result as the relatively more complex and awkward Fitts' IP with the accuracy adjustment expression. There are clearly *some* differences. Preliminary analysis shows a relationship between these differences and task accuracy (or error) rate, with the largest differences for large error rates (e.g., 50%). It is not feasible, at present, to characterize such differences as "error", as that would require the assumption that the IP value is indeed a solid gold standard. The extensive Fitts'-related literature questioning aspects of conceptualization and proposing various tweaks, in part, argues against that premise.

We have noted that the well-known expressions of Fitts law and IP do not explicitly contain speed or accuracy variables. However, we also note that 1/MT has the units of speed. In $CP_{S\text{-}A}$, speed is expressed as motions/s. Comparing expressions for IP with the accuracy adjustment and $CP_{S\text{-}A}$ leads to a focus on the equivalency of [Accuracy (%) • $\log_2$ (A/W + 1)] and [$\log_2$ (A/W_e + 1)]. Data presented here illustrates a very high equivalency level. Expounding on this via both conceptual and mathematical avenues is likely.

Equation (5) contains a performance index ($CP_{S\text{-}A}$), speed, accuracy, and task index of difficulty (ID). Given any three, the fourth can be computed. Of course, there have been similar interests in the use of Fitts' law. One can argue they contribute to the existence of an "accuracy adjustment" to allow consideration of arbitrary accuracy rates (i.e., not just 96%). With such motives, Wobbrock and colleagues (2008) proposed what they termed an "error model for pointing based on Fitts' law". With Accuracy(%) defined as 1 - Error(%), our preliminary review suggests that an "accuracy model for pointing", based on GSPT and Fitts' law, will provide similar results with a simpler expression (i.e., Accuracy (%) = $CP_{S\text{-}A}$/(ID • Speed), where $CP_{S\text{-}A}$ is equivalent to Fitts' IP). Further analysis is warranted.

Fitts' contribution with regard to the definition of ID is not only useful but elegant in its simplicity. MacKenzie (1992) discusses this type of appeal with regard to Fitts' law. While an apparently sound conceptual basis can be argued for the accuracy adjustment, there are some initial assumptions involved that lead to the computational complexity present in the adjustment and a detraction from the simplicity appeal. Our results suggest a review of such assumptions and their impact in defining the speed-accuracy tradeoff in human pointing motions.

The powerful idea of PCEs can be traced to an aerospace context, where the dimensions of performance (i.e., speed, altitude, and range) and the metrics used (i.e., a larger value means "more" of that quantity) *naturally* lead to a PCE. This is not the case or so clear for many systems, where the commonly employed metrics (e.g., error vs. accuracy; time as a *speed-related* measure) do not result in an envelope! One might wonder about his modeling efforts if Fitts incorporated the notion of a speed-accuracy PCE. We emphasize that the speed-accuracy PCE can apply to not only human movement, but also information processing in general. It is perhaps unfortunate that work in those areas relies on time and error measures instead speed and accuracy, subverting identification of a PCE. It is also useful to observe that PCEs of greater dimensionality that incorporate dimensions of performance other than speed and accuracy are likely to be of great interest in human performance characterizations and modeling efforts.

References

Fitts, P. (1954). The information capacity of the human motor system in controlling the amplitude of movement. *Journal of Experimental Psychology*, 47(6), 381–391.

Kondraske, G. V. (1990). A PC-based performance measurement laboratory system. *Journal of Clinical Engineering*, 15(6), 467-478.

Kondraske, G. V. (1999). Determination of Fitts' Index of Performance using constructs of General Systems Performance Theory and implications for motor control performance capacity measurement. In A.M. Sherwood (Ed), *Seventeenth Annual Houston Conference on Biomedical Engineering Research*, Houston (p 86).

Kondraske, G.V. (2011). General Systems Performance Theory and its application to understanding complex system performance. *Information • Knowledge • Systems Management*, 10 (1-4), 235-259.

Kondraske, G. V., Potvin, A. R., Tourtellotte, W. W., & Syndulko, K. (1984). A computer-based system for automated quantitation of neurologic function. *IEEE Transactions on Biomedical Engineering*, 31(5), 401-414.

Kondraske, G. V., & Stewart, R. M. (2008). Web-based evaluation of Parkinson's Disease subjects: objective performance capacity measurements and subjective characterization profiles. In *Proceedings of the 30th Int. Conf. of the IEEE Engineering in Medicine and Biology Society*, Vancouver, 799-802.

MacKenzie, I. S. (1992). Fitts' law as a research and design tool in human-computer interaction. *Human-Computer Interaction*, 7, 91-139.

Saganis, X., Goreczny, A. J., Kondraske, G., Berman, M., Comnick, I., Allen, T., & Nussbaum, P. D. (2020) Test–retest reliability of RC21X: a web-based cognitive and neuromotor performance measurement tool. *Archives of Clinical Neuropsychology*, 00 (2020), 1–7. doi.org/10.1093/arclin/acaa056

Soukoreff, R. W., & MacKenzie, I. S. (2004). Towards a standard for pointing device evaluation, perspectives on 27 years of Fitts' law research in HCI. *Int. J. Human-Computer Studies*, 61 (2004), 751–789. doi:10.1016/j.ijhcs.2004.09.001

Wobbrock, J. O., Cutrell, E., Harada, S., & MacKenzie, I. S. (2008). An Error Model for Pointing Based on Fitts' Law. In *CHI '08: Proceedings of the SIGCHI Conference on Human Factors in Computing Systems*, 1613–1622. doi.org/10.1145/1357054.1357306

Zimmerman, M.E. (2011) Speed–Accuracy Tradeoff. In: J.S. Kreutzer, J. DeLuca, B. Caplan (Eds) *Encyclopedia of Clinical Neuropsychology*. Springer, New York, NY. doi.org/10.1007/978-0-387-79948-3_12

A SALUTE TO ROBERT KEY DISMUKES: A MENTOR FOR TRANSLATIONAL RESEARCH IN HUMAN FACTORS

Janeen Adrion Kochan
Aviation Research, Training, and Services, Inc.
Winter Haven, FL

As the fields of study associated with human factors (aviation psychology, cognitive systems engineering, engineering psychology, etc.) become broader in scope, the drive to bring the findings from academic research to those who can benefit from study findings must also expand. This paper honors Robert Key Dismukes, Ph.D., through a case study that illustrates how the bridge from research to practice (and back to research) can be built and how human factors professionals can translate and share what they know with new scientists, target populations, and the public at large. This review of Dr. Dismukes' work demonstrates how the findings from human factors research can be brought to the operational world with a focus on his mentorship and modeling of ethical science.

The year was 1997 and the airline had just suffered their first fatal accident the previous December. The mandate for US FAA Crew Resource Management (CRM) training had not yet taken effect. The impending requirement had been published in the Federal Register on 14 June 1996 to become effective 19 March 1998 (14 CFR 121.404). Although the airline had a voluntary "Cockpit Resource Management" class, it was still in the day where CRM was considered charm school (Helmreich & Foushee, 2010). It was obvious that there was little interest in the topic from the lack of luster in the few-hour class presentation and the pilots' lack of interest. Furthermore, it was clear that there was a need to respond to circumstances noted in the accident report (NTSB, 1997) and to prepare for the new CRM training requirement. The training team decided to leverage the use of more Line Oriented Flight Training (LOFT) which required effective debriefings with the crews. Although published guidance (Advisory Circular 120-51E) includes information about LOFT debriefings, additional resources were needed to adequately train the instructor pilots on *how to* conduct an effective LOFT debriefing. Additional resources were also needed to create research and science-based training materials for the CRM classroom and simulator scenario development. Fortunately, there were publications with just the information needed for facilitating these debriefings (Dismukes et al., 2000).

Translational Research

Many useful findings are brought to the operational world through agencies and organizations that have contracted for research. However, this is not necessarily *translating* the findings directly into suggestions, instructions, procedures, or practices that are useful to end users. Although formal translational research methods were first mentioned in the medical field in the 1970s (Wolf, 1974), the identification and definition of the construct is still illusive (Austin, 2018; Krueger et al., 2019). Little interest was shown in the topic until the 2000s which is still contained primarily to the biomedical and associated clinical arenas, with little or no activity in direct human factors research (Krueger et al., 2019). However, as the vingette above suggests, there can be infomal avenues to disseminate needed information. It is these informal pathways that were utilized by Dr. Dismukes and his collegues thus making critical materials available to the training team at the airline.

Fleming et al. (2008) expanded upon the traditional 'from bench to beside' definition of translational medicine to include a feedback loop from the users, in this case the community and public health practitioners. In their framework for translational research, the inclusion of follow-up with the users of the clinical research findings through evidence-based practice and patient utilization provides for assessment of the entire process. Thus, the formalized process of translation in biomedical models is moving beyond the linear path from basic research, applied research, preclinical, clinical, and standard practice of care to include aspects of traditional human factors (Pettibone et al., 2018). As applied to human factors research, the idea would be to bring the research findings to the field, and then assess whether the findings accomplished the intended goal(s). The monitoring of unintended (positive and negative) consequences would be integral to the feedback loop. Knowing the audience and the needs of the end user is important in the assessment of success (or failure) of a program that is brought forth from the research community. Not only did Dr. Dismukes investigate human performance from a basic and applied perspective, he put forth the effort to know his audience, the problem space, and the conditions surrounding the humans and teams that could benefit from his research findings. In aviation, he earned an Airline Transport Pilot certificate for airplane multiengine land with type ratings in the Boeing 737 and Cessna 500 (Citation) and Commercial Pilot privileges for airplane single engine land and glider. He was also a tow pilot for gliders as well as a Certified Flight Instructor for glider and competed in (and won) numerous glider contests over the years. A colleague points out that Dr. Dismukes "embedded himself with the troops" by participating in airline pilot training so that he could understand the operational context and personally relate to the demands and pressures of the domain (I. Barshi, personal communication, March 5, 2021).

In an attempt to further clarify the terms associated with translational research and translational science, Austin (2018) explains that the definition of *translation* is "the process of turning observations in the laboratory, clinic and community into interventions that improve the health of individuals and the public – from diagnostics and therapeutics to medical procedures and behavioural changes" (p. 455). While *translational research* focuses on a specific case or disease, *translational science* is concerned with the general case or disease. Therefore, individual translational research projects can be aggregated to contribute to translational science as well as to test principles associated with translational science. Dr. Dismukes' body of research into pilot expertise and memory produced a number of studies with findings that he moved into the real-world of flying. One of the best examples (see also Table 1) was his academic work on prospective memory (Dismukes, 2006) that was translated to usable tips for pilots (Dismukes, 2015).

The medical model of formal translational research and translational science deals almost exclusively with bringing therapeutics and biomedical devices to the end users in healthcare. Similarly, a human factors model would consist of tools, equipment (hardware/software/interfaces), technologies, policies, procedures, processes, and training methodologies that could be applied (translated) to the applicable domain(s) of interest.

Indeed, the healthcare industry could benefit from human factors translational research as well. Though the concept of translational research has a variety of ethical perspectives to consider (DeRenzo, et al., 2020) and inherent gaps in the process as identified in the healthcare arena, it is clear that the concept is worthy of consideration in human factors research (Rubio et al., 2010). As early as 1977, Dr. Dismukes' writings communicated clearly about the ethics of research (Dismukes, 1977; 1979; 1980) and that thread was woven tightly throughout his research career making his mentoring of particular value.

Teaching Translational Research through Mentoring

So where does mentoring fit in? First, there is very little teaching of *how to do* translational research or translational science (Rubio, 2010). This is not to say that research findings cannot find their way to a target audience or end-user of the information. There are many avenues in place to contract for human factors work with the aim of providing information for use in a particular domain. However, those avenues are often long-and-winding roads where the information may not arrive intact and may be fraught with delays, potentially leaving a hazard unchecked. It is unknown what may be lost in translation without a clear path and someone who understands the spectrum of the journey. Dr. Dismukes made it part of his research methods to keep the usefulness of his studies in mind. One colleague notes that, "the most important idea that Key tried to impress upon any of us who worked with him in a collaborative setting in the research lab was to really think deeply about an issue or question being proposed, especially in terms of how it might be applied or used in as many other settings or applications as possible in order to get the most out of an investigation" (K. Jobe, personal communication, March 15, 2021). He was known to apply research findings in meetings such as using the "sixty-second rule" (to wait a full 60-seconds before answering a question) (Dismukes et al., 2000).

Through a unique and open style, Dr. Dismukes taught others the value of and techniques for translating research that could facilitate the usefulness of their work. This was not generally accomplished through a formal mentoring program or process, but was highly effective. Walk-the-walk, teach by example, practice what is preached, are all phrases that could be applied to Dr. Dismukes. Whereas mentoring is another construct that has a multitude of definitions and by definition is multi-faceted (Dominguez & Kochan, 2020), there is no doubt that Dr. Dismukes was a skilled mentor. Typical behaviors and processes used in mentoring such as taking time to hear the mentee (protégé), modeling good values and ethics, showing how to do things by example, and sharing tacit knowledge (Budd, 2007; Irby et. al., 2020) were his standard operating procedures. Mentoring research often focuses on the dynamics of the relationship between the mentor and mentee in a particular context (Budd, 2007; Janssen et al., 2016), temporal influences, culture, and developmental mechanisms (Irby et al., 2020; Janssen et al., 2016). But, there is only brief mention of *how to do* informal mentoring as demonstrated expertly by Dr. Dismukes whether at his office at NASA or at the Williams Soaring Center where he was known to be able to share his expertise with the pilots in a way that made sense to them.

The Case Study

An investigation into Dr. Dismukes publications was conducted to provide evidence of his work in translational research, mentoring, and ethical science. All publicly available information and documents were located through online searches on Google Scholar, ResearchGate, the Hunt Library, Swisscows, and the NASA Ames website using key words Robert Key Dismukes, R. Key Dismukes, and Key Dismukes. Publications with Dr. Dismukes as an author were reviewed and categorized as *strictly research*, *research and operational*, or *operationally focused* where content of strictly research papers had been translated to be of use to those in the field.

Results

A total of 127 publications (to include books, book chapters, papers, and other publications) were found. Of the 127 publications, 83 were strictly research, 30 operationally focused, and 14 were written for both the research and operational audience. This indicates that over one-third of the publications brought scientific research results to the operational world and over half of his publications were of use to an operational audience. Academic or basic research projects that were clearly translated into information for end users were also noted and examples of such are displayed in Table 1. It is noteworthy that the translation time from the academic research to the outlet for the end-users is very short which is illustrative of Dr. Dismukes' goal of making the findings from his work available to all.

Table 1.

Dismukes' Examples of Translating Human Factors Research

Academic Research Papers *Translated to:*	Operational Outlet or Another Domain
Dismukes, R. K. (2006). Concurrent Task Management and Prospective Memory: Pilot Error as a Model for the Vulnerability of Experts. *Proceedings of the Human Factors and Ergonomics Society 50th Annual Meeting* (pp. 903-114). San Francisco, CA: HFES.	Dismukes, R. K. (2010). Remembrance of things future: Prospective memory in laboratory, workplace, and everyday settings. In D. H. Harris (Ed.), *Reviews of human factors and ergonomics*, 6, 79-122. Santa Monica, CA: Human Factors and Ergonomics Society.
Dismukes, R. (2012). Prospective memory in workplace and everyday situations. *Current Directions in Psychological Science, 21*(4), 215-220.	Dismukes, R. K. (2015). Cognition, aging, and the soaring pilot. *Soaring* 79 (10) 35-37.
Dismukes, R. K. & Berman, B. (2010). *Checklists and monitoring in the cockpit: Why crucial defenses sometimes fail.* NASA Technical Memorandum (NASA TM-2010-216396). Moffett Field, CA: NASA Ames Research Center.	Berman, B. A. & Dismukes, R. K. (2010). Designing a better mousetrap. *AeroSafety World*, 12-17, July 2010.
McDonnell, L., Jobe, K., & Dismukes, R. K., (1997) Facilitating LOS Debriefings: A Training Manual (NASA Technical Memorandum 112192). *Moffett Field, CA: National Aeronautics and Space Administration.*	Dismukes R. K., & Smith, G. (2000). *Facilitation and debriefing in aviation training and operations.* Aldershot, UK: Ashgate.

Conclusion

As informal mentoring can be as helpful as formal mentoring for students and incoming colleagues (Irby et al., 2020: Janssen et al., 2020), so can the idea of informal roads to translational research for the human factors community. More formal methods and guidance on how to move our human factors and aviation psychology research into practice should prove to be useful, but in the meantime every effort to share our findings where they may best take root in a timely manner should be considered. As one of Dr. Dismukes later mentees, I found his model of ethical research and his style of mentoring to be precious gems in my research and flying journey. I know I speak for many that this world is a better place and aviation is safer because of the contributions of Dr. Robert Key Dismukes.

Acknowledgements

The views expressed in this paper are solely those of the author and are not associated with any other organization or entity. I would like to thank Dr. William Adrion, Dr. Barbara Burian, Dr. Jessica Cruit, and Michaela Satter for their reviews of earlier versions of this paper.

References

Austin, C. P. (2018). Translating translation. *Nature Reviews*, Volume 17, July 2018.

Budd, F. C. (2007). Mentoring in the U.S. Air Force: A cornerstone for success through organizational transformation. *Performance Improvement*, Volume 46, No. 3, March 2007. DOI: 10.1002/pfi.112

DeRenzo, E. G., Singer, E. A., & Moss, J. (2020). Special issues raised by evolving areas of clinical research. In E. G. DeRenzo, E. A. Singer & J. Moss (Eds.), *Ethical considerations when preparing a clinical research protocol* (2nd ed., pp. 271-298). Elsevier. https://doi.org/10.1016/B978-0-12-386935-7.09989-8

Dismukes, R. K. (2015). Cognition, aging, and the soaring pilot. *Soaring* 79 (10) 35-37.

Dismukes, R. K. (2006). Concurrent task management and prospective memory: Pilot error as a model for the vulnerability of experts. *Proceedings of the Human Factors and Ergonomics Society 50th Annual Meeting* (pp. 903-114).

Dismukes, R. K. (1980). The Supreme Court and patenting life: life is patently not human-made. *Hastings Center Report* 10(5):11.

Dismukes, R. K. (1979). What should society expect from scientists? *Bulletin of Atomic Scientists 35*(9):19. Reprinted in Current (1980) 220:23. DOI: 10.1080/00963402.1979.11458658

Dismukes, R. K. (1977). Recombinant DNA: a proposal for regulation. *Hastings Center Report* 7:25. Reprinted in Bioethics: Selected Readings. T. A. Shannon (1980) (revised). New York: Paulist Press.

Dominguez, N. & Kochan, F. (2020). Defining mentoring: An elusive serch for meaning and a path for the future. In (Eds.) B. Irby, J. N. Boswell, L. J. Searby, F. Kochan, R. Garza, & N. Abdelrahman *The Wiley international handbook of mentoring: Paradigms, practices, programs, and possibilities* (pp. 1-18). Wiley. DOI:10.1002/9781119142973.ch1

Federal Aviation Administration. (2004). Advisory Circular No. 120-51E, *Crew Resource Management Training*, 1/22/04, AFS-210.

Fleming, E. S., Perkins, J., Easa, D., Conde, J. G., Baker, R. S., Southerland, W. M., Dottin, R., Benabe, J. E., Ofili, E. O., Bond, V. C., McClure, S. A., Sayre, M. H., Beanan, M. J., & Noris, K. C. (2008). The role of translational research in addressing health disparities: A conceptual framework. *Ethnicity & Disease,* Volume 18, Spring 2008.

Helmreich, R. L, & Foushee, H. C. (2010). Why CRM? Empirical and theoretical bases of human factors training. In E. L. Wiener, B. G. Kanki, & R. L. Helmreich (Eds.), *Crew Resource Management* (2nd ed., pp. 3-57). Academic Press. https://doi.org/10.1016/B978-0-12-374946-8.10001-9

Irby, B., Boswell, J. N., Searby, L. J., Kochan, F., Garza, R., & Abdelrahman, N. (Ed.). (2020). *The Wiley international handbook of mentoring: Paradigms, Practices, Programs, and Possibilities.* Wiley.

Janssen, S. , van Vuuren, M., & de Jong, M. D. T. (2016). Informal mentoring at work: A review and suggestions for future research. *International Journal of Management Reviews*, Vol. 18, 498–517. DOI: 10.1111/ijmr.12069

Krueger, A. K., Hendriks, B., & Gauch, S. (2019). The multiple meanings of translational research in (bio)medical research. *History and Philosophy of the Life Sciences* 41, 57. https://doi-org.ezproxy.libproxy.db.erau.edu/10.1007/s40656-019-0293-7

National Transportation Safety Board (NTSB) (1997). Public Docket. *ABX Air (Airborne Express), Douglas DC-9-63, N827AX, Narrows, Virginia, December 22, 1996. NTSB/AAR-97/05.* Washington, D.C., National Transportation Safety Board.

Pettibone, K. G., Balshaw, D. M., Dilworth, C., Drew, C. H., Hall, J. E., Heacock, M., Latoni, A. R., McAllister, K. A., O'Fallon, L. R., Thompson, C., Walder, N. J., Wolfe, M. S., Wright, D. S., & Collman, G. W. (2018). Expanding the concept of translational research: making a place for environmental health sciences. *Environmental Health Perspectives* 126(7):07450.

Rubio, D. M., Schoenbaum, E. E., Lee, L. S., Schteingart, D. E., Marantz, P. R., Anderson, K. E., Platt, L. D., Baez, A., & Esposito, K. (2010). Defining translational research: Implications for training. *Academy of Medicine*, 85:470-5.

Wolf, S. (1974). Editorial: the real gap between bench and bedside. *New England Journal of Medicine*, 290 (1974), pp. 802-803.

SYSTEMATIC REVIEW OF WEATHER OBSERVATION AND FORECAST RESOURCES AVAILABLE TO GENERAL AVIATION PILOTS

J.W. Wallace
Oklahoma State University
Stillwater, OK

Nicoletta Fala
Oklahoma State University
Stillwater, OK

Weather continues to be a consistent hazard for pilots despite decades of progress in both pilot education and weather observation and forecasting technology. Much research has been done on the various facets of this problem, from pilot psychology to the weather information sources themselves. Weather-Intelligent Navigation Data and Models for Aviation Planning (WINDMAP) is a NASA University Leadership Initiative (ULI) that aims to use Unmanned Aerial Systems (UAS) to improve the accessibility and accuracy of weather information for General Aviation (GA) pilots and UAS operators. This paper aims to produce a systematic review of research on the topic using the Preferred Reporting Items for Systematic reviews and Meta Analyses (PRISMA) method that will then guide a further survey-based study of source utilization by GA pilots and UAS operators. Through the survey, we aim to evaluate satisfaction and need for improvements among weather products and education.

Weather continues to be a problem for pilots despite advances in both weather observation/forecasting technology and pilot education. According to the most recent complete Nall Report, of 42 weather related accidents in 2017, 32 proved fatal (AOPA Air Safety Institute, 2020). Despite making up only 4% of total accidents, the high fatality rate when compared to most other accident types makes weather accidents a problem worth investigating. Additionally, the growing Unmanned Aerial Systems (UAS) industry and budding development of Urban Air Mobility (UAM) will demand different or better weather observation and forecasting technology. To develop new and useful systems for pilots and UAS operators we must first review prior work and evaluate what products and information is available against the community's needs. The purpose of this review is to evaluate recent research in pilot education and weather observation and forecasting technology to better inform future work. Following the conclusions of this review, we will conduct a survey of both general aviation pilots and UAS operators to determine their respective aviation weather product awareness, use, and needs, as well as attempt to identify areas where new products could better serve UAS operators.

This research supports the WINDMAP NASA University Leadership Initiative (ULI), a four-year project which aims to address needs in real-time weather forecasting to improve the safety of low-altitude aircraft operations by integrating real-time observations from drones and other aircraft with weather prediction and flight management systems (Jacob, 2020). The literature review and survey introduced in this paper will provide customer requirements to inform system design and research within WINDMAP.

Methodology

The Preferred Reporting Items for Systematic reviews and Meta Analyses (PRISMA) method presents a systematic review method which allows us to simultaneously cast our literature net as wide as possible while also being able to narrow down to relevant literature in an efficient manner, as shown in Figure 1 (Moher, Liberati, Tetzlaff, & Altman, 2009). We included three search databases (Google Scholar, Scopus, and Microsoft Academic Graph) in the search. Each search used the same five sets of keywords: (1) 'aviation' or 'aircraft' or 'cockpit', (2) 'weather', (3) 'safety' or 'hazard' or 'risk' or 'decision making' or 'decision-making', (4) 'training' or 'education' or 'instruction' or 'information', and (5) 'pilot' or 'UAV' or 'drone'. Combining the items using Boolean operators yielded the following search criteria: "(aviation OR aircraft OR cockpit) AND weather AND (safety OR hazard OR risk OR 'decision making' OR 'decision-making') AND (training OR education OR instruction OR information) AND (pilot OR uav OR drone). Including 'UAV' or 'drone' proved to be more of a hindrance than a benefit as we ended up rejecting most of the papers with those keywords for failing to address the human UAS operators, instead focusing on the autonomous systems.

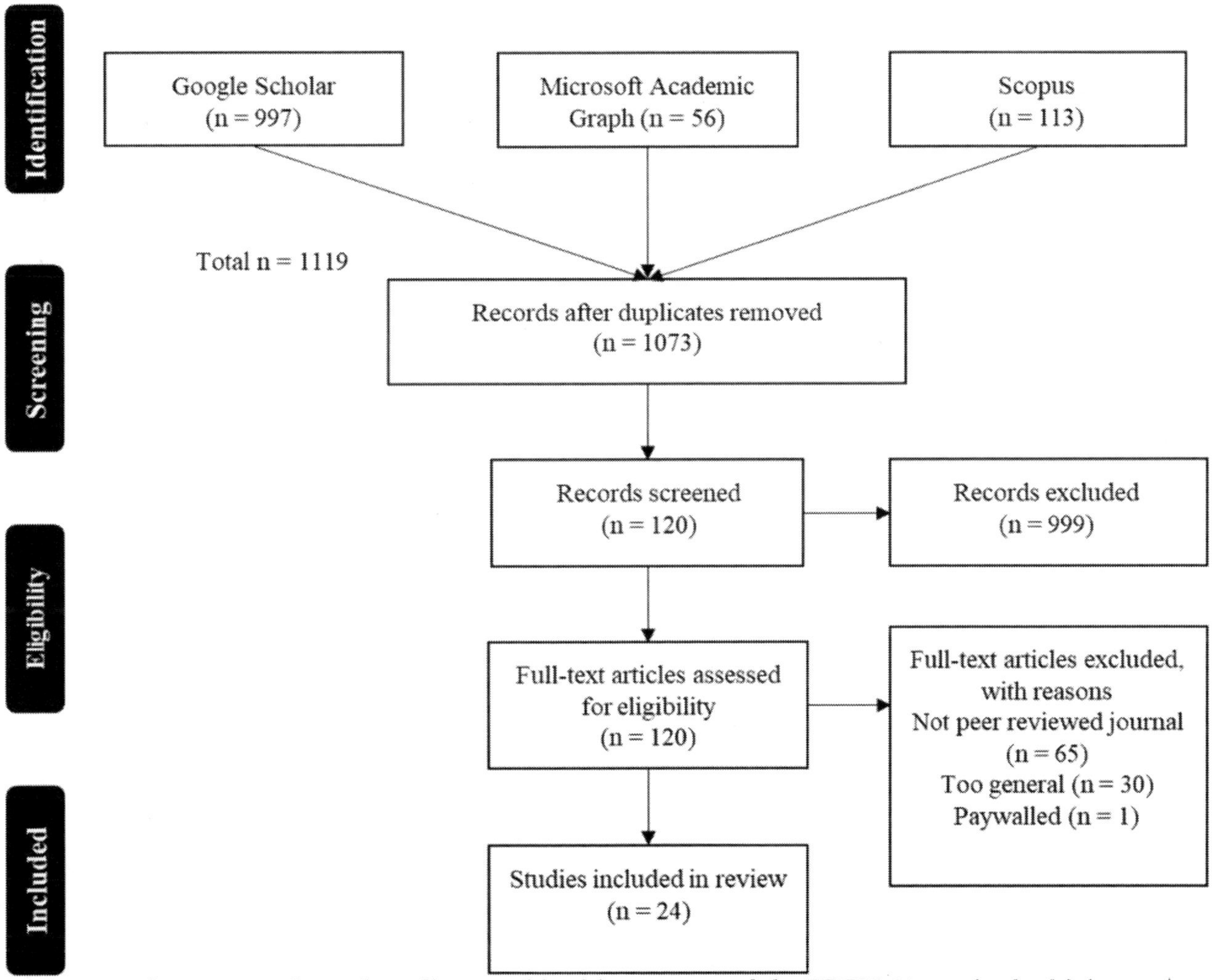

Figure 1. The systematic review discussed in this paper used the PRISMA method which consists of four steps that narrow down the identified papers based on relevance.

Removing duplicates resulted in a list of 1073 papers. Two reviewers (JW and NF) scanned titles and abstracts to determine eligibility and exclude papers which were not relevant to the subject. The two reviewers classified papers as *include*, *exclude*, and *maybe include*, and advanced any papers that belonged in the *include* and *maybe* categories to the full-text eligibility assessment. JW's review advanced 120 papers for full-access eligibility and NF's 130, with a conflict of 65 papers (5.6%). However, the conflict percentage includes disagreements where one reviewer classified a paper as *include* and the other as *maybe include*. Advancing both categories to full-text assessment eligibility decreased the conflict. We used the Rayyan web app to do this review (Ouzzani, Hammady, & Elmagarmid, 2016). At the full-text assessment stage, we evaluated papers for eligibility using two inclusion criteria: the papers had to be published in a peer-reviewed journal and have an adequate focus on pilot-weather interactions.

Results

We selected 24 articles that meet the inclusion criteria. Articles selected covered a range of topics from methods for educating pilots about new weather products to the development and implementation of the weather products themselves. While three selected papers do not directly address pilot-weather product interactions, they contain relevant information for design and technology implementations. We identified three themes in the reviewed literature. The accuracy and interpretation of weather products by pilots was the primary focus of most of the papers reviewed, some focusing more on the weather products and others more on the pilots. Papers focusing on the products themselves frequently addressed the symbology used by the product to convey weather information, while those focusing on pilots examined the use, effects, and education considerations for different weather products. A third theme emerged focusing on the pilots' biases and experiences with poor weather. This section describes the prior research in the literature in the context of the three themes.

Theme 1: Weather Products

Weather products are a central theme in ten of the papers reviewed. Within this theme there emerged two subthemes: symbology, and non-graphical modes of communicating information.

While papers on symbology were not definitive in their recommendations with respect to display symbology, they indicated that the graphical language used impacted pilot interpretability. Weather display symbology impacted pilot behavior and decision making in both VMC and IMC simulated flights (Ahlstrom, 2015). However, rather than recommending specific symbology for weather displays, Ahlstrom recommended that the development and assessment of a cockpit application which would automatically track and alert the pilot to weather conflicts or changes.

Papers with design as a central theme researched additional modes of conveying information. Pilot aids, in the form of either general digital copilots or more specialized tools, were featured in two papers. A digital copilot decreased head-down time in all tasks except determining the weather communication frequency (Wilkins, 2018). While the tasks assessed do not relate to pilot interpretation of weather information in the cockpit, the technology shows promise. A Risk Situation Awareness Tool (RSAT) also shows promise as an additional input source for pilots making decisions about how to best route around thunderstorms and other hazardous weather (Parmar & Thomas, 2020). The study presented pilots with NEXRAD loops with flight paths overlaid with and without the RSAT calculated risk and asked them to determine if the path was safe, or to determine which of two paths was safest. The study found that pilots who used RSAT were more likely to choose safer flight paths than the control group.

NEXRAD has been in use in GA for some time, but the topic of its reliability is not settled, with some researchers arguing that the current NEXRAD cannot reliably enable safe flight around heavy weather (Knecht, 2016). Knecht developed a study using a storm model to generate a looping NEXRAD-type simulation, and found that weather movement greatly degraded safety while weather depth had no effect. Knecht recommends adding future predicted weather and a range ring to NEXRAD to improve safety.

Theme 2: Education

Education played a large role across the literature reviewed. With many new technologies becoming available, research needs to evaluate 1) whether (or how much) education is required on how to use these new technologies and 2) if education is needed, integration of new technology education into existing training for new pilots.

A two-hour course on NEXRAD for GA pilots improved the subjects' knowledge scores and ability to apply concepts in paper-based scenarios (Blickensderfer, et al., 2015). However, the study did not employ a simulation or flight evaluation of pilot knowledge. This study affirms findings by a similar study on NEXRAD education, where a short course provided similar benefits to pilots (Cobbett et al, 2014).

The introduction of Electronic Flight Bags (EFB) improved preflight skill development and aeronautical decision making in student and private pilots with under 100 total flight hours (Misra & Halleran, 2019). In this study, participants not given EFBs were less likely to detect weather-related hazards. However, while EFBs proved useful, it is important for ab-initio pilots' interpretation, analysis, and decision making skills to be able to make accurate decisions without the assistance of an EFB (Misra & Halleran, 2019). An analysis of instrument approach accidents between 2002 and 2012, found that instrument approach accidents peak around 120 days after the last Instrument Proficiency Check (IPC) (Fanjoy & Keller, 2013). However more accidents occurred closer to the IPC date than further out. Current FAA IPC regulations do not mandate what training is required for IPCs, only giving a recommendation instead (Fanjoy & Keller, 2013). A more recent FAA Advisory Circular provides additional information on how to conduct an IPC, including guidelines for an IPC conducted in an approved simulator, but Advisory Circulars are not regulatory (Federal Aviation Administration, 2018). Evaluating the effectiveness of the newest updates to IPC guideline could have research potential.

Evaluating thunderstorm-related accidents from the NTSB database from 1996 to 2014 determined that the majority of flights resulting in accidents violated FAA-recommended separation distance from extreme convection (Boyd, 2017). Boyd argues for additional emphasis on thunderstorm hazards and safe practices during ab-initio and recurrent pilot training.

Theme 3: Pilot attitudes, biases, and experiences

To design effective weather products for pilots we need to know how pilots behave as humans. Papers that address pilot attitudes with respect to hazardous weather as well as cognitive biases in the general aviation pilot population help investigate how pilots think and make decisions and the research has applications in weather decision making. Developing tactics to combat risk-prone attitudes and de-bias pilots may prove helpful in reducing weather related fatalities in general aviation.

Common cognitive heuristics such as anchoring and adjustment, confirmation, and outcome bias, can lead to cognitive biases with adverse effects in three different studies of weather-related decision making (Walmsley & Gilbey, 2016). Weather reports obtained pre-flight affect pilots' interpretation of weather in-flight, evidence of anchoring bias. In one of the reported studies, pilots interpreted the decisions of pilots who flew into deteriorating weather more favorably when the outcome was positive than when it was negative, evidence of outcome bias. Another study found no evidence that pilots favored disconfirmatory evidence over confirmatory evidence when deciding which environmental cues were most useful in deciding whether to continue a flight. Using the "considering the alternative" technique to reduce the effect of the two negative biases identified in previous studies and de-bias weather-related decision making was ineffective at countering both biases (Walmsely & Gilbey, 2017).

Research on pilot attitudes may also point to differences between pilots who avoid adverse weather and those who do not (O'Hare, Hunter, Martinussen, & Wiggins, 2011). Pilots with more recent flight time may be more likely to be involved in adverse weather encounters, and pilots who are risk intolerant less likely. Experienced pilots with instrument ratings and high levels of instrument flight time were more likely to have not flown "VFR into IMC," though they have encountered weather conditions of significant concern during flight. Flight training hours nor number of flight safety seminars attended in the past year were not helpful in discriminating the three groups of pilots, casting doubt on the efficacy of flight safety seminars and flight instruction. Given enough exposure nearly all pilots will encounter weather conditions, some will emerge emboldened and optimistic about their skills while others will emerge more cautious and unwilling to encounter such conditions again (O'Hare, Hunter, Martinussen, & Wiggins, 2011).

Conclusion and Future Work

General aviation weather products, training, and pilots represent a complex system which spans many disciplines and industries. In this paper, we did a systematic review of the literature on weather information and products and how pilots use them. The review did not identify any research on what information UAS operators require or how they use it. WINDMAP aims to use drones to add to our weather observation, forecasting, and reporting capabilities for all low-level flying operations. While the literature review did not result in UAS weather decision making requirements, our future work includes developing and disseminating a survey to General Aviation pilots and UAS operators to identify their weather information needs. The needs identification from this literature review and upcoming survey will help WINDMAP develop new and improved weather products.

Acknowledgements

This work is funded by NASA Aeronautics Research Mission Directorate (ARMD) University Leadership Initiative Grant Number 80NSSC20M0162.

References

Ahlstrom, U. (2015). Weather display symbology affects pilot behavior and decision-making. *International Journal of Industrial Ergonomics, 50*, 73-96.

AOPA Air Safety Institute. (2020). *30th Joseph T. Nall Report.* Richard G. McSpadden Jr.

Blickensderfer, E. L., Lanicci, J. M., Vincent, M. J., Thomas, R. L., Smith, M., & Cruit, J. K. (2015). Training General Aviation Pilots for Convective Weather Situations. *Aerospace Medicine and Human Performance, 86*(10), 881-888.

Boyd, D. D. (2017). In-Flight Decision-Making by General Aviation Pilots Operating in Areas of Extreme Thunderstorms. *Aerospace Medicine and Human Performance, 88*(12), 1066-1072.

Fanjoy, R. O., & Keller, J. C. (2013). Flight Skill Proficiency Issues in Instrument Approach Accidents. *Journal of Aviation Technology and Engineering, 3*(1), 17-23.

Federal Aviation Administration. (2018, 4 30). Advisory Circular 61-98D, Currency Requirements and Guidance for the Flight Review and Instrrument Proficiency Check. Retrieved from https://www.faa.gov/documentlibrary/media/advisory_circular/ac_61-98d.pdf

Jacob, J. (2020). *Weather Intelligent Navigation Data and Models for Aviation Planning*. Retrieved from WINDMAP: https://uasweather.com/

Knecht, W. R. (2016). Current NEXRAD Cannot Reliably Enable Safe Flight Around Heavy Weather. *The International Journal of Aviation Psychology, 26*(1-2), 46-61.

Misra, S., & Halleran, M. (2019). The Effect of Electronic Flight Bags in Flight Training on Preflight Skill Development and Aeronautical Decision Making. *Collegiate Aviation Review International, 37*(2), 16-37.

Moher, D., Liberati, A., Tetzlaff, J., & Altman, D. G. (2009). Preferred Reporting Items for Systematic Reviews and Meta-Analyses: The PRISMA Statement.

O'Hare, D., Hunter, D. R., Martinussen, M., & Wiggins, M. (2011). Classification and Prediction of Pilot Weather Encounters: A Discriminant Function Analysis. *Aviation, Space, and Environmental Medicine, 82*(5), 543-549.

Ouzzani, M., Hammady, H. F., & Elmagarmid, A. (2016). Rayyan-a web and mobile app for systematic reviews. *Systemetic Reviews*.

Parmar, S., & Thomas, R. P. (2020). Effects of Probabilistic Risk Situation Awareness Tool (RSAT) on Aeronautical Weather-Hazard Decision Making. *Frontiers in Psychology, 11*.

Walmsely, S., & Gilbey, A. (2017). Debiasing visual pilots' weather-related decision making. *Applied Ergonomics, 65*, 200-208.

Walmsley, S., & Gilbey, A. (2016). Cognitive Biases in Visual Pilots' Weather-Related Decision Making. *Applied Cognitive Psychology, 30*, 532-543.

Wilkins, S. A. (2018). Examining Head-Down Time in Transportation: Case Study in Single-Pilot General Aviation Operations. *Transportation Research Record, 2672*(23), 137-145.

PILOT IS A PILOT IS A PILOT: EXPLORATION OF EFFECTS OF NATIONAL CULTURE IN HELICOPTER PILOTS

Anna Kaminska, Amy Irwin, Devin Ray
University of Aberdeen
Aberdeen, UK
Rhona Flin
Robert Gordon University
Aberdeen, UK

Culture has been identified as a factor influencing the way people communicate and behave. Though often imperceptible by its members, cross-cultural interactions can lead to misunderstandings and conflicts. The current study explored how national culture interacts in the cockpit and affects pilots' safety behaviours. The study used in-depth semi-structured interviews with 15 participants (14 helicopter pilots) to collect data on national culture's impact. The data were analysed using conventional content analysis. Content analysis indicated two categories relevant to working with others irrespective of culture and three categories relevant to perceptions of national culture. The findings indicate that pilots acknowledge the cultural differences present between themselves and others, and that culture can have an effect on their and other's safety behaviours. The participants also highlighted the importance of standardisation to overcome cultural influences. This research highlights the various ways in which culture affects pilots' safety behaviours and interactions with one another.

Helmreich and Merritt (1998) suggest that there are three main types of culture: national, organisational and professional. National culture is based on geographic country of origin and is thought to shape people's attitudes and behaviours and influence social interactions. The potential influence of national culture on behaviour has become a growing focus within the safety research as organisations, and teams, become more multi-cultural. Specifically, within aviation, the potential importance of culture has been recognised within crew resource management training (CRM) provision (Anca, 2019). For example, when non-adapted CRM training was first brought to Asia, most pilots did not actively participate in group activities because they viewed the instructor as an authority who should be listened to, rather than somebody to actively discuss material with, crucially reducing training effectiveness (Helmreich & Merritt, 1998). Despite this, 'culture' as a component has only recently been added to the list of CRM curriculum (Flin, 2019) and there is a general lack of up-to-date empirical research exploring the impact of culture in aviation crews. Thus, the current study aims to bridge that gap by exploring how national culture interacts in the cockpit and how it affects pilots' safety behaviours.

National culture

Hofstede's (1984) dimensions of national culture are a commonly used method of studying influences of national culture on various factors, e.g. leadership, accident rates in aviation, etc. The dimensions are Power Distance, Individualism-Collectivism, Uncertainty Avoidance and Masculinity-Femininity. Soeters and Boer (2000) examined NATO air forces data on total losses from military aviation accident reports and compared them to the individual country scores on Hofstede's four dimensions of culture. The authors found evidence of cultural differences across various national cultures involved in military action (and patrol). They found that air forces with (1) more individualist (more oriented toward work itself) culture have relatively less accidents as they allow themselves to be led by profession-related motives in decision making, rather than organisation-related motives (i.e., less bureaucratic characteristics controlling thoughts and behaviours); (2) greater level of uncertainty avoidance (or regulation orientation) have greater chance of accidents, perhaps due to unwillingness to 'improvise' in a potentially dangerous situation; and (3) greater power distance have relatively more accidents, perhaps due to those in relatively lower power positions not daring to speak up to others in higher power.

A more recent follow up by Enomoto and Geisler (2017) also accounted for weather conditions, GDP per-capita and number of flights in the country. They were able to support Soeters and Boer (2000) findings that countries with higher power distance have a higher amount of plane accidents even when weather conditions and GDP is considered. Thus, the authors underline the importance of training pilots and co-pilots in communication to overcome (national) cultural barriers, such as the feeling of inability to speak out against someone in power or criticise their actions, especially when life is at risk.

Research examining effects of cross-cultural interactions is scarce, and non-existent in regard to helicopter pilots (to author's best knowledge). This is surprising because helicopter flight crews are often comprised of multicultural teams that fly as part of multinational companies. Given the potential issues outlined above, in addition to communication difficulties arising due to a language barrier or different communication styles (van Glinow et al., 2004), it is vital that we improve our understanding of cultural impacts within helicopter crews, particularly in reference to flight safety.

The Current Study

The current exploratory study examined how national culture[1] affects helicopter pilots' safety-related behaviours. Semi-structured interviews were chosen to suit the exploratory nature of the first study and to allow for more in-depth examination of all three culture types. The study had three main aims: (1) to explore pilots' views on the effects of culture on safety behaviours; (2) to determine which aspects of culture are perceived as potential factors that might influence safety behaviours, performance and training; and (3) to determine which aspect of culture is perceived as the most important and / or most likely to influence safety behaviours and performance.

Methods

Participants

Three groups of oil and gas pilots (pilots, trainers and management team) were contacted internally by the company's training lead, and an invitation poster was hung in the break room at

[1] The effects of organisational and professional culture were also explored but are not reported here due to space constraints. Professional culture findings are reported elsewhere.

the heliport. In total 14 participants (2 female) were interviewed: 4 pilots, 6 trainers and 4 managers. Remaining participants (n = 14) age ranged from 36 to 64 (M = 47.20, SD = 7.98). Interviews were conducted both in person (n = 11), over video call (n = 2) and over the phone (n = 2). Most participants (all but one who was only involved in training pilots) were current pilots with varying flight experience. Nine participants were trained in the UK, and six pilots received their training in other countries (e.g., USA, Netherlands, etc.)

The study was approved by the University of Aberdeen, Psychology Ethics committee.

Interviews and Analysis Strategy

Semi-structured interviews took place took place between January and July 2020. The in-person interviews took place in private meeting rooms at the company's training offices, over-the-phone interviews were conducted at the University of Aberdeen, and video call interviews were conducted from home (both researcher and the participants).

In each interview participant demographic information was sampled, then participants were asked pre-prepared questions in 3 sections, each relevant to a culture type, and one overall question. Participants were encouraged to give full answers and provide examples, where appropriate. This procedure was followed until all questions were covered, whereby participants were asked if there was anything else that they would like to bring up that had not been covered by the interview questions. Throughout the interview process, the researcher remained neutral and inviting, being aware as to not provide physical or verbal (dis)approval to the answers given, apart from context specific facial expressions.

For the analysis, content analysis (Hsieh & Shannon, 2005) was performed. Codes were generated in primarily inductive coding (i.e., the analysis was data-driven (bottom-up) rather than theory-driven (top-down)) with some aspects of deductive coding (i.e., only information related to culture and safety was coded). Data saturation, the point at which no new categories were developed (Guest, Bunce & Johnson, 2006), was reached by the 14th interview.

Results

Working with other people irrespective of culture

Content analysis generated two overarching themes relevant to working with others.

Individual character exerts bigger influence than culture. Half of the pilots (n = 7) mentioned that rather than a culture influencing the way a person behaves, they only had issues with certain individuals.

'You can have problems with people from the UK... it's irrelevant.' (Participant #8)

Importance of standardisation for elimination of cultural influences. Many pilots (n = 6) observed that numerous cultural issues that could come up are eliminated by strong internal standardisation procedures within the organisation.

'When the company has little or no, um, standardization or very little standardisation, it ends up being up or down to each individual culture.' (Participant #6)

National culture

Content analysis generated three themes relevant to national culture (Table 3).

Table 1.
Themes and codes relevant to national culture.

Theme (definition)	Code
Language barrier can present difficulties: Pilots highlight language barrier as the most prominent national difference that leads to actual in-flight difficulties.	Speaking slower and clearer with non-native speakers ($n = 4$)
	Language can lead to misunderstandings ($n = 2$)
	Language issues become apparent in emergencies ($n = 2$)
	Standardisation mitigates language issues ($n = 1$)
	Working in a second language is more cognitively taxing ($n = 1$)
Minimal differences in Western world: Pilots describe small differences between European or Western pilots, and stress that these differences are minimal due to largely standardized CRM training that was developed in the Western world.	Rules and procedures in UK are strict and safe ($n = 7$)
	Pilots from European/Western countries are very similar ($n = 9$)
	Higher Power Distance in Southern and Eastern countries ($n = 5$)
	Standardised Western training ($n = 4$)
	Low Power Distance in Western cultures ($n = 3$)
Pilots from outside of EU are different: Pilots describe larger cultural differences impacting flights with non-European pilots.	American differences ($n = 5$)
	Arab differences ($n = 3$)
	Relaxed attitude to safety in Africa ($n = 3$)
	Non-European flying can be riskier ($n = 5$)

Language barrier can present difficulties. Most participants mentioned that the biggest national culture influence is the language barrier with non-native English pilots. Some pilots mentioned that it mostly is an issue with understanding accents and having to slow down:

> *'I can talk quickly and, you know, some guys have asked to just slow down and, erm, struggling to understand.' (Participant #1)*

Others also mentioned that it can be an issue due to technical language used in aviation:

> *'Then he was involved in an emergency... and it turned out he couldn't read the checklist... because A he was under a bit of pressure, and B it's quite technical language using some long words that are very aircraft specific [..]erm, and essentially he was completely unable to manage the emergency.' (Participant #3)*

Participants highlighted that non-native English speakers occasionally need to put in more effort into even simple tasks.

Minimal differences in Western world. Pilots described EU pilots as having fairly uniform flight behaviours and attitudes towards safety. Participants characterised power distance in Western Europe to be predominantly low, meaning that co-pilots can challenge their captains without trouble, while more Southern and more Eastern nations have a slightly higher PD. Pilots also noted that most pilots receive the same, standardised Western CRM training:

'Most of the people we fly with, regardless of where they have grown up, tend to have trained in similar areas, just because that's the way it works in aviation.' (Participant #5)

Pilots from outside of EU are different. Pilots described non-EU pilot differences as more significant due to lower standards and regulatory frameworks that are less strict:

'If you trained with regulatory framework being relatively lax, then there's a lot of scope to fly the aircraft however you want to basically and have some fun. [..] Erm, so, for people who are transiting from one area of the world into another, that can be a bit of a challenge.' (Participant #5)

Western pilots were described as being similar in terms of team interactions, but having slight differences in their approach to flying. The higher power distance and 'saving face' (i.e., avoiding embarrassing seniors) was noted in Eastern cultures:

'I flew some senior, erm, Arab officers around... erm, and it really required a huge amount more tact because you couldn't, especially if they had their own guys in the back, you couldn't show them up in any way, shape or form' (Participant #3)

Discussion

The qualitative data from this study provides insight into helicopter pilots' perceptions of culture and its influence on performance, safety behaviours and training. Key themes discussed importance of standardisation, international differences and language barriers.

Previous literature suggests that national culture can have a negative effect on flight safety (Helmreich & Merritt, 1998; Soeters & Boer, 2000), however, the helicopter pilots in our sample did not strongly support this idea. This may be due to the fact that all participants came from the Western world and had limited (if any) experience of interacting or flying with non-Western pilots. The pilots mentioned that there are very small differences between pilots from European countries (and most of the Western world) because of the standardised Western training styles and techniques, along with similar routes to becoming a pilot. This is largely in line with previous literature suggesting that many Western pilots (European, American and Australian) have similar characteristics (Helmreich & Merritt, 1998).

Pilots that did have experience of flying with others from non-Western countries, mentioned that main differences came down to team interactions, rather than actual skills of flying the aircraft. Participants reported instances in which they experienced the higher power distance when flying with Middle Eastern pilots, meaning that if they were in a junior role, they could not challenge the authority of their captain, even when the captain was wrong. Pilots also spoke of the 'saving face' culture in Eastern countries whereby junior pilots cannot embarrass their senior colleagues. These observations are in line with Hofstede's research comparing power distance between Western world and Eastern countries, where it has been found that power distance is higher in the latter. In line with conclusions of researchers (Soeters & Boer, 2000), some pilots observed that the higher power distance seemed to go hand in hand with a higher accident rate in those parts of the world.

The key aspect of national culture that was thought to have an effect on flight safety, and has caused in-flight difficulties, was the language barrier faced by some non-native English speakers. Pilots mentioned that standardised language in the cockpit helps to a certain extent but can also hinder communication in some instances – due to specialised terminology and 'difficult words' that can be harder to pronounce.

A potential limitation of the current study lies in the almost exclusively European (and predominantly British) sample, the majority of whom had only flown with other pilots from European countries. In future studies, we hope to expand out recruitment to such areas of the world as Asia, Latin America and Middle East. The literature discussed earlier (e.g., Helmreich & Merritt, 1998) suggests that these regions differ vastly on Hofstede's dimensions of culture, with power distance and uncertainty avoidance being of particular interest in their effects on flight safety. Thus, expanding the data collection to these areas will be of benefit to determine if these cultural differences persist and can be distinguished.

Acknowledgements

This work was supported by the Economic and Social Research Council through a collaborative studentship (ES/P000681/1). The research was conducted as part of a doctoral dissertation.

References

Anca, J. (2019). Cultural issues and crew resource management training. In B.G. Kanki, J. Anca, & T.R. Chidester (Eds.) *Crew resource management (3rd ed.)* (pp. 539-552). San Diego, CA: Academic Press. https://doi.org/10.1016/B978-0-12-812995-1.00020-8

Enomoto, C. E., & Geisler, K. R. (2017). Culture and plane crashes: A cross-country test of the Gladwell hypothesis. *Economics & Sociology*, *10*(3), 281-293. doi:10.14254/2071-789X.2017/10-3/20

Flin, R. (2019). CRM (nontechnical) skills: A European perspective. In B.G. Kanki, J. Anca, & T.R. Chidester (Eds.) *Crew resource management (3rd ed.)* (pp. 185-206). San Diego, CA: Academic Press. https://doi.org/10.1016/B978-0-12-812995-1.00006-3

Guest, G., Bunce, A., & Johnson, L. (2006). How many interviews are enough? An experiment with data saturation and variability. *Field Methods*, *18*(1), 59-82. doi:10.1177/1525822X05279903

Helmreich, R. L., & Merritt, A. C. (1998). *Culture at work in aviation and medicine: National, organizational and professional influences.* Aldershot, UK: Ashgate.

Hofstede, G. (1984). *Culture's consequences: International differences in work-related values.* Newbury Park, CA: SAGE.

Hsieh, H. F., & Shannon, S. E. (2005). Three approaches to qualitative content analysis. *Qualitative Health Research*, *15*(9), 1277-1288. doi:10.1177/1049732305276687

Soeters, J. L., & Boer, P. C. (2000). Culture and flight safety in military aviation. *The International Journal of Aviation Psychology, 10*(2), 111-133. doi: 10.1207/ S15327108IJAP1002_1

Von Glinow, M. A., Shapiro, D. L., & Brett, J. M. (2004). Can we talk, and should we? Managing emotional conflict in multicultural teams. *The Academy of Management Review*, *29*(4), 578-592. https://doi.org/10.5465/amr.2004.14497611

UTILIZING IMMERSIVE TECHNOLOGIES IN THE AIR TRAFFIC CONTROL DOMAIN

Shivani Bajaj
Grado Department of Industrial & Systems Engineering, Virginia Tech
Blacksburg VA
Andrew Jorquera
Department of Computer Science, Virginia Tech
Blacksburg VA
Charlie Duff
Grado Department of Industrial & Systems Engineering, Virginia Tech
Blacksburg VA
Rafael N.C. Patrick
Grado Department of Industrial & Systems Engineering, Virginia Tech
Blacksburg VA

The Federal Aviation Administration (FAA) holds a vital role in the United States, employing over 14,000 Air Traffic Control/Management (ATC/ATM) specialists responsible for managing roughly 43,000 flights each day. ATC education "wash-out" rates have shown that there is a disconnect between the training process and the implementation of cognitively demanding, safety-critical ATC duties. The purpose of this research was to investigate if, how, and where immersive technologies (i.e., augmented, virtual, and mixed reality) could be helpful within the ATC/ATM educational domain. To accomplish the overall research goal, subject matter expert (SME) interviews were conducted and a potential educational tool was developed and tested in two distinct research phases. Eighteen (N = 18) subjects volunteered to participate throughout both phases, and the tool was rated to be above average meaning the tool is usable in its current form; however, further development is suggested and expected.

In order to become an air traffic controller (ATC), potential candidates undergo a rigorous training process to prepare for the fast-paced, cognitively demanding, and high stressed safety-critical work environment. Despite a large number of interested applicants, training facilities experience "washout rates" as high as 70% (FAA, 2018). This inevitably increases the demand for newly trained ATCs who have successfully completed training; therefore, timely and effective training is imperative in order to meet the current demand.

During the early stages of immersive technology development, Kozak, et. al (1993) was noted as the first to investigate the use of VR within the air traffic control/management (ATC/ATM) domain. Although immersive technologies showed immense potential, insufficiencies in VR displays prevented these formats from being accepted as useful training tools. In recent years, immersive technologies have proven to be an incredibly beneficial learning tool in other complex domains (i.e., military (Bhagat, 2016), medical (Loukas, 2013), and engineering (Wickens, 2018)). This information combined with additional insights from Mackay, et. al (1999), Hoc, et. al. (1998), and the National Research Council's (1998) report on the future of air traffic control led the current research to adopt an exploratory research design aimed at confirming the statement made by Akselsson, et. al. (2000), that immersive technologies now possess the necessary capabilities to serve as effective tools within the ATC domain.

Methods

Research Design. The primary purpose of this research was to investigate the use of immersive technologies (i.e., augmented, virtual, and mixed reality) within the ATC/ATM domain and to determine where these technologies can be best integrated into an educational environment. To accomplish this, two distinct research phases were adopted. First, Phase I sought to gather contextual information about the potential of such tools; while Phase II sought to provide a potential solution in the form of a technological tool.

Participants. In total, eighteen (N = 18) subjects volunteered to participate throughout both phases. Ten (n = 10) participants engaged as subject matter experts (SMEs) interviewees while eight (n = 8) engaged in user-testing consisting of three males and five females with an average age of 23.5 years old (± 3.7). The only restriction was that all qualified participants must be of legal age (i.e., 18 years old) or above during the time of participation. It should be noted that due to COVID-19, special considerations were required.

Apparatus and Equipment. During Phase I & II semi-structured interviews, computer-based notes were taken and cross-referenced with audio recordings to ensure accuracy. During the data collection portion of Phase II, a web-based pre-survey tool was used to gather basic demographic information and gauge familiarity with the ATC/ATM domain and immersive technologies. A post-study questionnaire was administered upon completion of the interaction, with the incorporation of the System Usability Scale (SUS). Bangor (2008) and Lewis (2009, 2018) states that the SUS allows for the evaluation of a wide variety of products and services. With respect to the VR immersive tool development, an open-source 3D modeling engine, Blender, and a game engine, Unreal Engine 4 (UE4), were used. While an Oculus Quest 2 head mounted display (HMD) was used to display the developed tool to users in Phase II.

Procedure

The following components of the research will be divided into sections to provide sufficient detail related to specific aspects of the research.

Phase I. Prior to performing Phase I activities, a literature review was conducted to inform questions presented to SMEs during semi-structured interviews. It was necessary to develop a thorough understanding of the required approaches in designing and developing novel technology solutions for use in ATC trainee usability studies. Subsequently, technical experts who specialize in human-computer interaction and cognitive engineering were selected for engagement. Interviewees were asked to openly discuss their thoughts on the use of immersive technologies in the ATC/ATM domain.

Phase II. First, a second round of semi-structured SME interviews were performed to gain a rich understanding of what a future educational tool should do to help the ATC/ATM domain. Through consultation, it was clear that developing a map memorization tool would be the most beneficial to the ATC/ATM educational domain, and VR Avenue would be most suitable for testing spatial cognition This also provided greater control of software scalability and expandability.

Tool Development. A cross-platform modular approach provided the greatest flexibility for development. This workflow allowed for interaction techniques and mechanics developed for the initial prototype to be extended across future iterations of the tool.

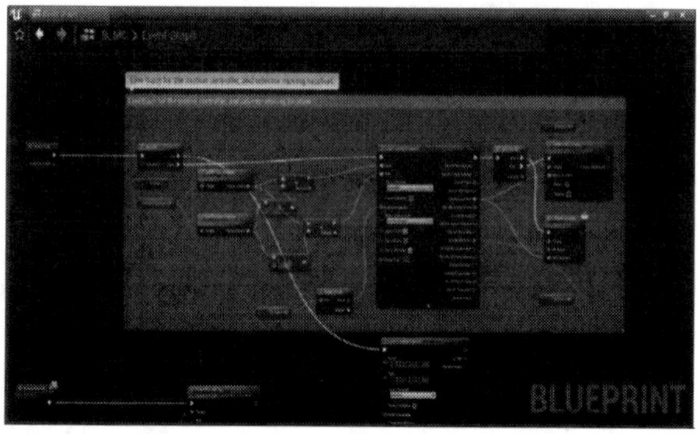

***Figure 1*. Unreal Engine Development**

Several design and user considerations (i.e. color, feedback, and intended population) impacted development. The color palette implemented coincides with the colors used by the FAA on official IFR maps ATC industry standards without overstimulating the user. As for feedback, visual feedback was chosen to guide and caution the user. Considering intended-populations, the interface is predominantly geared towards corrective lens users and right-handed users although users with glasses and left-handed individuals can still utilize the tool with minor adjustments.

With respect to software development, a selected sectional (provided by the *FAA's Aeronautical Information Services Aeronautical Chart Users' Guide)* was uploaded and populated with critical information in layered form. Figure 1 above represents visual snapshots of development within the Unreal Engine software. To maintain the accuracy of the data embedded in the prototype, a multi-step workflow was used to extract symbols from a geographic information system (GIS) and import into the game engine. Many data points, such as intersection points, were identified by target points in the software. Since these could not be extracted, 3D models were developed that could be placed on the markers so that their relative coordinate space was identical to the GIS imported markers. Finally, the layers were exported from Blender as 3D models (.fbx) and imported to UE4 for game engine manipulation and packaging. UE4 is a necessary choice for creating interactive immersive reality applications using the desired geographical data, and does not allow for GIS plugins, as it references its unique coordinate system that defines the architecture of the engine. However, the object models exported from Blender can be imported and scaled in UE4; and mechanics can be developed for data interaction.

***Table 1*. Usability Study Tasks**

Task Number	Task
1	Rotate controls left-right move up-down, move forward-back
2	Find and select the help menu, then tell us what the controller says
3	Toggle the laser pointer off then on
4	Point the motion controller at map data and confirm that the map data illuminates
5	Toggle on/off the layers menu
6	Turn off victors, intersections, and NAVAIDs
7	Turn off ALL layers and Turn on ALL layers
8	Find and name all 6 Airspaces
9	Find and name 5 airports in the Columbus 3 MOA airspace
10	Name the 4 victor lines coming from Greenville
11	Find Airport Columbus AFB, Tell us its information

Usability Tasks: A pre-study survey asked users to rate their ATC-relevant skills such as memorization and the ability to understand new technologies quickly on a 5-point Likert scale. Participants were also asked if they had any specific ATC experience. Users performed a task analysis intended to explore and test the functionality of the platform, control manipulation, and interface design and interaction. Table 1 outlines the tasks presented to each user within the ATC VR HMD which increased in difficulty as the evaluation progressed while Figure 2 represents a user interacting with the tool within the experimental environment and point-of-view perspectives of the actions performed.

Figure 2. **VR ATC Point of View and User Interaction**

Results

A collection of information was obtained in various forms to draw meaningful conclusions with respect to confirming that immersive technologies now possess the necessary capabilities to serve as effective tools within the ATC domain. Feedback from SME interviews and design recommendations have been discussed in previous sections; therefore, the following sections will report results pertaining to the user-testing and usability aspects of the research. Using the Systems Usability Scale (SUS), participants were asked to score 10 items with one of five responses that ranged from Strongly Agree to Strongly Disagree.

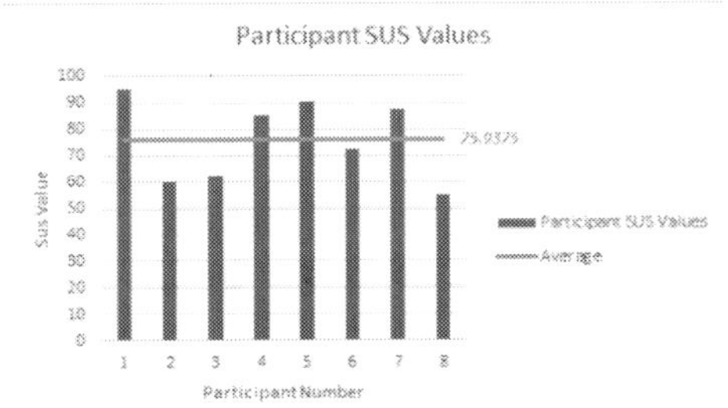

SUS SCORE	GRADE	RATING	
95	A	Excellent	P1
60	D	Poor	P2
62.5	D	Poor	P3
85	A	Excellent	P4
90	A	Excellent	P5
72.5	B	Good	P6
87.5	A	Excellent	P7
55	D	Poor	P8

Figure 3. **Participant SUS values**

Interpretation of SUS scores required calculations to normalize raw scores to produce a percentile ranking. Research states that a SUS score above 68 is considered above average while a score below that threshold would be considered below average indicating the user rated the platform to be unusable to some extent. As can be seen in Figure 3, the average SUS score for the developed VR HMD ATC map visualization tool was 75.93 with a standard deviation of 14.41. This indicates that the tool was rated to be above average. Higher scores on odd questions add to a SUS score while higher scores on even questions result in decreased SUS values. As can be seen in Figure 4, odd-numbered questions scored higher while even-numbered questions scored less on average, resulting in a higher overall SUS score. Of particular note, Question 4 asks

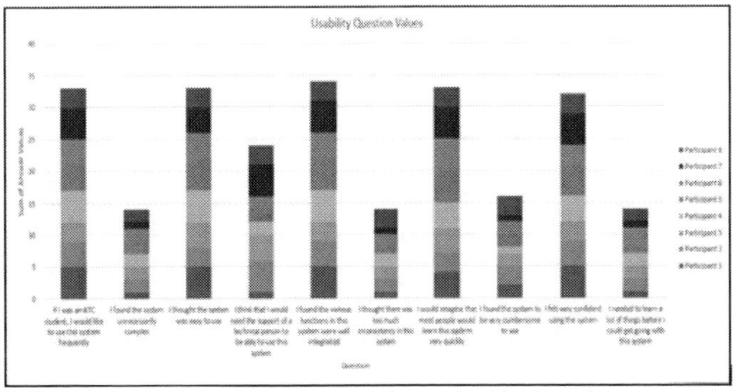

Figure 4: **Stacked Bar graph with SUS values**

whether users feel as if they would need the support of a technical person when using a system. This question scored around twice as much than its corresponding even-numbered questions. It can be inferred that there were some problems with explaining how the tool worked and/or it did not have enough built-in to be self-explanatory.

Discussion

Analysis of the post-survey demonstrates that the visualization tool is beneficial. All users liked the look of the product and six rated a positive interaction rating with the product, while the other two users gave a three out of five rating. Users found the tool easy to use, not unnecessarily complex, and felt the features were well-integrated. The GUI was found to be informative and clear to understand. One user asked to have "*a mutual guide to scaffold learning in the environment.*" Implementing a training/instructional element to the visualization tool can help satisfy this comment. Additionally, when asked about their overall interaction with the visualization tool, users had relatively positive comments.

Many users found the visualization tool, after a brief learning process, to be fairly easy to use and helpful. The question: "*Would you need the support of a technical person to use this product*" was the only one met with a mixed response. Three users felt that they needed the help of a technical person while using the tool while four users did not feel this need. It can be determined that the tool would benefit from a tutorial program that clearly shows how to use the visualization tool. Many users suggested that a search bar be implemented into the interface. Other suggestions included a map legend and bigger text. Majority users asked for brighter, friendlier colors, however, air traffic products do not use bright colors and many SMEs point to not using color at all. The improvements made in this iteration of the visualization tool will help bring a more effective tool to the intended population of users (i.e. intermediate and expert ATC/ATM specialists) which will then aid in creating a more effective final product.

Conclusion

The current exploration yielded information that will be extremely useful in the continuation of this tool's development and the development of future modules (i.e., AR map visualization and other tools for early ATC training). The team was able to build and test a VR educational tool with a novice population and gain a better understanding of the usability of such a tool within this context. This stage consists of the overall exploration of utilizing immersive tools within the ATC domain. Subsequently, the research team hopes to begin investigating the use of this educational tool and other immersive technologies (i.e. AR) with an ATC/ATM intermediate and expert population.

Acknowledgements

The research team would like to acknowledge and thank the Northrop Grumman Corporation for support of the project through the Northrop Grumman Undergraduate Research Experience in Industrial & Systems Engineering at Virginia Tech.

References

Akselsson, R., Källqvist, C., Bednarek, V., Cepciansky, M., Trollås, A., Davies, R., & Johansson, G. (2000). Virtual reality in air traffic control. In *Proceedings of the Human Factors and Ergonomics Society Annual Meeting* (Vol. 44, No. 33, pp. 6-273). Sage CA: Los Angeles, CA: SAGE Publications.

Bangor, A., Kortum, P. T., & Miller, J. T. (2008). An empirical evaluation of the system usability scale. *Intl. Journal of Human–Computer Interaction, 24*(6), 574-594.

Bhagat, K. K., Liou, W. K., & Chang, C. Y. (2016). A cost-effective interactive 3D virtual reality system applied to military live firing training. *Virtual Reality, 20*(2), 127-140.

Federal Aviation Administration (2018). Air traffic by the numbers. Retrieved from https://www.faa.gov/air_traffic/by_the_numbers/

Hoc, J. M., & Lemoine, M. P. (1998). Cognitive evaluation of human-human and human-machine cooperation modes in air traffic control. *The International Journal of Aviation Psychology, 8*(1), 1-32.

Kozak, J. J., Hancock, P. A., Arthur, E. J., & Chrysler, S. T. (1993). Transfer of training from virtual reality. *Ergonomics, 36*(7), 777-784.

Lewis, J. R., & Sauro, J. (2009). The factor structure of the system usability scale. In *International conference on human centered design* (pp. 94-103). Springer, Berlin, Heidelberg.

Lewis, J. R. (2018). The system usability scale: past, present, and future. *International Journal of Human–Computer Interaction, 34*(7), 577-590.

Loukas, C., Lahanas, V., & Georgiou, E. (2013). An integrated approach to endoscopic instrument tracking for augmented reality applications in surgical simulation training. *The International Journal of Medical Robotics and Computer Assisted Surgery, 9*(4), e34-e51.

Mackay, W. E., & Fayard, A. L. (1999). Designing interactive paper: lessons from three augmented reality projects. In *Proceedings of IWAR* (Vol. 98, pp. 81-90).

National Research Council. (1998). *The future of air traffic control: Human operators and automation*. National Academies Press.

Wickens, C., Dempsey, G., Pringle, A., Kazansky, L., & Hutka, S. (2018). Developing and Evaluating an Augmented Reality Interface to Assist the Joint Tactical Air Controller by Applying Human Performance Models. In *Proceedings of the Human Factors and Ergonomics Society Annual Meeting* (Vol. 62, No. 1, pp. 686-690). Sage CA: Los Angeles, CA: SAGE Publications.

TASK MODELING IN AIR TRAFFIC CONTROL WITH TRAJECTORIZED EN-ROUTE TRAFFIC DATA PROCESSING SYSTEM

Sungju Maeng, Makoto Itoh
University of Tsukuba
Tsukuba, Japan
Hiroko Hirabayashi, Atsushi Senoguchi
Electronic Navigation Research Institute
Tokyo, Japan

As the volume of flight is extended, it is expected that the task complexity of air traffic controllers will increase. In Japan, air traffic control in en-route airspace is operating with TEPS (Trajectorized En-route Traffic Data Processing System), which has functions to display information necessary for air traffic controllers' tasks. It has evolved to support the controllers with less workload, resulting in escalated interaction between controllers. However, paradoxically this means more information is provided and more workload would be required. In order to manage mental workload of air traffic controllers, detailed analyses of their tasks with TEPS are needed. In this study, we tried to develop a model of the air traffic control tasks conducted on TEPS by applying the Multilevel Flow Model (MFM). Based on the model, we clarified the task structure in which several controllers participated and assess contributing factors in workload.

Introduction

Since the COVID-19 crisis caused a great loss of airline industry, IATA (2020) reported that it will take a long time to return to pre-crisis levels. However, airline financial performance in Asia Pacific region is forecasted to recover faster than other regions because of large domestic markets and expected that the demands of flights will increase once recovery begins. Before the crisis, air traffic volume in global was in rising tendency, and the importance of performance and workload assessment of air traffic controllers have discussed for decades.

For supporting controllers, automation technologies have been developed continuously, and direct mediation task has been decreased but monitoring task being more substantial. In accordance with this recent change in tasks, Japanese en-route air traffic control operations are conducted with TEPS (Trajectorized En-route Traffic Data Processing System), which is the air traffic control system terminal used by controllers. To sustain the safety and efficiency of air traffic flow, controllers are working as a team, and their tasks are distributed by considering the characteristics. A set of a radar controller who mainly focuses on the radar screen and takes direct communication with pilots (called "R-seat" in TEPS) and a controller who takes coordinate tasks with other facilities such as airport and the controller of side sectors (called "C-seat" in TEPS) conduct en-route air traffic control operations on en-route airspace sector. Accordingly, to evaluate the expected workload and performance in the use of TEPS, the relationship and composition between the two seats needed to be considered. Conventionally, Modified Messerschmidt, Bölkow und Blohm (MMBB) Method (ICAO, 1984) was utilized for workload assessment of air traffic controllers in Japan, but this method was applied only for radar controllers.

Gregory et al. (2012) reminded the status that there are fewer studies related to team workload than the individuals, and suggested theory and assessment methods related to team workload. Their proposed Multilevel Team Workload Model implies work environment and task characteristics required to be identified.

For use in team workload research of air traffic controllers, this research aims to suggest the entire task flow on the utilized system, TEPS, supports performing air traffic control tasks and sharing information through the entire Flight Information Region (FIR) especially in en-route operations, at the point of means-end relationships using Multilevel Flow Model (MFM).

Trajectorized En-route Traffic Data Processing System

The entire system consists of multiple displays and input devices. In the monitor positioned at the center of a desk-type system, the system information region with the radar domain and the support domain, provide information required for entire tasks as shown in Figure 1. Above the center monitor, two screens are arranged and each of them is a sub-display screen with reference information and a total information display unit showing meteorological information and notice to airmen (NOTAM).

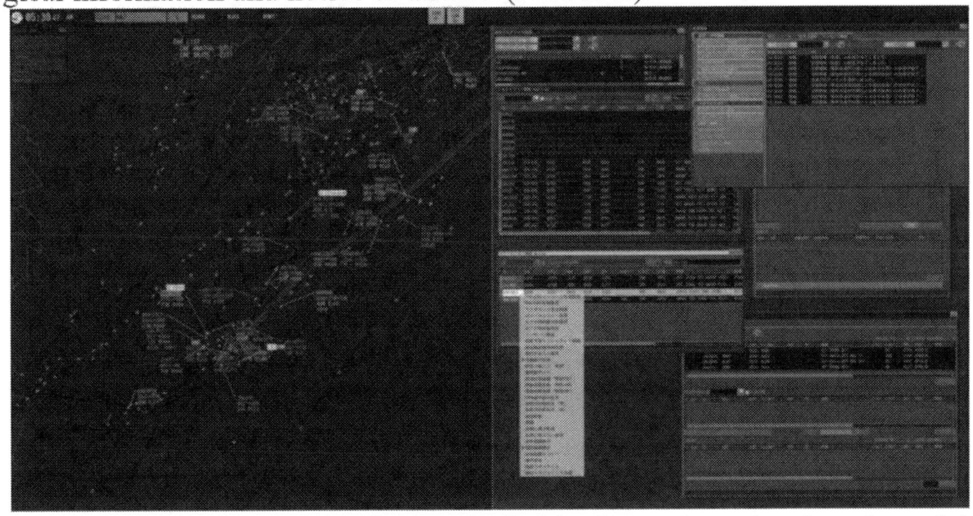

Figure 1. An example of TEPS main display. Left side shows the radar domain and the other side is support domain. The position information served by air route surveillance radar combined with flight planning information. (Ministry of Land, Infrastructure, Transport and Tourism, Overview of TEPS, Retrieved from https://www.mlit.go.jp/koku/content/001358999.pdf (In Japanese))

In the radar domain, data blocks in which simplified flight information containing call sign, altitude, ground speed, are deployed for each aircraft. To transfer and share the indication to pilots and whole system users, controllers perform tasks with input devices, such as a mouse, a keyboard, and a footswitch. They can take memos on the designated data block. The side of the radar domain is the support domain, selectable coordination windows pop up to show details of a situation and adjust the settings of the display. This domain contains the screen of flight list, departure clearance, sector coordination, AIDC (Air Traffic Services Interfacility Data Communications) transfer, etc. The downside of the main display, touch panels for taking communication with pilots and other facilities as other sectors or airports are prepared. Since whole displays are connected and have been automated, controllers who use the system are required to understand entire relationships and be trained for a considerable period to be used to.

Multilevel Flow Model

LIND (2011) introduced Multilevel Flow Model (MFM) as a qualitative modeling method for presenting the entire procedure of industrial system, representatively, Nuclear Power Plants. While the original step of MFM development was presenting Human-Machine Interfaces for complex systems with supervisory control, the model has been actively utilized in industrial areas because of the characteristics that identifying interconnected levels of means-end, part-whole abstraction, goals, and functions. Lind et al. (2014) suggested methodology for building MFM. Means-end hierarchy and relation diagram will have to be drawn first to survey the whole interconnected flow intending means-end relationship and between functions. The embodied MFM will have to be introduced with symbols shown in Figure 2.

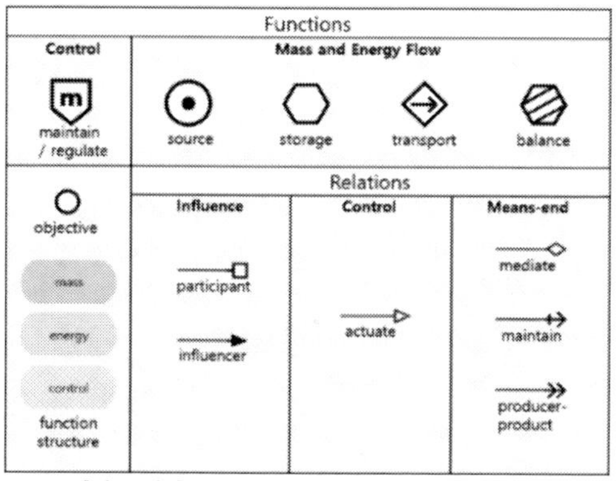

Figure 2. Basic symbols mainly used in this paper. TEPS comprises various functions and these are interconnected by relations symbols as participant and influencer.

Analysis and Results

Behavior analysis

Before undertaking the modeling work, we observed simulator trainings and analyzed the behavior of air traffic controllers using TEPS in need. Three typical scenarios performed daily were prepared, 6 participants (who are active-duty controllers) simulated scenarios 2 of them each, and the entire procedure was recorded as video data. Overall scenes were taken from behind for and the head-camera took the controllers' point of view. The first scenario was focused on the sector including traffic flow of descending phase to the Tokyo International airport from west. The second scenario included the cases of flights coming from the other FIRs, for example, from Incheon FIR of Republic of Korea to Fukuoka FIR of Japan. The last scenario was traffic flow of north of Tokyo area airports, and includes both departure and arrival phase.

To grasp the entire flow of tasks of R-seat controller and C-seat controller, video data were analyzed. Task categorized into three parts as Verbal, Behavior, and Visual. The verbal part was primarily recorded and focused on in this study. Behavior parts include actions using a keyboard, a mouse, and communication panels and Visual parts mean a rough record of where controllers' gaze is paying attention per designated time block. Recorded contents are sorted chronologically.

Multilevel Flow Model

Means-end Hierarchy and Relation Diagram Figure 3 shows the means-end hierarchy of TEPS. The general goal of TEPS is to achieve safe, efficient, and smooth air traffic management. Priorities (Abstract Functions) are divided into the management of aircraft represented inside of the sector and the coordinate work on the boundary of the sector in charge because the tendency of the tasks on each of priorities is different. Functions (Generalized Functions) accomplishes Priorities consisting of communication with other ACCs and Pilots, flight plan and intent check, instructions issue, real time position grasp, sector transition takeover and information sharing. Processes and Objects, as Physical functions and forms mounted on TEPS, directly represented. Based on the first hierarchy diagram, the means-end relation diagram was built within the range of the sector in charge. As the cornerstone of MFM, relation diagram focused on Processes.

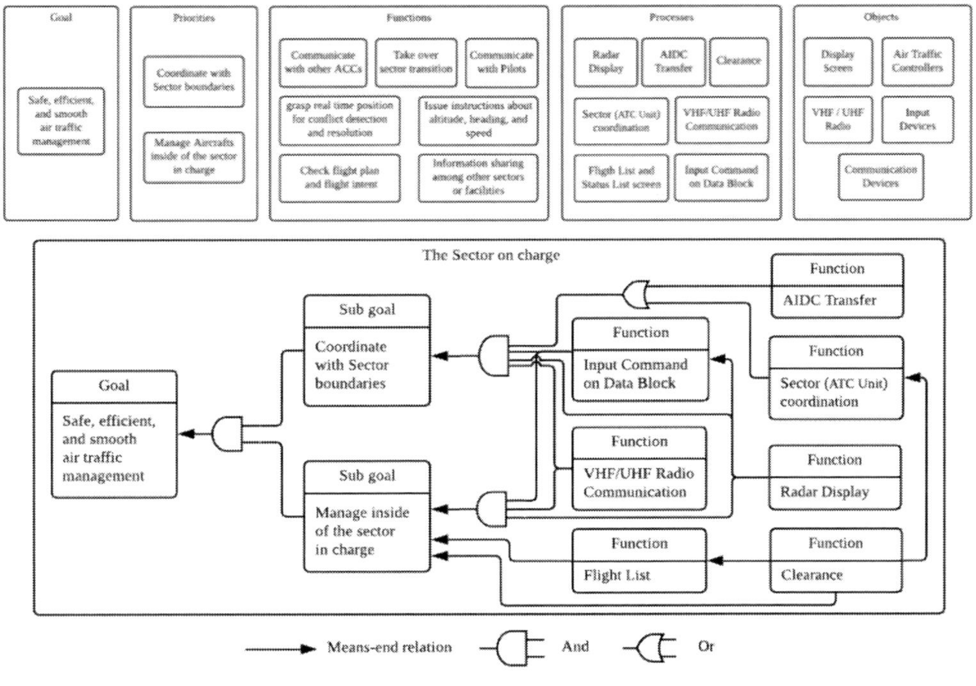

Figure 3. Means-end hierarchy and relation diagram of the sector in charge on TEPS.

The important point of coordination tasks at sector boundaries is safely taking over heading flights from or to the next sector or FIRs. In the case of "AIDC Transfer", almost tasks are automated. When taking over the flight, accepted clearance from another sector is needed. "Radar Display" function in which controllers can select the specific data block and input the command and "VHF / UHF Radio Communication" function are used for achieving both sub-goals.

On the side of management tasks in the sector in charge, controllers need to check "Flight List" which contains all flight information even not shown in radar display, and

"Clearance" offers the list of flights awaiting or accepting clearance. The issued flights by clearance function are renewed on the flight list.

Building MFM The MFM used for showing the entire flow of the system usually consists of a mass, energy, and control function structures. Figure 4 describes the MFM of TEPS. The information shown in the screen interface is treated as mass, and utterance content is assumed as energy. "Bal1" and "Bal2" in this model mean that normally tasks are processed to "Issued" storage automatically, but if the sector is in the situation with special circumstances such as sudden increase in volume, controllers are requested to manage takeover task manually and the aircrafts are temporally dropped in "Request" and "Input" storage to process.

The circle at the center contained in control function structure is the objective as the goal. The objective is maintained by 5 Function Structures, and "Bal1" and "Bal2" are regulated from that objective. "Input CMD" function conducted on "Radar Display" function. Therefore, two functions are presented in the same function structure. Most of Storage symbols represents windows for coordination in the support domain, except for "RadDis" means radar domain itself. "Tra10", connected with "RadDis" storage shows flights on radar display updated by intervened information mediated by radio communication.

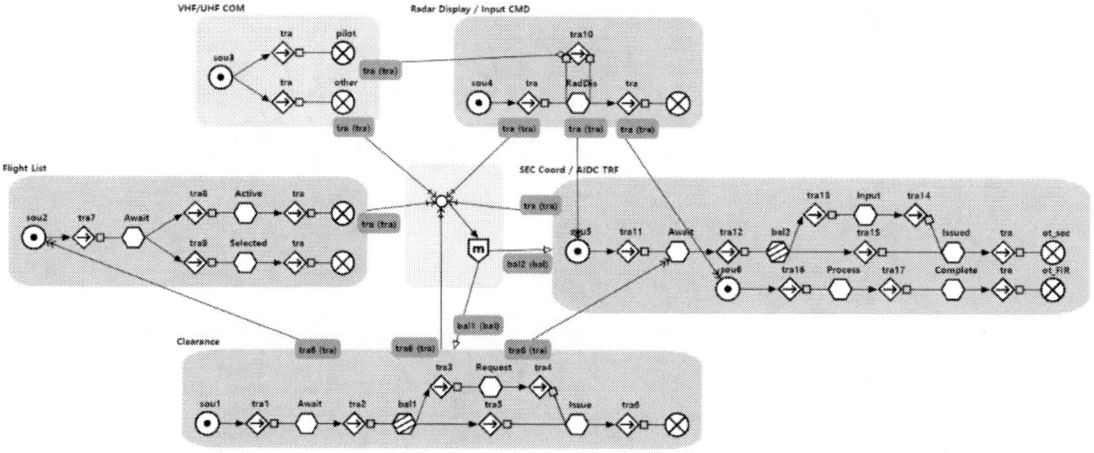

Figure 4. Multilevel Flow model of TEPS. Mass, energy and control flow are contained in the Function structure. VHF/UHF Radio communication is described as an Energy flow function. They are interconnected by maintaining, producing, and meditating relationships.

Limitation of represent who uses the Means One of the purposes of this research is to analyze the task as a team and used it for future work to clarify the relation which could affect team workload. With that point, we realize the limitation of MFM, not includes the detail of human operators who uses the Means as a function. The example of Kim and Seong (2018) are questioned the lack of representation of time to effect and detect in MFM and suggest the way to solve the limitation by writing the time under the arrow line. With this reference, our modeling diagram also includes the human operators, in this case, R-seat and C-seat.

Conclusion and Future Work

The paper presented a Multilevel flow model (MFM) of Trajectorized En-route Data Processing System (TEPS), the system supporting air traffic control tasks, especially on en-route traffic management in Japan ACCs, for illustrating interconnected means-end and Part-whole relationship of functions. The model combined with the means-end hierarchy and relation diagram and shows an entire flow of information transferring between functions and view of almost processes contained in TEPS briefly.

Although the complexity and connectivity of the system are identified comprehensively, there is the limitation that the original MFM cannot explains controllers who participate in the process. In the future work, it would be with the information of controllers who are involved in relations arrow and guess which function could be the potential of the workload.

References

Bedwell, W. L., Salas, E., Funke, G. J., & Knott, B. A. (2014). Team workload: A multilevel perspective. *Organizational Psychology Review*, 4(2), 99-123.

Glantz, E. J. (2006). Challenges supporting cognitive activities in dynamic work environments: Application to policing (Ph.D. thesis). University Part, PA: The Pennsylvania State University.

Gregory J. F., Eduardo, S., & Adam, S. (2012). Conceptualization and Measurement of Team Workload A Critical Need, *Human Factors*, 54(1), 36-51.

Hirabayashi, H., Senoguchi, A., & Toratani, D. (2020), A Study on Regional Characteristics of En-Route Air Traffic Control Sector Traffic Flow. In *The 58th Aircraft Symposium*, JSASS-2020-5159-3C02 (in Japanese)

IATA (2020), Economic Performance of the Airline Industry, *2020 End-year report,* Retrieved from https://www.iata.org/en/iata-repository/publications/economic-reports/airline-industry-economic-performance---november-2020---report/

ICAO (1984). ICAO Doc 9426 Air traffic services planning manual - part 2, section 1, chapter 1, appendix c.

Kim, S. G., & Seong, P. H. (2018). Enhanced reasoning with multilevel flow modeling based on time-to-detect and time-to-effect concepts. *Nuclear Engineering and Technology,* 50(4), 553-561.

Lind, M. (2011). An introduction to multilevel flow modeling. *Journal of Nuclear Safety and Simulation,* 2(1), 22-32.

Lind, M., & Zhang, X. (2014). Functional modelling for fault diagnosis and its application for NPP. *Nuclear Engineering and Technology,* 46(6), 753-772.

Suarez, N., López, P., Puntero, E., & Rodriguez, S. (2014). Quantifying air traffic controller mental workload. In *Fourth SESAR Innovation Days*, 1-6.

PSYCHOLOGICAL ASPECTS IN PILOT TRAINING:
COGNITION AND HUMAN FACTORS

Shagun Gupta

University of Delhi
New Delhi, India

Introduction- International Civil Aviation Organization's (ICAO) SHELL model was designed to study and investigate the way the pilot interacted with the various dimensions of the model. Aim-The present study aimed to explore psychological aspects while flying using the dimensions of ICAO's SHELL model of aviation. Methods- The study followed the qualitative research paradigm. The sample consisted of 9 (Males =4, Female =5) student trainee pilots in the age range of 18 – 21 years. They were subjected to in-depth interviews which lasted for around 30 minutes. The data were analyzed using the thematic network of analysis. Results-The results show that in the interaction of each dimension of the SHELL model various cognitive and human factors are involved in flying. Conclusion-This research is highly applicable to understand the psychological aspects that can be used to improve the efficiency of pilots and ensure safety measures in the aviation sector.

Pilots have a picture of being brave and allure, which incorporates that they are solidified experts. Psychologists and analysts have tried to explore and learned about it from the ideal mental qualities through research in the civil and military flight deck team. The SHELL model was created first by Edwards in 1972, with an adjusted chart to outline the model created by Hawkins in 1975. The model has four measurements which are the software, hardware, liveware and environment. The SHELL model was intended to contemplate and research the manner in which the information with the pilot's collaborated with different SHELL model factors and have any kind of effect in their proficiency. The SHELL model has four dimensions which are hardware, software liveware and environment.

The researchers Yu-Hern Chang and Chung-Hsing Yeh (2010) in the study of the human performance interfaces in air traffic control aimed to find using the SHELL model of ergonomics the ATC system performance interface. The research hypotheses were about the relationship between human performances. Interfaces of the system were developed and tested on the basis of the data collected from the air traffic controller using structural equation modeling. The research findings suggest that the organization plays a significant role than the individual differences in how the controllers interact with the software, hardware, and the environment of the ATC system. The conclusion of the study was that there are mutual influences in all dimensions but there is an exception of the controller–controller interface. .

The study conducted by T. K. Matsuoka was to propose a human factors classification framework. The SHELL model was adopted for this study. The purpose of the study is to provide a framework intended to be applicable to the circumstances .The data was collected using the method of questionnaire surveys. They used a quantitative approach in order to analyze the data. Using statistical analysis and discussion on the profiles show that relation exists between safety attributes and others. The conclusion helps in classification and provides better understanding of human factors and their contribution in reducing future collisions and human errors.

Results

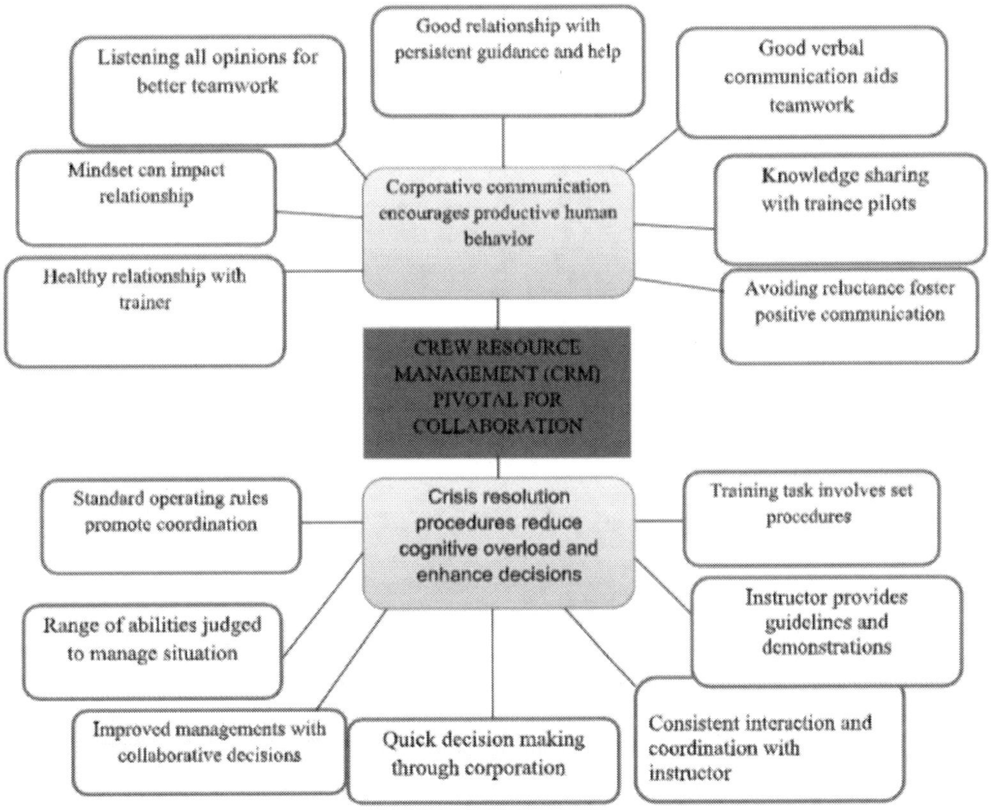

Figure 1 Diagrammatic representation of the thematic network analysis for the global theme "Crew resource management pivotal for collaboration"

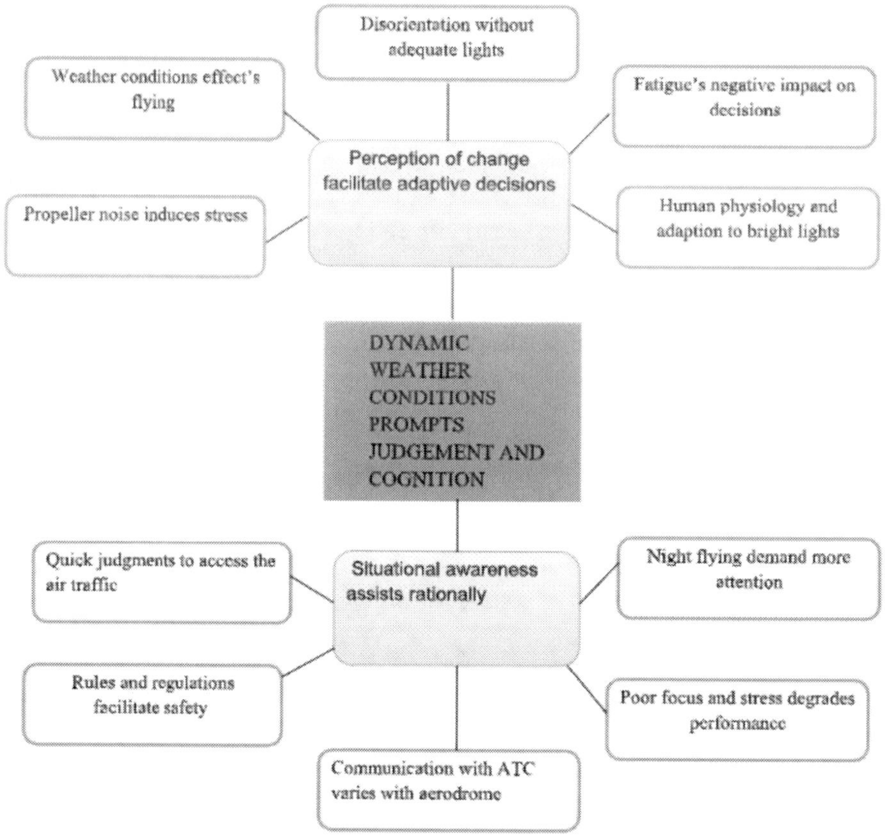

Figure 2 Diagrammatic representation of the thematic network analysis for the global theme "Dynamic weather conditions prompts judgment and cognition"

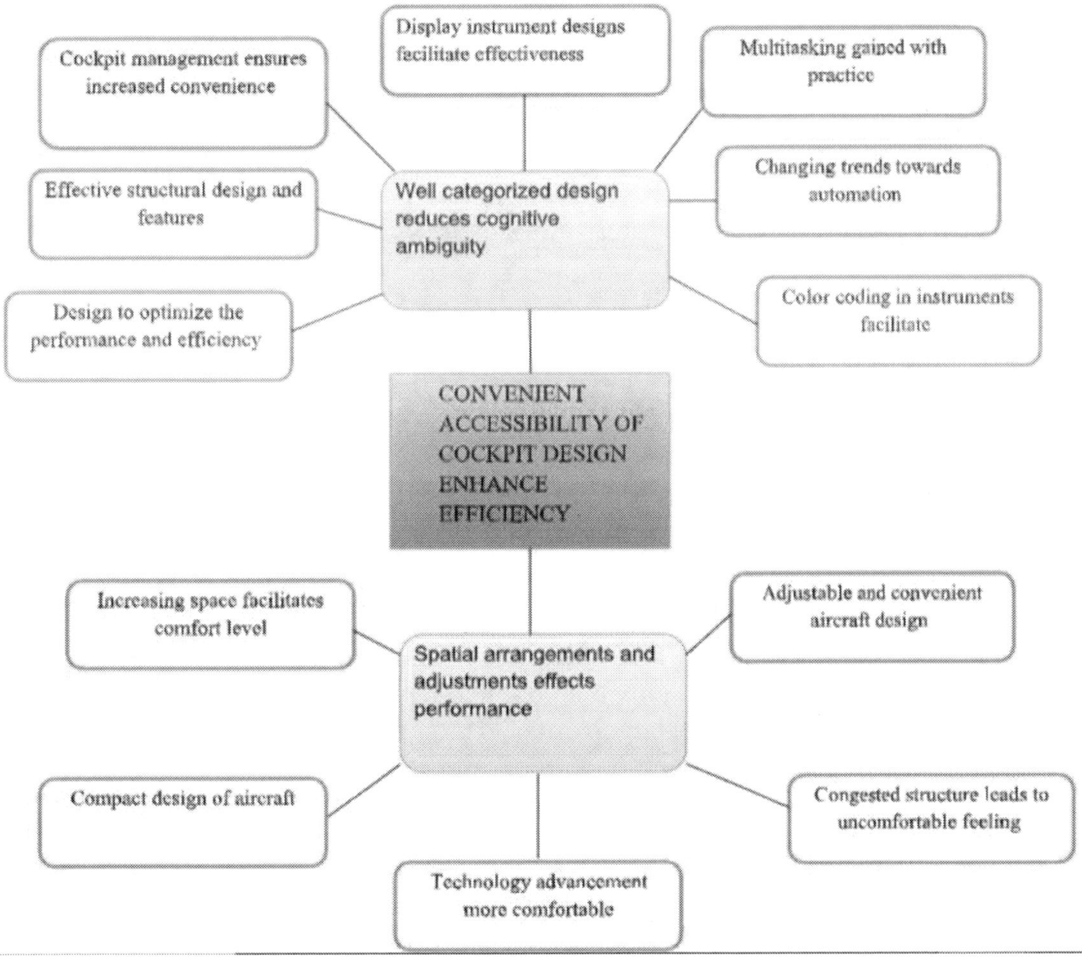

Figure 3 Diagrammatic representation of the thematic network analysis for the global them "Convenient accessibility of cockpit design enhance efficiency"

- Proper use of resource management helps to better coordinate and increase efficiency.
- Procedures provide a set course of action that reduces the cognitive overload.
- Situational awareness is crucial due to the dynamic nature of events.
- The perception may have cognitive biases thus can be reduced with proper knowledge and skills.
- The night flight demands more focus and attention considering the nature of human physiology.
- The design of the cockpit has an impact on human performance.
- The human factors consideration in an aircraft system is important due to constant interaction between display and pilots.
- The rules and regulations provide guidelines and facilitate safety.

The research study helps us to understand the various psychological aspects in flying and interaction of the human component with aircraft operation using the SHELL model. In order to reduce human error accidents it is necessary to understand them and help through improved decision making training this study can be highly applicable. Firstly, it will help to improve safety management in the aviation sector. It helps in building safety policies as well as help in critical planning and achievement of maximum safety. Secondly, it can be used for building better training procedures that focus on central processing and can be used to reduce the errors through training. Thirdly, the enhanced level of conditioning and reinforcement in all the dimensions can lead to improved decision making as well as the level of efficiency. Lastly, the enhanced level of awareness through this study can significantly lead to an improvement in the safety and reduction in human errors.

The limitation associated with the study was that the sample consisted of training of pilots from different demographic areas including the United States of America, Canada, New Zealand and India, so there were slight differences in their training due to the environmental conditions and geographical barriers.

Conclusion

There are human factors involved in flying. The SHELL model of aviation was designed to understand all the dimensions that interact with the human so that human errors can be reduced through a better understanding of things. Human errors, accidents and need to increase the performance of pilots led to this research on analyzing the cognitive processes. The events which comprise this accident aptly illustrate the dire need for reliance on the pilot's cognitive powers of perception, procedural knowledge, evaluative and predictive judgment. It is commonly realized that the greater part of the air mishaps is identified with human blunders, while the mechanical disappointments in airplane upkeep today has hugely been on the lessening with various new high innovative types of technical developments. The research provides an in-depth knowledge of the psychological aspects. Thus it is really useful to understand the human behavior, the cognitive skills involved. Further, the information processing systems, the impacting and facilitating factors for the efficiency can better be comprehended from this research.

REFERENCES

Anderson, J. R. (1985). Cognitive Psychology and its implications. New York: W. H. Freeman and Company.

Diehl, A. E., &Buch, G. D. (1986). Developing an international program to Improve Pilot Decision making.Vancouver: Aight Safety Foundation.

Ericksen, C.W.(1955).Stimulus Differences and Accuracy of Discrimination.Journal of Experimental Psychology.50,153-601.

Garner, W.R., & Creel,C.G.(1964) .Effect of Redundancy and Duration on Absolute Judgments of Visual Stimuli. Journal of Experimental Psychology. 7, 168-172.

Glaser, R. (1986). Human memory and cognitive capabilities: Mechanisms and performances.The Netherlands: Elsevier North Holland Publishers.

Glaser, R. (1987).Cognitive functioning and social structure over the life course. Norwood, NJ:Ablex Publishing

Gordon, S. E. (1990). Implications of cognitive theory for knowledge acquisition. Journal ofCognitive psychology.65,325-355.

Guilford, J. P, (1967). The nature of human intelligence. New York: McGraw- Hill.

Green, B.F.,&Anderson,L.K.(1967).Color Coding in a Visual Search Task. Journal of Experimental Psychology. 19(5), 19-24.

Holyoak, K. J. (1990). Symbolic connectionism: toward third-generation theories of expertise.New York: John Wiley.

A MODEL OF SPACE OPERATOR TRAINING MOTIVATION USING SERIOUS GAMES

Foster E. Davis
Michael E. Miller
Air Force Institute of Technology, Wright Patterson Air Force Base, OH, USA

Services provided by spacecraft, including communications and global positioning, are integral to small businesses, multinational corporations, and the United States Department of Defense. United States rivals recognize the advantage provided by the space domain and are exploring ways to degrade these services in their warfare doctrine. In response, the United States requires space systems suited to counter these threats and personnel who are trained to respond to the newly contested environment. Training research has shown that trainee characteristics, including motivation, can significantly impact training outcomes. Beyond the training literature, guidelines have been developed for motivating game play which might apply to the development of serious games to support training. This paper reviews the academic literature and written guidelines from each of these domains and proposes a model of motivation to guide development of future interactive training environments for space operator training.

Ensuring common domains remain open and free is a key objective of the United States (U.S.) Department of Defense (DoD) (Mattis, 2018). One of the key domains the U.S. has sought to keep free is space. Space is integral to small businesses and multinational corporations, as space-based capabilities enable commerce by providing worldwide communications, logistics support, and monitoring of space-based threats to earth-based communication and electronic dissemination. These capabilities are essential to the global economy. Simultaneously, space provides the DoD a medium for global information collection, communications, logistics support, and command and control of forces. These capabilities provide military commanders intelligence, surveillance, and reconnaissance allowing them to understand the security environment (Department of Defense, 2019). Adversaries recognize a need to undermine the advantage the space domain offers the U.S. and are exploring ways to degrade these capabilities .

To promote space as an open and free domain, the existing U.S. space architecture was not created to counter these threats. Further, the significant cost of each system made redundancy hard to justify in an uncontested environment. This combination resulted in a fragile U.S. space architecture which is vulnerable to exploitation. The United States Space Force (USSF) formation creates a force with the goal of seeking to regain secure access to and freedom to operate in space (McCall, 2019). This requires increases in the number and quality of space-rated personnel (Department of Defense, 2018). These personnel must be trained to the new reality, including becoming skilled in rendezvous and proximity operations (RPO), where satellites are maneuvered near other satellites. Students should be motivated to learn as new problems are presented by adversaries, requiring the students to become self-learners dedicated to continually learning the threats encountered and responses, which are non-intuitive some orbital regimes. Thus, training is required to prepare these personnel to deal with the complexities of the control and protection of space assets to provide an effective system by enabling human integration (Grossman, Oglesby, & Salas, 2015).

Unfortunately, students often lack the motivation to learn (Alsawaier, 2018) and therefore, lack the desire to actively partake in training (Eyal, 2019). A rising solution to this problem is properly implemented gamification (Seaborn & Fels, 2015); the "game-based mechanics, aesthetics, and game thinking to engage people, motivate action, promote learning, and solve problems" (Kapp, 2013). However, the literature on motivation in training and motivation in gaming are discussed in separate bodies of literature. This paper reviews each of these bodies of literature and proposes an integrated model.

Motivation in Training

Training often begins with an analysis of the trainees to baseline knowledge, skills, and abilities (KSAs) of the individuals entering training. This analysis can include assessment of cognitive ability, self-efficacy, goal orientation, and motivation. Goal orientation can vary from mastery to performance focused where someone with a mastery or learning orientation is more interested in obtaining new knowledge or skills and a person with a performance orientation is more focused on appearing to acquire training content to obtain high grades (Grossman, Oglesby, & Salas, 2015). Fink mentions a similar concept, referred to as a person's sense of self as a learner. He states a person's sense of self will impact their training. Those with a weak sense of self as a learner often fail to form a clear understanding of what they need or want to learn, while those with a strong sense of self are proactive about their training experience and actively seek to discover what and why they need to learn. The strong sense of self may correspond to the mastery orientation and the weak sense of self may correspond to the performance orientation. Fink suggests that to improve training one must help students learn something that is significant about the subject matter to help students develop a strong and proactive sense of self as a learner (Fink, 2003). Influencing the motivation level requires the manipulation of what Fogg calls the core motivators. Fogg states that the core motivators are to seek pleasure and avoid pain, seek hope and avoid fear, and seek social acceptance and avoid rejection (Fogg, 2009).

Internal and external factors can impact trainee motivation to learn and transfer learning to the workplace. Internal factors, such as perceived utility, is high when trainees believe the training will provide value. This value includes the belief that the KSAs being taught will enable them to perform a job they find value in or enjoy. In this instance they feel a desire to improve their performance, and there will be some level of return on investment from improving performance (Berkling & Thomas, 2013; Grossman, Oglesby, & Salas, 2015). These internal factors provide intrinsic motivation as the individual finds these KSA interesting and performs without conditioning, for the pleasure of learning (Alsawaier, 2018). External factors include items such as organizational climate, organizational commitment, supervisor support, peer support, and subordinate support generate high motivation when buy-in is established. Organizational climate was found to be the most important factor in a study focused primarily on motivation to transfer. Motivation to transfer refers to the trainee's intent to utilize the skills and knowledge obtained from the training environment in the real world (Seyler, Holton III, Bates, Burnett, & Carvalho, 1998). A student can have different levels of motivation depending on the phase of training. Before training motivation is influenced by reputation of the training (Seyler, Holton III, Bates, Burnett, & Carvalho, 1998; Grossman, Oglesby, & Salas, 2015). During training, training features impact training effectiveness; including presentation format of information about the KSAs, proper demonstration of the KSAs, opportunities to practice the KSAs, and feedback (Grossman, Oglesby, & Salas, 2015). After training, the transfer climate of the gaining organization will largely determine a trainee's motivation to apply the KSAs. Transfer climate refers to attitudes and processes in the organization that hinder or facilitate the use of newly acquired KSAs (Grossman, Oglesby, & Salas, 2015).

Motivation in Gaming

Motivation and engagement are central to the self-determination theory. This theory lists three principles: autonomy, competence, and relatedness as being important to motivation and engagement. Competence is related to the motivation to persevere through difficulties and attain success. Autonomy relates to the need to make choices in pursuit of being responsible for one's actions. Relatedness is about social status and connections with others based on mutual respect and interdependence. These three elements fulfill the human psychological requirements to feel confident about their abilities, make choices, and compete or collaborate with others (Alsawaier, 2018). To obey the self-determination theory competence principle, the gamified course should have a good feedback system, provide an appropriate challenge, provide sufficient build up for complex topics, have intuitive controls, variable rewards, and social engagement (Seaborn & Fels, 2015). To adhere to autonomy the system should allow customization of items such s profiles, avatars, interface elements, alternate activities, privacy settings, and notification controls (Seaborn & Fels, 2015). Providing autonomy helps maintain interest by keeping what psychologist term reactance, the instinctive response to threats on free choice, at bay (Eyal, 2019). To provide a sense of relatedness the system should provide a way to compete or cooperate with others by forming groups, employing blogs, messaging options, and chat functions (Seaborn & Fels, 2015). The ability to share encouragement, exchange advice, and receive praise satisfies the human need for social acceptance (Eyal, 2019). Social acceptance can be catered to through relatedness elements and leaderboards (Alsawaier, 2018; Fogg, 2009).

Bartle's Test of Game Psychology differentiates among four types of gamers. Individuals who enjoy competing and playing against other players, i.e., *Killers*, are motivated by public recognition. Individuals who seek status with a high level of performance, i.e., *Achievers*, are motivated by tracking their achievement and progress. *Explorers* enjoy collecting virtual goods and discovering new things, thus are interested in pursuing quest rather than impressing others. Individuals who seek to collaborate with others, i.e., *Socializers*, seek to interact with others through mutual support. The various player types illuminate factors that motivate them to play (Alsawaier, 2018). Games developed for training may benefit self-determination theory by incorporating these considerations during course design to improve intrinsic motivation. With gamification, it is important to reward the effort through timely feedback, allowing them to learn from their mistakes. This feature should motivate students to put in more effort while conquering different learning challenges (Alsawaier, 2018).

Method

To develop an integrated model of these two bodies of literature, a Concept Map (CMAP) was developed to represent the motivating factors and their relationships. The question used to create the CMAP was "what is involved in motivation in training?" While this model captured the primary factors, it did not provide insight into the dynamics of the process. To capture these dynamics, the concepts were incorporated into a Causal Loop Diagram (CLD) from systems dynamics. The goal of this model is to understand the influence and feedback structure among the concepts. The loops created help link concepts to one another, giving direction and polarity to the relationship. The CLD integrates these feedback loops to aid in understanding the dynamic influences among the motivating factors (Sterman, 2001).

Results

The CMAP shown in Figure I depicts the core motivators, self-determination theory, and effect of gamer types as methods to understand motivation. Further, it differentiates between intrinsic and extrinsic motivation, with many of the extrinsic factors arising from organizational influences beyond the training organization. Finally, it includes attributes to guide game design.

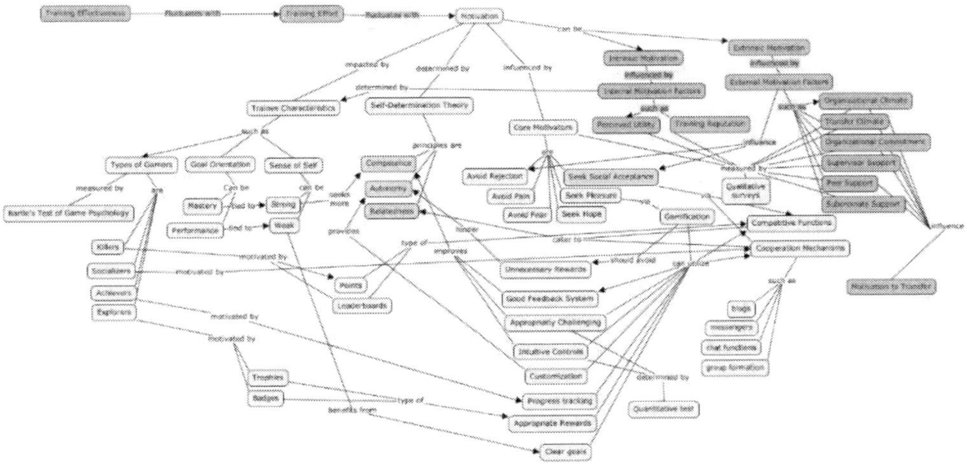

Figure I. Concept map linking motivation theories and ideas together.

Among desirable attributes are providing appropriate rewards, without unnecessary rewards for actions they perform freely as these rewards can be perceived as controlling; negatively impacting intrinsic motivation. The game should provide appropriate challenges with feedback, permitting students to experiment and increasing engagement through improving autonomy or self-determination. Frequent and immediate feedback further supports experimentation (Hanus & Fox, 2014). Students should also be provided avenues to help, challenge, and congratulate each other to foster relatedness and adherence to the social acceptance core motivator. Through these attributes, *Killers* are motivated by competing with their peers, *Socializers* are permitted to seek and provide aid as well as receive praise from their fellow students, *Achievers* can track their progress, and *Explorers* can enjoy discovering all that the game has to offer. *Explorers* are further supported by way of badges or extra information. The ability to repeat levels permits students to attempt higher scores, which supports *Killers* and *Socializers* and permits *Explorers* to attempt new avenues of play.

The CLD in Figure II**Error! Reference source not found.** illustrates the proposed model of the dynamic relationships among the motivators. The CLD represents both organizational, i.e., long-term, influences shown with red connectors, as well as influences internal to training, i.e., short-term influences, shown with blue connectors. The long-term influences begin with organizational commitment in which the organization both seeks to improve training effectiveness and transfer the knowledge to operations. As the transfer climate improves the students become increasingly motivated to transfer new knowledge to the workplace. This improves the perceived utility of the training for new students, improving their motivation. The improved motivation improves training effort, improving training effectiveness. Improvements in training effectiveness improve both training reputation and achieves the goal of improved system performance. Finally, improvements in training effectiveness improves training

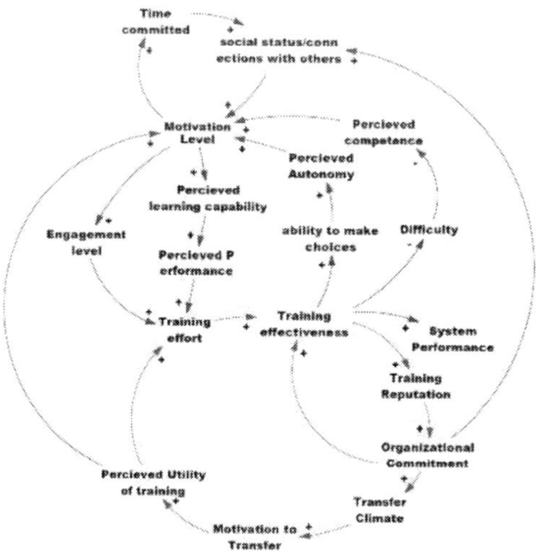

Figure II . Causal loop diagram explaining the interactions between various motivation factors.

reputation which improves organizational commitment. Generally, the long-term loop is a reinforcing loop.

Short-term influences within the training arise predominantly from increases in the student's perceived autonomy, competence, connections with others, as well as training utility. Although increased motivation is shown as increasing learning capability, leading to improved perceived performance, and improved training effort, not all of the short-term influences are reinforcing. For example, if the difficulty is too great, student will experience loss of perceived competence, reducing motivation (Alsawaier, 2018). Difficulty can be adjusted through many forms, including providing hints on how to complete task. By keeping the course at an appropriately challenging level, Hope, a core motivator, can be created while building competence, a tenet of self-determination theory (Fogg, 2009). Unlocking perks and leveling up satisfy a student's desire for competency by rewarding development and achievement (Eyal, 2019). *Socializers* will find the improved social connection appealing and will experience an increased motivation level and therefore put forth more training effort. *Killers* will experience improved motivation and increase training effort as their perceived performance rises. Providing students freedom to make choices increases perceived autonomy, increasing motivation.

Conclusion

Training in the space environment needs to develop the intrinsic motivation necessary to create lifetime learners (Fink, 2003) and appropriate gamification of training may offer an avenue for achieving that objective, if properly implemented. This paper proposes a potential model of human motivation within gamified training with the goal of establishing requirements or evaluation criteria space operator training. Future research will be needed to assess the utility of this model for designing and assessing serious games for space operator training.

Acknowledgments and Disclaimer

I would like to thank to Dr. Michael Miller and Dr. Richard Cobb for the aid and insight provided in completing this paper. The views expressed in this thesis are those of the author and do not reflect the official policy or position of the United States Air Force, Department of Defense, or the United States Government.

References

Alsawaier, R. (2018). The effect of gamification on motivation and engagement. *International Journal of Information and Learning Technology*, 56-79.

Berkling, K., & Thomas, C. (2013). Gamification of a Software Engineering Course and a detailed analysis of the factors that lead to it's failure. *International Conference on Interactive Collaborative Learning* (pp. 525-530). Karlsruhe: Cooperative State University Baden Württemberg.

Colby, E. (2016). *From Sanctuary to Battlefield: A Framework for a U.S. Defense and Deterrence Strategy for Space.* Washington: Center for a New American Security.

Department of Defense. (2018). *Final Report on Organizational and Management Structure for the National Security Space Components of the Department of Defense.* Department of Defense.

Department of Defense. (2019). *United States Space Force.* Department of Defense.

Eyal, N. (2019). *Hooked.* Penguin Random House LLC.

Fink, D. L. (2003). *Creating Signifiant Learning Experiences.* San Francisco: Jossey-Bass.

Fogg, B. J. (2009, April). A Behavior Model for Persuasive Design. *4th international Conference on Persuasive Technology* (pp. 1-7). Claremont: Stanford University.

Grossman, R., Oglesby, J., & Salas, E. (2015). The Training Process: Using The Science Each Step of the Way. In D. A. Boehm-Davis, F. T. Durso, & J. D. Lee, *APA Handbook of Human Systems Integration* (pp. 501-516). American Psychological Association.

Hanus, M. D., & Fox, J. (2014). Assessing the effects of gamification in the classroom: A longitudinal study on intrinsic motivation, social comparison, satisfaction, effort, and academic performance. *Computers & Education*, 152-161.

Kapp, K. M. (2013). *The gamification of learning and instruction fieldbook: Ideas into practice.* John Wiley & Sons.

Mattis, J. (2018). *Summary of the 2018 National Defense Strategy of the United States of America.* Department of Defense.

McCall, S. M. (2019). *Military Space Reform: FY2020 NDAA Legislative Proposals.* Congressional Research Service.

Seaborn, K., & Fels, D. (2015). Gamification in Theory and Action: a Survey. *International Journal of Human-Computer Studies*, 14-31. Retrieved from Research Gate: https://www.researchgate.net/publication/266398023_Gamification_in_Theory_and_Action_A_Survey/citation/download

Seyler, D. L., Holton III, E. F., Bates, R. A., Burnett, M. F., & Carvalho, M. A. (1998). Factors Affecting Motivation to Transfer Training. *International Journal of Training and Development 2.1*, 2-16.

Sterman, J. D. (2001). Fine-Tuning Your Casual Loop Diagrams-Part I. *The Systems Thinker*, 6-7.

TRANSITIONING FROM FACE-TO-FACE TO VIRTUAL TRAINING: TRAINEE PERCEPTIONS OF VIRTUAL AIR TRAFFIC TRAINING

Brett S. Torrence
Federal Aviation Administration
Oklahoma City, OK
Chanda Sanders
Federal Aviation Administration
Oklahoma City, OK
Charla Long
Federal Aviation Administration
Oklahoma City, OK

The public health emergency has adversely impacted the aviation community, resulting in reduced air traffic operations and challenges for the workforce. The Air Traffic Controller workforce has experienced interruptions to initial and on-the-job training due to social distancing practices and extended periods of reduced traffic. In response, the Federal Aviation Administration (FAA) transitioned components of its air traffic training to an online (virtual) platform to continue training and reduce delays. An initial evaluation of the FAA's Virtual Air Traffic Basics and Virtual Initial Lessons courses was conducted by examining air traffic control trainee ($N = 180$) perceptions of the virtual format. Preliminary findings suggest that trainees were satisfied with the virtual courses and found the virtual environment to be supportive of their learning and helpful for developing air traffic knowledge. Challenges of the virtual format were noted as well. Implications of the current findings for virtual training are discussed.

Effective Air Traffic Controller (ATC) training is vital for ensuring that controllers have the necessary technical knowledge and skills to manage a safe and expeditious flow of air traffic in the National Airspace System (NAS). Newly hired controllers participate in a standardized training program consisting of academic courses, lab exercises, simulations, and on-the-job training through the Federal Aviation Administration's (FAA) Air Traffic Training program. This program consists of two phases: FAA Academy and Field Qualification training. The FAA Academy serves as the initial phase of training, teaching newly hired controllers foundational air traffic principles and procedures needed for the second phase of training conducted at an assigned field location. Training newly hired ATCs is of significant interest due to the complexity of the job, the mission of providing a safe and efficient NAS, and the costly investment by the agency. Effective training ensures that the FAA can maintain a highly efficient and effective workforce that meets the skill demands of the air traffic environment.

However, the public health emergency has caused several disruptions to air traffic, and the aviation community as a whole. The International Civil Aviation Authority (ICAO; 2021) reported a 40% reduction in domestic passenger flights across North America in 2020 compared to 2019, as well as billions of dollars in lost revenue. In addition to the economic impacts, the crisis has produced unprecedented impacts on the aviation workforce, particularly ATCs. Some of the impacts on ATCs include extended periods of inactivity at work, changes to staffing procedures, temporary closure of facilities for cleaning, and changes to training requirements and procedures. Changes to ATC training included a temporary stoppage of training at the FAA

Academy, a transition from instructor-led classes to a virtual learning environment, and reduced class sizes for in-person FAA Academy training.

The FAA Academy consists of two courses, Air Traffic (AT) Basics and Initial Qualification training. AT Basics teaches trainees basic aviation and air traffic concepts, and provides an introduction to air traffic control procedures. Initial Qualification training provides option-specific training for en route and terminal (tower) controller positions. Prior to the changes, the FAA Academy utilized in-person, instructor-led training to train newly hired ATCs. However, the temporary stoppage in training resulted in a shift; transitioning parts of training to a virtual format. The entire AT Basics course was moved online to enable the delivery of basic training to newly hired controllers. Similarly, the academic components of Initial Qualification training were developed into a virtual course to maintain knowledge retention of trainees waiting to be assigned to in-person Initial Qualification training at the FAA Academy. Both Virtual AT Basics and Virtual Initial Lessons are taught synchronously over a virtual learning platform with training materials delivered primarily through instructor-led lectures and breakout room sessions (i.e., online group activities). The curriculum of the virtual courses is equivalent to the respective in-person courses. Prior to training, trainees are provided a device (i.e., iPad) that contains the software applications and learning materials needed for accessing, and participating in, training.

The FAA responded quickly by developing the virtual training courses to minimize training delays, meet staffing requirements, and maintain the safety of employees. The purpose of this study is to provide an initial evaluation of the Virtual AT Basics and Virtual Initial Lessons course through an examination of trainee perceptions of the virtual environment. Specifically, trainee satisfaction with the course(s), their perception of virtual learning, the benefits and challenges of virtual learning, and technological challenges faced by the trainees was evaluated. The data reported here provide a preliminary look at the new virtual courses. This study is a part of a larger, ongoing effort exploring the effectiveness of virtual training for ATCs.

Method

Participants and Procedure

We collected data from 180 air traffic control trainees enrolled in Initial Qualification training at the FAA Academy. Participants (M_{age} = 26.47 years, SD_{age} = 2.99 years) included trainees assigned to the En Route (n = 55), Tower Cab (n = 113), and Terminal Radar (n = 12) training track. Seventy-two percent (n = 130) of the sample had previously taken an online course (e.g., high school, college) prior to Virtual AT Basics. All participants had successfully completed Virtual AT Basics and finished Virtual Initial Lessons prior to completing the survey.

Trainees' perceptions were collected using an online survey. Using a cross-sectional design, trainees completed the survey on their first day of in-person Initial Qualification training. Participants provided consent to participate in the study prior to completing demographic questions, a training evaluation questionnaire for Virtual AT Basics and Initial Lessons, and questions about their use of technology during training. The training evaluation questions focused on different elements of the virtual environment, such as engagement, learning activities, and interactions. Participants were also asked to report their satisfaction with the course and confidence following the training. Additionally, participants described the benefits and challenges of virtual training. The survey took approximately 30 minutes to complete.

Results

Trainee Perceptions

The Virtual AT Basics evaluation questionnaire asked participants to rate items using a 4-point Likert scale (1- *strongly disagree*, 4- *strongly agree*) to indicate the extent to which they agreed or disagreed with statements about the course. Trainees agreed or strongly agreed that the online environment was easy to navigate (90%; $M = 3.11$, $SD = 0.58$), supported their learning (84%; $M = 3.02$, $SD = 0.66$), and was moderately engaging (70%; $M = 2.81$, $SD = 0.77$). Trainees also agreed or strongly agreed the training applications (e.g., learning platforms and software) used during the course were easy to use (88%; $M = 3.12$, $SD = 0.60$), supported their learning (92%; $M = 3.15$, $SD = 0.55$), and were moderately engaging (72%; $M = 2.86$, $SD = 0.77$). Finally, trainees agreed or strongly agreed the learning activities, which included lectures and individual/group exercises, were helpful for developing ATC knowledge (91%; $M = 3.12$, $SD = 0.60$), provided an opportunity to practice what they had learned (88%; $M = 3.10$, $SD = 0.65$), were engaging (82%; $M = 2.91$, $SD = 0.65$), and prepared them for the end-of-course test (82%; $M = 3.12$, $SD = 0.60$). However, roughly half of the trainees agreed the learning activities promoted interactions with other classmates (50%; $M = 2.48$, $SD = 0.80$).

Trainees, on average, were satisfied with Virtual AT Basics and rated their overall learning experience as positive on a 0 to 10 scale ($M = 7.19$, $SD = 1.84$). Additionally, after completing the Virtual AT Basics, trainees felt moderately confident or very confident about their knowledge of ATC job responsibilities (89%; $M = 3.22$, $SD = 0.65$) and ability to be successful in Initial Qualification training (84%; $M = 3.10$, $SD = 0.68$).

The Virtual Initial Lessons evaluation questionnaire also used a 4-point Likert scale (1- *strongly disagree*, 4- *strongly agree*). Trainees agreed or strongly agreed the online environment for Virtual Initial Lessons was easy to navigate (96%; $M = 3.23$, $SD = 0.50$), supported their learning (91%; $M = 3.15$, $SD = 0.65$), and was engaging (88%; $M = 3.13$, $SD = 0.68$). Trainees agreed or strongly agreed the training applications were easy to use (96%; $M = 3.18$, $SD = 0.48$), engaging (86%; $M = 3.02$, $SD = 0.64$), and supported learning (92%; $M = 3.12$, $SD = 0.59$). Finally, trainees agreed or strongly agreed the learning activities, which consisted of lecture and practice exercises (e.g., flight strip, maps), were helpful for developing ATC knowledge (97%; $M = 3.33$, $SD = 0.53$), provided an opportunity to practice what they had learned (96%; $M = 3.32$, $SD = 0.57$), were engaging (87%; $M = 3.07$, $SD = 0.63$), and promoted interactions with other classmates (82%; $M = 2.99$, $SD = 0.71$).

Trainees, on average, were satisfied with the Virtual Initial Lessons course and rated their overall learning experience as positive on a 0 to 10 scale ($M = 7.38$, $SD = 1.78$). Following the completion of Initial Virtual Lessons, trainees felt moderately confident or very confident about their knowledge of ATC duties and responsibilities (94%; $M = 3.38$, $SD = 0.59$) and ability to be successful in Initial Qualification training (94%; $M = 3.34$, $SD = 0.59$).

Benefits and Challenges

Qualitative responses provided by the participants were reviewed to identify common benefits and challenges of the virtual training. Almost 45% of trainees that responded reported training from their home as a top benefit of Virtual AT Basics. Additionally, trainees described training at home as a benefit because it offered a convenient (15%) and/or comfortable learning environment (20%) to learn ATC material. Virtual training also afforded trainees additional time to study and/or be with family (17%) and stay safe during the public health emergency (6%). Sixty-five percent of trainees that responded, listed additional exposure to learning material before attending the FAA Academy as a top benefit of Virtual Initial Lessons, in addition to the quality of instructors (21%), learning experiences (18%), and completing training at home (14%).

The top reported challenges for Virtual AT Basics included a lack of interaction with instructors and students (34%), understanding abstract material (19%), at-home distractions and stressors (17%), disengagement (17%), and connectivity or technology issues (15%). Trainees also reported a number of challenges for Virtual Initial Lessons. Thirty-three percent of responding trainees cited lack of interaction with instructors and students as a challenge. Other difficulties included feeling disengaged and unable to pay attention to online content (20%), understanding the content (20%), and instructor-related issues (16%) that ranged from changing instructors too often to conflicting information across instructors.

Discussion

The mission-critical nature of Air Traffic Control underscores the need for effectively designed training. Trainees are required to learn foundational information to successfully advance in ATC training, and it is crucial that trainees graduate with the knowledge needed for the next phase of training and to, ultimately, control live air traffic. The purpose of this evaluation was to investigate trainee perceptions of the new virtual air traffic training. Although challenges were noted, trainee responses offer preliminary evidence that the virtual courses provide a satisfactory learning experience and meet the training requirements of newly hired ATCs. Noteworthy, however, are the inconsistencies between the survey data and the challenges described by trainees (e.g., interaction between instructors and students, understanding material, disengagement). Potential explanations for the discrepancies could be individual differences in learning preferences among trainees. Further investigation is needed to clarify and provide a better understanding of this finding.

Technology can be beneficial for training as it provides flexibility and affords the opportunity to continue training when face-to-face delivery is not viable. Prior research suggests properly designed online training can be as effective as classroom training and tends to be most effective when the course incorporates active learning, provides practice opportunities, encourages interactions, offers learners control over their learning experience, and blends content with face-to-face instruction (e.g., Sitzmann et al., 2006). Virtual training, therefore, must be designed with the right learning principles to support knowledge acquisition and retention. The design and content of training will influence trainees' information processing, attentional focus, metacognition, motivation, and emotional responses (Gully & Chen, 2010). The results obtained from this research study, in combination with recommendations from the scientific literature on training (e.g., Goldstein & Ford, 2002; Kraiger, 2003; Salas et al., 2012), were used to develop a brief list of recommendations, shown in Table 1, for virtual training.

However, as this study is an initial evaluation, some limitations should be kept in mind when interpreting the results. First, only trainee attitudes toward the virtual training environment are presented in this report. While understanding how well trainees liked the training is important, the larger research effort will also evaluate other types of subjective and objective learning outcomes (e.g., knowledge tests). Additionally, instructors' perceptions of training quality may provide insights that differ from the trainee viewpoint. Second, the data reported in this study are from the first wave of classes that participated the virtual Air Traffic courses. As such, responses may reflect trends and attitudes unique to a newly implemented training course.

Conclusion

Training is crucial to ensuring the continued success of the ATC workforce and technology enables the delivery of training to be flexible and adaptable. As the use of training technologies continues to evolve in the air traffic domain, ongoing evaluation is needed to ensure the design and development of training provide trainees with the needed knowledge and learning experiences.

Table 1.
Recommendations for Virtual Training and Example Practices

Recommendations	Example Practices
Address instructional design before technology	Ensure training material can be taught and learned using technology (Salas et al., 2012) Use instructional principles and learning objectives to drive training design before selecting technology
Encourage interactions	Implement active learning activities (e.g., breakout rooms, discussion boards) that provide a variety of interactions Promote collaborative learning through planned interactions among trainees and instructors
Consider instructor needs	Provide training and resources to instructors for virtual learning techniques, assessing learning in virtual settings, and keeping trainees engaged Emphasize the use of feedback as a key interaction between instructors and students (Tannenbaum, Beard, McNall, & Salas, 2010)
Account for learner preferences and individual differences	Provide hands-on activities to provide trainees the opportunity to learn through doing (Gully & Chen, 2010) Offer the option of electronic or hard-copy learning materials (e.g., strip marking boards, maps)
Focus on trainee engagement	Encourage instructor sharing of real-world uses and cases of learning material (Garrison & Cleveland-Innes, 2005) Emphasize job relevance and learning outcomes to trainees throughout learning activities (Bell & Kozlowski, 2010) Leverage virtual learning features to mirror in-person classroom activities (Cannon-Bowers & Bowers, 2010)

Acknowledgements

This report was developed under the Air Traffic/Technical Operations Human Factors Program Directive/Level of Effort Agreement between the Human Factors Division (ANG-C1), FAA Headquarters, and the FAA Aerospace Human Factors Research Division (AAM-500), Civil Aerospace Medical Institute (CAMI).

References

Bell, B. S., & Kozlowski, S. W. J. (2010). Toward a theory of learner-centered training design: An integrative framework of active learning. In S. W. J. Kozlowski & E. Salas (Eds.), *Learning, Training, and Development in Organizations* (pp. 263-293). Routledge Taylor & Francis Group.

Cannon-Bowers, J., & Bowers, C., (2010). Synthetic learning environments. On developing a science of simulation, games, and virtual worlds for training. In S. W. J. Kozlowski & E. Salas (Eds.), *Learning, Training, and Development in Organizations* (pp. 229-254). Routledge Taylor & Francis Group.

Garrison, D. R., & Cleveland-Innes, M. (2005). Facilitating cognitive presence in online learning: Interaction is not enough. *The American Journal of Distance Education*, *19*, 133-148.

Goldstein, I. L., & Ford, J. K. (2002). *Training in organizations*. Wadsworth.

Gully, S., & Chen, G. (2010). Individual differences, attribute-treatment interactions, and training outcomes. In S. W. J. Kozlowski & E. Salas (Eds.), *Learning, Training, and Development in Organizations* (pp. 3-49). Routledge Taylor & Francis Group.

International Civil Aviation Organization. (2021). *Effects of novel coronavirus (COVID-19) on civil aviation: Economic impact analysis*. https://www.icao.int/sustainability/Pages/Economic-Impacts-of-COVID-19.aspx

Kraiger, K. (2003). Perspectives on training. In W. C. Borman, D. R. Ilgen, & R. J. Klimoski (Eds), *Comprehensive handbook of psychology: Volume 12. Industrial and organizational psychology* (pp. 171-192). John Wiley.

Salas, E., Tannenbaum, S. I., Kraiger, K., & Smith-Jentsch, K. A. (2012). The science of training and development in organizations: What matters in practice. *Psychological Science in the Public Interest*, *13*, 74-101.

Sitzmann, T., Kraiger, K., Stewart, D., & Wisher, R. (2006). The comparative effectiveness of web-based, and classroom instruction: A meta-analysis. *Personnel Psychology*, *59*, 623-664.

Tannenbaum, S. I., Beard, R. L., McNall, L. A., & Salas, E., (2010). Informal learning and development in organizations. In S. W. J. Kozlowski & E. Salas (Eds.), *Learning, Training, and Development in Organizations* (pp. 303-326). Routledge Taylor & Francis Group.

DOES ATTENTION TRAINING ENHANCE STRESS RESILIENCE AND PERFORMANCE IN UNFORESEEN SAFETY-CRITICAL SITUATIONS?

Christine M. Talker
ChrisTalkeR Coaching Training Research e.U.
Vienna, Austria

Dealing successfully with unforeseen safety-critical situations is a prerequisite for save pilot performance. Studies applying new attention-based training approaches have revealed positive effects on emotion regulation and on concentration abilities. Hence, the question arises to what extent attention training would facilitate cognitive adaptation processes, thereby attenuating emotional stress responses and reducing performance decrements in unforeseen flight situations. Twenty-four pilots will be randomly assigned to two groups and will either be trained in attention regulation or in relaxation techniques. "Home training" will be followed by training in the flight simulator. Performance ratings, video and audio recordings, subjective data, and EDA data will be collected. It is expected that the experimental group "Attention" will show lower stress responses and better performance compared to the control group "Relaxation" when faced with an unexpected situation in the final simulator test. It is suggested, that attention training positively influences cognitive appraisal processes and cognitive flexibility.

Unforeseen safety-critical situations with high complexity are among the most stressful challenges in high-risk environments such as aviation (Fornette, Bourgy, Jollans, Roumes, & Darses, 2014) and can severely hamper an operator's performance (Casner, Geven, & Williams, 2013). Recent incidents and accidents in civil aviation, classified as "loss of control in flight", have provoked safety experts to sound the alarm (Landman, Groen, van Paassen, Bronkhorst, & Mulder, 2017). They strongly emphasize that there is a need to reinforce a pilot's skills in dealing successfully with unforeseen safety-critical situations. Cognitive adaptation training could be a valuable supplement to the conventional training. In this regard, Fornette, et al. (2014) stressed the implementation of new training approaches based on attention regulation. However, the effects of these training techniques have not yet been evaluated in detail. Hence, the question arises to what extent attention training would facilitate cognitive adaptation processes which may enhance stress resilience and performance in unexpected flight situations.

Safety in all flight situations is a challenging demand. Operators, once selected, have to be intensively trained in order to manage the challenges faced in the time-dynamic working environment. Training so far mainly concentrated on improving a (student) pilot's anticipatory abilities (e.g., Talker, 2017) to successfully apply the knowledge and skills in time when faced with expected situations (Fornette et al., 2014). However, as accident reports revealed, a flight situation can rapidly change from "manageable" to "extremely challenging" if the safety-critical situation is unforeseen. The breakdown of anticipation might require a change from an automatic mental mode (= state of mind that is predominant in well-trained situations) to an adapted mental mode (essential in new, unforeseen situations) in order to respond flexibly and adequately to the changed conditions (Fornette et al., 2014). These additional cognitive processes, however, might pose the risk of losing valuable time in a life-threatening and highly time-critical situation (Burian, Barshi, & Dismukes, 2005; Fornette et al., 2014). In this regard, the study of Casner et al. (2013) revealed a significant increase in response times when pilots were faced with an abnormal in-flight event in unfamiliar cirmcumstances. Were these effects of an anticipation failure?

A main contributing factor to anticipation lapses can be seen in the increasing complexity of automated aircraft systems in the last decades (Landman et al., 2017). The increased pilot reliance on aircraft automation (European Aviation Safety Agency [EASA], 2017) and/or the less transparent flying process might increase the probability of a mismatch of the anticipated flight situation and the actual event. As recent studies revealed, a breakdown of anticipation may manifest itself in a considerable increase in emotional stress responses (Talker, 2017) and can negatively affect a pilot's performance (Casner et al., 2013; Landman et al., 2017).

How can pilots be cognitively trained in order to be prepared for the unexpected? Promising results from studies in a combat aviation population (Meland, Fonne, Wagstaff, & Pensgaard, 2015) revealed positive effects of cognitive adaptation training on concentration abilities as well as on arousal regulation. This new training approach is based on attention regulation and may overcome some limitations of previous training methods that are based on cognitive control (cf. Fornette et al., 2014).

But little is known about the impact of attention training on pilot performance and stress responses in unforeseen situations. Hence, the present study aims to elucidate to what extent attention training facilitates cognitive adaptation processes. These processes allow for the instantaneous adaptation to unforeseen safety-relevant changes in the environment and make use of the "on-line" mechanism of anticipation (i.e., closely related to the actual stimuli; Pezzulo, Butz, & Castelfranchi, 2008) in order to flexibly and appropriately respond to the current situational requirements, while keeping unnecessary stress activation low. In order to shed light on this issue, the effects of attention training on stress resilience and performance in unexpected flight situations will be experimentally examined in a FNPT-II simulator. Stress responses and performance will be assessed by collecting subjective data (performance ratings, questionnaires, and interviews), video and audio recordings of cockpit communication, and psychophysiological data (electrodermal activity).

Method

Participants

Twenty-four active pilots holding an Airline Transport Pilot Licence (ATPL) will be recruited for the study. They will not have to meet requirements in regard to a pilot's completed flying hours and the type rating a pilot holds. Each participant will take part in the experiment voluntarily. They will have to sign an informed consent and will be given the opportunity to quit the experiment whenever they wish, without giving any reasons. The participants will be naïve to the purpose of the experiment.

Design and Procedure

The study will comprise three main experimental phases (P): (I) training outside the flight simulator, (II) training in the simulator, and (III) the final simulator test. P I is scheduled for three months, P II and P III for about one hour, each.

Participants will be randomly assigned to two groups. In P I and P II, the experimental group (n = 12) will undergo an attention training, while the control group (n = 12) will do a muscle relaxation training in order to control for possible relaxing or restorative effects of the attention training procedure. Both groups will do the same final simulator test in P III where they will be faced with an unforeseen safety-critical flight situation.

The experiment will start with a 6-hour classroom seminar, performed separately for each group. The experimental group will be introduced to the theoretical background of attention training followed by the practical training session, where they will learn to deliberately regulate the allocation and the focus of attention. The practical training will include the following exercises (cf. Kabat-Zinn, 2003; Wagner, 2011; Williams & Kabat-Zinn, 2013): (1) Changing the focus of attention in the sense modalities seeing, hearing, and feeling, (2) Sitting upright with eyes closed with a "narrow" focus on the breath, and (3) sitting upright with eyes closed with a "broad" focus on thoughts, feelings, body sensations plus a "constant focus" on the breath. Participants will be instructed to observe arising sensations, thoughts, and feelings without judging them or wanting to change them. The exercises 1 and 2 will be for preliminary practice. The exercise 3 will be for further training outside the classroom (i.e., "home training"). The control group will be introduced to the theoretical background of relaxation followed by relaxation training in practice (Jacobson, 1934).

The participants of both groups will have to practice for 30 minutes three times a week, in a time frame of three months. Once a week, the participants will take part in a five-minute online one-to-one supervision session with the instructor where they will have the possibility to report their progress and to get support in case of problems.

After this training phase, the participants will undergo two simulator sessions – the simulator training and the final simulator test. The cockpit crew will consist of the pilot flying (= participant) and the copilot (= an experienced pilot who will be a member of the experimental team). The copilot will only take actions if instructed by the pilot flying.

Prior to the first simulator session, participants will have to complete the first questionnaire package. Thereafter, the electrodes for recording the participant's electrodermal activity (EDA) will be applied. EDA baseline measurements will be taken in an upright sitting position in the dark flight simulator cabin, with eyes closed. The simulation will be switched off during the baseline measurement.

Immediately before the simulator training, participants will do a 10-minute familiarization flight. In the simulator training phase, both groups will conduct an instruction flight. In order to simulate a real flight, the maneuvers will also include a takeoff and a landing procedure. At the beginning of each maneuver, the experimental group will be instructed to keep attention in the "here and now". The control group will will be instructed to keep their muscles relaxed.

After a break, where the participants will complete the second questionnaire package, both groups will undergo the final simulator test. Other than in the simulator training, the participants will not get any instructions in regard to attention or relaxation. Towards the end of the final simulator test, the participants will be faced with an unforeseen safety-critical situation. During the simulator test, video and audio recordings of the cockpit crew will be taken.

After the final simulator test, the third questionnaire package will be presented and the participants will attend a post-task reconstruction interview.

Apparatus

In order to fulfill the requirements of the planned study, a FNPT-II MCC (Flight and Navigation Procedures Trainer Type II Multi-Crew Co-operation) will be used. Offering a totally integrated system, the FNPT-II is fully instrumented for pilot and co-pilot stations. With a full autopilot capability, the autopilot can be controlled by either the pilot or the co-pilot (ELITE Simulations Solutions AG / S923 FNPT II MCC, 2021). The flight model of a Beech King Air B200 Twin Engine Turbine Aircraft will be used.

Dependent Variables

Performance will be assessed by using pre-defined criteria checked by a qualified instructor pilot. In order to evaluate different aspects of physical well-being, the Multidimensional Physical Symptom Check-List (MKSL – 24 – ak; Erdmann & Janke, 1978) will be used. The questionnaire includes 24 items which are aggregated into the four subscales: (1) nausea/cholinergic physical arousal, (2) adrenergic physical arousal, (3) pain, and (4) physical relaxation. Video and audio recordings in the cockpit during the final simulator test should reveal special aspects of a pariticipant's behaviour and his/her commands to the copilot. A post-task reconstruction interview after the simulator test will focus on the participant's perception of the unforseen situation as well as his/her thoughts, emotions, and self-described behavior before, during, and after the safety-critical situation. During both simulator sessions, electrodermal activity (EDA) will be recorded by using the method of exosomatic recording. Baseline measurements of 60 seconds will be taken at the beginning of each simulator session.

Statistical Analyses

Questionnaire data and EDA data will be analyzed using the procedure of mixed-design univariate ANOVAs with "group" as between-subject factor and "time" as within-subject factor. Independent samples t-tests (main effect of "group") and paired-samples t-tests (main effect of "time") will be used for post hoc analyses. In case of statistically significant "group" x "time" effects, post hoc analyses will be done by means of repeated-measures ANOVAs (and post hoc paired-samples t-tests) and by using independent samples t-tests. A significance level of $\alpha \leq .05$ will be adopted for the statistical tests. The assumption of normal distribution will be checked by means of the Kolmogorov-Smirnov Test, the premise of variance homogeneity will be evaluated by means of Levene Test, and the sphericity assumption will be evaluated by means of the Mauchly's Test. In case of violation, the Greenhouse-Geisser Test will be used in order to correct the degrees of freedom. Because of the explorative character of the study, no correction for type-I-error will be conducted.

EDA data (SCL, NS.SCRfreq) will be baseline-corrected and will be analyzed in three successive time intervals of 10 s, i.e., before, during and after the unforeseen situation (= "anticipation", "unforeseen effect", and "post effect").

Results

The main objective of the experiment is to reveal the effects of attention training, suspected to facilitate cognitive adaptation processes, on performance and stress responses in unexpected flight situations.

It is expected that the pilots trained in attention regulation will show lower emotional stress responses during the unexpected safety-critical situation, will get higher (i.e., better) performance ratings, and will show less decrements in physiological well-being after the final simulator test compared to the control group trained in relaxation techniques.

Discussion

Complex and unforeseen situations in flight can be extremely challenging even for experienced pilots and pose the risks of severe decrements in pilot performance (Casner, et al., 2013). For save pilot performance, anticipation of the near future is stressed to play a pivotal role (cf. SA, Endsley, 1995). Training approaches developed so far aim to improve anticipatory abilities (e.g., anticipation-based training, Talker, 2017) in order to facilitate anticipatory learning processes, suggested to build up strengthened mental representations of expected flight situations. However, these approaches might have limitations in case a pilot is faced with an unforeseen situation.

Current training approaches try to handle this problem by standardized trainings of abnormal events in the flight simulator. However, as Casner et al. (2013) could show, abnormal events become predictable when the flight scenarios are presented in the same sequence under the same circumstances. This procedure poses the risk of a low transfer of skills from training to the varying situations in the real flight environment. The question arises if a tested pilot really meets the requirements of an expert in managing abnormal and unforeseen events. An alarming answer has been provided by the findings of Casner et al., which revealed severe pilot performance decrements only in unfamiliar safety-critical flight situations in the simulator. These findings might reflect a breakdown of anticipation.

Does cognitive adaptation training based on attention provide an answer to this problem? It can be assumed that attention training facilitates cognitive processes which allow for the formation of continuously updated mental representations of the current flight situation. The proposed state of mind of being in the "here and now" (Kabat-Zinn, 2003; Williams & Kabat-Zinn, 2013) might play a key role in a pilot's ability to stay in immediate touch with the special aspects of the ongoing flight situation and might facilitate the "online-usage" of anticipation (Pezzulo et al., 2008). Because of the nonjudgmental attitude, attention training might also influence cognitive appraisal processes which might have positive effects on the occurrence of unnecessary emotional stress responses. The positive effects of attention training might manifest themselves as higher levels of stress resilience and as save pilot performance in unforseen flight situations.

Conclusions

Cognitive adaptation training based on attention might be a promising approach to improve flight safety – especially in complex and unforseen situations. Pilots trained to deliberately regulate their attention might have considerably improved skills for identifying safety-relevant cues from the flood of information and might be better equipped to flexibly and appropriately respond to unforeseeable safety-critical flight situations, while experencing low emotional stress responses.

Acknowledgements

The project is planned to be submitted for funding by the Federal Ministry for Climate Action, Environment, Energy, Mobility, Innovation and Technology, in The Austrian Aeronautics Programme TAKE OFF, 2021. The views of the research reported do not reflect the views of the granting organization.

References

Burian, B. K., Barshi, I., & Dismukes, K. (2005). *The challenge of aviation emergency and abnormal situations.* (NASA Technical Memorandum 2005–213462). Moffett Field, CA: NASA Ames Research Center.

Casner, S. M., Geven, R. W., & Williams, K. T. (2013).The effectiveness of airline pilot training for abnormal events. *Human Factors, 55*, 477–485. doi:10.1177/0018720812466893.

ELITE Simulations Solutions AG / S923 FNPT II MCC (2021, March 27). Retrieved from https://www.flyelite.ch/en/s923-fnpt-ii-mcc/

Endsley, M. R. (1995). Toward a theory of situation awareness in dynamic systems. *Human Factors, 37*(1), 32–64.

Erdmann, G., & Janke, W. (1978). *The Multidimensionl Physical Symptom Check-List* [Unpublished manuscript]. Technische Universität Berlin and Julius-Maximilians-Universität Würzburg.

European Aviation Safety Agency (2017). *Loss of control prevention and recovery training* (RELATED NPA/CRD 2015-13 — RMT.0581). Retrieved from https://www.easa.europa.eu/sites/default/files/dfu/Opinion%20No%2006-2017.pdf

Fornette, M.-P., Bourgy, M., Jollans, J.-Y., Roumes, C., & Darses, F. (2014). Enhancing management of complex and unforseen situations among pilots: new trends in cognitive-adaptation Training. In M. A. Vidulich, P. S. Tsang, & J. M. Flach (Eds.), *Advances in Aviation Psychology* (Vol. 1, pp. 229–246). Burlington, VT: Ashgate.

Jacobson, E. (1934). *You must relax. A practical method of reducing the strains of modern living.* New York, NY: McGraw-Hill Book Company Inc.

Kabat-Zinn, J. (2003). Mindfulness-based interventions in context: past, present, and future. *Clinical Psychology: Science and Practice, 10*(2), 144–156. doi:1093/clipsy.bpg016

Landman, A., Groen, E. L., van Paassen, M. M., Bronkhorst, A. W., & Mulder, M. (2017). The influence of surprise on upset recovery performance in airline pilots. *The International Journal of Aerospace Psychology, 27*(1–2), 2–14. doi:10.1080/10508414.2017.1365610

Meland, A., Fonne, V., Wagstaff, A., & Pensgaard, A. M. (2015). Mindfulness-based mental training in a high-performance combat aviation population: a one-year intervention study and a two-year follow-up. *The International Journal of Aviation Psychology, 25*(1), 48–61. doi:10.1080/10508414.2015.995572

Pezzulo, G., Butz, M. V., & Castelfranchi, C. (2008). The anicipatory approach: definitions and taxonomies. In G. Pezzulo, M. V. Butz, C. Castelfranchi, & R. Falcone (Eds.), *The challenge of anticipation: a unifying framework for the analysis and design of articificial cognitive systems* (pp. 23–43). Berlin: Springer.

Talker, C. M. (2017). *Anticipation in flight, The effect of experience on the impact of Coriolis effects examined in a motion-base flight simulator* [Unpublished doctoral dissertation]. University of Graz.

Wagner, A. C. (2011). *Gelassenheit durch Auflösung innerer Konflikte, Mentale Selbstregulation und Introvision* [Serenity Through Resolution of Inner Conflicts, Mental Self-Regulation, and Introvision] (2nd ed.). Stuttgart: Kohlhammer.

Williams, M. G., & Kabat-Zinn, J. (2013). *Mindfulness, Diverse Perspectives on its Meanings, Origins and Applications.* New York, NY: Routledge.

HAZARD ANALYSIS FOR HUMAN SUPERVISORY CONTROL OF MULTIPLE UNMANNED AIRCRAFT SYSTEMS

Elias B. Johnson, Andrew N. Kopeikin, Nancy G. Leveson
Massachusetts Institute of Technology
Cambridge, MA

Andrew W. Drysdale
U.S. Army Combat Capabilities and Development Command - Data and Analysis Center
Aberdeen Proving Ground, MD

Unmanned Aircraft Systems (UAS) operations are shifting from multiple operators controlling a single-UAS to a single operator supervising multiple-UAS engaged in complex mission sets. To enable this, there is wide consensus in literature that limitations in human cognitive capacity require shifting low-level control responsibilities to automation so that human operators can focus on supervisory control. However, hazard analyses to identify related safety concerns have largely been unexplored. To address this shortfall, this paper applies System-Theoretic Process Analysis (STPA) on an abstracted model of a multi-UAS system. This hazard analysis approach handles complex systems and human-machine control interactions together. The paper describes both how to execute the analysis, and provides examples related to an operator approving or denying plans developed by the automation. Numerous traceable causal scenarios are systematically identified and generate both design recommendations and questions that must be addressed to ensure the system is designed to be safe.

Control of Unmanned Aircraft Systems (UAS) is undergoing a paradigm shift from multiple operators remotely *controlling* a single-UAS to a single operator *supervising* multiple-UAS (Belecastro et al., 2017). In this context, the difference between operator *control* and *supervision* is characterized by a shift in delineation of control responsibilities between the operator and the UAS automation. Operators that control UAS are responsible for providing lower-level control inputs directly to UAS flight, navigation, and payload sub-systems to achieve the flight and mission objectives. In contrast, when operators perform supervision of UAS, the responsibility for lower-level control is delegated to the UAS automation (Porat et al., 2016). The operator becomes responsible for providing higher level control actions to the UAS decision making automation entity. In examples of supervisory control in several multi-UAS implementations, the operator will input mission planning parameters into the autonomy so that it can develop courses of actions and present them to the operator for review (Porat et al., 2016).

The allocation of more control responsibilities to automated controllers has the potential to increase the mission reach without increasing human operator resource requirements. For example, early studies showed that a single operator could only control 4-5 vehicles (Cummings, 2007a), but they could supervise around 12 UAS at a time (Cummings and Guerlain, 2007). However, increase use of automation also introduces new human factors concerns which have been raised extensively in the literature (Belecastro et al., 2017). For example, the skills and training required for operators to perform supervisory control may be considerably different than those previously required in lower-level control. Furthermore, in certain conditions, the UAS operator may have to override the automation and revert to lower-level control, potentially leading to cognitive overload if the system is not designed to account for these situations (Leveson, 2012).

To ensure operators and automation can work together to safely control multi-UAS requires a rigorous safety guided design process. A large body of work points to numerous studies related to the design of control algorithm (Saif et al., 2019) or the human factors implications of various design and interface choices (Levulis et al., 2018). However, the two domains are often considered separately in initial design, rather than taking a holistic approach that integrates them from the onset. This leads to potential hazards that may emerge later in the lifecycle of the system.

Few hazard analyses have been performed on these systems, and the ones performed (Belecastro et al., 2017) assumed linear causality which limits the results and opportunity to address safety through design recommendations. In addition, much of the human factors research is centered on simulation, which while important, should not be the only tool used in early system development. Simulation only reveals what is being specifically tested, and relies on assumptions that limit their scope, such as: set configurations, limited adverse factors, simplified dynamics, and reliable automation (Levulis et al., 2018). In reality, these systems will face unforeseen scenarios that will challenge the brittle autonomy in ways not detected in simulation.

To begin to address this shortfall, this paper applies a System-Theoretic approach centered on human-machine control interactions for such systems (Leveson, 2012). It demonstrates examples from a larger analysis of how human factors and control system design can be integrated in early concept development, modelling, and analysis. This ensures the multi-UAS system designers consider strengths and limitations of the operator at the onset design. The example explores hazardous supervisory control actions associated with approving or denying plans developed by the automation. The results of this abstracted modelling approach (1) provide design recommendations that enable safety features to be designed early into the system when most effective, (2) are applicable to a wide range of multi-UAS systems. The approach allows more design details to be refined using STPA for iterative safety guided design.

System Modelling and Hazard Analysis Process

The System-Theoretic Process Analysis (STPA) is a top-down hazard analysis approach which treats safety as a control problem rather than just considering component failures. As a result, the method is effective at handling complex systems with unsafe interactions between components, software, and human controllers. Complexity is managed through abstraction, and the analysis is initiated at a high level, as illustrated below, and can then be iteratively refined by adding design details. The following subsections demonstrate the process.

Purpose of the Analysis and Description of the System

The first step is to define the purpose the analysis, and the assumptions about the system and the environment. For this paper, the purpose is to analyze safety hazards for an abstracted model of a multi-UAS system with supervisory control to provide early design recommendations. In the system under consideration, an operator provides high-level planning guidance, the UAS automation develops courses of action (COAs) to control multiple UAS, and an operator is responsible for approving or denying them. No restrictive assumptions are made about the environment of the UAS or the operator.

STPA begins by identifying the system losses unacceptable to the stakeholders (Leveson and Thomas, 2018). For this multi-UAS system, these may include (L-1) loss of mission, (L-2) loss of life or permanent disabling injury, and (L-3) loss or damage to UAS or equipment. Next, system level hazards are identified. A hazard is "a system state or set of conditions that, together with a particular set of worst-case environmental conditions, will lead to a loss" (Leveson and Thomas, 2018). Table 1 presents a sub-set of the hazards considered in this analysis, and traceability to the losses.

Table 1.
Example Multi-UAS System hazards.

Hazard ID	Hazard Description	Loss Traceability
H-1	UAS does not complete mission objectives and tasks	L-1, L-2
H-2	Structural integrity of UAS is violated	L-3
H-3	Violation of UAS separation standards (min & max)	L-1, L-2, L-3

Hierarchical Control Structure

The second step in STPA is to build a hierarchical control structure of the system. This is a conceptual functional model composed of feedback control loops that shows responsibilities, control actions, feedback and mental models of each element within the system boundary. The control structure enables a hazard analysis on the interactions between elements.

The abstracted control structure for the multi-UAS system with control responsibilities split between the pilot and UAS automation is shown in Figure 1. The operator provides high-level guidance on the mission objectives and constraints. The Multi-UAS Fleet Controller generates a COA plan based on its process models of the environment, mission objectives and physical UAS systems. The operator can then "Approve" or "Deny" the COA as guided by their mental models of the environment, mission objectives, and feedback provided by the Fleet Controller.

Unsafe Control Actions

The third step of STPA is to identify unsafe control actions (UCAs), which are control actions that, in a particular context, and worse-case environment, will lead to a hazardous state (Leveson and Thomas, 2018). There are four possible ways to consider how each control action in the control structure can lead to a hazard: (1) not providing the control action, (2) providing the control action, (3) providing a safe control action but too early, too late, or in the wrong order, and (4) providing a control action that last too long or is stopped too soon. Table 2 provides examples of some of the UCAs that are identified for the "Approve COA" control action from the operator. Additional UCAs may exist in each UCA Type, and additional UCAs are similarly identified for the other control actions in the control structure.

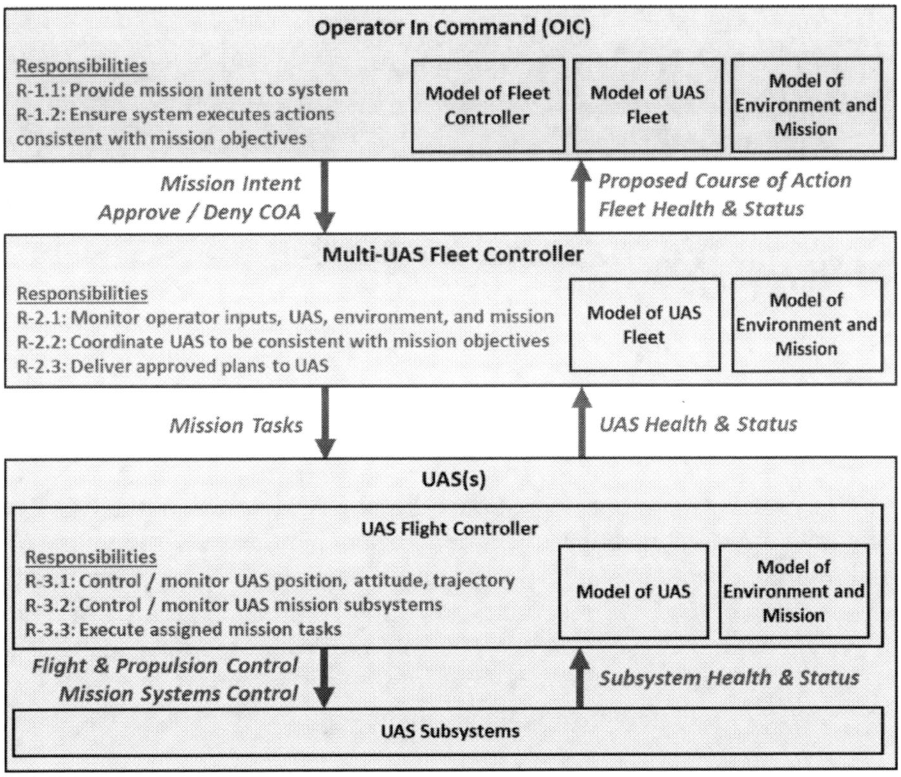

Figure 1. Safety hierarchical control structure of an abstracted multi-UAS system.

Table 2.
Example Unsafe Control Actions (UCA) for the "Approve COA" operator control action.

UCA Type	UCA	Hazard Traceability
Not Providing	[UCA-1] Operator *does not provide* "Approve COA" when the COA fulfills the flight or mission objectives	H-1, H-2, H-3
Providing	[UCA-2] Pilot *provides* "Approve COA" when the COA does not fulfill the flight or mission objectives	H-1, H-2, H-3
Too Early / Late / Wrong Order	[UCA 3] Pilot provides "Approve COA" *too late* when the COA will no longer fulfill the flight or mission objectives	H-1, H-2, H-3
Applied too long / Stopped too short	*Not applicable for this analysis because "Approve COA" is a discrete command*	

Causal Scenarios

The fourth step of STPA is to identify loss scenarios that describe the casual factors that can lead to the unsafe control actions and to the hazardous state. Scenarios help discover early design recommendations and questions that must be addressed to enforce safety constraints and refine the design. Causal scenarios consider potential breakdowns in feedback control loops as a result of unsafe interactions between elements of the control structure and component failures.

Of the 130 causal scenarios (CS) identified in the multi-UAS system analysis, three examples are highlighted that are traceable to Table 2 UCA-3: *The operator provides "Approve COA" too late and the COA which originally satisfied flight objectives will no longer fulfill the flight or mission objectives* [H-1, H-2, H-3]. Scenarios can potentially also trace to other UCAs.

CS-1: The Operator does not know that a proposed COA request is time critical. The COA was not originally time sensitive when the request was sent from the UAS Fleet Controller to the Operator, but became time sensitive because of dynamics in the mission or environment. The system is not designed to alert the operator when this occurs. [UCA-3]

CS-2: The Fleet Controller updates the COA request so frequently that the operator cannot assess its validity before it is replanned. Thus, the operator is in a perpetual cycle of reviewing proposed COAs. [UCA-3]

CS-3: In the time between operator approval and UAS execution, the COA becomes no longer consistent with mission objectives. Reasons for this include the following: (CS-3.1) The system design allows the operator to approve commands preemptively or with long time horizon; (CS-3.2) There is a delay in the UAS receiving execution commands because of environmental interference of system degradations; (CS-3.3) The Operator cannot modify the COAs once they are approved; (CS-3.4) The Fleet Controller generates an infeasible plan; (CS-3.5) The system is not designed to detect changes that may invalidate an already approved plan. [UCA-3]

Safety Guided Design Recommendations and Questions

Next begins an iterative cycle of safety guided design where the results of the hazard analysis are used to develop both design recommendations and questions to be addressed in refinement of the system. Recommendations are traceable directly to causal scenarios to provide critical context. The questions raise valuable insights to consider in the design. The full analysis revealed 65 design recommendations and 64 questions. The following are examples of Design Recommendations (DR) and their resulting questions (DR-Q) that illustrate how human factors considerations related to multi-UAS supervisory control were generated through analysis of the Causal Scenarios listed in the previous section. STPA is an iterative process. After design recommendations are implemented, changes must be reexamined using STPA to ensure they do not introduce sources of hazards themselves.

DR-1: There must be a feedback mechanism to alert the pilot when a non-time sensitive tasks becomes time sensitive [UCA 3, CS-1]. (DR-Q-1.1) How should the operator be alerted when a task becomes time critical? (DR-Q-1.2) How should the feedback for non-time critical tasks differ from time critical tasks?

DR-2: The system must not enter a state where the operator cannot provide input because the UAS perpetually updates the COA [UCA 3, CS-2]. (DR-Q-1.1) If there is [TBD] time gap in between approval and execution, which controller(s) is responsible for ensuring the command is still appropriate? (DR-Q-1.2) Which controller(s) is responsible for monitoring which tasks have been completed? (DR-Q-1.3) When is it appropriate for an operator to approve a COA in advance? (DR-Q-1.4) When is it not appropriate?

Conclusions and Recommendations

Multi-UAS supervisory control is a shift in the delineation of responsibilities between human operators and the automation. To date, few hazard analyses have been conducted on these systems to allocate responsibilities for safe operations. This paper demonstrated how to apply the STPA hazard analysis and safety guided design method on an abstracted model of a multi-UAS system. STPA specifically considers interactions within complex systems, in which components may or may not have failed, and that are controlled by both humans and software controllers. The analysis provides both design recommendations and questions, that if addressed, can help ensure safety is built into the system from the early design phases.

Acknowledgments

This research is based upon work supported by the U.S. Army Combat Capabilities Development Command (CCDC) Data and Analysis Center (DAC) under Contract W911NF-19-2-0124. Any opinions, findings, conclusions or recommendations expressed in this material are those of the authors and do not necessarily reflect the views of CCDC DAC.

References

Belcastro, C. M., Newman, R. L., Evans, J., Klyde, D. H., Barr, L. C., & Ancel, E. (2017, June 5). Hazards identification and analysis for unmanned aircraft system operations. 17th AIAA Aviation Technology, Integration, and Operations Conference, Denver, Colorado.

Cummings, M. L., Bruni, S., Mercier, S., and Mitchell, P. J. (2007a). Automation architecture for single operator, multiple UAV command and control. Int. C2 J. 1, 1–24.

Cummings, M. L., & Guerlain, S. (2007). Developing operator capacity estimates for supervisory control of autonomous vehicles. Human Factors: The Journal of the Human Factors and Ergonomics Society, 49(1), 1–15.

Leveson, N. G. (2012). *Engineering a Safer World: Systems Thinking Applied to Safety*. The MIT Press. Retrieved from: mitpress.mit.edu/books/engineering-safer-world

Leveson N. G., Thomas J. P. (2018). *STPA Handbook*. Retrieved from: psas.scripts.mit.edu/home/get_file.php?name=STPA_handbook.pdf

Levulis, S. J., DeLucia, P. R., & Kim, S. Y. (2018). Effects of touch, voice, and multimodal input, and task load on multiple-uav monitoring performance during simulated manned-unmanned teaming in a military helicopter. Human Factors: The Journal of the Human Factors and Ergonomics Society, 60(8), 1117–1129.

Porat, T., Oron-Gilad, T., Rottem-Hovev, M., & Silbiger, J. (2016). Supervising and controlling unmanned systems: A multi-phase study with subject matter experts. *Frontiers in Psychology*, *7*. https://doi.org/10.3389/fpsyg.2016.00568

Saif, O., Fantoni, I., & Zavala-Río, A. (2019). Distributed integral control of multiple UAVs: Precise flocking and navigation. IET Control Theory & Applications, 13(13), 2008–2017.

IDENTIFICATION OF A FAILED ENGINE IN TWIN-ENGINE PROPELLER AIRCRAFT: PILOT SURVEYS

Andrey K. Babin
Embry-Riddle Aeronautical University
Daytona Beach, FL

Andrew R. Dattel
Embry-Riddle Aeronautical University
Daytona Beach, FL

Twin-engine propeller aircraft accidents occur for many reasons including misidentifying a failed engine. Pilots learn to use a procedure called dead leg-dead engine to identify the failed engine; however, misidentification of the failed engine still occurs, questioning the effectiveness of this procedure. Two surveys were created. Survey 1 was completed by 49 airline pilots operating twin-engine turboprop aircraft; Survey 2 was completed by 22 instructor pilots operating twin-engine piston aircraft. Survey 1: Average flight time was 6,230 hours. Approximately 19% of respondents reported using the Engine-Out procedure at least once. Twenty-nine percent agreed that there could be a better method for failed engine identification. Survey 2: Average flight time was 420 hours. Half of respondents reported using the Engine-Out procedure at least once. Fourteen percent agreed that there could be a better method for failed engine identification. Forty percent of all respondents who suggested improvements recommended adding a visual indicator.

Engine failure is not a rare occurrence in aviation. A review of the National Transportation Safety Board (NTSB) database showed that in visual conditions, engine failure and incorrect handling caused one-third of all accidents in twin-engine piston aircraft (Boyd, 2015). Although a second engine provides additional power and reliability, twin-engine propeller aircraft require special handling in case of an engine failure to ensure the safe outcome of the situation. Not only does a failed engine stop providing power, it also can add significant drag in flight as its propeller starts windmilling, which is followed by a notable yaw toward the failed engine due to thrust asymmetry. An engine failure on takeoff, combined with the propeller drag, may result in a power loss as high as 80% (Federal Aviation Administration [FAA], 2016). Such power loss would be most detrimental at climbout immediately after takeoff when the aircraft is at full power and at a high angle of attack. Hence, a significant portion of multi-engine pilot training is devoted to single-engine operation of twin-engine aircraft and a successful recovery, especially if the failure occurs on takeoff.

Since at least the 1970s (Bramson & Birch, 1973), multi-engine pilots operating propeller aircraft have been trained to utilize the Identify-Verify-Feather (I-V-F) procedure as a response to an engine failure in flight, particularly on takeoff. Per the procedure, as a pilot depresses one rudder pedal to compensate for the yaw from the thrust asymmetry in an effort to stabilize the aircraft, he or she identifies the failed engine by determining which leg is not pushing the rudder pedal (dead leg). The dead leg is on the side of the dead engine, hence this method is called "dead leg – dead engine." Identification is verified by pulling back the throttle of the presumably dead engine and expecting no change in engine sound or power. Finally, the propeller is feathered, i.e., propeller blades are turned parallel to the airflow to minimize drag.

Data collected for a period of 12 years (Sallee & Gibbons, 1999) showed that almost half of all inflight shutdowns in turboprop multi-engine aircraft involved the good (i.e., working) engine. Investigations from several past fatal accidents involving an engine failure on takeoff in a twin-engine propeller aircraft suggested that a working engine was shut down in error (Aviation Safety Council, 2016; National Transportation Safety Board, n.d.). The method currently used for identifying a failed engine may not be effective in all circumstances and may, on the contrary, create confusion in a situation where a startled and preoccupied pilot has little room for error. On takeoff, when pilot workload is at elevated levels, the mental capacity and time available to make a decision are limited, and thus the identification of the failed engine and the action to feather it shall be quick and accurate to avoid a catastrophic outcome. The "dead leg – dead engine" method, however, is reliant on one's sensation of leg movement and requires mental resources to process that information. Hence, pilots operating twin-engine aircraft may benefit from a simpler and more straight-forward identification method using other sensory channels. Babin et al. (2020) introduced and tested a visual indicator of a failed engine and compared it to the "dead leg – dead engine" method. The visual indicator was designed to provide accurate information at a glance and consisted of a panel with two circles imitating aircraft annunciator lights (one for each engine), colored either in green (engine working) or red (engine not working). The color changed based on the corresponding engine parameters. The results revealed that, in a simulated scenario involving an engine failure on takeoff, pilots who used the visual indicator were able to identify a failed engine significantly faster than those who used the traditional method.

Although the data in the simulated environment highlights the benefits of using a visual indicator, it is important to learn the perspective from operators who have had to deal with real-life engine failures. Even with past accident data and research findings as supporting evidence, reluctance to change exists among the general population, especially when it comes to an FAA endorsed procedure (dead leg-dead engine) commonly taught, practiced, and used. Eliciting pilots' experiences and opinions would be beneficial to understand the general attitude and receptiveness to potential changes to the current procedures of how pilots identify a failed engine. Two surveys were conducted on twin-engine propeller aircraft pilot opinions of procedure, identification, and verification of a failed engine.

Method

Two surveys were created and distributed to two different pilot groups. Survey 1 was distributed to pilots of a US regional airline operating twin-engine turboprop aircraft. Survey 2 was distributed to instructor pilots at a US aeronautical university.

Participants

Survey 1. Forty-nine airline pilots participated in Survey 1. All participants were employed as pilots (either captain or first officer) at the time of participation and had prior or current experience in operating twin-engine piston and turboprop aircraft.

Survey 2. Twenty-three instructor pilots participated in Survey 2. All pilots had at least a Certified Flight Instructor (CFI) rating and were actively engaged in flight instruction at the time of participating in the survey.

Materials and Apparatus

Survey 1 contained 10 questions, with four open-ended questions, four categorized questions (Yes/No), and one scaled item. Survey 2 contained 11 questions, with six open-ended questions, three categorized questions, and one scaled item. The questions in both surveys asked pilots about their experience flying twin-engine aircraft (and twin-engine turboprop aircraft for airline pilots), difficulties handling an engine failure during simulator training, engine problems encountered in real-life operations, and their opinions on the current method of identification of a failed engine, including how comfortable they were with the I-V-F procedure (scaled from 1 to 5), any positive and negative aspects of the method, and if they had any suggestions for improvement to the current method of identifying a failed engine. Some categorized questions had additional comment fields for participants to provide additional information. Some questions had to be modified between surveys to account for the difference in experience between the two participant groups. Both surveys were created through the https://www.surveymonkey.com website (SurveyMonkey). The surveys had unique links that could be used by participants to access the survey and answer questions. Microsoft Excel and IBM Statistical Package for Social Sciences (SPSS) software were used for data analysis.

Procedure

Each Survey was distributed to pilot groups via an internal email (sent from the Safety Department for Survey 1 and Training Department for Survey 2) asking for their participation and providing a direct link to the survey. Upon following the link, each participant was provided a consent form. Individuals who volunteered to participate were redirected to the next page containing the survey. Individuals who did not agree to participate were redirected to the last page of the survey and prompted to close the browser window. All data were automatically collected and scored by SurveyMonkey and later exported into a spreadsheet for analysis.

Results

Survey 1

The average experience in flying twin-engine turboprop aircraft was 8.97 years (SD = 11.21) and 6,230 flight hours (SD = 8,695.11). The average experience flying all types of multi-engine aircraft was 13.91 years (SD = 12.53) and 7,229 flight hours (SD = 8,924.87). The most experienced participants in the sample had 40 years as a pilot and over 30,000 flight hours.

Almost a fifth (18.75%) of all respondents reported utilizing the Engine-Out procedure when operating the twin-engine turboprop aircraft in their capacity as a pilot with the airline. For past simulator training, 23% of respondents admitted having problems with identifying a failed engine at least once, 5.71% of respondents had problems with feathering an engine, and the rest did not report having any problems. Fifty-three percent of respondents reported encountering engine problems at least once in their real-life experience flying all types of aircraft. Although most pilots (71.43%) indicated that they were very comfortable with the I-V-F procedure, 24.49% were somewhat comfortable, 2% felt neutral, and 2% felt somewhat uncomfortable. The most commonly reported benefits of the I-V-F procedure were categorized as "redundant," "accurate," and "simple," and the most mentioned negative aspects of the I-V-F procedure included opportunity for error, high workload that the procedure may introduce, and long time required to complete it. Of the one-third of participants who provided their suggestions for improvement of the current method, 34% suggested adding visual indication (e.g., a light), 22% proposed audio indication, 22% suggested other improvements to the indications, and the other 22% proposed improving aircraft automation to better handle a failure (see Figure 1).

Survey 2

The average flying experience in operating twin-engine aircraft was 4.0 years ($SD = 7.2$) and 419.64 flight hours ($SD = 631.31$). The most experienced participant had 25 years as a pilot and 2,500 flight hours.

Half of the participants reported previously using the Engine-Out procedure in their experience operating twin-engine aircraft. Regarding simulator training experience, 9% reported difficulties identifying a failed engine, 9% reported difficulties verifying a failed engine, and 18% reported problems feathering the failed engine. Forty-one percent of participants reported having had engine problems in their real-life experience. Regarding the current I-V-F procedure, 59% reported that they were very comfortable with the current method, 32% were somewhat comfortable, and 9% were neutral. Most common reported benefits of the I-V-F procedure were described as simplicity, reliability, and ease of remembering, while the most mentioned negative aspects included opportunity for an error if the method is rushed or pilot stress levels are high in an emergency which could potentially cause loss of aircraft control due to a pilot fixating on the procedure. Three pilots (14%) provided suggestions for improvement to the current method. The suggestions included a visual indicator of a failed engine, both an aural or a visual indicator, and an aural indicator that plays a signal on the side of the failed engine (see Figure 1).

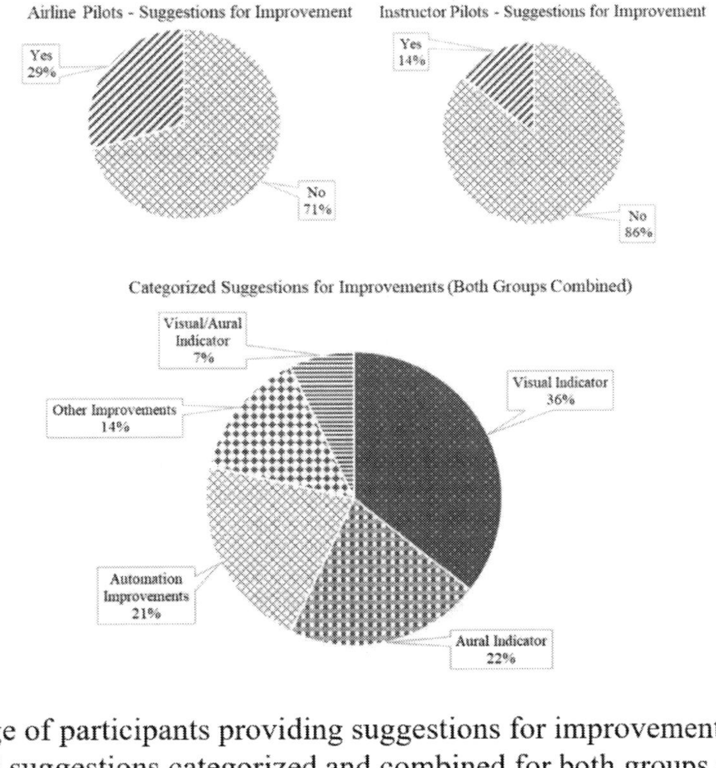

Figure 1. Percentage of participants providing suggestions for improvements to the current method (top) and all suggestions categorized and combined for both groups (bottom).

Discussion and Conclusion

The two pilot groups had different experience, with participants in Survey 1 having more years and hours of flying and aircraft type ratings than participants in Survey 2. This difference provides insight into perspectives and opinions from various representatives of the pilot population, from aspiring pilots at the beginning of their airline careers to seasoned captains. It is notable that both groups had a similar number of pilots who reported experiencing engine troubles before, showing some consistency despite certain difference in aircraft types operated.

A greater number of instructor pilots from Survey 2 reported using the Engine-Out procedure in past experiences compared to the airline pilots from Survey 1. This finding is surprising, considering that overall, participants from Survey 2 had accumulated less flight hours than the participants from Survey 1. Although this greater experience with more Engine-Out experience (despite fewer flight hours) is possible, it could also be caused by the misinterpretation of the questions. In-flight simulation of an engine failure (completed by reducing its power to idle but not shutting it down) is part of the twin-engine pilot training, thus an Engine-Out procedure must be utilized before one becomes a CFI. Additionally, for someone who teaches other pilots to fly twin-engine aircraft, it would be a common practice to utilize the Engine-Out procedure as part of the training curriculum.

Another interesting similarity between the two groups was how comfortable pilots felt with the I-V-F procedure that includes the "dead leg – dead engine" method. Of all respondents (both surveys combined), only one pilot admitted being somewhat uncomfortable with the current method while the majority felt either neutral or comfortable, with most saying they were very comfortable. We believe that several factors may have contributed to this opinion. The "dead leg – dead engine" method is widely common, applicable to most twin-engine propeller aircraft types, hence it is then not unusual that pilots would feel comfortable using it. It is a practice which is recommended by the FAA and is extensively used in pilot training (personal experience revealed that many pilots rated for single-engine aircraft are also familiar with the I-V-F method). Additionally, as pilots undergo periodic proficiency checks to maintain their license, social desirability may have been an additional factor that contributed to their response. Despite the high comfort levels of using the dead leg-dead engine procedure, both groups listed multiple negative issues to the method, including opportunities for error and increased workload, further supporting the potential for a better method of identification of a failed engine.

Possibly the most important findings were in the suggestions provided by the pilot groups. Among all suggested improvements, the overall majority of participants proposed a visual indicator to help in the identification of a failed engine, a trend seen across both more and less experienced pilots. This recommendation can be explained by the fact that 80% of information we receive comes visually (Geruschat & Smith, 2010) and humans tend to prioritize and trust visual information over audio and haptic when it is received at the same time (Xu et al., 2012). Hence it may feel more natural to receive timely and critical information through the visual channel, especially if it is placed in a fashion that it is not intrusive yet remains within the operator's field of view. These suggestions further corroborate findings by Babin et al. (2020) and show that not only a method that relies on the visual channel is more effective in a simulated environment, but its implementation would most likely be accepted and acknowledged by trained and experienced pilots who operate twin-engine propeller aircraft and who would benefit from it.

The results of this study provide a good overview of the opinions of pilots who have been in the profession for years as well as those who may be just starting their careers. It is intriguing that despite the differences in age, flight hours accumulated, and aircraft types operated, one can see similarities in experiences with engine failures and pilot opinions on how to handle them. These results highlight the potential for a better method of identifying a failed engine as shown by the recommendations from those who are most likely to encounter these failures.

Acknowledgments

The authors of this study would like to thank the Safety department of one U.S. regional airline and the Training department of one U.S. aeronautical university for their overall cooperation and help with preparation of the surveys and distribution of the survey links to their pilot groups. The names of these organizations shall remain hidden for confidentiality purposes. The authors would also like to thank everyone who has provided feedback to the surveys, thus helping them achieve reliable and valuable results.

References

Aviation Safety Council (2016). *Aviation occurrence report: 4 February, 2015, TransAsia Airways Flight GE235 ATR72-212A, loss of control and crashed into Keelung river three nautical miles east of Songshan airport* (Report No.: ASC-AOR-16-06-001). Author.

Babin, A. K., Dattel, A. R., & Klemm, M. F. (2020). An alternative method for identifying a failed engine in twin-engine propeller aircraft. *Aviation Psychology and Applied Human Factors*, 10(2), 103–111. doi: 10.1027/2192-0923/a000195

Boyd, D. D. (2015). Causes and risk factors for fatal accidents in non-commercial twin engine piston general aviation aircraft. *Accident Analysis & Prevention,* 77, 113-119.

Bramson, A. E., & Birch, N. H. (1973). *Pilot's guide to flight emergency procedures.* Garden City, NY: Doubleday & Company, Inc.

Federal Aviation Administration (2016). *Airplane Flying Handbook.* Oklahoma City, OK: Author.

Geruschat, D. R., & Smith, A. J. (2010). Low vision for orientation and mobility. In Wiener, W. R., Welsch, R. L., Blasch, B. B. (Eds), *Foundations of orientation and mobility (3rd ed., Vol. 1),* pp. 63-83. New York, NY: AFB Press.

National Transportation Safety Board (n.d.). *National Transportation Safety Board aviation accident final report* (Accident No. LAX92MA1183). Retrieved from https: //aviation-safety.net/database/record.php?id=19920422-1

Sallee, G. P., & Gibbons, D. M. (1999). Propulsion system malfunction plus inappropriate crew response (PSM+ICR*). Flight Safety Digest*, 18, 1-193.

Xu, Y., O'Keefe, S., Suzuki, S., & Franconeri, S. L. (2012). Visual influence on haptic torque perception. *Perception*, *41*(7), 862-870. doi: 10.1068/p7090

Loss of Situational Awareness during Commercial Flight Over Colombian Mountains

OSCAR SARMIENTO MEJIA PSY.D CIVIL AVIATION AUTHORITY OF COLOMBIA BOGOTA, COLOMBIA

FLIGHT HISTORY:

On March 27, 2018, an Airbus A 318 took off from Eldorado International Airport in the city of Bogotá (IATA: BOG - ICAO: SKBO) with destination Alfonso Bonilla Aragón International Airport in the city of Cali (IATA: CLO - ICAO: SKCL) with 75 passengers, 4 cabin crew members and 2 pilots. The aircraft was scheduled to depart BOG at 21:25 local time and land at CLO at 22:33.

According to the Operator's information, the First Officer (PF) was conducting Initial Operational Training in the right seat. The flight commander was a senior captain and Instructor Pilot serving as Pilot Monitoring (PM) in the left seat.

During the descent phase in a Standard Terminal Arrival procedure (STAR) and when the aircraft was about 40.5 miles from Cali VOR, the Ground Proximity Warning System (GPWS) was activated. The aircraft was descending to an altitude of 13,980 feet, falling below the established MEA of 17,000 feet. The flight commander performed the corresponding evasive maneuver, immediately climbing to 20,640 feet, exceeding the authorized flight level.

The ATC Radar Minimum Safe Altitude Warning system (MSAW) at Cali had a visual and aural terrain proximity alert system; however, the audio system was down due to configuration problems. Even so, the visual alert AW (Altitude Warning) was activated on the radar screen display when the aircraft left the MEA on R564 but the situation was not noticed by the CLO Approach Controller; the pilot deviation and subsequent evasive action were also not reported by the flight crew to ATC.

An analysis of the radar video recordings revealed that the radar return signal and Mode C on the radar screen disappeared at 38.5 NM, shortly before the aircraft crossed 14.000 feet on its descent. Moments later, the radar signal and Mode C reappeared on the controller's radar display, showing the aircraft at 19,400 feet.

Subsequently, the aircraft resumed the arrival procedure (MANGA8) and landed on RWY 02 at the Alfonso Bonilla Aragón Airport in Cali without further incident.

Location of GPWS alarm activation over mountainous terrain

HUMAN FACTORS

Loss of Situational Awareness (SAW) by the crew, by wrongly programming a descent altitude limit established in the MANGA 8 arrival procedure, not noticing the error and descending below the MEA on mountainous terrain; this circumstance brought the aircraft closer to the ground and triggered the GPWS "PULL UP, TERRAIN" alarm.

The investigation on this case revealed that first officer who was flying the aircraft at that time had wrongly programmed 14,000 ft on the STAR MANGA 8 route when the level established by the MEA was 17,000 ft.

On the other hand, the Pilot Monitoring (PM) did not notice this situation until the moment when the GPWS "pull up - terrain" was activated, forcing him to take command of the airplane and perform an evasive maneuver with a steep climb and ending above the authorized flight level 20640 feet.

There was a lack of assertive communication between both pilots, given that when programming the FMS, the arrival procedure was not crosschecked and was not verified and confirmed by both of them. The lack of supervision by the PM to his First Officer (PF) whom, due to his little experience and while being in training stage should have been under strict control and feedback **(CRM)**.

The Flight Commander's actions – Aeronautical Decision Making **(ADM)** - were quick and appropriate after hearing and verifying the GPWS alarm, managing the situation according to the **TEM** model for threat and error management and prompting a maneuver which avoided an Undesired Aircraft State **(UAS)**.

CORRECTION OF BAD PRACTICES

- Errors in the selection of altimetry adjustment.
- There was no cross-check between the PF and PM.
- Human - Machine interface (automation).
- Overconfidence.
- Direct effort to important tasks 80/20.
- Instruction and Training.

CORRECTIVE ACTIONS – AIRLINE

- Two periods of flight simulator with emphasis on the following topics:
-

  GPWS Memory items.

  EGPWS use and modes.

- LOFT with a scenario for the correct application of procedures in the vertical handling of the aircraft and descent calculations, as well as workload management.
- Cockpit Systems related to terrain. MEA – MORA - MOCA.
- Cockpit Systems related to position of aircraft.

PROBABLE CAUSES:

- Flight Crew's Loss of Situational Awareness (SAW).
- Not verifying the descent restrictions established in the MANGA 8 STANDARD ARRIVAL PROCEDURE.
- Performing a continuous descent (**"Open Descent"**) and not noticing the error while descending below the MEA over mountainous terrain; this action

brought the aircraft closer to the ground and triggered the activation of the GPWS with the call out «PULL UP, TERRAIN»

- Occurrence Category - Serious Incident

- TAXONOMY ICAO:
- CFIT - Controlled flight into or toward terrain.
- NAV - Navigation error.
- ATM - ATM / CNS - Air Traffic Management (ATM) / Communication, navigation, or surveillance service (CNS).

LESSONS LEARNED:

- Citing Captain Enrique Piñeiro, producer of WRZ a documentary film, HUMAN ERROR is an inherent part of human nature; and one of the defenses to minimize it, is to generate mechanisms that would detect it at an early stage to break the chain and prevent it from setting up an accident or incident because of it.
- Improve the SOPs: in this case, the airline updated its procedures in IOE cases.
- Factors to consider:

  Human physiology
  Flying discipline
  Training - TEM
  Technology interface
  Technical briefing SOPs
  Crew-ATC communication
  Workload
  Judgment and criteria.

THE IMPACTS ON THE FLIGHT CREW WITH THE IMPLEMENTATION OF THE AIRBORNE IMAGE RECORDER SYSTEM (AIRs) IN THE COCKPIT.

Tatiana Lícia da Silva Rangel
São Paulo University
Brasília/ Brazil
Marcelo Fernandes da Costa
São Paulo University
São Paulo/ Brazil

The technological resources used in aviation are widely used in the occurrence investigations process. However, they present technical-operational limitations, mainly regarding the reliable reproduction of the information that the flight crew really has. The implantation proposal of the Airborne Image Recording System (AIRs) arises to overcome these technical limitations presented by other technologies. Thus, this study aimed to verify if the implantation of AIRs in the cockpit could affect the pilot's perception, behavior and performance during flight. Preliminary results with ten volunteer pilots performing in a flight simulator in Brazil pointed to a series of behavioral and performance changes when the cockpit environment was being filmed. On-site observations allowed us to identify behaviors such as delay in response time, improvisation of procedures, lack of perception of stimuli, errors of judgment and communication failure.

Keywords: Human Factors, Psychophysics, Human performance, aerospace, pilots, AIRs and safety.

Through Cockpit Voice Recorders - CVR and Flight Data Recorders - FDR aviation has technological resources to record flight data. However, such voice and data recording tools are not always sufficiently reliable to clarify the contributing factors of aeronautical events. Such tools have technical-operational limitations, especially regarding their ability to reproduce accurately the information actually disclosed to operational crewmembers (pilots).

The proposal for the implementation of the Airborne Image Recorder System - (AIRs) in the cockpit by the International Civil Aviation Organization (ICAO.2016) emerged as a possible alternative for the provision of complementary data to the CRV and FDR. It would allow for the recovery of a pilot's actions and physical reactions by cameras and their interactions with buttons and switches on the instrument panels, in order to clarify aeronautical occurrences. It is noteworthy that AIRs are currently not required on any aircraft by any ICAO member state.

Thus, this study aimed to verify if the implantation of the AIRS in the aircraft cockpit would affect a pilot's perception, behavior and performance in flight.

Discussion

The presence of cameras to record images of the professional in a work situation, even if these recover only part of the professional's body, technically symbolizes a way of monitoring and surveillance of the work context. In researches related to surveillance in the work context, the effects of this monitoring on work-related attitudes are still poorly studied. However, in the context of aviation, in which the cockpit is already monitored by CVR and

FDR, it is reasonable to assume that the inclusion of cameras as just another monitoring tool would be easily accepted.

It is part of the activity to make the sequence of movements, gestures, and the subjective management of a work situation more flexible according to the technical capacity, professional experience, requisite of the operating environment and operational procedures related to the job. The actual working condition often impose requirements that overlap or are beyond the standard operating procedures (Guérin et al, 2001; Ferreira, 2012).

The presence of a camera that records movements and actions could inhibit a pilot's natural behavior, adding permanent tension and concern, possibly making the pilot's work even more complex, enhancing cognitive and psychomotor limits of the operational activity, inducing a decrease in crew performance, contrary to the interests of flight safety (Belletier et al., 2015).

Methods

We used the psychophysical method of the theory of signal detection as a methodological basis in this study, in which discriminability was measured as well as the response criterion of the pilot .We established a reaction time criterion of ten (10) seconds, and measured the response time between the presented stimulus and the executed response (Mori et al. 2002) for each pilot.

As a comparative basis, we utilized the task prescribed for simulator training (Oliveira, P.A.B. 2011) and the pilot's performance during piloting activity recorded by the camera. We highlight that discriminability in this context was the ability to detect or not the stimulus (Costa, M.F. 2011) and the stimuli presented to the pilots were sounds and lights for a determined period of time and situation. In order to compare the pilot's discriminability and the criterion during the flight, the AIRs started recording at the beginning of the training but the pilots were only informed in a pre - determined stage of the flight and during the state of emergency of the flight.

Results

As preliminary results, the on-site observations of the flight simulator training of ten (10) pilots at Azul Airline' training center in Brazil pointed out a series of behavioral and performance changes during the filming of the cabin environment. We highlight behaviors such as delay in response time, improvisation of procedures, lack of perception of stimuli, errors of judgment and failure in communication. This analysis allowed for the verification of the pilots' mediation and adaptation strategies to the real work conditions when facing different stimuli and threats.

Final Consideration

This study highlights and provides important empirical evidence that invite the scientific and aviation community to reflect on the theme and foretells an important scientific contribution to world aviation.

Acknowledgements

I am particularly grateful for the assistance given by my family.

I would like to thank the following companies for their assistance: Azul Airlines, for their help in collecting the data. USP/ CAPES for the financial support to conduct this project.

I wish to acknowledge the help provided by to the volunteer pilots, without whom it would not have been possible to conduct this project.

I would like to express my very great appreciation to Ms. Lilian Moraes and Mr. Maurício de Lucio for their contribution assisting in critiquing the paper.

Special thanks should be given to Dr. Marcelo F Costa, my research project supervisor for his professional guidance and valuable support.

Referências

Belletier, C., Davranche, K., Tellier, I.S., Dumas, F., Vidal, F., Hasbroucq, T., & (2015). Choking under monitoring pressure. Being watched by the experimenter reduces executive attention. *Psychon Bull, 22*, 1410-1416. doi: 10.3758/s13423-015-0804-9

Costa, M.F 2011. A clínica da psicofísica. *Revista Psicologia USP*, 22(1),15-44.

Guérin, F., Laville, A., Daniellou, F., Duraffourg, J., & Kerguelen, A. (2001). Compreender o trabalho para transformá-lo. *A prática da ergonomia*. Tradução Giliane M.J. I, & Marcos, M. São Paulo: Blucher.

Holland, P.J., Cooper, B., & Hecker, R. (2015). Electronic monitoring and surveillance in the workplace. The effects on trust in management, and the moderating role of occupational type. *Personel Review*, 44, 161-175. doi: 10.1108/PR-11-2013-0211

Internacional Civil Aviation Organization, Assemby – 36th Session – *Agenda Iten 35: Aviation Safety and air navigation standardization, Airborne Image Recorders*, 2016 in https://www.icao.int/Meetings/a39/Documents/WP/wp_307_en.pdf.

Link, S. W. Rediscovering the Past: Gustav Fechner and Signal Detection Theory. Psychological Science, 5(6), 335-340.

Mori, S; Yoshio, O; Imanaka, K. Reactions times and antecipatory skills of karate athletes. *Human Movemente Science*. V. 21, p.213.2002

Oliveira, P.A.B. (2011). Trabalho prescrito e trabalho real. In A.D. Cattani & L. Holzmann (Orgs.), *Dicionário de trabalho e tecnologia* (pp. 462). Porto Alegre -RS: Zouk.

Verghese, P., Visual Search and Attention: A Signal Detection Theory Approach. *Neuron*, Vol. 31, 523–535.

TOWARD MODELING PILOT WORKLOAD IN A COGNITIVE ARCHITECTURE

Christopher Stevens
Air Force Research Laboratory
Wright Patterson, OH USA
Christopher R. Fisher
Cubic Defense
Beavercreek, OH USA
Megan B. Morris
Air Force Research Laboratory
Wright Patterson, OH USA

Characterizing and predicting pilot cognitive workload remains a formidable challenge, especially in tasks with a high perceptual/motor demand like aerial refueling. Cognitive models are useful tools for this, as they offer the potential to derive both performance and workload simulations before a test is conducted. We conducted a task analysis of a C-17 aerial refueling mission and developed a low-fidelity Atomic Components of Thought – Rational (ACT-R) model and environment to simulate the task. ACT-R models have been successful in predicting workload in other domains, such as menu navigation and problem solving. Eight aerial maneuvers were examined, including takeoff, climb, cruise, descent, refueling, contact station keeping, and landing. The exercise revealed two subtasks not currently described in great detail by workload modeling methods: trajectory estimation and collision avoidance. We identify gaps in research on workload modeling approaches and explore preliminary predictions made by the model using default ACT-R parameters.

Aerial refueling is among the most cognitively demanding maneuvers that pilots perform, requiring sustained attention, planning, perceptual-motor coordination, and strategy adjustment in the face of changing environmental conditions. If cognitive workload becomes excessive during flight in general—and aerial refueling in particular—the chance of error increases, potentially leading to catastrophic consequences. A deeper understanding of the factors that contribute to pilot workload could improve training and risk mitigation efforts. We propose using cognitive architectures to understand how workload changes with task demands and cognitive moderators such as fatigue. A cognitive architecture is a computational instantiation of an integrative theory of cognition, detailing how memory, attention, and perceptual-motor processes operate as a coherent system capable of performing complex tasks (Newell, 1990). Cognitive architectures are well-suited for understanding pilot workload because (1) they provide quantitative workload and performance predictions based on sound theoretical principles and (2) they have the potential to scale up to complex tasks, such as aerial refueling. We report an initial effort to use a cognitive architecture to understand workload dynamics during aerial refueling.

Workload estimates from our model show a high degree of correspondence to subjective workload ratings collected during various maneuvers of an aerial refueling exercise, such as takeoff, approach/refueling and descent. In what follows, we will describe the aerial refueling exercise, introduce a model of aerial refueling based on a cognitive architecture, and show that the model's workload predictions correlate with subjective workload assessments.

Aerial Refueling Study

We observed seven test pilots in the Air Force 418th Flight Test Squadron at Edwards Air Force Base during a routine aerial refueling maneuver. The pilots were flying a C-17 aircraft being refueled by a KC-135 tanker. During flight, we asked the pilots to complete a NASA-TLX subjective workload scale (Hart & Staveland, 1988). Pilots completed nine total maneuvers during this exercise. We report three of those maneuvers here because these are the most interesting for present purposes and further research is needed to successfully model the other six. The first maneuver, takeoff, involved a standard takeoff procedure in which the pilot and co-pilot were required to complete a pre-flight checklist, taxi to the runway, and achieve liftoff. The second maneuver, approach/refuel, required the pilots to approach the tanker and establish contact with the refueling boom. Finally, descent involved the pilot gradually decreasing the altitude of the aircraft.

Simulated Refueling Task

Cognitive architectures are computer simulations that operate in a simulated task environment designed to be analogous to real world tasks in terms of the cognitive demands they impose. Figure 1 provides a screenshot of the task environment in which the model operated, featuring a simplified flight deck with keyboard navigation controls for speed, climb, and direction, and indicators for position, speed, climb, and fuel level. The two panels in the top center jointly indicate the model's position relative to the tanker in 3D space. In both panels, the model's position is represented as a fixed central cross and the tanker is represented as an unfilled red circle. The panel on the left displays the model's altitude relative to the ground, which is represented as a dashed horizontal line. The panel on the right displays the model's position in the remaining two dimensions (forward-backward and left-right). As the model approaches the tanker, the unfilled red circle will move closer to the central cross. The task environment also features a basic communication center, located at the bottom left of Figure 1, where it can send and receive simple messages. The grid of buttons to the left represents a gauge checklist, which is used in preparation of takeoff. The model's visual attention is represented by the filled yellow circle.

In the aerial refueling task, the model must perform a series of aerial maneuvers: takeoff, approach/refuel, and descent. The model begins the takeoff maneuver at ground level in a stationary position with its heading oriented towards the tanker. During takeoff, the model must increase its speed and climb rate to specified values. As soon as the tanker becomes visible in one of the radar panels, the model initializes the approach/refuel maneuver. During approach, the model makes necessary adjustments to the speed, climb, and angular speed in order to align with the tanker. Fuel can be transferred from the tanker as long as the distance is within a predefined tolerance. Finally, once the target fuel level has been achieved, the model begins the descent maneuver in which the tanker and receiving plane depart.

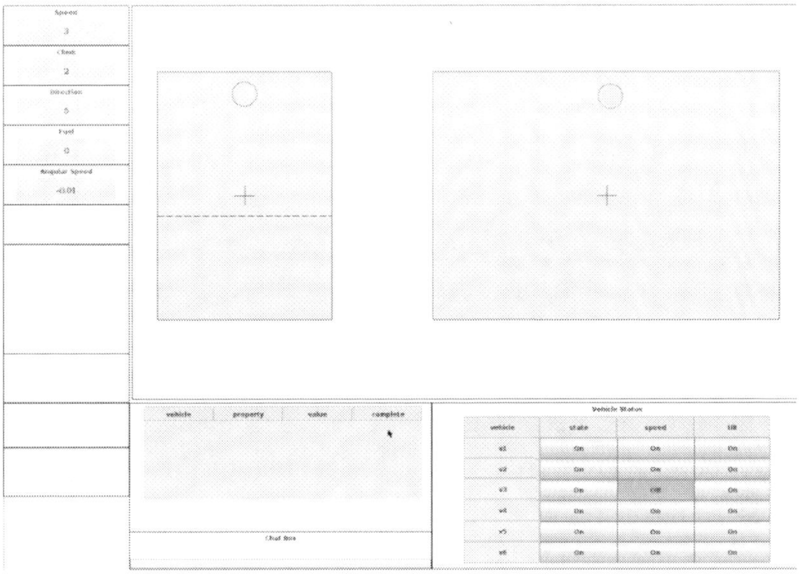

Figure 1. An illustration of the flight deck in the aerial refueling task. The instrument panel is located on the far left. Aerial position indicators for altitude and position are located in the left and right grey boxes, respectively. A message box is located in the bottom left and an information panel is located at the bottom right.

Model

We developed the aerial refueling model in the cognitive architecture Adaptive Control of Thought-Rational (ACT-R; Anderson, 2007; Anderson et al., 2004). The architecture is organized as a set of specialized information processing units called modules, which are dedicated to functions such as goal directed-behavior, procedural memory, declarative memory, tracking the problem state, visual and auditory perception, and motor control. Each module can process only one request at a time, leading to a processing bottleneck within the architecture that mimics limitations found in humans. The procedural memory module functions as the "engine" of the architecture, which uses production rules to issue processing requests to other modules and control the flow of information within the architecture. Production rules specify the conditions under which modules process information. When translated to natural language, a production rule might specify "if the goal is to refuel, and the tanker is in front and moving away, then issue a command to the motor module to press the arrow up key to increase speed." Each production rule is associated with a utility value that represents its ability to accomplish a goal, and is a function of the match between the conditions in the production rules and the state of the architecture. The production rule with the highest utility (i.e. match to conditions) is selected and executed. This process of selecting and executing production rules is known as a production cycle and is responsible for producing complex cognition.

Model Strategy

In this section, we describe the high level strategies the model uses during different maneuvers throughout the aerial refueling task. During the takeoff maneuver, the model cycles through six goals: (1) achieve target speed, (2) achieve target climb rate, (3) inspect altitude panel for the tanker, (4) inspect position panel for the tanker, (5) inspect message list, and (6) go into temporary standby. The approach/refuel maneuver begins as soon as the model identifies the tanker in one of the position panels. The model's primary goal during this maneuver is to align its position with the tanker to enable fuel transfer. In order to accomplish this goal, the model must continually estimate and adjust its trajectory to achieve alignment while avoiding collision. The model cycles through three phases: adjust trajectory, address communications, and temporary standby. Trajectory adjustment involves iterative adjustments to climb, speed, and angular speed until the correct trajectory is achieved. Adjustments to the trajectory must be made within safe parameter ranges. For example, the model attempts to avoid approaching the tanker with excessive speed. In the next phase, the model addresses communications in the message box. Finally, in the last phase, the model goes into a brief standby period before beginning a new cycle.

Once the model positions itself with the tanker, refueling will begin. The strategy for refueling is similar to approach, except the model also monitors the fuel level. The model makes adjustments to its speed, climb, or angular speed if it loses proper alignment with the tanker. Once the model recognizes that the target fuel level has been achieved, it will enter the descent phase where it will decrease its altitude to depart with the tanker.

Workload Predictions

We generated workload predictions in the aerial refueling task using an approach called cognitive metrics profiling (CMP) (Gray, Schoelles, & Sims, 2005; Jo, Myung, & Yoon, 2012). In CMP, workload is measured as activity within each module over a time interval. The basic idea is that workload increases with increased use of a given module (i.e. vision), making less of the resource available for competing demands. Workload can be analyzed for individual modules (i.e. memory) or can be combined into a composite workload index. Prior research has found that composite workload— defined as a weighted sum of activity across modules—predicts NASA-TLX ratings across a variety of laboratory cognitive tasks (Jo et al., 2012) in addition to high fidelity unmanned vehicle management tasks (Stevens, Morris, Fisher, & Myers, 2019). We used this composite workload measure to estimate workload during different maneuvers of the task.

Results

Figure 2 provides a side-by-side comparison of subjective workload as measured by the NASA-TLX and the model workload predictions using the linear regression results equation in Jo et al. (2012) to transform model workload in NASA-TLX units. The model captures the rank order across maneuvers, but underestimates workload during approach/refueling.

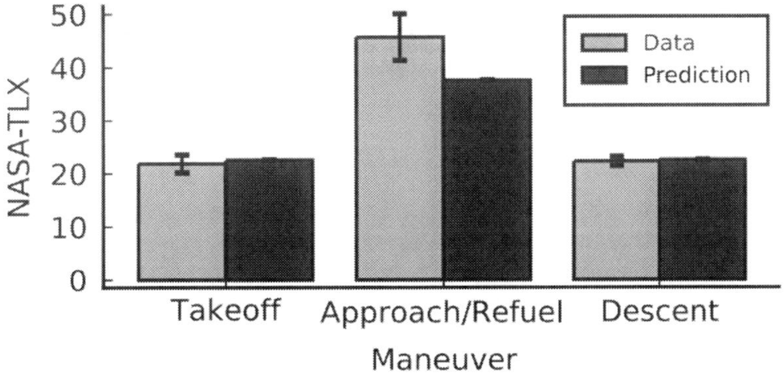

Figure 2. A comparison of subjective workload ratings (grey) and model workload predictions (red) for different maneuvers.

Discussion

Our goal was to demonstrate how cognitive architectures can be used to understand and predict workload in aerial refueling. As a proof of concept, we developed a cognitive model of aerial refueling and showed that its workload predictions agreed with the rank ordering of workload across aerial maneuvers. According to the model, the high level of workload found in the approach/refuel maneuver is due to continual monitoring, trajectory estimation, and collision avoidance.

Although the rank order of workload predictions was correct, the model underestimated workload during the demanding refueling/approach maneuver. It is possible that some sources of workload were omitted in the model, leading to underestimation. For example, some cognitive operations may have been abstracted away during the development of the model. Alternatively, emotions such as anxiety could have contributed to workload judgments, which is currently outside the purview of ACT-R. Nonetheless, this initial effort highlights the potential for using cognitive architectures to predict and mitigate pilot workload in complex flight maneuvers, such as aerial refueling.

In future research, we plan to extend the model in several ways. First, we want to compare the accuracy and robustness of different approach and refueling strategies. In the current strategy, there are some cases in which the model fails to align with the tanker or devolves into a tailspin. Although this might be consistent with the performance of novices, it likely underestimates the performance of pilots who have acquired at least some training.

Second, we would like to use more realistic controls, such as a control stick or a flight simulator, to produce a more accurate model of the pilot.

Cognitive architectures provide a theoretically grounded approach for understanding and predicting pilot workload. Our simulation serves as an initial demonstration of the potential use of cognitive architectures in pilot workload prediction and assessment. We believe that the potential of cognitive architectures remains largely untapped. Unlike direct measures, such as subjective workload ratings, it is possible to generate predictions under a variety of hypothetical scenarios with cognitive architectures. For example, the space of strategies could be explored to inform training regimens, or the design space of flight decks could be explored to understand the implications for usability and workload.

Acknowledgements

The views expressed in this paper are those of the authors and do not reflect the official policy or position of the Department of Defense or the US Government. Approved for public release; distribution unlimited. Cleared 03/24/2021; Case Number MSC/PA-2021-0050. CF's contributions to this work were supported in part by a postdoctoral research associateship, administered by the Oak Ridge Institute for Science and Education (ORISE). MM completed a portion of this work as a contractor with Ball Aerospace. The authors thank Glenn Gunzelmann, Justin Estepp, James Christensen, Allen Dukes, and Maj. Clifford Johnson for their efforts to design the study and collect the data. We are also deeply grateful to the 418th Flight Test Squadron for providing the opportunity to conduct the study.

References

Anderson, J. R. (2007). How can the human mind occur in the physical universe? Oxford University Press.

Anderson, J. R., Bothell, D., Byrne, M. D., Douglass, S., Lebiere, C., & Qin, Y. (2004). An integrated theory of the mind. Psychological review, 111(4), 1036-1060.

Gray, W. D., Schoelles, M. J., & Sims, C. (2005). Cognitive metrics profiling. In Proceedings of the human factors and ergonomics society annual meeting (Vol. 49, pp. 1144–1148).

Hart, S. G., & Staveland, L. E. (1988). Development of nasa-tlx (task load index): Results of empirical and theoretical research. In Advances in psychology (Vol. 52, pp. 139–183). Elsevier.

Jo, S., Myung, R., & Yoon, D. (2012). Quantitative prediction of mental workload with the act-r cognitive architecture. International Journal of Industrial Ergonomics , 42 (4), 359–370.

Newell, A. (1990). Unified theories of cognition. Harvard University Press.

Stevens, C. A., Morris, M. B., Fisher, C. R., & Myers, C. (2019). Cognitive metrics profiling of a complex task: Toward convergent validity with behavioral and EEG workload indicators. Paper presented at the 17th International Conference on Cognitive Modeling, Montreal, QC

GO-AROUND NONCOMPLIANCE DURING UNSTABILIZED APPROACHES AND LANDINGS IN COMMERCIAL AVIATION: A HUMAN FACTORS ANALYSIS

Garrin Ross
Embry-Riddle Aeronautical University
Lafayette, Indiana

Linda Tomko
Purdue University
Lafayette, Indiana

Informed by findings and recommendations from the Flight Safety Foundation's Approach and Landing Accident Reduction Task Force, we examined and analyzed Aviation Safety Reporting System (ASRS) incident report data from unstabilized approach and landing events. The aim of this study was to investigate the human factors reported as contributing to operational incidents of unstabilized approaches and landings in United States-based commercial aviation. Results showed the unstabilized approaches were significantly less likely to be responded to with go-around compliance. Binomial logistic regression analysis revealed descriptive differences in the associations of the ASRS-coded human factors with the likelihood of unstabilized approaches being continued to landing rather than go-around compliance. Content analysis of flight crew incident report narratives may allow for identification of other contributory human factors not explicitly coded by ASRS, such as decision making. Results from such investigations have the potential to inform effective go-around compliance training designs.

Although U. S. commercial aviation has long been classified as the safest mode of passenger transportation, safety remains a primary focus in NextGen airspace developments. Approximately 65 percent of commercial aviation accidents occur during the approach and landing phases of flight (FSF, 2017; International Air Transportation Association [IATA], 2016). According to the Flight Safety Foundation study, 83 percent of those approach and landing accidents were avoidable if flight crews had intervened on their unstabilized approaches and initiated a go-around. Thus, following proper operational procedures of initiating a go-around in response to an unstabilized approach could potentially avoid 54 percent of commercial aviation accidents. However, despite commercial aviation industry go-around policies, it is estimated that only approximately 3 to 5 percent of unstabilized approaches are met with go-around policy compliance (FSF, 2017).

Unstabilized approaches and landings are persistent and pervasive risks to commercial aviation safety, and they have been identified as a top current safety threat. Echoing the earlier recommendations by the National Transportation Safety Board, the Flight Safety Foundation's Approach and Landing Accident Reduction (FSF ALAR) Task Force called for increased efforts improving flight crew training in order to promote go-around compliance. The ALAR Task Force concluded that go-around policies and procedures have not been sufficient for ensuring aviation safety during approaches and landings (FSF, 2017). Deficiencies in flight crew training for the appropriate operational decision making during unstabilized approaches and landings were identified. According to the ALAR Task Force findings, improvements to flight crew training for go-around compliance need to be informed by the lessons learned from the review and analysis of operational events and incidents of unstabilized approaches and landings (FSF, 2017). To date, there have been no documented efforts reviewing and analyzing operational incidents of unstabilized approaches and landings in commercial aviation toward the end of understanding the psychology of go-around noncompliance and improving effective commercial pilot training for go-around execution.

At the onset of the descent phase of flight, commercial flight crews aim to continuously manage the aircraft configuration of speed and attitude for a stabilized approach to a safe landing. Although occupying less than 14 percent of total commercial flight time, more than half of all fatal accidents in worldwide commercial aviation operations occur during the approach and landing phases of flight (Boeing, 2017). Unstabilized approaches are the primary risk factor in approach and landing accidents, and nearly 97 percent of unstabilized approaches are voluntarily continued to landing (FSF, 2017) in conditions that unnecessarily jeopardize commercial aviation safety. In other words, flight crew continuation of an unstabilized approach was the causal factor, and attributable to human factors. Despite go-around policies and procedural training designed to mitigate needless risks to aviation safety, the tendency for highly trained flight crews to continue with an unstabilized approach persists.

In response to the pervasiveness of go-around noncompliance, the FSF ALAR Task Force conducted an extensive study of the psychology of go-around noncompliance as part of the FSF Go-Around Safety Initiative of 2011 (FSF, 2017). The results of the study revealed that there were differences between commercial pilots who had continued an unstabilized approach to landing and commercial pilots who executed a go-around in response to an unstabilized approach. It was found that a pilot's ability to correctly perceive and assess risk during unstabilized approaches was directly affected by the pilot's situational awareness competencies (FSF, 2017). Pilots who executed a go-around scored higher across all nine factors of situational awareness compared to pilots who landed during unstabilized approaches. As for human factors associated with go-around noncompliance, there were also differences (FSF, 2017). Compared to pilots who executed a go-around, it was revealed that pilots who landed during unstabilized approaches experienced greater influence of human factors associated with a perceived pressure to land, lack of crew support for a go-around, discomfort in being challenged or challenging others, and inhibitions about calling for a go-around due to a perceived authority imbalance in the flight deck (FSF, 2017). Further, the ALAR Task Force interpreted from the results a concerning risk to the commercial aviation culture. Commercial pilots who do not comply with go-around policies and procedures appear to have normalized an attitude of go-around noncompliance (FSF, 2017).

The ALAR Task Force recommendations included the need to understand the psychology of go-around noncompliance, and the lessons learned need to be applied to commercial pilot training programs. Go-around training needs to incorporate lessons learned from operational incidents in order to appropriately reflect typical and atypical go-around execution risk scenarios, and training scenarios should involve realistic simulation (FSF, 2017). The assumption is that training in a wide range of typical and atypical operational conditions may facilitate increased awareness of the risks inherent in those conditions that pose risk to stabilized approaches and warrant execution of a go-around. According to the ALAR Task Force, realistic training scenarios are needed for validation of recommended strategies for improved go-around compliance training (FSF, 2017). In sum, understanding the attitudes and conditions of noncompliance with go-around policies begins with understanding the characteristics of unstabilized approach and landing incidents.

The purpose of this study was to investigate human factors identified and coded as contributing to reported operational incidents of unstabilized approaches and landings in commercial aviation. Understanding the attitudes and conditions of flight crew noncompliance with go-around policies and procedures begins with understanding characteristics of unstabilized approach and landing incidents. Thus, the aims of this study were three-fold: (1) identify the human factors that are coded in Aviation Safety Reporting System (ASRS) reports as contributing to aviation incidents of unstabilized approaches; (2) assess to what extent, if any, the ASRS-coded human factors are associated with unstabilized approaches reported in the ASRS database; and, (3) determine if there was a relationship of the human factors in the likelihood that the reported incident was an unstabilized approach continued to landing versus go-around. This study had the potential of identifying human factors associated with and contributing to reported incidents of flight crew go-around noncompliance during unstabilized approaches and landings and informing effective go-around compliance training designs.

Method

The reports of interest in this study were from commercial passenger air carriers operating under Federal Aviation Regulations Part 121. A study sample pool of incident reports was gathered from the ASRS online reporting system database using the following criteria:

- Date of incident: 01 January 2012 to 31 December 2016
- Federal aviation regulations: Part 121
- Reporting organization: air carrier
- Reporter function: captain, first officer, pilot flying, pilot not flying
- Phase of flight: initial approach, final approach, landing
- Event type: unstabilized approach
- Contributing factors: human factors

The database query output resulted in a return of 444 reports meeting this initial study sample criteria. Following exclusions (e.g., incomplete fields, sole human factor was "other/unknown"), a final study sample of 95 reports was randomly selected, based on an *a priori* power analysis to achieve .95 statistical power in detecting a medium sized effect.

The ALAR Task Force report was consulted for the "situational awareness constructs" and "key psychosocial factors" that were assessed as part of the prior FSF 2017 study, since the ALAR Task Force report was informing this current study. The ALAR Task Force situational awareness constructs and key psychosocial factors were carefully mapped to the ASRS-coded human factors (see Figure 1). Taking this informed approach, eight ASRS human factors were identified for the current study: communication breakdown, confusion, fatigue, human-machine interface, situational awareness, time pressure, training/qualifications, and workload. Since these ASRS-coded human factors map to the constructs and factors identified by the ALAR Task Force, these eight human factors were identified as IVs for this current study.

Given that the overall goal of this study was to inform aviation training designs, the remaining four human factors were reviewed for reconsideration as an IV in the current study. Of those remaining human factors, distraction was identified for inclusion. It was assumed that training designs can impose distractions, and distractions have been found to influence overall flight crew performance (Barnes & Monan, 1990; Foyle et al., 2005; Strayer & Cooper, 2015). This resulted in a total of nine human factors used for this current study.

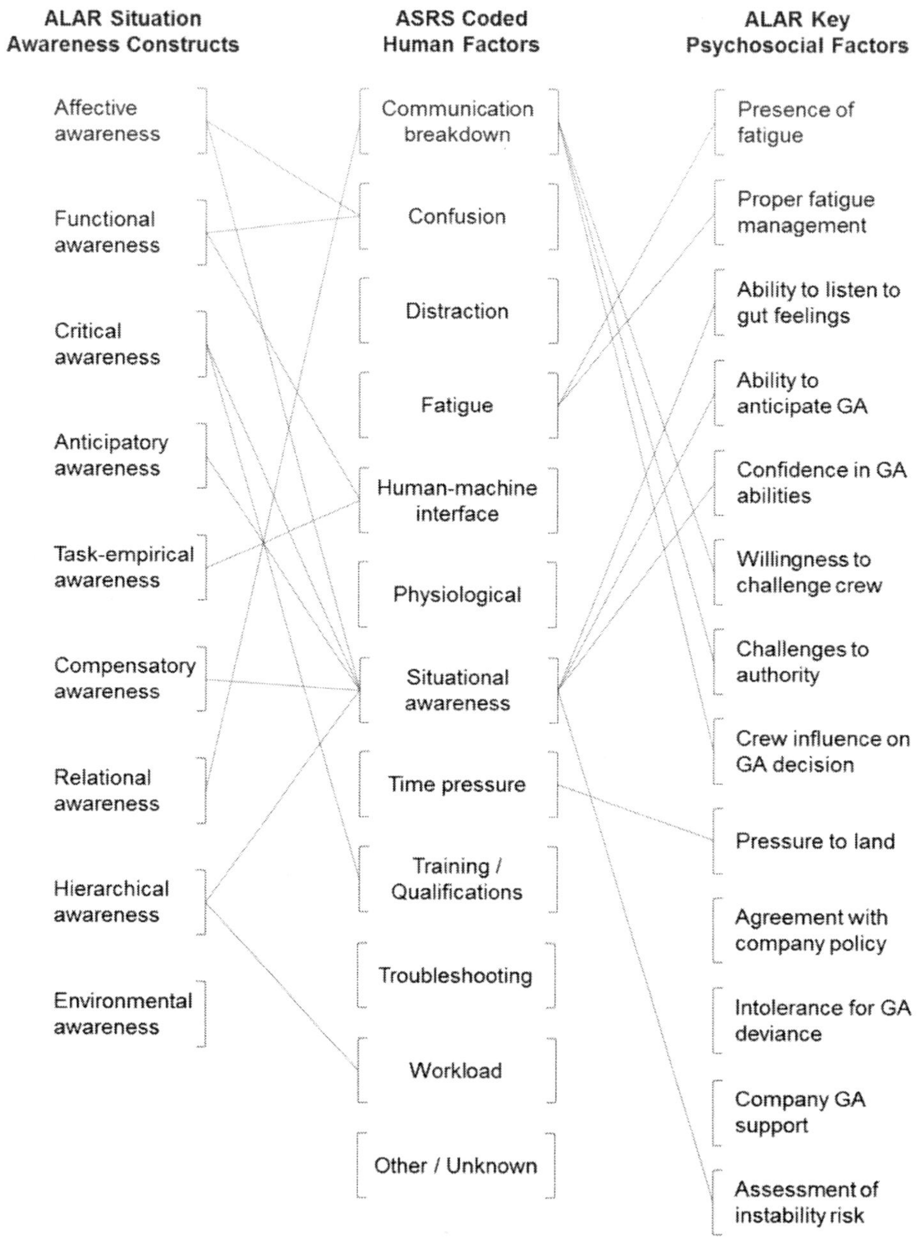

Figure 1. Diagram mapping the ALAR Task Force "situational awareness constructs" and "psychosocial factors" to the ASRS-coded human factors. Each line indicates a mapping from the situational construct or psychosocial factor to the human factor.

Results

A binomial logistic regression was used to test the associations and relationships of ASRS-coded human factors on the likelihood that the reported event was an unstabilized approach continued to landing versus go-around. Preliminary chi-square test analysis of the study sample revealed statistically significant differences in the outcome of reported unstabilized approaches ($\chi^2(1) = 6.58$, $p = .01$, $w = .26$), with more than 63% of the reported unstabilized approaches continued to landing and less than 37% responded to with go-around compliance. Nine of the twelve ASRS-coded human factors were used as the independent variables in the logistic regression: communication breakdown, confusion, distraction, fatigue, human-machine interface, situational awareness, time pressure, training/qualifications, and workload. The dependent variable was the reported unstabilized approach event outcome, either *continued to landing* or *go-around*. The model explained between 7.9% and 10.7% of the variance in event outcome, depending on the method used in calculating the explained variance (Cox & Snell R^2 or Naglekerke R^2, respectively). The model sensitivity was 88.3%, specificity was 31.4%, positive predictive value was 68.8%, and negative predictive value was 61.1%. However, the logistic regression model was not statistically significant ($\chi^2(9) = 7.78$, $p = .56$).

Although there were associations of the ASRS-coded human factors with reported unstabilized approaches, the relationships of these associations were not statistically significant. Three human factors – communication breakdown, confusion, and time pressure – were associated with decreased odds of the report being one of an unstabilized approach continued to landing when the human factor was coded as contributing to the event outcome (see Table 1). The remaining six human factors – distraction, fatigue, human-machine interface, situational awareness, training/qualifications, and workload – were associated with increased odds of the report being one of an unstabilized approach continued to landing when the human factor was coded as contributing to the event outcome.

Table 1.
Logistic Regression Predicting Likelihood of Unstabilized Approach to Landing Report

Variable entered at Step 1	*B*	SE	Wald χ^2	*df*	*p*	OR	95% CI for OR Lower	95% CI for OR Upper
Communication breakdown	-.210	.561	.140	1	.709	.811	.270	2.435
Confusion	-.650	.562	1.337	1	.248	.522	.173	1.571
Distraction	.092	.520	.031	1	.860	1.096	.396	3.036
Fatigue	.076	.700	.012	1	.913	1.079	.274	4.257
Human-machine interface	.191	.573	.111	1	.739	1.210	.394	3.722
Situational awareness	1.067	.571	3.487	1	.062	2.906	.949	8.906
Time pressure	-.275	.721	.145	1	.703	.760	.185	3.124
Training/qualifications	.227	.559	.165	1	.684	1.255	.420	3.754
Workload	.541	.520	1.084	1	.298	1.718	.620	4.757
Constant	-.264	1.033	.065	1	.798	.768		

Note. $\alpha = .05$. OR = Odds Ratio.

Using the Wald χ^2 test to determine the statistical significance of the contribution for each human factor to the model, the results indicate that none of the human factors added significantly to the model (all *ps* > .06). Although situational awareness was expected to improve the fit of the model as indicated during the baseline analysis, it did not result in a statistically significant contribution to the model when added (Wald $\chi^2(1) = 3.49$, $p = .06$, OR = 2.91, 95% CI [.95, 8.91]). This result suggests that when situational awareness is coded as a contributing factor, the reported event is 2.91 times more likely to be an unstabilized approach continued to landing than when the factor is not coded as contributing.

Discussion

This study was an analysis of human factors identified as contributing factors in unsafe acts and attitudes, operational errors, and flight crew behaviors during unstabilized approaches in commercial aviation incidents reported to ASRS. The primary aim was to assess if there was an association of the human factors with reported unstabilized approaches, such that the relationship of the human factors influenced the likelihood that the reported event was an unstabilized approach continued to landing versus go-around compliance. The results revealed that there is a statistically significant difference in the outcome of reported unstabilized approaches, in which it is more likely the unstabilized approach will be continued to landing. The influence of decrements in flight crew situational awareness approached the threshold of being a significant contribution to the likelihood that the reported unstabilized approach was continued to landing. However, results from the binomial logistic regression of this study do not support a claim of the outcome likelihood being influenced by the contribution of any sole or combination of human factors. A recommendation is to analyze associations of the different combinations of human factors coded by ASRS as contributing to reported unstabilized approaches. It may be that certain combinations of human factors are associated with an increased likelihood in the outcome of unstabilized approaches. Human factors may indeed have an influence on the likelihood of unstabilized approaches continued to landing rather than go-around compliance, and these human factors may be interacting with other non-human contributing factors. Analyses of these other contributing factors, human factors, and other flight characteristics is warranted.

References

Barnes, V. E., & Monan, W. P. (1990). Cockpit distractions: Precursors to emergencies. *Proceedings of the Human Factors and Ergonomics Society Annual Meeting, 34*, 1142-1144.

Boeing Commercial Airplanes, Aviation Safety. (2017). *Statistical summary of commercial jet airplane accidents. Worldwide operations 1959–2016.* Retrieved from www.boeing.com/news/techissues/pdf/statsum.pdf

Flight Safety Foundation. (2017). *Go-around decision-making and execution project* (Final Report to Flight Safety Foundation). Retrieved from https://flightsafety.org/toolkits-resources/go-around-project-final-report/

Foyle, D. C., Hooey, B. L., Byrne, M. D., Corker, K. M., Deutsch, S., Lebiere, C., . . . Wickens, C. D. (2005). Human performance models of pilot behavior. *Proceedings of the Human Factors and Ergonomics Society Annual Meeting, 49*, 1109-1113.

International Air Transportation Association. (2016). *Unstable approaches: Risk mitigation policies, procedures and best practices* (2nd ed.). Montreal, Quebec: Author.

Strayer, D. L., & Cooper, J. M. (2015). Driven to distraction. *Human Factors, 57*, 1343-1347.

MODERN PRACTICES FOR FLIGHTCREW TRAINING OF PROCEDURAL KNOWLEDGE

Nathan Sonnenfeld, Blake Nguyen, Class Tido Boesser, Florian Jentsch
University of Central Florida
Orlando FL

Instructional systems utilizing electronic and distance learning approaches (E&DL) have advanced the accessibility and capabilities of training in aviation and other industries. The aviation industry is particularly interested in how to extend the use of E&DL from training facts and information to training procedures, and how to integrate E&DL into existing training. We reviewed a corpus of literature including over 1,200 scientific, regulatory, and technical documents—across domains including aviation, defense, healthcare, and education—focusing on the effectiveness of E&DL for training procedures and the design, development, and evaluation of E&DL. We received input from subject matter experts with respect to contemporary and near-future aviation training practices, and also created a glossary of over 5,600 terms related to E&DL. In this paper, we discuss our findings and provide suggestions for their application to flightcrew training.

A priority for the aviation industry is training flightcrews on procedures, with an emphasis on those pertaining to corrective actions in response to non-routine and emergency situations on the flight deck (Human Factors & Aviation Safety, 2019). Training of procedures requires both the conveying of knowledge (declarative and procedural) and the practice of skills (Matton et al., 2018), and it is traditionally provided through either live training or high-fidelity simulation (Dunne et al., 2010). However, organizations increasingly use E&DL to satisfy learning needs (Martins, 2019), and this shift has only accelerated recently. The goal of our work was to identify guidance for the potential application of E&DL configurations to procedures training within the aviation industry. In support of this research, we also sought input from subject matter experts with respect to contemporary and near-future aviation training practices.

Findings & Recommendations

Within the space of E&DL in aviation, there are many approaches that employ recent developments in training design and technology. Like others (eg., ACT ARC 1-6, 2016), we found that the terminology is used inconsistently (Moore et al., 2011) and often conflates technologies with approaches, and vice versa. Relevant approaches (e.g., eLearning, online learning, computer-based training, technology-based training, simulation-based training) may each be combined with various hardware delivery mediums [e.g., personal computers (PCs), mobile devices, simulators, virtual reality (VR) or augmented reality (AR) displays and peripherals], as well as hardware and software technologies—ranging from common tools such as internet, multimedia, and simulation, to modern technologies including physiological sensors, learning analytics, and virtual agents. These instructional and technical choices are further crossed with pedagogic/didactic choices; training may vary in the location of delivery and the level of instructor involvement; most of this training may be conducted with or without instructors. Given the numerous ways relevant approaches and technologies may be combined or blended to form a specific configuration with which to deliver training, in the following, we will refer to the overall set as E&DL and describe specific facets as needed to distinguish different approaches.

E&DL without Task Simulation

Traditional eLearning/CBT. Executing procedures requires both knowledge and skills, and it is conventional wisdom in training that skills will only be learned if they are actively practiced – and the best practice opportunities in aviation involve on-the-job training or simulation. Despite this, training for procedures that uses E&DL without a simulation component still has been found to be more effective than no training (e.g., Potter et al., 2014). Furthermore, the literature suggests that this type of training can be equally or more effective than similarly focused classroom-based instruction—effectiveness of non-simulation E&DL largely depends on the design of each training. Evidence of this comes from (a) computer-based training (e.g., Kearns, 2011); (b) web-based training (e.g., Sitzmann et al., 2006); (c) multimedia/video (e.g., Keller et al., 2019); (d) webinars/video conferencing (e.eg., Abbot et al., 2017); and (e) intelligent tutoring systems (ITS; e.g., Kulik & Fletcher, 2016). In all of the aforementioned examples, learning tools such as video content and intelligent tutoring functionality were part of the training, and we therefore suggest them for inclusion in training of procedures using modern training approaches such as E&DL.

Mobile learning and microlearning. The literature provides some evidence that mobile learning, i.e., the use of mobile devices to access and display training content "*anywhere/anytime*", may be an effective training approach (Kearns, 2018). However, at this time, it is the least proven of all non-simulation training approaches. That said, some aviation training is already conducted using tablets and other mobile devices, and mobile devices have unique functionalities to support procedures training, including tactile interfacing, display of animated and interactive multimedia content, and collaborative Web 2.0 features (Tucker, 2010). While there are limitations to the use of mobile devices for training, most notably (small) screen size (Park et al., 2018), most of these could feasibly be resolved by adapting content to mobile devices (Bhuttoo et al., 2017) and using instructional methods suitable for mobile devices such as microlearning (Kearns, 2018). Further, mobile devices are potentially a low-cost solution to providing trainees with "anywhere/anytime" access to (limited) task simulation, as they can form the hardware basis for rudimentary VR (HMD) and AR (fusion of camera and training content) capability (see also below). Aviation practitioners who wish to include mobile learning as a component of procedures training should ensure that: (a) mobile instruction is accessible by other means; (b) mobile instruction is designed to utilize the unique capabilities of the medium; (c) mobile content is usable, accessible, and compatible to learners' devices; and (d) instructional content is designed with empirically supported strategies.

E&DL with Task Simulation

Scenario-based training (SBT). Based on our review of the literature, we determined that E&DL configurations using components of task simulations are generally more suitable for procedures training than those without it. Whether implemented as actual simulations (Reweti, et al., 2017; Walker, 2014) or more basic part-task scenario-based training (SBT; Blickensderfer et al., 2012), embedding relevant practice components into E&DL will improve the likelihood that trainees can acquire and apply procedural knowledge[1].

Game-based training. Potential exemplars for extending SBT further include intelligent tutoring, adaptive training, and automated scenario generation (Nicholson et al., 2009), and the used of game-based approaches. The latter, in particular, can be effective for improving cognitive skills (Sitzmann, 2011), although this evidence is more limited than for other simulation approaches, particularly in aviation contexts (cf. ACT ARC, 2016). Balancing engaging game elements with effective instruction is a difficult endeavor, however, necessitation the use of specialized personnel and development approaches (Hirumi & Stapleton, 2009), which may be infeasible for smaller carriers. Overall, aviation practitioners should consider leveraging SBT approaches when conducting procedural training and could further innovate this space by incorporating adaptive training, intelligent tutoring, automated scenario generation, and game-based elements within SBT approaches—while sharing the results of those efforts to strengthen and further advance E&DL research across and within the aviation community.

Virtual (and Augmented) Reality (VR/AR). VR promises to be a particularly suitable delivery medium for procedures training in contexts applicable to aviation. Extant studies generally show equivalence between VR and desktop configurations (Taylor & Barnett, 2011), although some VR configurations have been more effective than traditional training (Aggarwal et al., 2007) or computerized training on desktop displays, especially for novices, stationary operators, and for tasks involving contexts congruent with the flight deck (Taylor & Barnett, 2011). In a meta-analysis, Kaplan et al. (2020) found that Extended Reality (XR) platforms are more suitable for training physical tasks but are otherwise equally effective to traditional approaches. There is also some evidence that incorporating other modalities, specifically, emphasis on haptic interfaces, may improve the efficacy of VR training (Sigrist et al., 2013; Matton et al., 2018). AR has so far been limited to individual cases studies in the literature, but these have demonstrated that AR can be an effective medium for training knowledge, procedures, and spatial tasks (Hatfield et al., 2019). We believe that there are the following four future AR use cases: (1) using mobile technologies to capture field data (e.g., external views) and incorporate it back into the training environment (e.g., Medford et al., 2017); (2) visually overlaying training content such as text and video (Fehling et al., 2020; Keebler et al., 2017); (3) providing interactive procedural training across platforms (HMD, mobile, tablet, PC) (Hatfield et al., 2019); and (4) spatially overlaying procedural guidance (Limbu et al., 2018). In aggregate, while aviation practitioners should leverage immersive VR technologies in procedures training, we believe that although AR is a promising emerging technology, it remains unproven in the literature.

Evaluation of E&DL

To provide guidance on which training effectiveness evaluation (TEE) framework(s) may be suitable for the training evaluations when training and evaluation are conducted at a distance or without the presence of an instructor, we reviewed and qualitatively compared the most relevant, unique, and representative frameworks across identified categories, adapting criteria suggested by Dessinger and Mossely (2006): *feasibility*, *utility*, *propriety*, *accuracy*—and a fifth criterion, the *suitability* of each framework with respect to the evaluation of flightcrew training delivered using

[1] See also the discussion on levels 1 through 4 of eLearning in ACT ARC Recommendation 16-6 (pp.6-8); https://www.faa.gov/about/office_org/headquarters_offices/avs/offices/afx/afs/afs200/afs280/act_arc/act_arc_reco/media/2016/ACT_ARC_Reco_16-6.pdf)

E&DL configurations. Overall, we found that the selection of an evaluation framework should be based on a comparison of the unique advantages of each against training needs and organizational requirements, given that each TEE framework has unique advantages and limitations (e.g., Alvarez et al., 2004; Goodwin et al., 2018; Holton, 1996; Sitzmann & Weinhardt, 2019; Sottilare et al., 2012; cf. Kirkpatrick, 1959). While smaller carriers may not have the resources to implement every framework, training practitioners may be able to obtain more actionable training data by (a) assessing a greater scope of training factors and system affordances, (b) measuring learning, performance, and transfer at multiple levels; and (c) leveraging the unique capabilities inherent to E&DL systems—such as automated assessment and systems interoperability.

Conclusion

From our review of the literature on modern E&DL training approaches relevant to flightcrew training, and specifically the training of procedures, we can conclude that a number of them have promise for procedures training. Specifically, we found sufficient evidence configurations using video and/or VR, intelligent tutoring, and scenario-based practice that provide trainees with interactive opportunities to apply and practice the training content can generally be effective for procedures training, given it is appropriately designed. The empirical evidence supporting the application of mobile devices, game-based approaches, and AR to flightcrews' procedural training, on the other hand, is still comparatively limited, however, and air carriers should cautiously approach implementing these training practices. Lastly, to support training outcomes, air carriers should consider adapting their training evaluation practices by selecting TEE frameworks that are more suitable for the evaluation of training conducted using E&DL. In this context, tying data collected during and after E&DL training to subsequent performance in simulator training and to more distal outcome data is particularly important.

Disclaimer & Acknowledgements

The views expressed in this article are those of the authors and do not necessarily reflect the official policy or position of the University of Central Florida or the Federal Aviation Administration (FAA). This work is sponsored by the FAA NextGen Human Factors Division (ANG-C1) under grant CRA 692M151940002.

References

Abbott, E. F., Thompson, W., Pandian, T. K., Zendejas, B., Farley, D. R., & Cook, D. A. (2017). Personalized video feedback and repeated task practice improve laparoscopic knot-tying skills: Two controlled trials. *Academic Medicine*, 92(11), 26-32. https://doi.org/10.1097/ACM.0000000000001924

Air Carrier Training Aviation Rulemaking Committee (2016). *Recommendation 16-6: Delivery of Cross-Functional CRM Training*. Federal Aviation Administration Flight Standards Service. Retrieved from https://www.faa.gov/about/office_org/headquarters_offices/avs/offices/afx/afs/afs200/afs280/act_arc/act_arc_reco/media/2016/ACT_ARC_Reco_16-6.pdf

Aggarwal, R., Ward, J., Balasundaram, I., Sains, P., Athanasiou, T., & Darzi, A. (2007). Proving the effectiveness of virtual reality simulation for training in laparoscopic surgery. *Annals of Surgery*, 246, 771–779. https://doi.org/10.1097/SLA.0b013e3180f61b09

Alvarez, K., Salas, E., & Garofano, C. M. (2004). An integrated model of training evaluation and effectiveness. *Human Resource Development Review*, 3(4), 385-416.

Bhuttoo, V., Soman, K., & Sungkur, R. K. (2017, July). Responsive design and content adaptation for e-learning on mobile devices. 1st International Conference on Next Generation Computing Applications (NextComp), Mauritius, 2017, 163-168. https://doi.org/10.1109/NEXTCOMP.2017.8016193

Blickensderfer, E., Strally, S., Doherty, S. (2012). The effects of scenario-based training on pilots' use of an emergency whole-plane parachute. *The International Journal of Aviation Psychology*, 22(2), 184-202. https://doi.org/10.1080/10508414.2012.663253

Dessinger, J. C., & Moseley, J. L. (2006). The full scoop on full-scope evaluation. In J. A. Pershing (Ed.), Handbook of Human Performance Technology: Principles, Practice, and Potential (3rd ed.) (pp. 312-330). Pfeiffer.

Dunne, R., Schatz, S., Fiore, S. M., Martin, G., & Nicholson, D. (2010, September). Scenario-based training: Scenario complexity. In *Proceedings of the Human Factors and Ergonomics Society Annual Meeting* (Vol. 54, No. 27, pp. 2238-2242). Sage CA: Los Angeles, CA: SAGE Publications. https://doi.org/10.1177%2F154193121005402704

Fehling, A., Fedorovich, M., Johnson, M., Cunningham, J., Ariotti, S., Beck, S. (2020, May 12). *Microsoft Hololens 2 Roadmap, Applications and Experiences*. Webinar presented by the Central Florida Chapter of the VR/AR Association.

Goodwin, M. S., Reinerman-Jones, L., Harris, S., & Arrieta, A. (2018). A taxonomy for the evaluation of training simulations and environments. *MODSIM World 2018*. Retrieved from http://www.modsimworld.com/papers/2018/MODSIM_2018_Paper_not_required_for_a_43.pdf

Medford, A. L., Diaz, G., Murphy, J. S., & Goodwin, G. A. (2017). *Support for training effectiveness assessment and data interoperability (STEADI)*. U.S. Army Research Laboratory. Retrieved from https://apps.dtic.mil/sti/pdfs/AD1033634.pdf

Hatfield, N., Padron, C., Skinner, A., Fidopiastis, C., & Stanney, K. (2019). Augmented reality for training and operational support solutions. *MODSIM World 2019*. Design Interactive, Inc. Retrieved from http://modsimworld.org/papers/2019/MODSIM_2019_paper_56.pdf

Hirumi, A., & Stapleton, C. (2009). Applying pedagogy during game development to enhance game-based learning. In *Games: Purpose and Potential in Education* (pp. 127-162). Springer, Boston, MA. https://doi.org/10.1007/978-0-387-09775-6_6

Hoffman, R. R., Ward, P., Feltovich, P. J., DiBello, L., Fiore, S. M., & Andrews, D. H. (2013). *Accelerated expertise: Training for high proficiency in a complex world*. Psychology Press.

Holton III, E. F. (1996). The flawed four-level evaluation model. Human Resource Development Quarterly, 7(1), 5-21. https://doi.org/10.1002/hrdq.3920070103

Human Factors & Aviation Safety. *Hearing on The Boeing 737 MAX: Examining the Federal Aviation Administration's Oversight of the Aircraft's Certification* before The House Committee on Transportation & Infrastructure, House of Representatives, 116th Cong. 1-17 (2019, December 11) (testimony of Mica Endsley). Retrieved from https://transportation.house.gov/imo/media/doc/Endsley%20Testimony.pdf

Kaplan, A. D., Cruit, J., Endsley, M., Beers, S. M., Sawyer, B. D., & Hancock, P. A. (2020). The effects of virtual reality, augmented reality, and mixed reality as training enhancement methods: A meta-analysis. *Human Factors*. https://doi.org/10.1177%2F0018720820904229

Kearns, S. K. (2018). Snap-courses: An instructional design strategy for aviation mobile learning. *The Collegiate Aviation Review International*, 31(1). Retrieved from https://search.proquest.com/scholarly-journals/snap-courses-instructional-design-strategy/docview/1502610549/se-2?accountid=10003

Kearns, S. (2011). Online single-pilot resource management: Assessing the feasibility of computer-based safety training. *The International Journal of Aviation Psychology*, 21(2), 175-190. https://doi.org/10.1080/10508414.2011.556499

Keebler, J. R., Patzer, B. S., Wiltshire, T. J., & Fiore, S. M. (2017). Augmented reality systems in training. In Brown, K. G. (Ed.), *The Cambridge Handbook of Workplace Training and Employee Development* (pp. 278-292). Cambridge University Press, New York, NY.

Keller, A., Langbauer, M., Fritsch, T., Lehner, F. (2019). Interactive video vs. hypertext documents - the effect on learning quality and time effort when acquiring procedural knowledge. *Proceedings of the 52nd Hawaii International Conference on System Sciences*, 22-31. https://doi.org/10.24251/HICSS.2019.005

Kirkpatrick, D. L. (1959). Techniques for evaluating training programs. *Journal of the American Society of Training Directors*, 13, 3–9.

Kulik, J. A., & Fletcher, J. D. (2016). Effectiveness of intelligent tutoring systems: A meta-analytic review. *Review of Educational Research*, 86(1), 42-78. https://doi.org/10.3102%2F0034654315581420

Limbu, B., Fominykh, M., Klemke, R., Specht, M., & Wild, F. (2018). Supporting training of expertise with wearable technologies: The WEKIT reference framework. In S. Yu, M. Ally, & A. Tsinakos (Eds.), *Mobile and Ubiquitous Learning: Perspectives on Rethinking and Reforming Education* (pp. 157-175). Springer Singapore. https://doi.org/10.1007/978-981-10-6144-8_10

Martins, L. B., Zerbini, T., & Medina, F. J. (2019). Impact of online training on behavioral transfer and job performance in a large organization. *Journal of Work and Organizational Psychology*, 35(1), 27-37. https://doi.org/10.5093/jwop2019a4

Matton, N., Vrignaud, C., Rouillard, Y., Lemarie, J. (2018). Learning flight procedures by enacting and receiving feedback. *Applied Ergonomics*, 70, 253-259. https://doi.org/10.1016/j.apergo.2018.03.009

Moore, J. L., Dickson-Deane, C., Galyen, K. (2011). E-learning, online learning, and distance learning environments: Are they the same? Internet and Higher Education, 14(2), 129-135. https://doi.org/10.1016/j.iheduc.2010.10.001

Nicholson, D., Schatz, S., Stanney, K., & Lackey, S. (2009). *Next-generation Expeditionary Warfare Intelligent Training (NEW-IT) Program Summary Booklet* (No. IST-CR-09-06). University of Central Florida, Institute for Simulation and Training. Retrieved from https://apps.dtic.mil/sti/pdfs/ADA514853.pdf

Park, E., Han, J., Kim, K. J., Cho, Y., & del Pobil, A. P. (2018, January). Effects of screen size in mobile learning over time. *Proceedings of the 12th International Conference on Ubiquitous Information Management and Communication*, 1-5. https://doi.org/10.1145/3164541.3164625

Potter, B. A., Blickensderfer, E. L., & Boquet, A. J. (2014). Training monitoring skills in helicopter pilots. *Aviation, Space, and Environmental Medicine*, 85(5), 543-549. https://doi.org/10.3357/ASEM.3771.2014

Reweti, S., Gilbey, A., Jeffrey, L. (2017). Efficacy of low-cost pc-based aviation training devices. *Journal of Information Technology Education: Research*, 16, 127-142. Retrieved from http://www.informingscience.org/Publications/3682

Sigrist, R., Rauter, G., Riener, R., & Wolf, P. (2013). Augmented visual, auditory, haptic, and multimodal feedback in motor learning: a review. *Psychonomic Bulletin & Review*, 20(1), 21-53. https://doi.org/10.3758/s13423-012-0333-8

Sitzmann, T. (2011). A meta-analytic examination of the instructional effectiveness of computer-based simulation games. *Personnel Psychology*, 64(2), 489-528. https://doi.org/10.1111/j.1744-6570.2011.01190.x

Sitzmann, T., Kraiger, K., Stewart, D., Wisher, R. (2006). The comparative effectiveness of web-based and classroom instruction: a meta-analysis. *Personnel Psychology*, 59, 623-664. https://doi.org/10.1111/j.1744-6570.2006.00049.x

Sitzmann, T., & Weinhardt, J. M. (2019). Approaching evaluation from a multilevel perspective: A comprehensive analysis of the indicators of training effectiveness. *Human Resource Management Review*, 29(2), 253-269. https://doi.org/10.1016/j.hrmr.2017.04.001

Sottilaire, R. A., Goldberg, B. S., Brawner, K. W., & Holden, H. K. (2012, December). A modular framework to support the authoring and assessment of adaptive computer-based tutoring systems (CBTS). Retrieved from https://gifttutoring.org/attachments/download/222/2012_09_Sottilare_ITSEC%20Presentation_Paper%2012017.pdf

Tucker, J. S. (2010). Mobile learning approaches for US Army training (No. ARI-RN-2010-07). Army Research Institute for the Behavioral and Social Science. Fort Benning, GA. Retrieved from https://apps.dtic.mil/sti/pdfs/ADA528742.pdf

Walker, D. R. (2014). A quantitative study evaluating the impact of training context on performance in simulator-based aviation training (Order No. 3666754). ProQuest Dissertations & Theses Global. Retrieved from https://search.proquest.com/dissertations-theses/quantitative-study-evaluating-impact-training/docview/1640918106/se-2

PROCEDURES AS AN ECOLOGICAL INTERFACE

Esa Rantanen[1] and Erik-Jan Huijbrechts[2]

[1]Rochester Institute of Technology,
Rochester, NY, USA

[2]Airline Pilot and Independent Aviation Safety Researcher,
Sassenheim, the Netherlands

The skills-, rules-, and knowledge (SRK) framework of Rasmussen (1983) is an immensely influential model of human behavior and performance and foundational to many other models. Ecological interface design (EID) seamlessly integrates analysis, design, and evaluation functions in complex sociotechnical, real-time, and dynamic systems. The design principles of EID fall into the SRK categories. In this paper we will examine airline flight deck procedures from the EID point-of-view. We argue that better procedures are needed to avoid many procedure-related accidents, including accidents resulting from deviations from procedures. An EID-based procedure/checklist, then, should support every kind of performance, skill-, rule-, and knowledge-based. In Karl Weick's notion of safety as a "dynamic non-event" safety results from continual compensations for external influences. In this sense, it is dynamic inputs that create stable outcomes, and therefore rigid procedures are antithetical to safety (i.e., stable outcomes). Another way of putting this idea is by applying C. S. Lewis' notion of looking at something vs. looking along something: Rather than looking at a given task (or just at a checklist or procedure), pilots should look along the checklist or procedure at a system itself, bringing knowledge-based behaviors to even otherwise routine (skill- or rule-based) performance.

In psychology the distinction between attention-demanding, controlled, reasoning and intuitive, automatic, behavior has been made at least since 1970s (Stanovich & West, 2000). The skills-, rules-, and knowledge (SRK) framework, first outline published in 1976 (Rasmussen, 1976), lays out this distinction in three ways (Rasmussen, 1983). The SRK framework is an immensely influential model of human behavior and performance and foundational to many other models (e.g., the Generic Error Modeling System; Reason, 1990).

In this framework, skill-based behavior consists of sensory-motor performance during acts that (after a statement of intention) take place without conscious control as smooth, automated, and highly integrated patterns of behavior. Diagnostic troubleshooting, for example, is done by a direct match between the features of the problem observed and patterns previously experienced and stored in the long-term memory (LTM). Skill-based performance is fast, requires little cognitive activity, and accurate. Rule-based behavior is controlled by a stored (in LTM) rule. Performance is goal-oriented, although the goal may not be explicitly formulated but is found implicitly in the situation that calls a stored rule. Diagnosis is done by applying sets of rules stored in LTM, e.g., sequence of steps, and the procedures for doing so. Finally, knowledge-based behavior is explicitly goal controlled performance in unfamiliar situations where no rules or know-how are available. Functional reasoning is done based on mental models. Knowledge-based processes are Iterative diagnostic testing and subsequent analyses necessary for problem-solving.

Ecological interface design (EID; Bennett & Flach, 2011; Burns & Hajdukiewicz, 2013) was introduced specifically for complex sociotechnical, real-time, and dynamic systems. EID seamlessly integrates analysis, design, and evaluation functions facilitating a truly user-centered design process. The goal of EID is to make constraints and complex relationships in the work environment perceptually evident (e.g., visible, audible) to the user. In this sense a checklist is an example of an EID-based interface. Such interfaces allow more of users' cognitive resources to be devoted to higher cognitive processes, for example, problem solving and decision making.\

The design principles of EID fall into the SRK categories. For example, for skill-based behavior the operator should be able to act directly on the display and thus the structure of the display should support skill-based performance; direct manipulation device is therefore preferred over command-language interface. For rule-based behavior the designers should provide consistent one-to-one mapping between the work domain constraints and the cues or signs provided by the interface (i.e., provide the operators signs they can use to select appropriate actions). This will allow operators operate by relying in perceptual cues instead of having to resort to knowledge-based behavior and to take advantage of the economy of rule-based behavior while preserving the wide applicability of knowledge-based behavior. Finally, designers should support knowledge-based performance by revealing the problem space in the form of an abstraction hierarchy presentation (Rasmussen, 1985), providing the operators a normative model of the work domain, supporting experimentation ("what if" scenarios), and relieving the operators of the burden of keeping track of causal nets within which they are reasoning.

Checklist as a Procedure and an Ecological Interface

Formal, written procedures are often the only "window" operators have into systems that are too complex for unaided human operation (i.e., most systems today, including modern aircraft). Typically, such procedures take the physical form of a checklist (Barshi, Mauro, Degani, & Loukopoulou, 2016). Once it is known what people have to do (from Cognitive Work Analysis; CWA), and it is determined whether how they do it is skill-based, rule-based, or knowledge-based (also by CWA), design of the relevant interface elements may be done accordingly. A procedural sequence will be some mix of skill-based and rule-based activity, with possibly some knowledge-based activity as well. From here on, the terms procedure and checklist are used interchangeably.

There are two complications, however. For experienced people, who have performed the procedure countless times and for whom the task has become routine, the activity will be more skill-based than is the case for less experienced people. It is a challenge to design procedures to accommodate those changes in the balance of the skills, rules, and knowledge. Another challenge for ecological interface design is to allow for different strategies within the same procedure, sometimes used by different people of different experience levels and sometimes situation dependent.

The conventional view of learning is top-down, that is, turning declarative knowledge into procedural knowledge through practice. Declarative knowledge corresponds to knowledge-based and procedural knowledge to skill-based, being largely automatic and effortless. Procedures are somewhere between knowledge-based and skill-based, providing simple rules, which, if followed, allow for successful completion of tasks. In this sense, rule- based—and procedures—are a "crutch" that allow for good task performance with imperfect knowledge and skills.

A good illustration may be cooking of baking: A recipe in a cookbook is a perfect example of a procedure, which allows for someone with limited expertise in cooking turn out a tasty dish on the first attempt. Expert chefs may create novel dishes without a recipe based on their knowledge of food chemistry and properties of different ingredients. Short- order cooks may prepare simple, fast food without recipes, too, but in a typical skill-based performance. The rest of us need cookbooks and to pay careful attention to following recipes (i.e., procedures) in a typical rule-based performance.

Development of better procedures and to avoid the many procedure-related accidents is another challenge. There are examples where following procedures has resulted in tragic accidents (e.g., Swissair flight 111 on Sep 2, 1998). Yet, most procedure-related accidents result from deviations from procedures. An EID-based procedure/checklist, then, should support every kind of performance, skill-based, rule-based, and knowledge-based.

For an expert operator performing at skill-based level, a checklist is just that, a check that all steps are taken. For a less experienced operator performing on rule-based level, a procedure/checklist is a necessary tool to get the job done right. For both, a well-designed, EID-based, checklist should provide a deeper view into the system, supporting knowledge- based performance. Another way of putting this idea is by applying C. S. Lewis' notion of looking at something vs. looking along something (Lewis, 1970): Rather than looking at a given task, or just at a checklist or procedure, and performing it just because it is part of the procedure, we should look along the checklist or procedure at a system itself, bringing knowledge-based reasoning to even otherwise routine (skill-based or rule-based) performance.

Procedures should be designed to allow much flexibility to operators to adapt to changing circumstances and even to be creative and innovate better ways of doing things. This notion is based on Karl Weick's idea of safety as a "dynamic non-event": It is dynamic because reliability "is an ongoing condition in which problems are momentarily under control due to compensating changes in components". In other words, "dynamic inputs create stable outcomes" (Weick, 1987). In this sense, rigid procedures are antithetical to safety (i.e., stable outcomes). Furthermore, looking along a procedure/checklist item to the system it pertains to is, by definition, essential for good situation awareness (SA; Endsley, 1988).

De-Icing Procedures

Before aircraft engine start aerodynamic surfaces have to be clear of snow and ice. If environmental conditions require so, aircraft have to be de-iced and sometimes anti-iced (ICAO, 2018). The Flight Crew Operation Manual (FCOM) delivered by an aircraft manufacturer for each aircraft type gives guidance on how crews have to perform de-icing/ anti-icing procedures. These procedures are established at the time of certification of the aircraft model. Through the years the practice of de-icing has changed, however. In the sixties and seventies it was common practice to have the aircraft de-iced at the gate before engine start. In the eighties remote de-icing on an apron platform became preferred. This offered a more efficient use of de-icing equipment, better use of Hold Over Times for anti-icing, and the spilled de-icing fluids could be better collected for environmental reasons.

Fokker Airplane Flight Manual (AFM) and Fokker FCOM cold weather procedures were primarily based on gate-de-Icing and type-specific de-icing and anti-icing systems. As de-icing is not a daily practice the procedure should be designed in a way it can be used as a do-list for novice and procedure-oriented pilots but left the freedom to skip unnecessary steps for experienced and task-oriented pilots. For example, if only fuel-induced ice is present in the wings, tail de-icing can be omitted to save on cost, time and environment. Explicitly stated recommended procedures allow for deviation from the procedure if circumstances require. Such recommended procedures allow pilots to look along the checklist item into the purpose of the procedure and use knowledge-based reasoning in deciding whether or not to heed the recommendation (i.e., will the environmental conditions and the purpose of the procedure warrant the action?). A well-designed procedure gives guidance for both de-icing before engine start (gate de-icing) and remote de-icing postponing the control- and flaps check to after de-icing treatment is finished. Hence, it is possible for an airline company to provide procedures that comply with manufacturers procedures yet are modified to adapt to changed practices and also offer flexibility for pilots to adapt to circumstances.

The FCOM procedures for the Boeing 737, including the later PG, NG and MAX models, have added remote de-icing procedures but still contain remnants of the original procedures that are not fully adapted to present way of operation. For example, the FCOM required that flaps and controls are checked before the de-icing treatment. However, the procedure originated from the time when gate de-icing was still common practice and was not effectively adapted to cover remote de-icing. Conversely, it is a matter of common sense to delay movement of controls and flaps to after the de-icing treatment to avoid damage that may result from moving ice- or snow-covered control surfaces. The electronic checklists in Boeing 777 and 787 aircraft offer the possibility to skip checklist steps and perform them later. Again, this kind of flexibility provides additional information about the purpose of the procedure and allows looking along it rather than just at it. Flight crews moreover prefer to adapt the standard procedure to perform control checks and flap settings after the de-icing treatment. Procedure design should be done by operators (pilots) with consent and approval from management and manufacturer. This, however, requires the consent of management and good communication between operators and procedure designers.

Decision Speed Calculation and Use

Before starting a take-off a flight crew makes a take-off performance calculation. Nowadays most of these performance calculations are made by computers. It is noteworthy that computer calculations, with opaque algorithms, will make it more difficult to look along the procedure to the system behind it than if doing the calculations by hand. In this performance calculation a decision-speed (V_1) is established. Before reaching V_1 a flight crew can reject the take-off in case of a serious malfunction or condition, after V_1 the crew must continue the take-off. There is a minimum decision speed, V_{mcg} -limited V_1, which must assure controllability in a continued take-off. This minimum speed is based on V_{mcg} (minimum control speed ground), a speed that is established by the manufacturer during the certification phase of an aircraft under certain conditions on a dry runway.

Through recommendations of aviation authorities many airline companies have chosen a policy in their performance software that will turn out the minimum allowable V_1, in most cases V_{mcg}-limited V_1, to use for contaminated runway operations. Research at Delft University of Technology shows that V_{mcg}-limited V_1, however, is not a safe speed to use on a contaminated runway (Huijbrechts, Koolstra, & Mulder, 2019; Koolstra, Huijbrechts, & Mulder, 2019). Authorities have recognized this and recommended adaptation of minimum speeds to use on contaminated runways. However, as there is no legal obligation yet to cater for a margin on V_{mcg}-limited V_1 on contaminated runways, many operators still get unsafe V_1 values from their performance calculations. In most cases a minor change in the software can increase the safety level without affecting payload.

Performance programs are available that give a graphic display about take-off distances available and comparing take-off performance with the runway length available. Such an interface closely corresponds to the definition of EID, that is, making constraints and complex relationships in the work environment perceptually evident. Furthermore, if the selection of V_1 is left to the pilot, such interfaces may indeeed improve safety. A Dutch charter company uses a performance program with the V_1-range policy that offers these features to pilots. With a good explanation about the hazards of low and high V_1 values safety will benefit and the responsibility of choice will also shift from the company to the crew.

Performance engineers in general understand the problem but are reluctant to give information on how their own companies handle V_1 calculations. For an individual pilot, working at the sharp end, it is difficult to get attention of the blunt end, that is, authorities and company management, for safety related problems. Information flow is often very slow between the "Sharp-" and "Blunt End", and slowed further by multiple actors there: airline companies, aviation authorities, and aircraft and systems manufacturers. The complexity described above, with most of the actors occupying the "Blunt End", and the opacity of computer-calculated performance parameters, force pilots to merely look at a checklist (in this case the V_1 value) without the benefit of looking along the procedure and using knowledge-based reasoning to determine whether the V_1 value may in fact be unsafe.

Conclusion

Procedures and checklists are typically considered to support rule-based performance. This is understandable, for they indeed look like clear rules to be followed for prescribed task performance. One may also argue that checklists are unnecessary for skill-based performance, and that knowledge-based performance, with its associated considerable "cognitive overhead" would be undesirable. However, as we hope our examples have illustrated, checklist/procedure supported knowledge-based performance is in many occasions critical to operations in the "Sharp End". Design of such procedures and their interfaces in checklists is a challenge. Accepting the challenge and responding to it requires much better coordination and communication between all the stakeholders, pilots and the triad of airline companies, aircraft manufacturers, and regulatory authorities, than what is presently apparent.

Flight deck procedures should be tailored to the circumstances within which the airline company works and adapted to changing practices (Barshi et al., 2016). Airline companies that develop their own procedures can offer some flexibility to their operators (pilots) through the design of company procedures. EID can be used as a philosophy to redesign flightdeck procedures. Examples of procedures that were adapted to cover the introduction of remote de-icing and offered the opportunity to use them on both rule- and a knowledge-based level could be found in a commuter company. If the Work-As-Imagined (Hollnagel, Woods, & Leveson, 2007) is presented as global directions instead of rigid procedures both company and operators can benefit. Manufacturers' procedures tend to remain unchanged after certification and are not adapted to changed practices like the introduction of remote de-icing (Huijbrechts & van Paassen, 2021). Unfortunately, many airline companies choose to present manufacturers' procedures without adaptation to their operators (pilots).

References

Barshi, I., Mauro, R., Degani, A., & Loukopoulou, L. (2016). *Designing flightdeck procedures*. Tech. Rep. No. NASA/TM-2016-219421. NASA.

Bennett, K. B., & Flach, J. M. (2011). *Display and interface design: Subtle science, exact art*. CRC Press.

Burns, C. M., & Hajdukiewicz, J. (2013). *Ecological interface design*. CRC Press.

Endsley, M. R. (1988). Situation awareness global assessment technique (SAGAT). In *Proceedings of the IEEE 1988 National Aerospace and Electronics Conference (NAECON)* (pp. 789–795).

Hollnagel, E., Woods, D. D., & Leveson, N. (2007). *Resilience engineering: Concepts and precepts*. Ashgate Publishing, Ltd.

Huijbrechts, E.-J. A., Koolstra, H. J., & Mulder, J. (2019). Using V_{mcg}-limited V_1, controllability issues on contaminated runways and in crosswind. *Journal of Aircraft, 56*(4), 1342–1352.

ICAO. (2018). *Manual of Ground De-icing/Anti-icing Operations* (No. DOC 9640 AN940).

Koolstra, H. J., Huijbrechts, E.-J., & Mulder, J. (2019). Analyzing aircraft controllability after engine failure during takeoff in adverse weather conditions. Journal of Aircraft, 56(4), 1330–1341.

Lewis, C. S. (1970). Meditation in a toolshed. In *God in the dock: Essays on theology and ethics* (pp. 212–215). William B. Eerdmans Publishing Co.

Rasmussen, J. (1976). Outlines of a hybrid model of the process plant operator. In T. B. Sheridan & G. Johannsen (Eds.), *Monitoring behavior and supervisory control* (pp. 371–383). Springer.

Rasmussen, J. (1983). Skills, rules, and knowledge; signals, signs, and symbols, and other distinctions in human performance models. *IEEE Transactions on Systems, Man and Cybernetics, 13*(3), 257–266.

Rasmussen, J. (1985). The role of hierarchical knowledge representation in decisionmaking and system management. *IEEE Transactions on Systems, Man, and Cybernetics, 2*, 234–243.

Reason, J. (1990). *Human error*. Cambridge university press.

Stanovich, K. E., & West, R. F. (2000). Individual differences in reasoning: Implications for the rationality debate? *Behavioral and Brain Sciences, 23*(05), 645–726.

Weick, K. E. (1987). Organizational culture as a source of high reliability. *California Management Review, 29*(2), 112–127.

VALIDATION OF COMPUTERISED APTITUDE SELECTION SYSTEM (COMPASS) IN PREDICTING SUCCESS OF UAV APPLICANTS IN THE REPUBLIC OF SINGAPORE AIR FORCE (RSAF)

Thevaki, Karunakarar
Yuhan, Lim
Christopher, De Roza
Janine, Loi En Qi
Grace, Yip Kam Luen
Republic of Singapore Air Force, Ministry of Defence, Singapore

The study examined the predictive validity of the Computerised Aptitude Selection System (COMPASS) that was set up to support the RSAF in its selection of pilots and other vocations. COMPASS measures cognitive abilities theoretically identified to be relevant to the vocation and was introduced for Unmanned Aerial Vehicle (UAV) pilot selection since 2003. With fast changing technological advancement of the UAVs, it is important that validation studies are regularly conducted to improve the effectiveness of the test suite in predicting training success. 219 UAV Pilot applicants' COMPASS scores were analysed against their actual training outcomes to determine a theoretically and statistically sound selection composite. Hierarchical multiple regression was done, and findings revealed that the current composite of tests remained to be significantly correlated with applicant success in UAV pilot training. The paper discusses the practical considerations in streamlining the tests to be included in the final assessment composite. Future studies should consider exploring non-cognitive assessment to improve the predictive validity of the overall selection system beyond COMPASS.

The UAV has evolved to play an increasingly critical role in modern warfare, with capabilities in tasks such as air intelligence, surveillance and reconnaissance. With increasing complexity of UAV operations that comes along with technological advancement, there is a need to study the ideal profile of a new generation of UAV operators. In addition, there is a need to ensure that existing selection systems continue to identify the appropriate abilities and traits expected in applicants to ensure the greatest likelihood of training success. This is particularly important, given the amount of resources invested in UAV pilot training, such as infrastructure support, aircraft maintenance, and highly-trained instructors. It is pertinent to develop a strong selection battery to identify candidates with higher potential of meeting the training criteria to improve overall organisational effectiveness (Carretta & Ree, 2003) and training efficiency.

Cognitive assessments are considered the gold standard in employee selection and assessment. They have been found to have comparable or better predictive validity over other selection tools (Schmidt & Hunter, 1998). The selection process for assessing UAV pilot applicants in the Republic of Singapore Air Force (RSAF) is multi-tiered and is comparable with established Air Forces around the world and has improved training efficiency and success for the RSAF. COMPASS forms one part of a four-stage selection process for applicants and is focused on assessing cognitive traits identified to be critical to UAV training success.

While the predictive validity of the selection composite for UAV pilots has remained largely stable in recent years, it is imperative for selection and training pipelines to pre-emptively evolve to meet shifting operational role of the platform. In anticipation of these changes, the RSAF selection system should be examined and reviewed to ensure that present assessment criterion continue to select candidates who are both willing and able to handle the demands of operating in a radically new operational environment.

The aim of the current study was to assess the predictive validity of the present cognitive assessment criterion based on the current operational environment and training demands. In addition, the study attempted to review and streamline the criterion for aptitudes that might no longer be relevant.

Methodology

Participants and Data

COMPASS and performance data of 219 UAV Pilot trainees from UAV Training School were examined for this analysis. The majority of the trainees were males, and their average age was 20.4 (SD = 2.1). COMPASS was administered at the point of candidate's application and the sub tests of COMPASS served as predictor variables for the criterion: trainees' performance data from UAV Training School. Regression analyses and Pearson correlations were used to determine best predictors of performance.

Procedure

The COMPASS test scores for the sample of UAV pilot candidates accepted into training were individually correlated with their end-of-course training results to determine the direction and strength between them. Tests that were significantly correlated and deemed to be measuring relevant abilities were entered into a hierarchical multiple regression model and the best-fit regression model was selected. The COMPASS composite score is therefore a weighted sum of the selected subtest scores. A multiple regression approach was adopted for two reasons. First, it allows for the tracking of performance of each predictor against the criterion defined, allowing for the determination and refinement of the selection composite. Second, it fits the recruitment requirements by allowing for the development of expectancy tables for HR and decision makers to easily understand candidate's probability of success in training (Tippins, Sackett, & Oswalkd, 2018).

Validation Methodology

Data cleaning and validation sample. Univariate descriptive statistics were obtained from the raw data. Repeated, missing, or outlying values were removed accordingly. Listwise deletion was also used to manage missing data. The spread of scores, skewness, kurtosis were checked to ensure normality. Skewness exceeding +/- 1 and kurtosis exceeding +/- 3 was considered unacceptable (George & Mallery, 2010). Checks were done on the validation sample to ensure the sample had the complete set of predictors (COMPASS score) and criterion data (end of course data).

Correlation and Regression. A correlation analysis was run with COMPASS test scores and training outcomes to determine the direction and strength between them. Tests that were significantly correlated and deemed to be measuring relevant abilities were identified. Subsequently, hierarchical multiple regression was run between the identified predictors and the criterion. Different combinations of multiple regressions were done to maximize the predictability of the composite scores.

Checks on statistical assumptions such as cases-to-Independent Variable ratio, multi-collinearity, and singularity among the Independent Variables were conducted. When determining the best regression equation, the following were considered: content validity of the test battery based on previous job analyses, correlation of individual tests with training outcome (*r*), low inter-correlations and high incremental validity (R^2 value), parsimony, stability of composite, accuracy of prediction, and distribution of applicant population that ensures a large enough selection pool.

Results

Correlation. Results show that all three tests of the current COMPASS selection composite remained moderately and significantly correlated to UAV training success (Cohen, 1988). The following test were identified to be entered based on their correlation with training outcome as well as the relevance to the training. The correlations are shown in Table 1.

Table 1.
COMPASS Tests with Significant Correlations with Training Success.

COMPASS Test Name	Ability Assessed	Correlations (*r*) with Success in UAV Training
SpatialO1	Estimation and spatial orientation	.31**
SpatialV1	Spatial Visualisation	.32**
MultiTask2	Multi-tasking test between psychomotor and auditory/mathematical processing	.25**

Note. ** indicates significance at $p < .01$.

Regression. Results from the hierarchical multiple regression show that the best-fit regression model was found to have a predictive validity, $R^2 = .39$. The three tests within the current model were retained within the proposed model, and their weightages were recalibrated based on the present validation analysis to form the proposed COMPASS selection composite. The regression equations are as shown in Table 2 below. Additional considerations in selecting the predictors include practical implications on the selection systems. This will be elaborated on in the *Discussion* section.

Table 2.
Current and Proposed Selection Composite.

	UAV Selection Composite	Correlations (R^2) with UAV Pilot Training Success
Current	Current Composite (SpatialO1 + SpatialV1 + MultiTask2)	.40** (N = 121)
Proposed	Retain Current Composite with Recalibrated Weights (SpatialO1 + SpatialV1 + MultiTask2)	.39** (N = 219)

Note. ** indicates significance at $p < .01$.

Discussion

In personnel selection, predictive validation tests for the inferences made during selection, especially if the inferences are not directly observable (e.g., psychological constructs) (Gatewood, Field, & Barrick, 2019). There needs to be balance between the strength of the intended validity inference and the practical limitations (Tippins, Sackett, & Oswalkd, 2018). As such, regular efforts are taken to consistently monitor the validity of the cognitive-test battery, and to further refine the COMPASS selection composite. In developing a selection composite, there were two key considerations. First, the selection composite should be able to reliably predict training outcomes. The composite score and the probability of success should also be positively correlated, such that a candidate's composite score increases vis-à-vis their probability of success at training. Second, the selection composite should be able to select sufficient pilots to meet pilot production demands. While developing a predictive composite, there is a need to balance this with practical consideration such that the supply pool continues to be sufficient to meet production targets.

One dilemma faced in determining the UAV selection composite was the difficulty in selecting between tests that were comparative in their statistical validity. In such cases, the face validity of tests was prioritised. For example, during initial analyses, multiple tests of multi-tasking in the COMPASS test battery were found to be significantly correlated to applicant training outcomes. In streamlining the tests to be included in the final assessment composite, the most demanding version of the Multi-tasking test was selected (MultiTask2), as it was determined that the test was more reflective of the high demands and complexity of UAV operations.

It was also noted that despite changes in operations and training in recent years, the predictive validity of the existing COMPASS composite has continued to be stable since its last validation in 2014. This finding suggests that the cognitive traits assessed by COMPASS continue to be essential despite advancements in technology and changes in operational requirements.

Nevertheless, it is clear that with the introduction of new work demands and the increasing complexity of UAV systems, there may be a need to supplement the COMPASS selection system with additional tests that measure relevant traits beyond the fundamental cognitive traits. This includes tests that assess soft skills and other non-cognitive traits that have been found to be related to UAV training success. One example would be the inclusion of personality testing as part of the overall selection composite. Studies examining the relationship between the Big Five personality traits and UAV training performance have suggested that Agreeableness, Extroversion and Conscientiousness are positively associated with UAV training success (Barron et al., 2016). To improve the predictive validity of the overall selection system beyond COMPASS, the introduction of such testing is essential.

The next generation of aptitude selection tests should incorporate the gaps (such as critical thinking) and also review testing methodology to also provide a realistic job performance preview to assessors, as well as to entice the right people for the job.

Conclusion

The reviewed COMPASS selection composite suggests that the RSAF continues to employ aptitude selection tools of good predictive validity for the recruitment of UAV Pilots. This allows for downstream benefits in optimising the production pipeline and deployment of UAV pilots.

References

Barron, L. G., Caretta, T. R., & Rose, M.R (2016). Aptitude and Trait Predictors of Manned and Unmanned Aircraft Pilot Job Performance. *Military Psychology 28(2), 65-77.*

Carretta, T. R. & Ree, M. J. (2003). Pilot selection methods In: *Human factors in transportation: Principles and Practices of Aviation Psychology* (seriesed. B. H. Kantowitz, vol. eds T. S. Tsang and M. A. Vidulich), pp.357–396. Erlbaum.

Cohen, J. (1988). Statistical Power Analysis for the Behavioral Sciences. (2nd ed.). Hillsdale, Erlbaum Associates.

Gatewood, R. D., Field, H. S., & Barrick, M. R. (2019). Human resource selection. Mason, Ohio: Thomson/South-Western.

George, D., & Mallery, M. (2010). *SPSS for Windows Step by Step: A Simple Guide and* Reference, 17.0 Update (10th ed.). Boston, USA: Allyn & Bacon.

Schmidt, F. L., & Hunter, J. E. (1998). The validity and utility of selection methods in personnel psychology: Practical and theoretical implications of 85 years of research findings. *Psychological Bulletin, 124(2)*, 262-274.

Schmidt, F. L., Hunter, J.E., & Outerbridge, A. N. (1986). Impact of job experience and ability on job knowledge, work sample performance, and supervisory ratings of work performance. *Journal of Applied Psychology, 71*, 432-439.

Steiner, D.D & Gilliland, S.W. (1996). Fairness reactions to personnel selection techniques in France and the United States. *Journal of Applied Psychology, 61 (1006),* 134-141.

Tippins, N., Sackett, P., & Oswalk, F. (2018). Principles for the validation and use of personnel selection procedures. *Industrial and Organizational Psychology, 11 (S1)*, 1-97.

Wheatcroft, J.M., Jump, M., Breckell, A.L., & Adams-White, J. (2017). Unmanned aerial systems (UAS) operators' accuracy and confidence of decisions: Professional pilots or video game players? *Cogent Psychology*, *4*(1).

THE SKILL ASSUMPTION - OVER-RELIANCE ON PERCEPTION SKILLS IN HAZARD ASSESSMENT

M. M. (René) van Paassen†, Jelmer R. Reitsma†, Erik-Jan A. M. Huijbrechts‡,
Clark Borst†, Annemarie Landman* and Max Mulder†
† Aerospace Engineering – Delft University of Technology
Delft, The Netherlands
‡ Airline pilot, KLM, Amsterdam, The Netherlands
* TNO Human Factors, Soesterberg, The Netherlands
and Aerospace Engineering – Delft University of Technology
Delft, The Netherlands

In the analysis of human performance and human error, considerable attention is given to the cognitive processes of actors involved in error or success scenarios. Even with awareness of hindsight bias, it takes effort to understand the actions of agents in later inspection of error scenarios. One such topic of heated discussion was the perceived poor performance of pilots in the two 737 MAX MCAS-related crashes in applying the "memory item" checklist pertaining to a runaway trim. In this paper, we argue that it is not so much the reproduction of the checklist that was lacking in these scenarios, but the trigger for even starting the checklist. Not only trim run-away problems, but several other issues likewise require an instant reaction from pilots, designated as "memory items". Rasmussen's simplified schematic for the "skill, rule and knowledge" taxonomy already provides the tools for properly analyzing this. The skill to provide the triggers for these reactions relies on pattern extraction from the available sensory input, and, importantly, it can only be learned in a valid training context. It is argued that re-appraisal of these items is needed, addressing explicitly the validity of the training environments that enable pilots to learn the required pattern recognition skills.

Introduction

With improving technical reliability of aircraft, and improved training and team concepts for flight crew, aviation safety has slowly but steadily improved (Pasztor & Martin, 2020). Each remaining accident or incident is carefully analyzed to learn lessons for further safety improvements, and to watch for trends in aviation safety. The twin accidents of the new Boeing 737 MAX aircraft certainly received their share of the aviation community's attention. In these accidents, a faulty sensor feeding its data into the Maneuvering Characteristics Augmentation System (MCAS), led to the MCAS system applying repeated nose-down pitch trim inputs, producing high forces on the control column, so that the pilots were not able to keep the aircraft from entering a dive. Analyses led to conclusions on failed oversight, a degradation in safety culture at Boeing, questioning of the practice of co-pilots with relatively little flight experience, and the safety process of low-budget airlines.

Research on improving flight safety investigates issues such as the role of startle and surprise (Landman et al., 2017), and training with variability (Landman et al., 2018). This work focuses on preparing pilots for failure diagnosis and recovery, also in the presence of distracting startle conditions, and when operating with a possibly wrong situational frame with respect to concepts on the aircraft and environment state. Other work focuses on safety culture, (Dekker, 2006; Roelen & Klompstra, 2012), investigating an organization's ability to instigate safe practices. Research on specific topics, such as the language and communication (Tajima, 2004), is work that can lead to recommendations by certifying authorities.

It is important to understand that safety in flight operations depends in a large part on proper preparation. Time to diagnose problems is often short, and the time pressure in flight impacts diagnostic capabilities. In some conditions, with flight near aircraft limits such as stall or overspeed, or near terrain, the time available to the first action is counted in seconds. Most incidents are also not new and unique, like the erroneous MCAS activation was, but can be reasonably prepared for. Hazard analysis and operational experience have resulted in a vast body of possible failure cases, and pilots train for known cases such as engine failures, inadvertent stalls, wind shear conditions, cabin leaks leading to depressurization, and a whole slew of technical malfunctions. Reference materials and checklists are prepared for these cases, providing additional pre-programmed responses to foreseen trouble.

In the aftermath of the 737 MAX crashes, several authors claimed that, even though the trim activation by the MCAS system was inappropriate, the flight crew could have responded as taught, by performing the "memory items" from the "Runaway Stabilizer" checklist. Of this checklist, the first five items are memory items, these are actions that a pilot should study, learn by heart, and be able to perform without referring to the physical paper checklist. They are also common with the items in checklists for previous 737 models, making them familiar to pilots transitioning from previous models.

Boeing's training materials for 737 MAX pilots do not even mention the MCAS system as a separate topic. From later investigation, it appears that Boeing's official consensus was that any malfunction of the MCAS system would be handled by the pilots as if it were a "normal" trim system malfunction. It has now been recognized that Boeing overly relied on the capability of pilots to provide the necessary responses to malfunctions (Anon., 2019).

Angle of attack sensor failure on the original 737 MAX

Both accident scenarios with the 737 MAX were triggered by the same failure of an angle of attack sensor on the captain's side. Signals from the angle of attack sensor affect a wide range of the aircraft's systems:

- The pilot's and co-pilot's speed and altitude indications started to differ in both scenarios. This is due to the use of the angle of attack value from the sensor for the correction of placement errors of the static and total pressure sensors, correcting the output values of these sensors.[1]
- The measured angle of attack forms the basis for the activation of the captain's stick shaker and low airspeed warnings.
- The different values for angle of attack affect the feel systems for the captain's and co-pilots control column differently, leading to a "feel differential" message.
- The angle of attack sensor forms the input for the MCAS system, leading to repeated and aggressive pitch down trim actions, but only for certain aircraft configuration, that is, with the autopilot disconnected and flaps up.

This all produces a situation that does not resemble a normal "old school" runaway stabilizer trim incident. We will subsequently look at the incidents, and use Rasmussen's Skill, Rule and Knowledge taxonomy to understand the pilots' failure to start the trim checklist.

The Skill-Rule-Knowledge taxonomy inspected

Last week, I (the first author) opened the family's dishwasher with the intention to fill it with the dirty items that had collected in the sink. As I noticed humid warm air coming off the machine, I changed my activity and cleaned out the dishwasher, since it had finished the cleaning cycle initiated by one of the other members of the household.

Rasmussen (Rasmussen, 1983; Rasmussen, 1986), presents a three-level taxonomy for human behavior, allowing a rough categorization into skill-based, rule-based and knowledge-base behavior, see Fig. 1. This distinction is useful, because each level of human behavior has its own strengths and weaknesses, and understanding which behavior is needed in which work situation enables an analyst to assess training needs and possibilities for error. A typical example of skill-based behavior would be manual flying of an aircraft, rule-based behavior would be performing the checklist, and knowledge-based behavior would be assessing whether a flight can be performed safely, given pilot skills, aircraft equipment and weather forecast. However, these examples do not do justice to the richness of the concepts in the SRK taxonomy. Most rule-based behavior does not correspond to activities where explicit rules have been formulated, and may equally be termed "habit-based", or "routine-based". Thus activities like cleaning out the dishwasher, should be classified as rule-based behavior. In this case, the activity of cleaning out the dishwasher was triggered by a familiar perception of humid warm air.

[1] Judging from the very large altitude and speed differences presented to the pilot and co-pilot, it might be assumed that this algorithm does not cap the correction value, allowing unrealistic measured values to drastically influence instrument presentation.

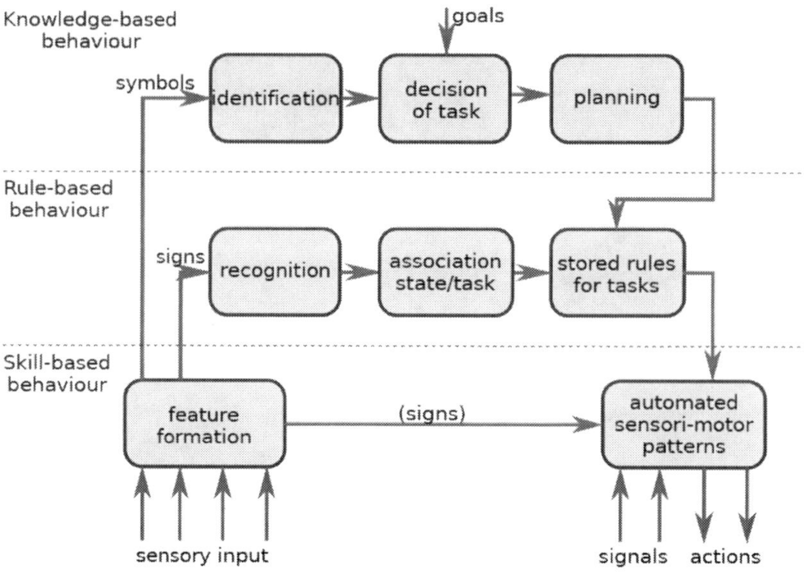

Figure 1: Simplified illustration of the skill-rule-knowledge taxonomy (Rasmussen, 1983)

Application to the MCAS incidents

To the reader or observer of transcripts, timelines, graphs and replays of aircraft accidents, it is often difficult to understand why the pilots did not initiate actions that would have led to a successful outcome. Of course, inspection of the incident after the fact is prone to hindsight bias; the knowledge of the outcome places all facts in a biased review, inviting the classification into wise and unwise decisions. It seems reasonable to assume that the pilots in both scenarios knew how to perform the memory items from the Runaway Stabilizer checklist, which makes it all the more curious why these actions were not performed swiftly or at all.

In Rasmussen's simplified model of SRK behavior, a significant clue is hidden in the connections between the blocks. Rasmussen labels three kinds of information: signal, sign and symbol. Signals are for perceptual communication. Signs are for triggering rule-based activities, and symbols are recognized morsels of information about the outside world that are used in cognition and knowledge-based behavior. In Fig. 1 Rasmussen's signs and symbol lines both originate in the "feature formation" block, explicitly depicted at the skill level. In the checklist for the 737 MAX (Fig. 2), the start of the procedure is simply listed as a condition: "Uncommanded stabilizer trim movement occurs continuously". This condition seems easy enough to detect; stabilizer trim movement is visible in the trim wheels and audible on the flight deck. Since memory items on the checklist must be performed quickly and skillfully, performing these items *should* be rule-based behavior. Rasmussen's simple model indicates that this rule-based behavior can be started by a "sign", which must be a recognized condition that triggers this pre-learned response. The recognition and production of the sign takes place *at the skill level* in Rasmussen's taxonomy. If we take the requirement that the Runaway Stabilizer checklist is executed swiftly and without delay, the pilots must therefore both have the skill to almost instinctively recognize the trigger, and the practice and routine to execute the required actions. The alternative option is to start these rule-based activities after knowledge-based evaluation, which is undesirable, since this is a possibly lengthy and effort-full process.

For the checklist trigger to occur, two other conditions are needed; the trim must be uncommanded and continuous. The Pilot Flying is supposed to recognize the condition, switch off autopilot and autothrottle, and, if needed, command the Pilot Not Flying to move the Trim Cutout Switches to Cutout by stating: "Runaway Stabilizer, memory items". To complicate matters, uncommanded trim input in the 737 (MAX as well as older models), is quite common. The Speed Trim System (STS), that is active in manual flight. can produce trim bursts of several second. As the third author has experienced, these trim actions can be counter-intuitive when the pilot wants to accelerate or decelerate.

Lion air 9.1

737 Flight Crew Operations Manual

Runaway Stabilizer

Condition: Uncommanded stabilizer trim movement occurs continuously.

1 Control column. . . . . Hold firmly

2 Autopilot (if engaged) . . . . .Disengage

Do **not** re-engage the autopilot.

Control airplane pitch attitude manually with control column and main electric trim as needed.

3 Autothrottle (if engaged). . . . . .Disengage

Do **not** re-engage the autothrottle.

4 **If** the runaway **stops** after the autopilot is disengaged:

■ ■ ■ ■

5 **If** the runaway **continues** after the autopilot is disengaged:

STAB TRIM CUTOUT switches (both) . . . . CUTOUT

If the runaway **continues**:

Stabilizer trim wheel . . . . Grasp and hold

- -

6 Stabilizer. . . . . .Trim manually

7 Anticipate trim requirements.

▼ Continued on next page ▼

June 15, 2017 D6-27370-MAX-MLI 9.1

Figure 2: First page from the "Runaway Stabilizer" checklist (Thanhjono, 2019)

To further compound matters, the trim activation was not the only activity at the flight deck at that time. As the Lion Air flight climbed out, pilots faced and became preoccupied with a "speed disagree" message, and differing airspeed on the captain's and co-pilot's speed instruments. In addition, the altitude indications on the left and right instruments disagreed. While confronted with the double problem, pilots were also questioning whether to turn back, and negotiating with air traffic control on where to enter a holding, and on the – erroneous – altitude indication. The co-pilot was focusing on stopping the climb and finding a waypoint to fly to when the MCAS system started its first ten-second nose down trim action. In response, the captain trimmed the aircraft nose up again. In between, a momentary "bank angle" warning was sounded by the Enhanced Ground Proximity Warning System (EGPWS), then "air speed low" warnings (while the air speed was actually high), and the stick shaker started to activate. On the captain's speed display, which was using data by the same faulty sensor that activated MCAS, the overspeed limits and the low speed limits merged. With navigation tasks, feel differential pressure, flight control low pressure, unreliable airspeed, stick shaker, the crew was busy on multiple tasks, and did not start the stabilizer runaway checklist.

Strictly speaking, the condition listed in the checklist was not even met. Common in older generations of aircraft, uncommanded stabilizer trim leading to a runaway was the result of either a trim switch on the control yoke that does not disengage, or of a solenoid that stays stuck. The event is usually after a trim commanded by the pilot, in the absence of other signals and events, and it is consistent; a trim that was initiated does not stop after the trim switch is released, but continues, steadily, in the original direction. This trigger is described in the checklist, and this scenario is implied in the checklist description. As a young trainee, the first author observed this phenomenon several times in the video recordings of a flight experiment with a Swearingen Metro. The Metro's pilots were accustomed to the event, quickly pulled out the circuit breaker disconnecting the trim, cycled the trim switches and turned the circuit breaker back on.

Because the Metro pilots were accustomed to the phenomenon, they instantly recognized the inappropriate trim actions of the aircraft. To the new Boeing 737 MAX pilots, conventional trim problems were a thing of the past, but the MCAS produced a-typical trim behavior in a non-typical context (i.e., not after one of the two pilots used the trim) and during a very busy time. The nominal description in the checklist, "continuous" activation, does not match what happened. The MCAS actions come in bursts, and untypically, in the opposite direction of the pilot's trim direction, instead of continuing in the same direction. The pilots in both accident scenarios only had received a tablet-based training before transitioning to the new aircraft variant, and before the first accident did not have any information on the MCAS system. It is thus clear that they did not have experience with the pattern and context in which the MCAS activates the trim system.

Considering Rasmussen's SRK taxonomy and the simplified model, it is clear that time-critical actions in response to a trim runaway should be implemented as a combination of skill and rule-based behavior; knowledge based behavior is too slow for this condition, and this justly motivates Boeing's choice to designate the checklist items as memory items. On the Lion Air flight preceding the accident flight with the same aircraft, the pilots did manage to switch off the electric trim and thereby stop the MCAS actions. However, an additional pilot was present on that flight in the observer's seat. Possibly due to not being in the thick of controlling the aircraft, he had time to detect the largest threat as coming from the uncommanded trim actions, and instructed the pilots to use the cutout switches.

Rasmussen's diagram places the triggering signals for a procedure clearly in the skill domain. But simply by not having experienced MCAS activation previously, there is no skill to recognize the start condition for the checklist. The pattern recognition is not possible, and the sign for activating a rule-based sequence of actions is not produced. With the short time available, and the multitude of sensory impressions, surviving an erroneous MCAS activation on the original 737 MAX design is only likely if the condition for action is trained and recognizable as a familiar pattern. This can only be achieved if detection and distinguishing of the trigger is practiced in the simulator using scenarios that present the trigger in the correct context, possibly with a multitude of masking signals and detractors as present in the accident scenario.

In this case, in the absence of the possibility to recognize a familiar pattern, the pilots are left with a large set of observations, from the various instruments and warnings, and with parallel checklists and actions that these indications require, and they lacked the time and capacity to determine which was the most urgent issue and attend to it. In Rasmussen's terms, in this scenario the pilots were forced to resort to knowledge-based behavior, which can be slow, error-prone and is always laborious.

Wider implications

In the aftermath of the two accidents, the design of the MCAS system was re-evaluated, and its main problems are addressed. The insight that oversight by certifying authorities and a degradation in safety culture at Boeing contributed to unsafe designs and practices has prompted a wider review, and the correction of other identified issues, e.g., the lack of capacity of one of the flight control computers. However, based on Rasmussen's proven insights, wider implications can be seen. Each time a (foreseeable) failure condition requires immediate or near immediate attention from pilots, we should ensure that not only the pilots have the opportunity to train and drill to execute their response, but that they are also trained in recognizing the sign pattern in the multitude of signals available on the flight deck, not just in isolated failure cases, but also in busy scenarios where multiple triggers for multiple checklists compete.

This recognition requires more than pictures on a tablet or paper materials. Signal timing, strength, noise, vibration, possibly heat, multiple cues need to be approximately correct. From people working on flight simulation devices, we know that in the aftermath of the two accidents the hardware behind the motors providing the trim feel in the 737 simulators had to be significantly upgraded, to approach the levels of forces present in the actual aircraft. Also with the airline pilot's job progression from an active, hands-on flight to supervisor and programmer of automation for flight, the need for simulator fidelity does not diminish. In all scenarios where quick action is required to avert danger, the recognition of pilots must be trained until the patterns become familiar and recognizable even in the presence of significant and distracting signals, and in unexpected situations. Constant innovation in training scenarios is also needed to prevent pilots from recognizing a situation in a simulator session on the basis of expectation (*"second scenario, likely an engine failure during take-off"*), because that would train for false patterns that are not present in real flights.

Conclusion

With the introduction of the 737 MAX and its MCAS system, which was intended to keep the aircraft from stall situations, Boeing assumed that the procedures for runaway trim would enable pilots to correct for potential malfunctions in the MCAS system. Boeing's assumption on the pilots' ability to perform the necessary actions might have been correct, but the assessment on whether pilots would readily recognize the condition for action was not. In this paper, we propose a more accurate perspective on pilot behavior based on Rasmussen's theory. This implies that pattern and trigger recognition for situations that require near-immediate response must be trained in relevant training environments, with the proper cues, and under a variety of conditions. This is not only needed for MCAS activation, but for all checklists with memory items, and for all other situations that require a quick response while triggering conditions may be ambiguous or vague. We should not force people at the sharp end to know when action is required through cognitive reasoning, but instead we should ensure that they can train their pattern recognition skills.

References

Anon. (2019, September). *Assumptions used in the safety assessment process and the effects of multiple alerts and indications on pilot performance* (Safety Recommendation Report ASR1901). National Transportation Safety Board. Washington, DC.

Dekker, S. (2006). *The field guide to understanding human error*. Ashgate.

Landman, A., Groen, E. L., van Paassen, M. M. (, Bronkhorst, A. W., & Mulder, M. (2017). Dealing with unexpected events on the flight deck: A conceptual model of startle and surprise. *Human Factors: The Journal of the Human Factors and Ergonomics Society*, *59*(8), 1161–1172.

Landman, A., van Oorschot, P., van Paassen, M. M. (, Groen, E. L., Bronkhorst, A. W., & Mulder, M. (2018). Training pilots for unexpected events: A simulator study on the advantage of unpredictable and variable scenarios. *Human Factors: The Journal of the Human Factors and Ergonomics Society*, *60*(6), 793–805.

Pasztor, B., & Martin, E. (2020, December). *Statistical summary of commercial jet airplane accidents*. The Boeing Company.

Rasmussen, J. (1983). Skills, rules, and knowledge; signals, signs, and symbols, and other distinctions in human performance models. *IEEE Transactions on Systems, Man and Cybernetics*, *SMC-13*(3), 257–266.

Rasmussen, J. (1986). *Information processing and human-machine interaction : An approach to cognitive engineering*. North-Holland.

Roelen, A. L. C., & Klompstra, M. B. (2012). The challenges in defining aviation safety performance indicators. *International Probabilistic Safety Assessment and Management Conference and the Annual European Safety and Reliability Conference. Helsink, Finland.*

Tajima, A. (2004). Fatal miscommunication: English in aviation safety [ISBN: 0883-2919 Publisher: Wiley Online Library]. *World Englishes*, *23*(3), 451–470.

Thanhjono, S. (2019, October). *PT. lion mentari airlines boeing 737-8(MAX); PK-LQPTanjung karawang, west java* (Aircraft Accident Investigation Report KNKT.18.10.35.04). Komite Nasional Keselamatan Transportasi. Jakarta (IN).

Knowns and Unknowns in Air Traffic Controller Safety Reports: Developing a New Method

Paul Krois, Ph.D., PMP[1]
Crown Consulting Inc.
Aurora, CO

Julia Pounds, Ph.D.[1]
TempoLogics
Mustang, OK

Air traffic controllers in the Federal Aviation Administration can submit voluntary safety reports when significant safety concerns and potential safety events are encountered during their everyday operations. We tested two questions: Can safety reports be classified according to whether the risk was known or unknown to the controller or the system; and would classifying reports in this manner yield useful safety information? A sample of 36 reports was assessed using this known-unknown method. 55% of the reports were classified as risks known to the controller but unknown to the system. 17% of the reports were scored as known to both the controller and the system. 14% were classified as unknown to the controller but known to the system, and 14% as unknown to both the controller and the system. Trends, limitations, and next steps are discussed.

The Federal Aviation Administration (FAA) continues in its mission of providing the safest National Airspace System (NAS) in the world. The Air Traffic Organization (ATO) uses policy, process, programs, and data to monitor safety in operations consistent with its Safety Management System (SMS) and includes safety-related performance targets in its annual business plan. The ATO seeks to mitigate known risks and to uncover unknown risk through its safety assurance and risk mitigation efforts. One of the avenues the ATO identifies and assesses risk, and improves safety culture, is through the Air Traffic Safety Action Program (ATSAP). The National Aeronautics and Space Administration (NASA) has a separate aviation voluntary safety reporting system called the Aviation Safety Reporting System (ASRS) used by pilots, controllers, and others to report safety concerns and issues (Billings et.al., 1976). A commercial airline may have its own Aviation Safety Action Program (ASAP) for a pilot to file a report.

Controller Safety Reporting System

Air traffic controllers in the ATO can submit voluntary safety reports when significant safety concerns and potential safety events are encountered during their everyday operations (FAA, 2017). These reports of hazards and risks are processed and, if appropriate, mitigations are developed and collected as part of safety data. The employee is responsible to ensure that all occurrences of which they are aware, through either direct involvement or observation, are reported. All personnel with knowledge of an occurrence are encouraged to report, even if this

---

[1] Retired from the Federal Aviation Administration. The opinions expressed in this paper are those of the authors and do not reflect any entity.

results in multiple submissions of the same occurrence. The Operations Supervisor, Operations Manager, and controller-in-charge (CIC) must also report occurrences. Reported occurrences are first reviewed at the facility level as a Mandatory Occurrence Report (MOR), e.g., airborne loss of separation. Facility points of contact review the MOR for possible inclusion of additional data before submission to Quality Assurance.

An ATSAP Event Review Committee (ERC) includes a member of FAA's Air Traffic Organization Management, a National Air Traffic Controllers Association (NATCA) representative, and a member of FAA's Air Traffic Safety Oversight Service. The ERC evaluates each report submitted and determines if it meets the requirements established through the FAA-NATCA Memorandum of Understanding. If the report meets the standards prescribed, the ERC accepts the report and logs it into the ATSAP. During the review process, the ERC also reviews each report to identify actual or potential safety issues and causal factors.

Between 2008, when ATSAP was established between the FAA and NATCA and 2018 over 160,000 reports were generated (NATCA, 2018). Certain ATSAP reports are shared with airlines through the Confidential Information Sharing Program (CISP) involving over 28 participants and over 98,000 reports.

By 2018 NATCA indicated there were over 185 formal Corrective Action Requests (CARs) issued to address serious system safety concerns, of which 112 had been closed/resolved. At least 805 systemic positive changes had been developed from voluntary reporting and informal changes taking place at FAA facilities as reported by NATCA (2018). Reports were used to develop recurrent training curricula and contributed to the development of the ATO's Top 5 safety issues. ATSAP Positive reports were use in this study (NATCA, 2015, 2016). Key to ATSAP and its relationship with safety culture are that reports are de-identified so the reporting employees are kept anonymous.

A Paradigm for Knowns and Unknowns

We tested two questions. First, can safety reports be classified according to whether the hazard or risk was known or unknown to the controller or the system? The system is broadly defined and encompasses the work environment including local facility management and operations, air traffic control procedures and airspace, and airlines. Second, would classifying reports yield useful safety information according to whether the reports represent safety issues either known or unknown to either the controller or the system?

For the purpose of this paper, in a safety management system (SMS), what can be considered as knowns and unknowns corresponds to what the front-line air traffic controller understands juxtaposed to what the system understands. The controller's understanding is based upon expertise built on cumulative years of experience, knowledge of automated systems, airspace, and operational procedures, controller training programs (initial and recurrent), reading safety-related and other ATC informational materials, and discussing operational situations with others. Controllers are professionals who continually take in and apply information to provide the safest ATC service.

Understanding of knowns and unknowns by the system is based on integration of complex types of information from many sources. Systems have been defined to "be people, hardware, software, information, procedures, facilities, services, and other support facets which are directly related to the organization's aviation safety activities" (FAA, 2015). What is known by one part of the system does not mean it is known throughout the system. People are part of the system and include the supervisor and operations manager, controller-in-charge, other controllers, airspace and training specialists, Technical Operations personnel (e.g., software specialists), and pilots.

How the system is expected to operate is prescribed through procedures and inter-facility letters of agreement with prescriptive instructions such as on airspace, communications, and flight restrictions. Automated radar and flight data systems perform functions the controllers use to ensure safe traffic flow and manage workload. Aircraft and avionics are designed, built, and integrated according to standards and certification requirements, and flown by pilots certified through training requirements.

Although the front-line controller is the person who first recognizes or deals with a safety issue, the situation may be emergent and heretofore not previously encountered. Its nature, origin, causal and contributing factors, and possible outcomes may not be understood especially if it has not been directly encountered it before, i.e., it does not fit any known pattern.

The ATO and NATCA categorized safety issues as knowns and unknowns for the controller and management based on over 130,000 ATSAP reports from 2010 through 2016 (2017). 100% of the reports were problems known to rank and file personnel. Of those, 75% were problems known to supervisors. Of those, only 9% were problems known to middle management, and of those, 4% were problems known to top management. Using our taxonomy decision rules, these results can be interpreted to mean that 25% of the problems were unknown to the system and 0% were unknown to controllers, as shown in Table 1.

Table 1. *NATCA Reports Classifications.*

		System		
		Known	Unknown	**Total**
Controller	Known	75%	25%	100%
	Unknown	0%	0%	0%
	Total	75%	25%	

Unknowns represent safety risks. What kinds of unknowns occur? Risk can result from the system performing a function in a manner unknown to the controller. For example, Fort Hill examined ATSAP reports associated with the en route automated handoff function (2012). Review of system design specifications showed there were 17 ways that the automated handoff feature could be manually discontinued for a flight being handed off from the transferring sector and many controllers were not aware of those conditions. The controller was also not informed that the aircraft would be handed off to the incorrect sector if the controller initiated the hand-off just after entering the altitude. This poses that what is known to one part of the system (e.g., software designers) may not be known to other parts (e.g., front-line supervisors and trainers).

Method

Our criteria for classification decisions are shown by our 2 x 2 table of knowns and unknowns as shown in Table 2.

Table 2. *Classification of Knowns and Unknowns Between Controllers and the System.*

		System	
		Known	**Unknown**
Air Traffic Controller	Known	• Known to the controller such as based on 7110.65, local agreements, or training. • Controller might say "I've seen that before," or "I've not seen that before but other controllers have told me about it." • Known to the system as part of design and operations (procedures, automation design documentation, automation expert knows of interaction in software design, airspace layout, airline flight operations information, etc.).	• Known to the controller such as based on 7110.65, local agreements, or training. • Unknown to the system (e.g., pilot not aware of NOTAM change, local operational workaround, unexpected behavior of system or equipment; unexpected outage, information not included in design or training documentation, etc.) • System unknowingly changes or removes information without understanding impacts or reverberations on the controller. • System creates a threat or hazard that the controller has to contend with. • Report states the condition was unknown to management or other system elements. • Error of omission or commission.
	Unknown	• Report states the condition was unknown to the controller. • Unknown to the controller (e.g., unexpected system action or response). • Unexpected surprise to the controller. • Known to the system as part of design (procedures, automation documentation, automation expert knows of interaction in software design, airspace layout, airline flight operations information, etc.). • Not everything known by the system is known by the controller, or known throughout the system.	• Unknown to the controller (e.g., unexpected system action or response). • Unknown to the system (e.g., pilot not aware of NOTAM change, local operational workaround, unexpected behavior of system or equipment; unexpected outage, information not included in design documentation, etc.)

Classification of an ATSAP Positive report as unknown to the controller was based on a lack of information in the report that the controller knew of the issue before encountering it or the report stated the controller did not know about it. Classification of an ATSAP Positive report as unknown to the system was based in part on a lack of information contained in the report that the facility knew of the issue until it was reported by the controller, the facility did not demonstrate awareness until prompted by the report, and avoiding assuming awareness by supervisor and other managers. The classification was based solely on the information contained in the report. Assumptions were avoided about what might have been implied in the reports or how the reports might have been prepared relative to policy, procedure, or process.

A sample of 36 ATSAP Positives reports were used in this study. There was no identifying information about the controller(s) involved with the reports. A pilot test of the method was applied to ten of the reports to assess the viability of the classification table and to develop agreed upon stopping rules for the actual classification by the authors. The reports were separately assessed and classified by the authors working independently. The classifications were then compared and coding differences were resolved by discussion. Final classification agreement was reached for all reports.

Results

Results of the classification are shown in Table 3 with 72% of the ATSAP Positives reports involved safety issues known to the controller. The system was aware of 31% of the issues.

Table 3. *Results of ASTAP Reports Classifications.*

		System		
		Known	Unknown	**Total**
Controller	Known	17%	55%	72%
	Unknown	14%	14%	28%
	Total	31%	69%	

Safety issues known to the controller but unknown to the system included different issues with terminal procedures (e.g., instrument flight procedures used by the terminal controller were refused by pilots because those procedures were not in their flight database, and a new missed approach procedure took aircraft directly into the flow of traffic at another airport) and areas of missing radio coverage.

Safety issues known to the system but unknown to the controller included controllers not being informed of equipment outages and en route controllers not informed about special approach procedures developed for one airline by Jeppesen and not being trained to read and interpret those procedures. An example of an unknown to both the controller and the system involved not fully understanding sector combine/de-combine en route automation so aircraft and data tags would not be seen in the proper sector.

Discussion

The purpose of this study was to develop and test a classification method for assessing safety data based on a pattern of knowns and unknowns between the controller (human operator) and the system (broadly defined). The data set used in this study was a set of safety reports called ATSAP Positives. The method used stopping rules for making classification decisions using information from the reports. Results showed that 17% of the safety issues shown in the ATSAP Positive reports were known to both the controller and the system. Controllers may need to be better informed about how automation works for different operational conditions as reflected by both 28% of the safety issues being unknown to them and findings from the Fort Hill study. The large percentages of safety issues unknown to the system reflect in part the complexity of interdependencies between different parts of the system.

Limitations of this study include that the ATSAP Positives reports used are not current and the procedures, automation, and training are different now so the identified issues and trends have most likely been mitigated, with perhaps new issues and trends emerging. The amount and quality of information found in the ATSAP Positives reports were considered in classification decisions relative to the stopping rules as well as coding reconciliation between the authors.

Further work is needed to assess the approach using a larger sample of reports such as from ASRS having more details. This method is not intended to replace current techniques for detailed safety analysis but rather to understand trends in safety data from a different perspective. Moreover, revealing patterns of unknowns can reveal potential system risks for mitigation.

References

Berry, K.A., & Sawyer, M.W. (2012). "Human Factors Assessment of ATC Handoff Process using ATSAP Reports." Fort Hill Group, LLC. Accessed at: https://www.forthillgroup.com/aggregate-safety-event-analysis

Billings, C.E., Lauber, J.K., Funkhouser, H.E., Lyman, G., and Huff, E.M. (1976). NASA Aviation Safety Reporting System Quarterly Report 76-1. NASA TM-X-3445. Accessed at: https://ntrs.nasa.gov/api/citations/19760026757/downloads/19760026757.pdf

Federal Aviation Administration (2017). Air Traffic Organization Policy: Voluntary Safety Reporting Programs. FAA Order 7200.20.

Federal Aviation Administration, Air Traffic Organization, and the National Air Traffic Controllers Association (2017). Voluntary Safety Reporting Program: Air Traffic Safety Action Program (ATSAP). APTA Rail Conference.

Federal Aviation Administration (2015). Safety Management Systems for Aviation Service Providers. AC No: 120-92B.

National Air Traffic Controllers Association (2018). CFS 2018: Safety Reporting. Accessed at: https://www.natca.org/2018/10/23/cfs-2018-safety-reporting/

National Air Traffic Controllers Association (2016). ATSAP Positives Bulletin: January 2016 through March 2016. Accessed at: https://www.natca.org/wp-content/uploads/2015/11/ATSAP_January2016throughMarch2016.pdf

National Air Traffic Controllers Association (2015). ATSAP Positives Bulletin: October 2015 through December 2015. Accessed at: https://www.natca.org/wp-content/uploads/2015/11/ATSAPOctoberthroughDecember2015.pdf

SYSTEM-THEORETIC REQUIREMENTS DEFINITION FOR HUMAN INTERACTIONS ON FUTURE ROTARY-WING AIRCRAFT

Sam M. Yoo, Andrew N. Kopeikin, Dro J. Gregorian, Adam T. Munekata,
John P. Thomas, Nancy G. Leveson
Massachusetts Institute of Technology
Cambridge, MA

Future rotary-wing aircraft designs are highly complex, optionally manned, and include advanced teaming concepts that create unknown human-automation interaction safety risks. System-Theoretic Process Analysis (STPA) enables analysis of hazards on these complex systems. This paper demonstrates how to apply STPA in future helicopters' early concept development to prevent unacceptable losses. The system is modeled as a hierarchical control structure to capture interactions between components, including human and software controllers. Unsafe control actions are identified from these relationships and are used to systematically derive causal scenarios that arise from both hazardous interactions between system components and component failures. System requirements are then generated to mitigate these scenarios. A subset of the scenarios and requirements that address human factors related concerns are highlighted. Early identification of these problems helps designers (1) refine the concept of operations and control responsibilities and (2) effectively design safety into the system.

Future Rotary-Wing Aircraft (RWA) concepts are highly complex and include technologies such as autonomous flight, optionally manned capability, and cooperative teaming with other Unmanned Aircraft Systems (UAS). Some of the challenges related to developing concrete user requirements for future RWA are well documented in recent literature (Sushereba *et al.*, 2019). The technological complexity that supports future capabilities creates vulnerabilities for unsafe interactions between system controllers, especially in environments where operators perform under stress, high workloads, and face conflicting control authority over systems. A hazard analysis method is required to systematically identify these potential issues early in development so that mitigations can be designed into the system to enforce safety.

The SAE International Aerospace Recommended Practice (ARP) 4761 outlines methods for conducting safety assessments on civil airborne systems, such as the Functional Hazard Assessment (FHA) (SAE, 1996). However, a recent UH-60MU helicopter hazard analysis found that FHA limited its hazards to component failures and omitted humans from the study, except in instances where humans were assumed to mitigate the effects of some failures (Albrecht *et al.*, 2016). Additionally, specific hazards such as "loss of altitude indication in a degraded visual environment" or "loss of internal/external communications" were categorized as marginal in severity. In some cases, these hazards can be far more severe. For example, lost communications were cited as a significant contributor in the 1994 friendly shootdown of two US Army helicopters (Leveson, 2012). Other traditional hazard analysis techniques such as Fault Tree Analysis (FTA) or Failure Modes and Effects Analysis (FMEA) also emphasize failures

(Wasson, 2016). These methods are difficult to apply at the system level and are not recommended for complex human causality analysis (Cabosky, 2020).

The System-Theoretic Process Analysis (STPA) is a relatively new hazard analysis approach that is well suited to effectively handle complex systems like future RWA (Leveson and Thomas, 2018). Unlike traditional hazard analyses, STPA considers interactions between system entities, including software and human controllers. The top-down process begins at a high-level of abstraction and is then refined by iteratively adding design detail. This higher view enables STPA to provide early insights, even at the conceptual design stage, into potential causes of losses not typically discovered until much later in the engineering lifecycle. The results provide a critical opportunity to design safety features in early system development. This paper explains how STPA can be applied to future RWA to provide a top-down approach to hazard analysis. The subset of causal scenarios derived through the analysis highlights some of the human factors related challenges that need to be addressed in the program. The causal scenarios and requirements discussed in this paper represent a small portion of the completed STPA on future RWA.

STPA Applied to Future Rotary Wing Aircraft

STPA defines safety as a control problem rather than a component failure problem. The goal is to identify and design controls that enforce safety constraints uncovered through the analysis. The process systematically follows four steps described in the following subsections. The process can be used to rigorously derive design requirements that ensure the system behavior is safe and that the requirements are end-to-end traceable to the hazards they mitigate.

STPA Step 1: Define the Purpose of the Analysis

The STPA process begins by identifying the system losses unacceptable to the stakeholders (Leveson and Thomas, 2018). Safety is defined as the absence of such losses. In future RWA systems, unacceptable losses include (L-1) loss of life or permanent disabling injury, (L-2) loss or damage to aircraft or equipment, and (L-3) loss of mission. Next, the analysis identifies the system hazards. These are system states that will lead to a loss under a particular set of worst-case environmental conditions (Leveson and Thomas, 2018). Table 1 lists some of the hazards identified for the aircraft and traces each of them to the loss(es) they can lead to. High-level system safety constraints (SC) can be developed to address each of these hazards. For example, SC-1 can be derived as follows with traceability to H-1: *SC-1 the aircraft must remain controllable during all manned/unmanned operations [H-1].* Many more traceable safety constraints with increasing details will be derived as the analysis unfolds.

STPA Step 2: Model the Control Structure

The next step of STPA is to model the hierarchal control structure. The model comprises feedback control loops and captures the relationships between various controllers and processes within the system (Leveson and Thomas, 2018). An effective control structure will enforce constraints on the behavior of the overall system. Each feedback control loop typically consists of five elements: *controllers* (in Figure 1, boxes at the top of each loop), *control actions* (down arrows), *feedback* (up arrows), *other inputs/outputs from components* (side arrows), *and controlled processes* (boxes at the bottom of each loop). Generally, the control structure starts at an abstract level and is iteratively refined to incorporate more system details. For example, the Operator(s) element might be refined into manned, remote, and autonomous configurations.

Table 1. Future rotary-wing aircraft system hazards.

Hazard ID	Hazard Description	Loss Link
H-1	Aircraft is uncontrollable (manned/unmanned)	L-1, L-2, L-3
H-2	Structural integrity of aircraft is violated	L-1, L-2, L-3
H-3	Minimum aircraft separation standards are violated	L-1, L-2, L-3
H-4	Aircraft environment is harmful to human health	L-1
H-5	Aircraft is unable to conduct mission tasks	L-1, L-2, L-3

STPA Step 3: Identify Unsafe Control Actions

The third step of STPA is to identify Unsafe Control Actions (UCAs). A UCA is a control action that will lead to a hazard in a particular context and worst-case environment (Leveson and Thomas, 2018). Each UCA consists of four parts: (1) the controller issuing the control action, (2) the **type** of control action, (3) the control action itself, and (4) the *context* under which it becomes hazardous (see Table 2). Each controller and control action in the control structure is considered. For each control action, there are four *types* of ways that each need to be considered on how a control action could cause a hazard: (1) not providing it, (2) providing it (incorrectly or in the wrong context), (3) providing it too early, too late, or out of order, and (4) providing for too long or short a time. Table 2 illustrates how a subset of the UCAs are developed for future RWA and how traceability is maintained to the hazards they cause.

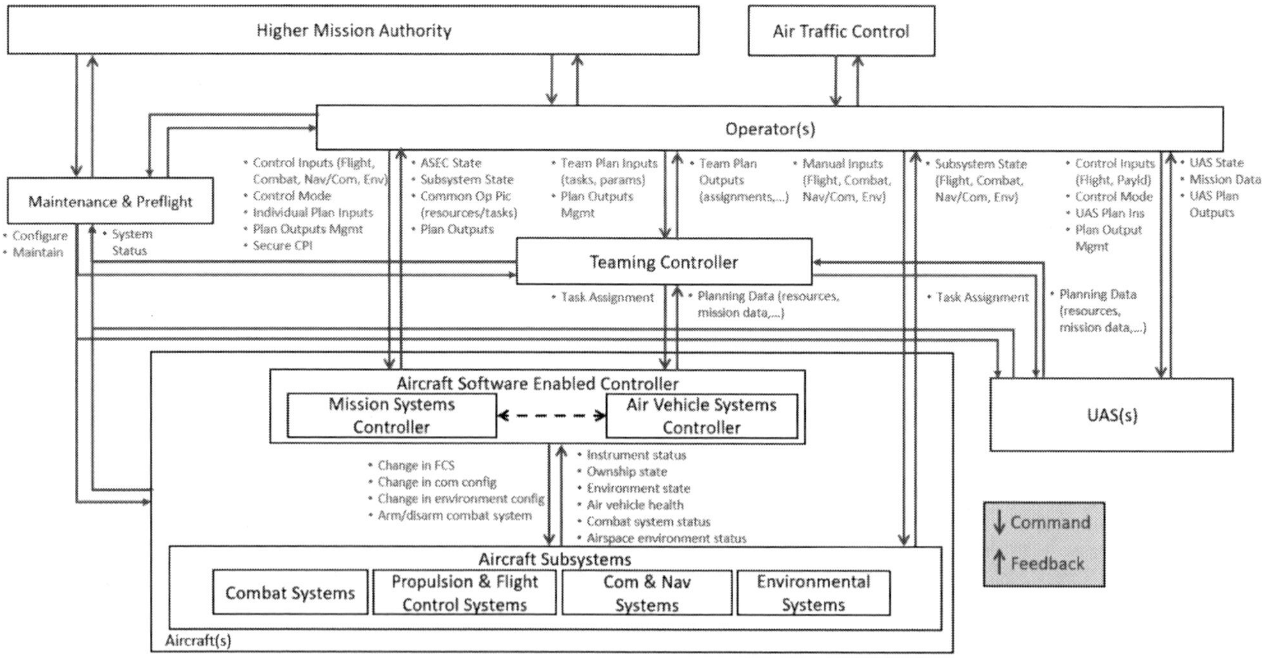

Figure 1. Safety Control Structure of the Future Rotary-Wing Aircraft System.

DISTRIBUTION STATEMENT A: Approved for public release; distribution unlimited.

Table 2. Subset of UCAs for the "Flight Control Inputs" operator control action.

UCA Type	UCA	Hazard Traceability
Not Providing	[UCA-1] Operator **does not provide** flight control inputs *when needed during high power maneuvers (e.g., takeoff, formation, hover, ...)*	H-1, H-2, H-3, H-5
Providing	[UCA-2] Operator **provides** flight control inputs *when aircraft is in autonomous operation*	H-1, H-2, H-3, H-5
Too Early / Late / Wrong Order	[UCA 3] Operator provides flight control inputs **too early** *before the autonomous mode is disengaged*	H-1, H-2, H-3, H-5
Applied too long / Stopped too short	[UCA-4] Operator provides flight control inputs for **too long** *during high power maneuvers*	H-1, H-2, H-3, H-5

STPA Step 4: Identify Causal Scenarios

The final step in STPA is to identify causal scenarios for each UCA. Four potential flaws in the feedback control loop are systematically analyzed for each UCA by exploring reasons why (1) the controller would make an unsafe decision, (2) feedback would be inadequate, (3) the controlled process would not receive the command, or (4) the controlled process behavior would be unsafe despite receiving the command. Interactions between these elements of the feedback control loop and other control entities in the control structure are considered. The following are examples of each of these instances for UCA-1 in Table 2: *Operator does not provide flight control inputs when needed during high power maneuvers (e.g., takeoff, landing, hover, formation flight, ...).* Many more scenarios can be systematically created for this UCA using this method. Traceability is provided back to the UCA for each scenario.

Causal Scenario CS-1: The Operator has adequate feedback that a high-power maneuver is needed. However, confusion regarding the current operational mode of the Aircraft Software Enabled Controller (ASEC) leads the Operator to believe no inputs are necessary and that the ASEC will accomplish the behavior. This mode-confusion may result from maintenance personnel uploading new firmware into the vehicle that alters the modes or a remote operator performing teaming with the RWA and changes the mode remotely to manual flight. [UCA-1]

Causal Scenario CS-2: The Operator does not have adequate feedback that a high-power maneuver is needed. The aircraft is being operated in a degraded visual environment (DVE) enabled by an onboard sensor suite. However, the operator interface is devoted to a separate high workload mission operation, such as teaming with multiple UAS, and does not alert the Operator with sufficient time. [UCA-1]

Causal Scenario CS-3: The Operator does provide flight control inputs when needed for a high-power maneuver. However, the aircraft is being operated remotely, and insufficient communication bandwidth, potentially due to degraded channel capacity, is available to send the command. Or alternatively, the remote Operator inadvertently sends thc command to a different aircraft. [UCA-1]

Causal Scenario CS-4: The ASEC does receive the flight control inputs from the Operator. However, the ASEC detects a different potential trajectory constraint, such as another aircraft, that could violate minimum separation standards. The ASEC overrules the Operator's control inputs, and the command is not issued to the aircraft's power and flight control system. Note that the other aircraft's detection may be caused by a malicious actor spoofing a transponder signal at a given location without a physical aircraft being there. [UCA-1]

Design Requirements

After scenarios have been identified, design requirements can then be generated to prevent those scenarios or UCAs from occurring or to mitigate their impact should they occur. For example, CS-1 may lead to the following design requirements. (R-1) The ASEC must inform the Operator about its control mode [CS-1]. (R-2) The ASEC must inform the Operator of any changes in control mode, actions taken by the ASEC as a result of that change, and the reason for the change [CS-1]. (R-3) The ASEC must be programmed with software consistent with operator tactics techniques and procedures [CS-1]. (R-4) Remote operators must not override onboard operators when they are actively controlling or supervising the aircraft.

The process described in the previous section provides end-to-end traceability between design requirements, scenarios, and back up to the unacceptable losses that should be prevented. The traceability provides an opportunity to document the rationale for each design requirement. The high-level abstraction of the presented analysis leads to the systematic development of high-level requirements in the early development stages. The early insight provides a new and unique opportunity to highlight the design trade-offs. As assessed through the analysis, features with significant risk may be candidates for removal from the architecture. The process can then be iterated by adding refinement in the design's details so that additional requirements can be uncovered. In addition to iterating with STPA, R-1 and R-2 might benefit from a more specific application of related human factors research in presenting critical information to operators at the right time using the Alerting and Reasoning Management System (ALARMS) framework (Saffell *et al.*, 2011). Additional details to R-3 and R-4 would benefit from the lessons learned through DARPA's Aircrew Labor In-Cockpit Automation System (ALIAS) program, as it works with Sikorsky to explore communication protocols between autonomously operated helicopters.

Conclusions

Future RWA will be increasingly complex and will challenge the traditional delineation between software and human controllers' responsibilities. This complexity creates new hazards that need to be identified and addressed early in design. Traditional hazard analysis methods are not capable of addressing complex systems with human interactions such as future RWA. However, STPA is well suited for this problem and is applied in this paper to demonstrate systematic identification of a subset of potential causal loss scenarios that emphasize human factors design elements. System requirements are then derived from the causal loss scenarios to design controls into the system to mitigate those scenarios and enforce safety. The process can be repeated for all control actions identified in the control structure to derive a rich set of safety requirements at this level of abstraction. As design decisions are made throughout the engineering lifecycle, additional detail can be incorporated into the model as refinement. The analysis can then be continued at that level to generate lower-level system requirements.

Acknowledgments

DISTRIBUTION STATEMENT A. Approved for public release; distribution is unlimited. This research is based upon work supported by the U.S. Army Combat Capabilities Development Command Aviation & Missile Center (DEVCOM AvMC) under Air Force Contract No. FA8702-15-D-0001. Any opinions, findings, conclusions or recommendations expressed in this material are those of the authors and do not necessarily reflect the views of DEVCOM AvMC.

References

Abrecht, B., Arterburn, D., Horney, D., Schneider, J., Abel, B., & Leveson, N. (2016). *A new approach to hazard analysis for rotorcraft.* Cambridge, MA: Massachusetts Institute of Technology. Retrieved from: sunnyday.mit.edu/papers/AHS-final.pdf

Cabosky, R. (2020). *A human factors study on vehicle automation*. Cambridge, MA: Massachusetts Institute of Technology.

Leveson, N. (2012). *Engineering a Safer World: Systems Thinking Applied to Safety*. The MIT Press. Retrieved from: mitpress.mit.edu/books/engineering-safer-world

Leveson, N., Thomas J. P. (2018). *STPA Handbook*. Retrieved from: psas.scripts.mit.edu/home/get_file.php?name=STPA_handbook.pdf

Reim, G. (2021). Sikorsky plans to autonomously fly multiple UH-60 Black Hawks in formation in 2021. Flight Global. Retrieved from: www.flightglobal.com/helicopters/sikorsky-plans-to-autonomously-fly-multiple-uh-60-black-hawks-in-formation-in-2021/141128.article

Saffell, T., Alexander, A., Carlin, A., Chang, A., & Schurr, N. (2011). An integrated alerting and notification system utilizing stages of automation and uncertainty visualization. In *16th International Symposium on Aviation Psychology*, 197-202. Retrieved from: corescholar.libraries.wright.edu/isap_2011/82

Society of Automotive Engineers. (1996). ARP 4761 - Guidelines and Methods for Conducting the Safety Assessment Process on a Civil Airborne Systems and Equipment. Retrieved from: www.sae.org/standards/content/arp4761/

Sushereba, C., Militello, L., Ernst, K., Diiulio, J., Roth, E., Scheff, S., Huff IV, W. (2019). Envisioning user requirements for first-of-a-kind future rotorcraft. In *20th International Symposium on Aviation Psychology*, 403-408. Retrieved from: corescholar.libraries.wright.edu/isap_2019/68

Wasson, C. (2016). *System Engineering – Analysis, Design, and Development*. 2nd ed. Wiley

COMPARING HUMAN AND MACHINE LEARNING CLASSIFICATION OF HUMAN FACTORS IN INCIDENT REPORTS FROM AVIATION

Claas Tido Boesser & Florian Jentsch
University of Central Florida
Orlando, FL

Machine learning algorithms have become popular tools for automated classification of text; however, performance of such algorithms varies and depends on several factors. We examined how a subjective labeling process based on a human factors taxonomy can influence human, as well as automated, classification of safety incident reports from aviation. In order to evaluate these challenges, we trained a machine learning classifier on a subset of 17,253 incident reports from the NASA Aviation Safety Reporting System using multi-label classification, and collected labels from six human annotators for a representative subset of 400 incident reports each, resulting in a total of 2,400 individual annotations. Results showed that, in general, reliability of human annotation for the set of incident reports selected in this study was comparatively low. Performance of machine learning annotation followed patterns of human agreement on labels. Suggestions on how to improve the data collection and labeling process are provided.

Continuous advances in computing power, in algorithms, as well as research in the fields of Artificial Intelligence and Machine Learning, have led to an increased application of these tools to Human Factors. What once were laborious tasks that had to be performed by humans are becoming increasingly automated. As such, an increasing number of studies are being conducted that seek to use a variety of computational methods for the analysis of incident reports with text narratives; studies are spanning across several industries, such as aviation, medicine, construction, and the railroad industry, among others. In the field of aviation safety, valuable insight into inflight incidents can be gleaned by examining narratives provided by personnel involved in flight operations that are reported under the condition of confidentiality (e.g., Dekker, 2014; Wiegmann & Shappell, 2003). Using such incident reporting data, researchers have used a variety of techniques, including the usage of topic modeling/data reduction algorithms to identify latent structures in the data, assessing report similarity, automatically labeling and classifying reports, and visualizing the results (e.g., Irwin et al., 2017; Kuhn, 2018; Robinson, 2016; Robinson et al., 2015; Tanguy et al., 2016).

Analyzing and categorizing data such as text narratives presents unique challenges. Along with the sheer volume of available narratives and their text form comes the challenge of extracting trends and information from unstructured data. One way to gain insight and, in turn, reduce the complexity of the data, is through the categorization of such data according to a taxonomy (e.g., Bailey, 1994; Tanguy et al., 2016; Wiegmann & Shappell, 2003). For aviation safety and incident reports, one such implementation is the human factors taxonomy consisting of 12 different labels that is being used in the public self-reporting database of aviation incidents known as the *Aviation Safety Reporting System* (ASRS; see Federal Aviation Administration [FAA], 2011, for a description of the program).

In this study, we compared human and machine learning classification of human factors categories in aviation incident reports from the ASRS database. In the process, we identified the challenges with regards to human and automated annotation, beginning with examining the pertinent characteristics of incident narratives and taxonomies, evaluating different ways of annotating the data, assessing whether some human factors constructs are easier to label reliably than others, all while discussing the implications of what is learned with regards to automatic classification. A main focus of this study was on evaluation of the viability of automated text classification given a subjective classification process. We studied (a) whether human annotators would be reliable and consistent in assigning the same labels to reports, when compared to one another and to the codes given by the experts at ASRS, and (b) whether an automated machine learning classifier could be trained to do this task at better than chance level and/or at a similar performance as human raters. Arguably, if a machine learning classifier does not perform better than chance, or when human annotation of a taxonomy is at the chance level, the reliability of the whole approach is in question.

Method

Using purposeful sampling, six annotators were recruited for this study. Three of the annotators were required to have at least a 4-year undergraduate or master's degree in Human Factors, or an associated discipline such as Psychology. They also had to have commercial flying experience or familiarity with 14 CFR Part 121 Air Carrier operations (we called these the domain plus classification, or "D+C experts"). The three other annotators did not have any formal schooling in Human Factors, but they were required to have commercial flying experience as active or former pilots of 14 CFR Part 121 Air Carrier operations (we called these the "D experts").

The human annotation of the ASRS narratives was followed by a qualitative and quantitative data analysis using machine learning and applying a mixture of statistical analyses from various disciplines in order to evaluate reliability of human annotation, machine learning performance, as well as the resulting interdependencies. In summary, this study consisted of the following steps:

1. Extract data from the ASRS database for the training of a machine learning classifier (17,253 incident reports and their associated human factors labels).
2. Generate a representative subset of the extracted data for the purpose of human annotation (400 incident reports to submit to human annotation).
3. Collect data from annotators including human factors labels for incident reports, confidence measures for selected labels and overall comments, if any.
4. Analyze inter-rater reliability (IRR) measures between the existing labels (referent labels), the D experts, and the D+C experts.
5. Split the 17,253 ASRS reports into a training and a test set using stratified sampling. Extract features. Train machine learning classifiers. Measure performance of machine learning classifiers using 10-fold cross-validation. Compare and contrast performance between different classifiers and between classifiers and human annotation.
6. Evaluate results.

Results

We set out to compare human and machine learning classification of human factors in aviation incident reports. One influences the other—classification is required in order to train a supervised machine learning model. Therefore, we also examined the interaction between human and machine learning classification. Hypotheses were based on some premises, mainly that (a) reliability in human classification is above chance level, (b) reliability depends on annotator and report characteristics, and (c) training a machine learning model can, to some extent, be beneficial for the task of analysis and classification of incident reports. Throughout this study, it became evident that there was considerable variability in the labeling of incident reports. As such, some hypotheses were supported, whereas others were not.

As hypothesized, we found that IRR was dependent on the label. Some labels of the taxonomy were more agreed upon than others, and in fact by a fairly large margin. Figure 1 shows agreement on labels based on Krippendorff's (2004) α.

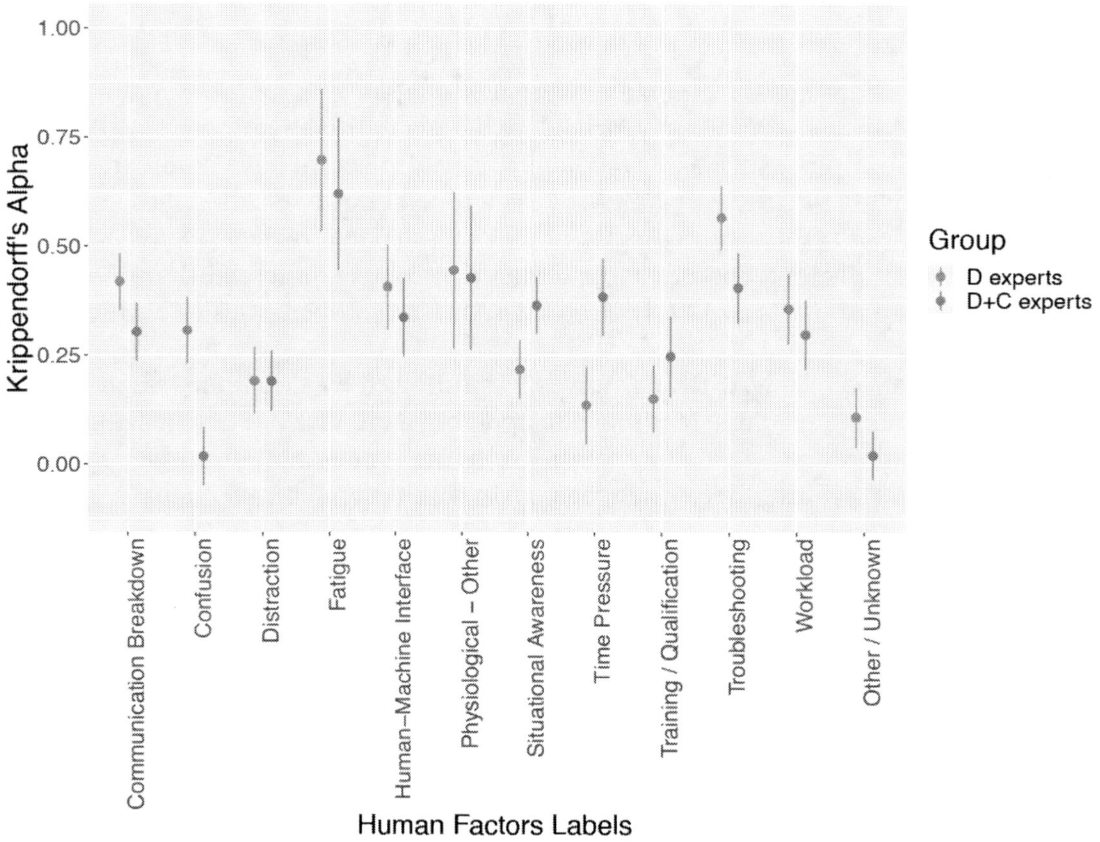

Figure 1. Krippendorff's α by label and group (D experts (left on each pair) vs. D+C experts (right on each pair)).

While, in general, agreement throughout the study seemed to be rather low, this is not necessarily unusual for the coding of raw incident reports. For example, Olsen and Shorrock (2010), as well as Olsen (2011) tested the reliability of the more widely researched HFACS taxonomy—the original HFACS taxonomy in one study and a derivative of it in the other study—with conditions that closely resemble the research herein in the sense that there was no extensive training, and the incident report narratives were presented as the raw narratives to participants (as opposed to coding causal factors that were already abstracted from the reports). In their studies, agreement also highly varied depending on the specific HFACS category, but average percentage agreement at the category level was as low as 34.5% in Olsen and 39.9% in Olsen and Shorrock. This shows that the results presented herein are not necessarily unusually low when similar tasks are considered.

With regards to machine learning performance (see Table 1 for results), we found that, while human agreement and machine learning performance on labels did not exactly correlate with each other, there were some notable trends. For example, *Fatigue*, while not exhibiting a large prevalence in the dataset, stood out as one of the labels that were most agreed upon. *Fatigue* was also most reliably labeled by the machine learning classifier. As the prevalence of *Fatigue* was fairly low (only 5% of the original dataset contained the label), we followed up with a measure of separate agreement on the positive and negative class (see Feinstein & Cicchetti, 1990, as well as Cicchetti & Feinstein, 1990) and found a similar pattern, indicating that annotators were good at discerning when reports included fatigue but also discerning when they did not.

When examining the coefficients for the model, it also was evident that, for the label *Fatigue*, by far the largest predictor of the label was the occurrence of the actual word "fatigue." This poses the question of hand-coding rules versus machine learning. If only a few rules might lead to acceptable performance, why use machine learning to begin with? In fact, Tixier et al. (2016) achieved very good results with hand-coded rules for assigning attributes and outcomes to injury reports. However, they also noted that the process is tedious, labor-intense, heavily based on domain-knowledge and does not scale well to problems outside of the domain for which the rules were coded.

Table 1.
Precision, Recall, and F1-Scores for Individual Labels and Averaged Scores.

Labels	Precision	Recall	F1-Score	Support
Individual				
Fatigue	0.66	0.67	0.66	166
Communication Breakdown	0.60	0.73	0.65	1,267
Situational Awareness	0.62	0.59	0.60	2,017
Troubleshooting	0.44	0.82	0.57	649
Confusion	0.49	0.59	0.54	1,143
Physiological – Other	0.37	0.66	0.48	131
Human-Machine Interface	0.39	0.56	0.46	714
Workload	0.36	0.49	0.41	774
Distraction	0.37	0.45	0.40	903
Time Pressure	0.30	0.50	0.38	560
Training / Qualification	0.27	0.51	0.35	554
Other / Unknown	0.10	0.18	0.13	188

Note. Labels are presented in order of decreasing F1-score.

Discussion

Overall, there are clear challenges to be met in order to improve the annotation process both on the human and the machine learning sides, with one side influencing the other. DiMaggio (2015) wrote about the paradox that task performance of humans and a machine learning classifier often suffers at similar tasks. The research herein to an extent supports this statement. A straightforward categorization of "*Fatigue*", often based on the words, *fatigue*, *fatigued*, *tired*, or *sleep*, was more consistent than for concept labels such as "*Distraction*."

Other challenges that were discovered in the research herein illustrate the complexity of the problem, while also leading to valuable lessons learned. For example, evaluating performance on an imbalanced dataset is not straightforward as regular measures of accuracy are not appropriate for imbalanced data (for an overview, see Weiss, 2013, or Sahu et al., 2017). A similar challenge presented itself for the evaluation of IRR measures. As most IRR measures are sensitive to trait prevalence (e.g., Feinstein & Cicchetti, 1990; Gwet, 2008), imbalance in the data also needed to be accounted for with regards to measures of IRR.

In summary, there is promise in using ML with regards to fairly routine and simple categorizations. On the other hand, a basic ML algorithm, as used in this study, seemed to perform worse at anything that required more context and deeper analysis; but so seemed the humans. With that being said, categorizing narratives in accordance with a human factors taxonomy is an inherently subjective process. This leads to the conclusion that the labels that are provided either by the ASRS experts or by other annotators should always be seen as "a" categorization and not "the" categorization. Finally, recognizing the influence of narrative content as a major source of annotation variability is crucial to improving both the narrative, as well as the annotation. To improve the underlying quality of the reports, it is suggested to investigate, inter alia, automated cognitive aids based on the idea of semi-structured interview processes (see Crandall et al., 2006, as well as Wiegmann & von Thaden, 2003 for related ideas). For people involved in the creation and maintenance of incident databases, working together closely with human factors practitioners, as well as leveraging knowledge of the field of computer science should help to greatly improve incident reporting systems.

Acknowledgments

Participant payment for this study was partially funded by a 2018 *Human Factors and Ergonomics Society* Training Technical Group Student Grant Award. The paper is based on the doctoral dissertation of the primary author. The views of the research report reflect the views of the authors and not the views of the employers, the granting organization, or the University of Central Florida.

References

Bailey, K. D. (1994). *Typologies and taxonomies: An introduction to classification techniques*. Sage Publications.

Cicchetti, D. V., & Feinstein, A. R. (1990). High agreement but low kappa: II. Resolving the paradoxes. *Journal of Clinical Epidemiology, 43*(6), 551–558.

Crandall, B., Klein, G., & Hoffman, R. R. (2006). *Working minds: A practitioner's guide to cognitive task analysis*. The MIT Press.

Dekker, S. (2014). *The field guide to understanding "human error"* (3rd ed.). Ashgate Publishing Company.

DiMaggio, P. (2015). Adapting computational text analysis to social science (and vice versa). *Big Data & Society*, *2*(2), 1–5.

Federal Aviation Administration [FAA]. (2011). AC 00-46E, Aviation safety reporting program.

Feinstein, A. R., & Cicchetti, D. V. (1990). High agreement but low kappa: I. The problems of two paradoxes. *Journal of Clinical Epidemiology*, *43*(6), 543–549.

Gwet, K. L. (2008). Computing inter-rater reliability and its variance in the presence of high agreement. *British Journal of Mathematical and Statistical Psychology*, *61*(1), 29–48.

Irwin, W. J., Robinson, S. D., & Belt, S. M. (2017). Visualization of large-scale narrative data describing human error. *Human Factors, 59*(4), 520–534.

Krippendorff, K. (2004). *Content analysis: An introduction to its methodology* (2nd ed.). Sage Publications.

Kuhn, K. D. (2018). Using structural topic modeling to identify latent topics and trends in aviation incident reports. *Transportation Research Part C: Emerging Technologies*, *87*, 105–122.

Olsen, N. S. (2011). Coding ATC incident data using HFACS: Inter-coder consensus. *Safety Science, 49*, 1365–1370.

Olsen, N. S., & Shorrock, S. T. (2010). Evaluation of the HFACS-ADF safety classification system: Inter-coder consensus and intra-coder consistency. *Accident Analysis and Prevention, 42*, 437–444.

Robinson, S. D., Irwin, W. J., Kelly, T. K., & Wu, X. O. (2015). Application of machine learning to mapping primary causal factors in self reported safety narratives. *Safety Science, 75*, 118–129. https://doi.org/10.1016/j.ssci.2015.02.003

Robinson, S. D. (2016). Visual representation of safety narratives. *Safety Science*, *88*, 123–128.

Sahu, M., Mukhopadhyay, A., Szengel, A., & Zachow, S. (2017). Addressing multi-label imbalance problem of surgical tool detection using CNN. *International Journal of Computer Assisted Radiology and Surgery, 12*(6), 1013–1020.

Tanguy, L., Tulechki, N., Urieli, A., Hermann, E., & Raynal, C. (2016). Natural language processing for aviation safety reports: From classification to interactive analysis. *Computers in Industry*, *78*, 80–95.

Tixier, A. J. P., Hallowell, M. R., Rajagopalan, B., & Bowman, D. (2016). Automated content analysis for construction safety: A natural language processing system to extract precursors and outcomes from unstructured injury reports. *Automation in Construction*, *62*, 45–56.

Weiss, G. M. (2013). Foundations of imbalanced learning. In H. He, & Y. Ma (Eds.), *Imbalanced Learning*. IEEE Press.

Wiegmann, D. A., & Shappell, S. (2003). *A human error approach to aviation accident analysis: The Human Factors Analysis and Classification System*. Burlington, Ashgate.

Wiegmann, D. A., & von Thaden, T. L. (2003). Using schematic aids to improve recall in incident reporting: the critical event reporting tool (CERT). *The International Journal of Aviation Psychology*, *13*(2), 153-171.

EXAMINING THE CHANGING ROLES AND RESPONSIBILITIES OF HUMANS IN ENVISIONED FUTURE IN-TIME AVIATION SAFETY MANAGEMENT SYSTEMS

Lawrence Prinzel
Kyle Ellis
John Koelling
NASA Langley Research Center
Hampton, VA

Paul Krois
Crown Consulting, Inc.
Aurora, CO

Misty Davies
Robert Mah
NASA Ames Research Center
Moffett Field, CA

Advances in technology are enabling new concepts of operations that will transform aviation including increasingly autonomous capabilities to handle evolving complex dynamic ecosystems like those associated with Advanced Aerial Mobility. A major challenge is how to ensure today's safety levels are maintained as the system scales for rapid detection and timely mitigation of safety issues. NASA has developed a concept of operation for In-Time Aviation Safety Management Systems (IASMS) that represents a system-of-system perspective on interconnected capabilities needed to proactively reduce risk in complex operational environments where unknown hazards may exist. As a result, NASA research priorities include understanding how the balance between humans and automation changes in such envisioned systems, which may lead to novel human-machine interaction paradigms and human-autonomy teaming for informed contingency management.

Advances in technology are enabling new concepts of operation that will transform aviation. The innovations for the future air transportation system will span increasingly autonomous capabilities to handle very complex, dynamic ecosystems comprised of a widening mix of vehicles and technologies, urban air mobility, and unmanned and traditional operations. The purpose of this paper is to discuss key considerations for the roles of human operators in the design of an In-time Aviation Safety Management System (IASMS) concept of operations (ConOps) to account for these innovations and their implications (Ellis et al., 2021).

The temporal parameter of in-time safety means quickly managing known operational risks in real- or near-real-time, quickly identifying unknown risks to be managed, and quickly informing system design as risk patterns are identified. As explained by the National Academies, "real time" pertains to events that occur at the same time or near real time, whereas other parts of a Safety Management System (SMS) operate over a longer period of time for identifying trends that cannot be identified in real time (2018), i.e., real-time safety assessment is but one aspect of a broader IASMS.

Future Vision of Emerging Aviation Domains

Demand for high-speed mobility, transformative advances in autonomous capabilities and emerging aviation sectors are enabling new future concepts. Advanced aerial mobility (AAM) is a concept of operation enabled by envisioned technological innovations imagined leading to on-demand, passenger carrying air taxi, small package delivery, autonomous cargo delivery, and emergency and disaster response (Patterson, 2021).

The FAA's Vision 2035 concept poses that in the future the vehicle and its missions will drive the services required and tailored for flight performance characteristics (MITRE, 2020). The Vision poses basic principles including that human-machine teaming with smart systems is pervasive.

Scalability for Autonomous Systems and Operations

A major challenge is how to ensure today's aviation safety levels are maintained as the system scales in volume and complexities (Holbrook et.al., 2020; Pritchett et.al.,2018; Shively, et.al., 2018). Another major challenge considers that as autonomy takes on increasing responsibilities, humans and machines will be required to work together in new and different ways. AAM's path forward is through design of responsible autonomy that embraces innovation while respecting its safety tradition. The digital transformation to assimilate AAM includes use of satellite, cell, and web capabilities for surveillance, required navigation performance, digital communications, machine learning and artificial intelligence for flight management replacing automated decision support tools, and prognostic safety assurance.

In-Time Aviation Safety Management System

As new entrants transition into the airspace system, maintaining safety will require more proactive risk mitigation of emerging safety issues before they become hazards. The IASMS concept of operations goes beyond today's SMS by addressing the design of new in-time safety systems and services, enhanced tools and technologies, increased access to data and data fusion, improved integrated data analytics, enhanced in-time risk monitoring and detection, hazard prioritization and mitigation, and safety assurance decision support.

The IASMS represents a system-of-system architecture of services, functions and capabilities (SFCs) for vehicle, airspace, and operators. SFCs monitor conditions, assesses data, and perform or inform an in-time mitigation action. A set of coordinated and collaborative in-time safety assurances together make up an in-time aviation safety management system. At the vehicle level, example SFCs include detect-and-avoid and contingency management. Example SFCs for the operator include people below flights as third-party risk, weather risk, and safety reports. Example SFCs for airspace include airspace conformance, constraints, and traffic dynamic density. Example SFCs at the infrastructure level include safety data repositories and vehicle post-flight operations data reduction and analysis.

Transitioning to increasingly complex AAM operations necessitates new roles for human operators. A possible evolution of AAM operations, shown in Figure 1, poses transitioning from today's simplified pilot operations to a future when the human remotely manages and later remotely supervises operations. Across AAM epochs, SFCs will improve situation awareness with vehicle-to-vehicle and vehicle-to-infrastructure communications across vehicles, systems, Unmanned Aircraft Systems (UAS) Service Suppliers (USS), and Supplemental Data Service Providers (SDSPs). Vehicles will exchange information about their current state and planned trajectory, and users will have access to information about the operating environment.

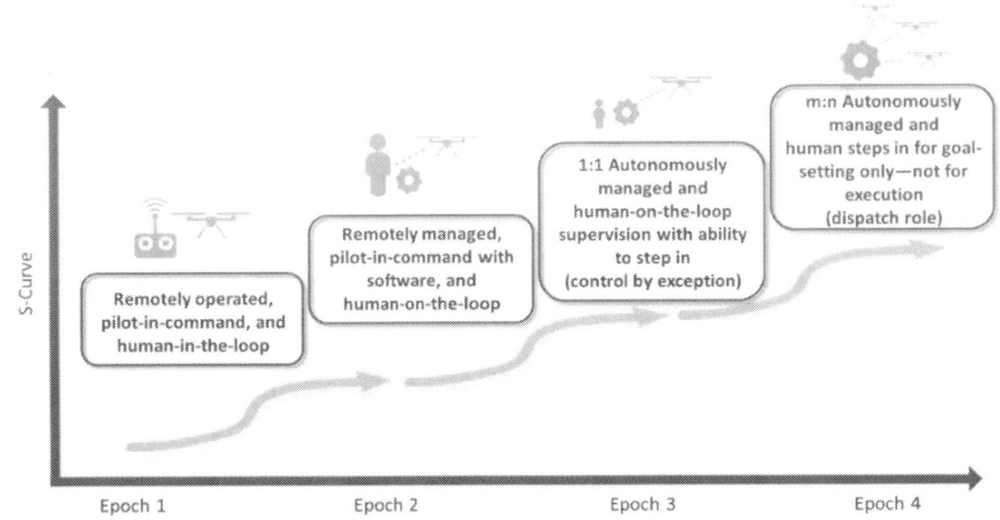

Figure 1. Evolution of AAM Operations and Safety.

Envisioned New Human Roles

AAM and the future National Airspace System (NAS) pose new roles for humans with shared responsibility for safety assured through continuous, real-time monitoring of operations and emergent risks. A key challenge is that today certification and safety assurance assume pilots and air traffic controllers are in-the-loop for operational safety. To realize AAM, innovations are needed to include paradigm changes in designing human-system architectures and considerations for safety assurance. Moving from simplified vehicle handling qualities to properly designed, resilient and capable automated systems may require phased implementation of autonomy that moves safely from today's highly capable pilot, to a pilot operating a simpler vehicle, to a passenger who can act during an emergency.

Defining human roles flows from the functions they perform. Sheridan and Parasuraman identified five human functions for supervisory control including planning off-line, monitoring the automation's execution of the plan, and intervening to abort or assume control as necessary (2005). Just as the IASMS uses the functions of monitor, assess, and mitigate to address safety risk, the human operator can be conceived to perceive, assess, and act on risk based on the human information processing model by Wickens (1992).

Automated systems capable of complete autonomy do not yet exist for aviation and future designs should avoid repeating problems and errors of the past. The Performance-Based Aviation Rulemaking Committee (PARC) and the Commercial Aviation Safety Team (CAST) examined modern flight deck systems for flight path management (2013). Automation concerns included that pilots sometimes rely too much on automated systems and may be reluctant to intervene and auto flight mode confusion errors continue to occur.

Managing safety risk across varied operations and services will require deep human-machine teaming to best use the increased volumes of information being exchanged (MITRE, 2020). These smart systems will analyze a situation and make decisions based on the available data in a predictive or adaptive manner. For instance, flight deck-based capabilities like Aircraft Health Monitoring Systems will proactively detect and mitigate undesirable aircraft states such as high energy on approach before they result in safety incidents.

Challenges and Mitigations in Human-Automation Teaming

Understanding the balance between humans and automation and how they seamlessly work together in increasingly complex operations may lead to novel paradigms for human-autonomy teaming (HAT) including contingency management. Part of the challenges in human-automation teaming resides with the capabilities and limitations of human operators assessed through cognitive models, and another part corresponds to the automation and the SFCs the operator relies upon. The in-time safety assurance SFCs were developed with these considerations in mind.

The key challenge for HAT is that the key critical enabler does not yet exist. It is "as envisioned" as shown in Table 1. The Monitor-Assess-Mitigate numbers signify notional increases in capability. This presents a challenge in identifying, developing, and implementing SFCs to provide for in-time safety assurance. This presents an opportunity for clean-slate design to possibly enable collaborative human-machine partnerships.

Table 1. *Maturation Levels of Human-System Interaction.*

Maturity Level	Maturity Description	Monitor	Assess	Mitigate
4	Fully Autonomous Functionality	3	3	3
3	Autonomous Functionality with Human Over-the-Loop	3	3	2
2	Automated Function with Human Fallback (On-the-Loop)	2	2	1
1	Alerting Function for Human	1	1	0

Endsley provided a human-autonomy system oversight model integrating past research (2018). This notional model integrates central design decisions with considerations for adaptive automation, granularity of control, key automation interface features such as information presentation and salience, attention allocation, task demands and competing tasks, automation robustness and reliability, and operator trust based on automation robustness and reliability.

Research has demonstrated the challenge when humans monitor and take over automated systems including when they do not fully understand these systems (Smith & Baumann, 2019). Potential negative effects from automation included reliance on automation as the primary agent to detect problems; reduced attentiveness to deal with automation degradation and failures; fatigue with false alarms; and degradations of skill, situation awareness and teamwork. Strategies for mitigating these effects included using data analytics to learn from past performance, using contingency procedures, designing technology that keeps the user informed and engaged, ensuring consistency across automated systems, and designing displays so the user can view the same data that the automation is using. Mitigations regarding alarm design included designing an interface so status indicators are salient, and information is easy to access. They cautioned that brittle systems can be due to scenarios that occur beyond what was anticipated for the automation to handle, gaps in the model used by automation, and unanticipated emergent behaviors resulting from mismatching interactions between multiple automated systems.

At least during earlier phases of AAM evolution human operators will have a supervisory role such as on-the-loop or over-the-loop. Vehicle SFCs will automatically execute the flight plan, provide health monitoring checks and alerts, and react quickly to trajectory and other deviations. The vehicle would make local decisions about the safety of flight such as loss of C2 link. Automation will develop the flight plan to marry expected flight time with battery life and airspace constraints. The human operator will validate the flight plan prior to departure, make strategic decisions to mitigate risks before departure, quickly respond to alerts and diagnose other exceptions, and re-route the flight because of new airspace constraints.

With AAM the balance between humans and automation goes beyond traditional function allocation. The human operator must remain vigilant when not in-the-loop to possible exceptions that can occur to ensure the automation responds appropriately. Handling an exception requires both sufficient time to intervene and a deep understanding of the system to avoid errors of omission and commission that could prevent mitigation of a risk or making it worse. Training on system design and operations requires a high level of proficiency, but without continual use those skills can degrade. Integrating cognitive modeling for the operator with machine learning and artificial intelligence poses further challenges for HAT. SFCs will provide data exchange as part of in-time risk management between flight operators and service providers.

Future challenges in HAT include understanding the information requirements for human operators and how those change with increasingly complex levels of autonomy and contingency management. Mathematical modeling of cognitive architectures is an approach to examine a hierarchy of autonomous systems (Bhattacharyya, 2015). Information requirements can be envisioned through questions about what the automation is currently doing compared to what it should be doing for particular operations. For example, NASA found in early UTM tests that missing details made it more difficult for crews to establish situation awareness and when more detailed information was available, it was needed more quickly. Results indicated more complex environments required more information to be presented to the pilot but there were too many messages to read. Some terminology was hard to interpret, different units of measurement required crews to interpret, and more time was needed for making complex contingency decisions. Cardosi and Lennertz examined human factors issues with trajectory-based operations and recommended that an assessment of functions performed by the pilot should include what information is needed, when it is needed, and in what format it should be presented (2020).

Summary

HAT offers an important design and operational safety perspective to the IASMS concept. Teaming approaches will scale as the architecture of SFCs, use of inter-dependent automated systems, and operational environments evolve toward greater complexity. This multi-dimensional space for design of an IASMS has implications for changing roles and responsibilities for human operators and fewer skilled operators. The Monitor-Assess-Mitigate functions can inform design decisions about what information the human operators should monitor, when they need to make assessments, and how they need to intervene.

References

Bhattacharyya, S., Davis, J.A., Vogl, T., and Smith-Velazques, L. (2015). Enhancing Autonomy with Trusted Cognitive Modeling. Association for Unmanned Vehicle Systems International.

Cardosi, K., and Lennertz, T. (2020). Flight Deck Human Factors Issues for Trajectory Based Operations. DOT-VNTSC-FAA-20-07. Cambridge, MA: John A. Volpe National Transportation Systems Center.

Ellis, K., Krois, P., Koelling, J., Prinzel, L., Davies, M., and Mah, R. (2021). "A Concept of Operations (ConOps) and Design Considerations for an In-time Aviation Safety Management System (IASMS) for Advanced Air Mobility (AAM)," AIAA Sci Tech Conference, 2021.

Endsley, M.R. Level of Automation Forms a Key Aspect of Autonomy Design (2018). *Journal of Cognitive Engineering and Decision Making*, 2018, Volume 12, Number 1, March 2018, pp. 29–34. doi: 10.1177/1555343417723432.

Holbrook, J.B., Prinzell, L.J., Chancey, E.T., Shively, R.J., Feary, M.S., Dao, Q.V., Ballin, M.G., and Teubert, C. (2020). Enabling Urban Air Mobility: Human-Autonomy Teaming Research Challenges and Recommendations. AIAA Aviation.

National Academies of Sciences, Engineering, and Medicine. (2018). In-time Aviation Safety Management: Challenges and Research for an Evolving Aviation System. Washington, DC: The National Academies Press. Retrieved from https://doi.org/10.17226/24962.

MITRE CORPORATION (2020). "The Future of Aerospace: Interconnected from Surface to Space." FAA Managers Association Managing the Skies. Retrieved from www.faama.org

PARC/CAST Flight Deck Automation Working Group (2013). Operational Use of Flight Path Management Systems. Commercial Aviation Safety Team (CAST)/Performance-Based Aviation Rulemaking Committee.

Patterson, M.D., Isaacson, D.R., Mendonca, N.L., Neogi, N.A., Goodrich, K.H., Metcalfe, M., Bastedo, B., Metts, C., Hill, B.P., DeCarme, D., Griffin, C., and Wiggens, S. (2021). "An Initial Concept for Intermediate-State, Passenger-Carrying Urban Air Mobility Operations," AIAA SciTech.

Pritchett, A., Portman, M., and Nolan, T. (2018). "Research & Technology Development for Human-Autonomy Teaming–Final Report: Literature Review," NASA Contract Number NNL16AA07C.

Sheridan, T., and Parasuraman, R. (2005). Human-Automation Interaction. *In HFES Reviews of Human Factors and Ergonomics.* Retrieved from https://doi.org/10.1518/155723405783703082

Shively, R.J., Lachter, J., Brandt, S.L., Matessa, M., Battiste, V., and Johnson, W.W. (2018). "Why human-autonomy teaming?" in Advances in Neuroergonomics and Cognitive Engineering, Switzerland: Springer, pp. 3-11.

Smith, P.J. and Baumann, E. (2020). Human-automation teaming: Unintended consequences of automation on user performance. *Proceedings of the 2020 Digital Avionics Systems Conference (DASC).*

Wickens, C. D. (1992). Engineering psychology and human performance (2nd ed.). Harper Collins Publishers.

MINIMIZING THE NEGATIVE IMPACTS OF AIRPORT CONSTRUCTION

Sarah Hubbard
Bryan Hubbard
Joseph Sobieralski
Purdue University
West Lafayette, Indiana

Airfield infrastructure projects are critical to ensure facilities are safe, in good condition and meet current standards. However, these airfield construction and reconstruction activities are usually conducted on an active airfield, which impose operational and human factors challenges for all users, including pilots, air traffic controllers, airport operations personnel, construction workers, and emergency responders. FAA recognizes the potential safety challenges, and provides supporting guidance and regulation as described in AC 150/5370-2G, Operational Safety on Airports During Construction. While this guidance is valuable and enhances safety, there remain human factors issues that are worthy of investigation and discussion.

Introduction

Airport infrastructure is critical to ensure mobility and safety for passengers and cargo in the US and worldwide. To ensure adequate infrastructure, capacity, current standards and condition, airfield construction (including reconstruction and maintenance) is vitally important. The National Plan for Integrated Airport Systems (NPIAS) identifies airport development that is needed and includes $34.3 billion dollars for projects related to reconstruction, standards, safety and capacity for 2021 through 2025 (2020). These allocations demonstrate the ongoing need for airfield construction related projects to ensure the integrity and reliability of our aviation system. Airfield construction is critical to our aviation system, however, it can create operational challenges for stakeholders, since aeronautical activities typically need to continue throughout construction. This paper presents a literature review regarding the impacts of construction, data related to airfield safety, and a discussion of the human factors considerations and mitigation measures that may be appropriate.

Literature Review

FAA recognizes the potential safety challenges, and provides supporting guidance and regulation as described in AC 150/5370-2G, Operational Safety on Airports During Construction (2017). This document provides information to support the development of a plan for safety throughout each phase of construction (referred to as Construction Safety and Phasing Plans, CSPP), checklists for daily inspections for airport operations personnel, examples of operational issues that my result from construction activities, and signs and barricades to identify the construction area.

There is limited information in the literature regarding safety during airport construction activities. There are a few publications related to construction safety at Denver International Airport, where there were 2,843 construction contracts and 4,634 injuries and illnesses (Glazner et al, 2005). These studies emphasize the significance of injuries for construction workers during airport construction activities. Despite this fact, these findings have limited applicability to most airport construction since this reflects construction at a new airport site rather than construction at an active airport.

Airfield construction may have operational and safety considerations that affect numerous airport stakeholders, including pilots, Air Traffic Control (ATC), airport operations, tenants, flight training, and emergency response. There are numerous characteristics of construction that have an impact, including the number of personnel, the kind of material and equipment being used, the nature of the construction activities, and the location of the construction site, material storage location, and access points. These characteristics may change throughout the project, and will affect the operations, safety, security (Khalafallah & El-Rayes, 2008) and cost. Activities not only affect the stakeholders, but also affect airport hazards such as wildlife (Khalafallah, & El-Rayes, 2006) and foreign object debris (FOD) (Khalafallah, & El-Rayes, 2006). Other issues that have been mentioned in the literature related to airside construction include security escort requirements, night work, short closures, segmenting of work, provision of barricades and fencing, maintaining operational surfaces free of FOD, maintaining operational surface zoning requirements, protecting workers from jet blast, the need for flexibility to adapt to changing circumstances, unusual weather and labor disputes (Stewart, 2001). Other considerations mentioned in the literature include construction contracts (Stewart, 2001), the benefits of partnering to reduce claims and improve schedules (Mollaoglu, et al, 2021), and the importance of communication and well defined roles and responsibilities (Stewart, 2001).

The limited analysis and publication regarding the safety impacts due to airport construction contrasts with other sectors, such as the roadway sector, where there have been numerous studies of the costs, risks, and characteristics associated with crashes in work zones (e.g., Saha, 2020; Schrock et al, 2014; Chen and Tarko, 2012; Li and Bi, 2009).

Results and Discussion

One way to assess the impact of airside construction activities is to investigate the incidence and cause of runway incursions when there are airside construction activities. Analysis of the FAA Runway Incursion Database indicates that there were 612 runway incursions that had "construction" in the narrative from 2001 to 2020. The associated incident type for construction related runway incursions and all runway incursions are shown in Figure 1. For construction events, 46% are vehicle or pedestrian deviations (VPD), 33% are operation error (OE, caused by ATC) and 20% are pilot deviation (PD). VPD and OE are much more likely for events with "construction" in the narrative than for all runway incursion events, which are dominated by pilot deviations (60%). This suggests that while we need to maintain the strong focus on ensuring operational safety for aircraft, there may be a need to provide additional consideration to the impact of runway construction on ATC, construction, and airport operations.

Additional information about the construction related runway incursions is shown in Table 1. Fortunately, severe runway incursions (A and B), are a rare event and represent only 1.4% of all construction runway incursions. Most of the runway incursions pose no risk of collision, with 15% Type C and 32% Type D incursions; approximately half of the construction runway incursions did not have a designated severity. Of the 612 construction runway incursions, 266 indicated a vehicle and 32 indicated a pedestrian (in the aircraft flight code columns); this suggests that vehicles may be a greater concern than pedestrians during airfield construction activities.

Figure 2 illustrates a barricade used to designate the construction area. The airfield construction barricade and construction signs are orange, which is consistent with the colors used in the roadway sector for signs and barricades, which provides consistency and reinforces cues associated with information presentation, which enhances performance through effective and consistent design, including colors. The low barricades (an evolution from railroad ties) provide a visual cue but do not present a hazard to aircraft.

Examples of Potential Impacts and Increased Risk Due to Airfield Construction

There are numerous ways to frame a discussion of human factors. One traditional framework for human factors in aviation is the ICAO SHELL model. The name is derived from the components Software, Hardware, Environment, and Liveware (International Civil Aviation Organization, 2012). This is a useful framework for the analysis of a single activity that is focused on a single unit or person (the central liveware).

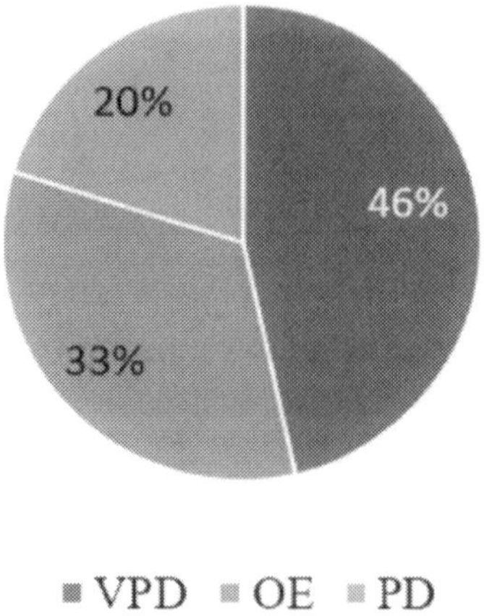

"Construction" in narrative (n = 612)

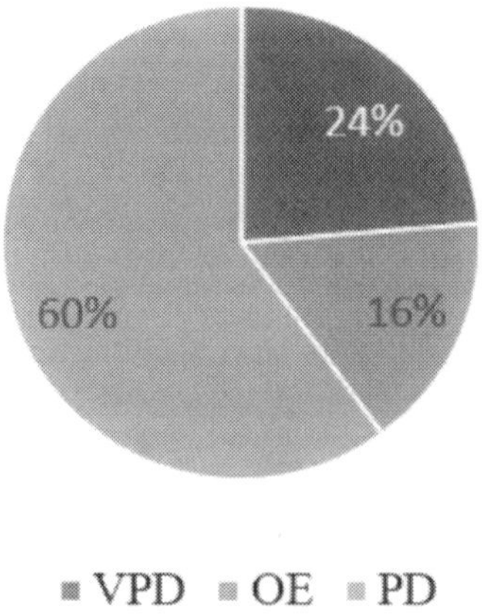

All events (n = 25,584)

Figure 1. Distribution of Incident Type for Runway Incursions from 2001 to 2020.

Table 1.
Characteristics of Runway Incursion Events with "Construction" in Narrative.

Severity	Total Number	Percent of all Events*	PD	OE	VPD
A	4	0.6%	100%	0%	0%
B	5	0.8%	20%	20%	60%
C	94	15.4%	38%	20%	41%
D	197	32.2%	28%	6%	66%
All events	612		20%	33%	46%

**Note.* All events is greater than the sum of A, B, C and D since many events did not indicate a severity rating. (Source: FAA Runway Incursion Database, https://www.asias.faa.gov/apex/f?p=100:28:::NO)

Figure 2. Low profile construction barricades provide a visual cue but may not prevent passage of pedestrians or vehicles (Source: OTW Safety, 2020).

Airfield construction is much more complex, with many people, activities, and organizations involved. Mapping out each of the required activities in the context of the SHELL model would be very challenging and may not support a comprehensive context for the wide variety of ongoing activities.

Another context for a discussion of the challenges associated with airfield construction is to consider the human factors areas as defined by FAA (2012). In this context, the impact of airfield construction may be considered both in general and as it may affect different users as shown in Table 2. The human factors focus areas related to the environment, error, situational awareness, workload, and staffing may be especially relevant for many affected users. Work space and safety and health are most relevant for constructors. An examination of these areas in the context of airfield construction suggest that some areas may be more relevant than others, especially considered in the context of standard practices, which reflect the fact that construction activities are of a limited duration at many airports. Example implications are provided in Table 2, and may be positive (+), negative (-) or neutral (o)in terms of the expected impact. Although not shown in Table 2, the human factors areas of documentation, training, and information are all supported by the development of the Construction Safety and Phasing Plan (CSPP).

Conclusion and Recommendations

The potential impacts of airfield construction are significant and the limited amount of relevant literature indicates that this may be an area that warrants further study. One of the challenges is access to relevant data, however, it may be possible to investigate the topic using case studies, considering data published by OSHA, through the use of the narratives associated with runway incursions, investigation of aircraft incidents and accidents, and development of a construction database by FAA. A better understanding of the most important issues related to airfield construction may provide insights that will translate to other airfield activities, including airport operations activities and construction activities in other sectors, such as the roadway sector.

While timing construction activities to occur when aeronautical activity is lower may be one possible strategy, other scheduling and contracting approaches are recommended for future investigation. Potential approaches to consider include accelerated construction schedules, and incentives for early completion of construction work, an approach that is commonly used in other sectors.

Table 2.
Human Factors Areas, Examples and Affected Users for Airfield Construction

Human Factors Area	Example	Affected Users: Pilots	ATC	Airport Ops	Constructors	Emergency Response
Environment	- Greater safety risks associated with construction at night or during low visibility conditions o Conducting construction activities at night may reduce impacts on and by aircraft operations, but may introduce additional hazards due to darkness	X	X	X	X	X
Workload	- Increased workload for pilots and emergency response due to changes associated with construction (e.g., different paths and routes) - Increased workload for controllers due to visual clutter associated with construction - Increased workload for airport ops due to additional inspection requirements - Increased workload for construction workers due to additional risks and distractions in airfield environment	X	X	X	X	X
Human Error	- Increased workload (and associated fatigue) may increase human error - Numerous NOTAMS at many airports may reduce the effectiveness of construction related NOTAMS for pilots	X	X	X	X	
Staffing	- Ops workers are often required to conduct additional construction inspections and other duties although additional staffing is usually not provided except at the largest airports			X		
Situational Awareness	+ Enhanced by visual cues such as signs and barricades + Enhanced by automated runway incursion warning systems	X	X	X	X	
Work Space	- Construction workers are in constrained environment - Space constraints affect material storage area, which may introduce additional risks associated with requirements for material movement o Although pilots may be required to land in a constrained space (e.g., a shorter runway), minimums ensure that the runway length is adequate				X	
Safety and Health	+ Construction workers use required PPE (e.g., safety vests and hearing protection) + Barricades around construction area with barricades supports worker health and safety.				X	
Information Presentation	- Changing taxiway nomenclature during construction violates consistency for pilots and air traffic control. + Use of standard construction signs and markings on all airfields	X				
Procedures	+ Construction inspection procedures support airport ops + Escort procedures help ensure safety for construction contractors			X		

References

Chen, E., & Tarko, A. P. (2012). Analysis of crash frequency in work zones with focus on police enforcement. *Transportation Research Record*, *2280*(1), 127-134.

Endsley, M. R., & Garland, D. I. (2000). Pilot situation awareness tralnlng in general aviation. *Proceedings of the Human Factors and Ergonomics Society Annual Meeting*, *44*(11), 357–360. https://doi.org/10.1177/154193120004401107

Federal Aviation Administrain (FAA) (2020, September). National Plan of Integrated Airport Systems 2021-2025. https://www.faa.gov/airports/planning_capacity/npias/current/media/NPIAS-2021-2025-Narrative.pdf

FAA (2017, December 13). Operational Safety on Airports During Construction. AC 150/5370-2G. https://www.faa.gov/documentLibrary/media/Advisory_Circular/150-5370-2G.pdf.

FAA (2012, June). Human Factors Acquisition Job Aid.

Federal Aviation Administrain (FAA). (n.d.). RWS search form. http://www.asias.faa.gov/pls/apex/f?p=100:28:0::NO:28::

Glazner, J., Bondy, J., Lezotte, D. C., Lipscomb, H., & Guarini, K. (2005). Factors contributing to construction injury at Denver International Airport. *American Journal of Industrial Medicine*, *47*(1), 27-36.

International Civil Aviation Organization (2012). *Safety Management Manual (SMM)*. Doc 9859. Third Edition.

Li, Y., & Bai, Y. (2009). Highway work zone risk factors and their impact on crash severity. *Journal of Transportation engineering*, *135*(10), 694-701.

Kennedy, J. (2016). Dublin Airport's runways are now on Google Street View. Retrieved February 11, 2017, from https://www.siliconrepublic.com/play/dublin-airport-street-view

Khalafallah, A., & El-Rayes, K. (2008). Minimizing construction-related security risks during airport expansion projects. *Journal of Construction Engineering and Management*, *134*(1), 40-48.

Khalafallah, A., & El-Rayes, K. (2006). Minimizing construction-related hazards in airport expansion projects. *Journal of Construction Engineering and Management*, *132*(6), 562-572.

Mollaoglu, S., Sparkling, A. E., Garcia, A., & Polkinghorn, B. D. (2021). Collaborative Partnering for Airport Construction Projects: State-of-Practice. *Journal of Management in Engineering*, *37*(3), 04021003.

OTW Safety (2020). The invention of Low-Profile Airport Barricades. https://otwsafety.com/blog/low-profile-airport-barricades-invention/ .

Saha, P. (2020). An optimization model to determine an appropriate budget for improving work zone safety. *IATSS Research*.

Schrock, S. D., Fitzsimmons, E. J., Lindheimer, T., Wang, M. H., & Bai, Y. (2014). Work zone positive protection policy guidance: Synthesis of devices and state of practice. In *Transportation Research Board 93rd Annual Meeting* (No. 14-5574).

Stewart, C. (2001). Airfield Safety: Developing Operational Procedures and Schedules for Airside Construction Projects. In *Advancing Airfield Pavements* (pp. 309-321).

IS OUR CURRENT CERTIFICATION PROCESS A THREAT TO SAFETY INNOVATION?

Erik-Jan A.M. Huijbrechts
Airline Pilot and Independent Aviation Safety Researcher
Sassenheim, the Netherlands
M.M. (René) van Paassen
Associate Professor Delft University of Technology
Delft, the Netherlands

Certification is an important process in the aviation industry. The certified status of aircraft, aircraft equipment and procedures is often regarded as a guarantee for safety. However, if shortcomings emerge during operation, this certified status can prevent improvement of the design. In addition, to develop and certify new equipment, it is often easier to modify existing, certified equipment than have a full certification of a new system. Doing so, safety problems may be overlooked. In this paper, a link is made between the certification process and organizational safety of both manufacturers of aircraft or aircraft equipment and airline companies.

To guarantee safety in aviation, equipment and procedures are certified by aviation authorities. In this process, the manufacturer needs to demonstrate that the equipment and procedures fulfil the prescribed airworthiness regulations and/or achieve the required safety level. Due to the cost and effort associated with certification, manufacturers choose to re-use previous certification efforts rather than seek new approval when developing newer versions of aircraft. At this point, the certification process can become a barrier to safety innovation. While the aircraft is modernized to meet new demands, its equipment, systems and procedures will be largely remain based on legacy versions, with only incremental improvement. New automation is introduced piecewise on the flight deck, which can result in situations where isolated systems provide counteracting inputs. When new systems are introduced it may be profitable to consider them as just a modification on an already certified system.

Pilots and other operational personnel may recognize the possible dangers in systems and procedures. They may ask for improvement, however, in general, company management is reluctant to deviate from prescribed procedures for liability reasons. Furthermore, manufacturers are wary of proposed changes and will try to maintain the certified status of existing equipment and procedures. This paper identifies situations where the certified status of equipment and procedures or the certification process has hindered potential safety improvements, and invites ideas for improvement of the certification process.

Influence of Certification on Systems and Procedures

Systems and procedures that have passed the certification stage tend to remain unchanged over time, even when shortcomings may become known. The certified status of a system or procedure is often used as an excuse not to correct imperfections or even known safety hazards. This applies also when new variants or aircraft types are developed and certified on the basis of older models. Examples can be found that have lasted for decades. Sometimes guidance of authorities is required in order to correct an unsafe situation. Gradual introduction of new automation may result in unexpected safety problems that are overlooked in the certification process. Economic pressures during the design process may shift the design goal from developing a new and safe system to, a development process where re-utilization of existing certified systems and components is maximized. This inevitably leads to compromises, as an already certified system is squeezed into fulfilling a new purpose.

Certified Systems

Systems that have passed the certification stage are often used in newer variants or types of aircraft without improvement on known deficiencies. The fuel crossfeed indication wiring in Fokker 70/100 and B737 aircraft can serve as an example.

The fuel crossfeed indication wiring. Both in the Fokker 70/100 and the Boeing 737 PG/NG, the fuel crossfeed indication light is wired over the Circuit Breaker (CB) that protects the Fuel Crossfeed valve. This design may cause confusion to pilots when a crossfeed operation is in place and the CB trips with the crossfeed valve not closed. In that case, the fuel transfer between tanks continues, while the indication light is extinguished suggesting that the crossfeed valve is closed. As a result, pilots have diverted their flights assuming there was a fuel leak. Manufacturers did not inform pilots about this in their Flight Crew Operations Manual (FCOM). Procedures for handling crossfeed problems were not suitable to cope with this situation either. Furthermore, the wiring design was not included in maintenance training manuals for technicians; it is only incorporated in wiring schematics. When Fokker was alerted to this safety related issue, they put some engineering effort into the design, however, this did not result in a modification as at that time the factory had already stopped the build of new aircraft. After alerting Boeing that the procedure was not correct, a change was made to the procedure, better clarifying the state of the system to pilots. Only in the B737 MAX, a separate CB for indication is installed, entirely solving the problem.

Certified Procedures

Procedures that have passed the certification stage and are incorporated in a manufacturer's Flight Crew Operations Manual (FCOM) or Aircraft flight Manual (AFM) remain unchanged over time, even if it may be clear to many operators (pilots) and the manufacturer that the procedure is not correct or optimal. Also when newer aircraft models are introduced, the procedure may remain unchanged. The Boeing stall recovery procedure can serve as an example.

The Boeing stall recovery procedure. The old Boeing recovery procedure for a stall or an approach to stall situation requires the operator (pilot) to first increase thrust and then reduce pitch attitude. When a stall is imminent, the stalled condition can be aggravated if the thrust on underwing mounted engines is increased. This is particularly prominent with the installation of new and large engines on later versions of the Boeing aircraft. Although the problem was recognized by Boeing, and a note to this effect was added in the Flight Crew Training Manual (FCTM), the procedure remained unchanged in the FCOM from the introduction of the first Boeing 737 models in 1968 until the Boeing 777. Only after several stall related crashes in 2009 and the issue of Advisory Circular 120-109 (Federal Aviation Administration, 2012), the procedure was corrected. The new stall recovery procedure for all Boeing aircraft models requires the operator (pilot) to decrease pitch attitude before increasing thrust.

Gradual introduction of new Systems

When new systems or automation is added to already certified equipment the combination may result in unexpected safety issues. The combination of actions of autothrottle and autopilot, that played a role in the Turkish Airlines crash at Amsterdam can serve as an example.

The Turkish airline crash. The crew of TK1951 was forced into a rushed approach. Through a defect of the radio altimeter, incompatible actions of the autothrottle and the autopilot systems lead to an aerodynamic stall and a thrust position that made it difficult to recover from this stalled condition (The Dutch Safety Board, May 2010). This crash also initiated a review of the stall recovery procedure.

Certification of new Systems and Procedures

When new systems are to be developed by a manufacturer, it is often easier to adapt an existing, certified system to serve a new purpose than to have a newly developed system certified. This is true for aircraft models, that are equipped with new engines and technology, to keep up with the demand for better efficiency and to comply with new regulations. It is also true for aircraft systems. Maintaining the current certification is often set as a constraint in the development of a system. This goal may invite for legal shortcuts rather than a thorough operational evaluation of the system. The recent safety problems with the Boeing 737 MAX MCAS (Maneuvering Characteristics Augmentation System) can serve as a good example.

The Boeing 737 MAX MCAS system. The Boeing 737 MCAS system was introduced on the B737 MAX to compensate for the additional pitch up effect that resulted from the modified placement of newer and larger engines on the 737 airframe. The system was supposed to make handling of the MAX aircraft similar to its predecessor, the B737 NG, and provide a safety catch in low speed situations when high thrust is delivered by the engines. To reduce efforts in the certification process, the MCAS was presented as a modification of the previously certified Speed Trim System (STS) in the B737 PG and NG variants. (DeFazio & Larsen, 2020) By using the previous system as a basis for certification, only a limited evaluation was needed, and a thorough evaluation of all safety aspects associated with a new system was avoided. Presenting the new MCAS as an incremental development of the STS also posed constraints on the design process, limiting the ways in which the two system could differ. Regarding procedures and pilot training, it was assumed that pilots would be able to compensate for possible malfunctions using the already established runaway stabilizer procedure also used in the older variants of the 737 airframe. (van Paassen et al., 2021) By stressing the similarity between the aircraft, the initial training requirements for transitioning to the new aircraft could be limited to computer-based instruction.

Influence of Certification on Organizational Safety

Companies in aviation, be they operators or manufacturers, generally have multiple stakeholders, and to each of these certification plays a different role. The management (blunt end) of an aviation related company is often focused on process and legal aspects, and to management the certified status of a system or procedure may be regarded as a guarantee for safety. The certified status may be used as an argument to quell concerns from actors at the sharp end (operators and designers) of the organization about safety. (Rantanen & Huijbrechts, 2021) There is a difference in how management attitude can affect organizational safety in manufacturing and airline companies.

Manufacturers

A manufacturer may use certification as a target in their design process of an aircraft or aircraft system. This invites shortcuts, like using the certified status of existing systems to facilitate the certification process of a new system. Effort in the design process must now be spent to re-use and adapt existing components and procedures, while at the same time the safety review is limited because existing certification efforts can be re-used. The cumulative effects of stepwise adaptation in multiple generations of aircraft on the operation can then easily be overlooked. Once certification is ensured, the accumulated safety record for older generations becomes part of the renewed airplane's reputation, making it difficult for concerns from technical and operational experts to gain traction.

Airline Companies

Airline companies may be adversely affected in their safety level when operating equipment and using procedures that were developed without a thorough safety screening during certification. Although flight deck procedures preferably must be tailor fit to the circumstances within which the airline company works (Barshi et al., 2016), many airline companies choose to trust and present manufacturers' procedures without adaptation to their operators (pilots). In smaller companies, the knowledge or assets to adapt procedures may not be available. In bigger companies, the fear for liability issues is often greater than the urge to improve safety by issuing company procedures. When a "process and legal" mindset is prominent in an airline company, the combination of unadapted procedures and a rigid procedure-oriented operation may impair organizational safety. (Rantanen & Huijbrechts, 2021)

Evaluation

Certification was intended to assure that systems and procedures fulfilled legal requirements and thus provided a certain safety level. In the past, the certified status of equipment and procedures has prevented improvement on the safety level. Clearly, the cost of certification and the effort invested in design of new procedures and systems must somehow be balanced with the yields from operation, both for manufacturers and operators. Re-using existing knowledge and certification is often key to profitability and ultimately success of the company. However, a means to identify and follow up on safety issues and prevent a slow drift into unsafety is important to long-term profitability. Indeed, as the saying goes, "if you think safety is expensive, try an accident…"[1]. With the practice of re-using designs, procedures and certification efforts, a means must be available to stop cumulation of small changes from resulting in real threats to safety. This drift is often first visible to the operators at the sharp end, but it is the responsibility of the operators at the blunt end, i.e., management, and certification authorities, to detect and amplify these alarms. When the management of a company is not product and operation oriented and instead focusses on processes and legal aspects, poor communication between sharp- and blunt end of the organization can have a detrimental effect on organizational safety (Rantanen & Huijbrechts, 2021). Hence, a mechanism must be found to convince management that the certified status does not relief a company from its responsibility for safety of a system, procedure or operation. This may require reviewing the certification process.

Conclusions

If a system or procedure is certified, it tends to remain unchanged although it is known that changes can improve the safety level. The certified status thus prevents safety improvement.

Gradual addition of new systems or automation to certified equipment may introduce new safety hazards that may remain concealed during the certification process. Attention must be given to signals from the operation when safety hazards emerge.

Certification can be used as a target in a design process. This goal invites for adaptation of the design process, possibly trading safety for a speedier and less costly certification.

The certified status of an aircraft, aircraft system or procedure does not absolve a manufacturer from its responsibility for safety of that aircraft, aircraft system or procedure.

Using certified manufacturers equipment and procedures does not absolve an airline company from its responsibility for a safe operation.

---

[1] Alternatively attributed to Stelios Haji-Ioannou or Trevor Kletz

Acknowledgements

I want to thank Esa Rantanen and Maartje Huijbrechts for critical reading, hints and tips. This report is not initiated by, nor does it reflect the views of my employer.

References

Barshi, I., Mauro, R., Degani, A., and Loukopoulou, L., *Designing Flightdeck Procedures,* NASA/TM-2016-219421, October 2016
§ 2.1 Step 1: Determining When Procedures Need to be Designed or Modified

DeFazio, P & Larsen, R, *Final Committee Report*
The design, Development & Certification of the Boeing 737 MAX, September 2020
Chapter 5 Maneuvering Characteristics Augmentation System (MCAS)

Federal Aviation Administration, *Stall and Stick Pusher Training,* Advisory Circular 120-109 (2012) (later replaced by:) *Stall Prevention and Recovery Training,* AC 120-109A (2015)

van Paassen, M.M. (René), Reitsma, J.R., Huijbrechts E-J.A.M., Borst, C. and Mulder, M.
The Skill Assumption – Over-Reliance on Perception In Error Analysis
Conference Paper, ISAP 21, may 2021

Rantanen, E. & Huijbrechts E-J.A.M., *Organizational Safety in Airline Operations*
Conference Paper, ISAP 21, may 2021

The Dutch Safety Board, *Crashed during approach, Boeing 737-800, near Amsterdam Schiphol airport, 25 February 2009*, May 2010 p.52

BEHAVIORAL INDICATORS IN AIR TRAFFIC CONTROL: DETECTING AND PREVENTING PERFORMANCE DECLINE

Tamsyn Edwards
San Jose State/NASA Ames
Moffett Field, CA
Rachel Seely
Federal Aviation Administration
Washington D.C.
Aaron Katz
National Air Traffic Control Association
Phoenix, AZ
Paul Lee,
NASA Ames
Moffett Field, CA

Air traffic controllers are responsible for the safety and efficiency of air traffic and therefore must maintain a consistently high standard of performance. However, performance can be negatively affected by factors such as workload and fatigue, potentially leading to performance decline and performance-related incidents. Real-time identification of negative influences would facilitate timely implementation of supportive strategies prior to performance decline. The current study aimed to explore the concept of 'behavioral indicators' to identify when a controller was reaching a performance limit. A second aim was to capture behavioral indicators associated with performance influencing factors. A total of 65 controllers spanning Tower, Approach and Enroute facilities across the United States of America were interviewed. Findings revealed that controllers were familiar with the concept of behavioral indicators, and that indicators were associated with specific performance-influencing factors. Implications for implementing behavioral indicators training in control environments are discussed.

Air traffic controllers are responsible for the safety of air traffic. It is essential that air traffic controllers maintain a consistently high standard of human performance in order to maintain flight safety and efficiency. Air Traffic Management is remarkably reliable (Amalberti & Wioland, 1997), however, controllers' performance can be negatively affected by performance-influencing human factors such as workload and fatigue (e.g. Cox-Fuenzalida, 2007), potentially leading to performance decline and performance-related incidents. Current mitigations to address these impacts on controller performance include various operational mechanisms, such as sector caps, traffic restrictions, and fatigue breaks. These techniques are very effective at supporting controller performance; however, less is known about preventing or mitigating these performance-related influences dynamically. Detecting the performance-related limits in real-time could allow for the implementation of supportive strategies prior to a performance decline or performance-related incident.

Real-time identification of indicators of potential performance decline is one approach that may permit identification and mitigation of potential performance influences to prevent performance decline. Edwards, Kirwan, Sharples, and Wilson (2016) explored the concept of behavioral indicators with 20 controllers from an Enroute facility in Maastricht, Netherlands. Behavioral indicators were identified that were common across all controllers interviewed. However, the sample was limited to European-based, Enroute controllers. The current research aimed to gain further insight into the concept of indicators and extend Edwards et al. (2016)'s findings by including controllers from Tower, Approach and Enroute control facilities across the United States of America.

Method

A total of 65, one-hour semi-structured interviews were conducted with controllers. Interviews were conducted in-person at three separate facilities: Tower Control, Terminal Radar Approach Control (TRACON), and Enroute. Facilities were selected by the FAA Human Performance team in association with a National Air Traffic Control Association (NATCA) national representative. The interviews included 10 open-ended questions which related to five areas of interest, including current use of indicators in an air traffic control settings, and generalization of indicators between controllers. At each interview, a NATCA representative was present in addition to the researcher. Interviews were transcribed orthographically, and thematic analysis was applied.

Out of a total of 65 controllers, 20 were Enroute controllers, 23 were Tower controllers and 22 were Terminal Radar Approach controllers (TRACON). Ages ranged from 21-56 years old. Years of experience post-certification ranged from 1-30 years, with 94% of participants certified professional controllers (CPCs). Four participants had been checked out of the academy but were not yet certified on their control positions (6%); for these participants, experience post-academy ranged from three months to two years. A total of 38 participants worked as On the Job Training Instructors (OJTIs), 14 from the Tower environment, 15 from TRACON and 9 from Enroute control. Years of experience as an OJTI ranged from three months to 25 years. In total, eight participants were also Operational Supervisors; three from the Tower environment, two from the TRACON environment and three from the Enroute environment.

Results

Controllers Perception and Use of 'Behavioral Indicators' of Performance

Nearly all of the controllers (64/65) were familiar with the concept of indicators and agreed that behavioral indicators occurred in the operations room; one new trainee, with three months post-academy experience, was the exception. In general, participants characterized indicators as cues that a controller (themselves or a colleague) was not completely comfortable with the control task, for example, when colleagues repeated 'say again' instructions to pilots, or when surprised by an aircraft on the radar screen. Indicators appear to serve as a mechanism to protect performance, and prevent performance decline during operations, cueing controllers to mitigate (such as through a change in control strategy) dynamic influences that can negatively affect performance. Controllers naturally monitored colleagues for indicators in addition to themselves, and once identified, applied a compensation strategy to mitigate the cause and support performance, for example, increasing the safety buffer between aircraft. The perception and use of indicators are therefore critical elements in maintaining a consistently high performance.

Indicators are Learned Through Experience

Indicators of potential performance decline are not formally taught but instead are learned through experience: *"The more you see, the more you know, 'ohh I'll never do that again'"* (Participant 23, TRACON). As a result, indicators are usually not discussed with other controllers and the opportunity to learn from other colleagues is limited. In addition, inexperienced controllers such as trainees are more vulnerable to performance decline without the learned experience that a performance limit is being reached.

Individual Differences in Observable Indicators

` Despite no formal training, findings showed that a majority of indicators were shared by every controller interviewed. Controllers' opinions regarding whether indicators were consistent between individuals were divided, however. While some believed indicators would be relatively similar between controllers, others believed that indicators were specific to the individual: *"Everyone is so different on how they interact with people. So, to generalize it, it'd be very tough.* (Participant 5, TRACON). The indicators used at the different facility types did not vary. The phase of control or a particular airspace may result in different compensation strategies employed, but the majority of the indicators were repeated in all facilities. This is an important finding, with implications for training and sharing of indicators.

Individual Differences in Awareness of Indicators

Awareness emerged as integral to the use of indicators; controllers needed to be aware of their own or colleagues' indicators in order to adapt to the situation and protect performance. Participants differed in the extent of conscious awareness of personal indicators. A majority of experienced controllers could identify personal indicators, although several other controllers suggested that they could 'sense' when they are reaching a performance limit, but not identify how they knew: *"I didn't even think about it myself until I just said it to you. I think I kinda knew it in the back of my mind"* (Participant 10, TRACON). It was reported to be easier to identify indicators in colleagues than self-indicators.

Indicators are Associated with Specific Performance-Influencing Factors

Participants were presented with a list of nine factors, including workload, fatigue, stress and situation awareness that are known to affect controller performance (e.g. Edwards et al., 2016). Participants were asked to identify internal and external indicators that were believed to be associated with each factor. Due to space constraints, three of the nine factors are presented below: workload (low and high), fatigue, and situation awareness.

High workload. Participants reported internal and external indicators of potential performance decline that were associated with high workload (Table 1). Changes to subjective feelings and performance changes were reported as important indicators that a controller may be reaching the edge of performance: *"The amount of times you hear, say again, the amount of uhs, you hear, the extremely loud typing, or the stomping of the foot pedal, they're all the same cues. And it doesn't matter if it's because of an internal factor or an external."* (Participant 7, Enroute). Because indicators were associated with specific factors (such as high workload), indicators provided controllers with information about effective mitigative compensation strategies. However, the specific compensation strategies would be specific to the airspace and the situation.

Table 1.
Internal and Observable Indicators of Performance Decline Associated with High Workload.

Cognitive Changes	Changes to control	Physiological changes	Performance changes
Don't know the next steps	Reactive	Faster heartbeat	Miss actions
Calls are a surprise	No back-up plan	Red face	Less negotiation
Mind racing/ 'busy in head'	No space for unexpected events	Sweating	Mixing call signs
Tunnel vision	Future plan reduces in minutes		Can't see solutions
Filtering out information; stop hearing readbacks	Prioritize ineffectively		Overlook aircraft

Low workload. In comparison to high workload, indicators related to low workload reflected a potential influence on performance through boredom or relaxation, leading to distraction: *"One of our tankers said they wanted an extra-long- a downwind because of a seat change. We said, 'Sure'. And then, we started talking.... And the next thing you know, this guy is 20 miles passed where he's supposed to be"* (Participant 7, Enroute). A particularly interesting finding was that controllers are more prepared to approve pilot requests in low workload situations, including shortcuts, which could create unfamiliar control situations: *"You're trying to be more expeditious when you don't have a lot of workload, and you end up putting aircraft where they aren't normally. It can put someone really out of place and get you in trouble"* (Participant 15, TRACON). Common indicators for low workload are presented in Table 2.

Table 2.
Internal and Observable Indicators af Performance Decline Associated with Low Workload.

Cognitive Changes	Control changes	Visible cues	Performance changes
Forgetting	Leave situations develop longer	Sit back	Overlooking aircraft
Easily distracted	Create more complex situations	Look away from radar screen	Forgetting aircraft
Reduced self - awareness	Less safety buffer	Talk to colleagues	Repeated mistakes

Fatigue. Controllers differentiated between tiredness, such as not sleeping well, and mental fatigue, resulting from the time and workload on session: *"Those are two completely different things. [Mental fatigue] You could hear the door open, and you're screaming for him to help you out"* (Participant 1, Tower). Sleepiness however, was largely felt to disappear after the first session: *"Once you get engaged in the operation, it'll go away pretty quickly."* (Participant 5, TRACON). Indicators of fatigue are presented in Table 3.

Table 3.
Internal and External Indicators of Performance Decline Associated with Fatigue.

Cognitive Changes	Control Changes	Visible cues	Performance changes
Slower	Less flexible	Less active	Multiple small mistakes
Not as sharp	Longer to see solutions	Quieter	Missing frequencies, transmissions
Mild confusion	Slower reactions	Yawning	Mixing call signs
Forgetting/surprised	Reactive control	Laid back in chair	Late on tasks
Extra time thinking			Incorrect plan without realization

Situation awareness. Controllers defined situation awareness as 'the picture'. As one controller described: *"You have to know where everybody's at, what they're doing... what they're gonna do in the next 10 minutes"* (Participant 14, Enroute). The loss of situation awareness was reported to be progressive and occur in stages, which were associated with different indicators: "*If you don't get catch it – it's easy to drown faster when you're already drowning–you get the first one [aircraft] and something happens. You're so focused on that, that when the other four get in you don't have time to sit there and do your plan.* (Participant 14, Enroute). Because of this progression, a distinction was made between losing the picture and having lost the picture. The progressive decline was only reported under conditions of high taskload. During low taskload, the loss of awareness was often instantaneous, potentially due to reduced task engagement and increased vulnerability to distraction.

Table 4.
Internal and External Indicators of Performance Decline Associated with Situation Awareness

Cognitive Changes	Control Changes	Visible cues	Performance changes
Running behind traffic	Reactive	Zig-Zag head movement	Falling behind
Thinking whilst giving clearance	Keep traffic static Build plan as go	Slow at task	Unsafe clearances
Tunnel vision	Reduce complexity	Silent	Missing calls
	Conservative clearance		Unexpected decisions

Discussion

Findings revealed that indicators were used in an air traffic control setting as an indication of when a controller was reaching the edge of performance, or a factor was negatively influencing performance. It was considered a natural process that controllers used. Participants confirmed that specific factor influences on performance were associated with specific internal and external indicators. Awareness emerged as an integral element in the use of indicators;

controllers needed to be aware of their own or colleagues' indicators in order to apply compensation strategies and therefore maintain performance. This study found evidence of individual differences in overall levels of awareness. This was especially true of inexperienced controllers who had not yet developed the awareness to identify indicator s and apply adaptive strategies. Indicators were found to be learned through experience rather than being formally taught. Because indicators are learned, there was an expectation that indicators are specific to the individual rather than similar between controllers. If controllers had greater awareness that indicators are used consistently, indicators and associated compensation strategies could be shared. Training on self- and colleague- indicators may support trainees to better protect performance whilst developing the required experience to identify additional indicators. In addition, a standardized list of generic indicators to look out for may be useful to trainees whist building awareness and experience. Awareness of common indicators would also be beneficial for new OJTIs and Supervisors who are still developing awareness of their colleagues' indicators (e.g., a new trainee, or a supervisor assigned to a new sector or facility).

These findings are particularly important given the current changes to the ATC environment during the pandemic. With low traffic levels, controllers face the risk associated with low workload, in addition to increased stress. Lower staffing levels may result in occasional spikes in workload. Controllers would benefit from training on the indicators and supportive strategies now, and as traffic increases. The unpredictability can lead to higher risk. Arming controllers to manage their response would be beneficial. Future research should explore program-specific training that would be most appropriate for specific roles to facilitate awareness and use of indicators to prevent performance decline and potential performance related incidents.

Acknowledgements

The authors would like to thank the Airspace Operations Laboratory, NASA Ames, and the Human Performance Team at the Federal Aviation Association for supporting this research. Thanks also goes to the representatives of the National Air Traffic Association (NATCA) for facilitation of data collection. Finally, the authors would like to acknowledge the three facilities that agreed to take part in this research, the many controllers who volunteered their time, and the operational supervisors who made the research possible.

References

Amalberti, R., & Wioland, L. I. E. N. (1997). Human error in aviation. *Aviation safety*, 91-108.

Cox-Fuenzalida, L. E. (2007). Effect of workload history on task performance. *Human Factors*, 49(2), 277-291.

Edwards, T., Sharples, S., Kirwan, B., Wilson, J.R. (2016). Identifying Markers of Performance Decline in Air Traffic Controllers. In: Di Bucchianico, G., Vallicelli, A., Stanton, N.A., Landry, S.J. (eds.) *Human Factors in Transportation: Social and Technological Evolution Across Maritime, Road, Rail, and Aviation Domains*

THE EFFECT TO HUMAN PERFORMANCE AND WELLBEING OF AIR TRAFFIC MANAGEMENT OPERATIONAL STAFF THROUGH THE COVID-19 PANDEMIC

Nathan Vink
Austro Control GmbH
Vienna, Austria

The Covid-19 pandemic reduced air traffic levels in Europe by up to 95% and the system had to respond quickly to preserve safety, maintain efficiency and performance. Operators were significantly affected both in terms of individual and team performance, as well as the longer-term impact to skills and attitudes. Human Performance data from Operators has been collected through this period. The impact to safety risk due to underloading of human performance, as well as the longer-term impact to wellbeing and competencies of operators was analysed. The largest impact to staff was reduced performance because of anxiety and uncertainty around the future as well as changing job roles. Results also demonstrated the positive effect of systems already place to protect human performance.

There has been almost no lives on earth left untouched by the Covid-19 pandemic (Zacher and Rudolph, 2020). Aside from the public health impact and the measures that have had to be put in place, the way we work and interact socially has shifted dramatically (Schieman et al., 2021) Human Performance is one of the essential key performance indicators for many industries and organisations as it produces outputs such as safety or business productivity. Individually, our human performance varies day to day and over time and is driven primarily by our *technical skills and experience* and also by our *non-technical skills* (also sometimes known as soft-skills) such as Confidence and Resilience (Wickens, et. al., 2015). The pandemic has had a significant impact to these non-technical skills, and if not treated seriously by employers may pose the larger threat to business outputs in the long term even as demand recovers (Murden et al., 2018)

Aviation as a specific example. The global pandemic was declared on March 11, 2020 by the World Health Organisation. This began a sharp decline in the amount of air traffic throughout Europe. Austria went into a first national lock down on 16 March 2020. By 30 March, with continued decline in air traffic globally, operations within Austria reduced to approximately 25% of normal traffic levels. By the end of May, Air Traffic Across Europe had reduced up to 95% (Eurocontrol Daily Traffic Variation, accessed Jan 31, 2021). The response to this significant loss of traffic, coupled with health and welfare responsibilities to impose physical distancing amongst technical and operational staff has seen an unprecedented shift in the task requirements not only of Air Traffic Controllers (ATCOs) but to all operational staff in almost every industry. Organisations have had to adapt at very short notice to new human performance demands that their staff were often not trained or prepared for (Vink, 2020a and Eurocontrol, 2020)

Focusing on just Air Traffic Control; a sustained loss of workload to Air Traffic Controllers posed a potential safety risk of 'underloading' due to skill fade, monotony and other human performance issues. In Austria, under guidance from the Eurocontrol Network Recovery Plan (Eurocontrol, 2020), measures were taken to off-set these risks and to study the effect of a pandemic on human performance. Initial risk assessment focused on the degradation of technical skills – being the actual skills required such as operating a radar screen, or radio communications. Several studies have been conducted from April 2020 until present looking at all facets of human performance. This paper discusses two key surveys and related occurrence data. Overall, it is concluded that the pandemic has led to a degradation in *non-technical skills* which has an effect to safety and performance. This is potentially a longer-term issue in the recovery from Covid-19 for Air Traffic Management. This means, for aviation the message is clear: *we must continue to focus on maintaining and improving staff wellbeing and non-technical skills just as we try to preserve technical skills, safety and business outcomes.*

Expanding this idea to society. In the immediate months following the pandemic, businesses and societies focused on acute solutions – economic, business and medical (Zacher & Rudolph, 2020) As the rest of 2020 unfolded, the impact to mental health was becoming clearer (Pereira-Sanchez et., al. 2020). Efforts are often made by human resource departments to measure it: absent days, sick days, reduced motivation and supervisor checks, however these do not show the true impact to performance because performance is often not very well defined (Patel et al., 2018) There is now a growing consensus that the pandemic and the shift to working from home, or other major changes to tasks of employees is having direct impact to job satisfaction, motivation and mental well-being (Zacher & Rudolph, 2020). In fact, it is likely that even once national and international measures to curb the spread fade away, the impact to employees may be felt for many years to come even as the global economy recovers (Schieman et al., 2021 and Polizzi, Lynn & Perry, 2020).

Within Europe, as in many other Air Navigation Service Providers globally, we call these skills *non-technical skills* and they include: Confidence, Resilience, Adaptability, Trust, Anxiety, Worry and Motivation. Taken together they represent a quantifiable output that directly contributes to human performance (Vink, 2020b). The SHELL Model (Edwards, 1972) is the basis of the human performance pyramid which is used to identify the *most important factors* for producing successful human performance (available upon request to the author). Generally, it is accepted that culture, infrastructure, training/experience, and individual daily variability are the keys to this performance. Individual factors consist primarily of workload, situational awareness, team interactions and non-technical skills (Vink, 2020a).

Even before the pandemic, burnout and other significant losses of human performance were being observed as operations were pushed to their theoretical limits (Vink, 2020b). For aviation, the pandemic has in some ways given some much-needed breathing room and crucially the opportunity to understand exactly how much impact non-technical skills has on our human performance. In their book, "Burnout," the Nagoski sisters discuss the idea of wellness as not being a state of safety and comfort, but as the ability to return to safety and comfort after adversity and difficult performance (Nagoski & Nagoski, 2020). But as the Nagoski sisters point out, we need to learn these skills alongside our day to day required skills. The Covid-19 pandemic has allowed us to capture a unique view into how these non-technical or 'through life-skills' mitigate and mediate our day to day performance. If society can adapt some of the concepts of the human performance pyramid and engineer these skills into sustainable living, then the recovery from this pandemic may be far more effective.

Surveying Human Performance in ATCOs and Operational Engineers

Participants. This paper focuses on the results of two subjective surveys which were carried out in July 2020 (for ATCOs) and January 2021 (for Operational Engineers). For the ATCO survey, n = 94 representing 28% of invited Controllers. For the Engineers n = 149 representing approximately 68% of invited Engineers. The majority of respondents had between 6- and 19-years' experience as operators. Respondents represented an even distribution of operational centres across the country.

Methodology. Two distinct but related questionnaires were produced each focusing on the more specific human performance requirements of the target groups. Both surveys were broken into three areas that asked human performance questions related to: 1 – perception and worry about skill fade, 2 – Monotony and general human performance and 3 – feedback and opinions on Covid-19 measures and impact. For section 1, the focus was on understanding what kinds of skills were impacted by the disruption to normal working patterns. Operators were asked to respond to statements using a 5-point Likert scale and questions included for example, "I am worried about skill fade as a result of the downturn in workload." For section 2, generic human performance measures were needed to understand the average impact to human performance across the reduced traffic period and determine whether boredom and monotony were serious safety threats. These included a variation on the NASA TLX workload indicator (Hart & Staveland, 1988) which asked operators on a 5-point Likert scale about mental and physical workload. Additionally, frustration, effort and self-rated performance were collected. Section 3 contained more generic subjective comment feedback from operators on the specific measures taken during the Covid-period. Using word frequency analysis, the day to day worries and anxieties as well as future psychological wellbeing and concerns could be captured.

Results. Overall, the results were similar between ATCOs and Operational Engineering staff with one notable exception – the difference in workload. In section 1, most operational staff did not need to adjust the techniques and methods for mentally and physically conducting their tasks. 40% of controllers were worried about skill fade, but many took personal initiatives to keep themselves sharp and active. Similarly, engineers were less concerned with general technical skill fade. Some specific skills not used were identified, especially those related to complex situations. But technical skill fade was shown to be less of a concern than first predicted.

Table 1.
Specific Human Performance indicators from ATCOs and Operational Engineers.

	Very Low	Low	Medium	High	Very High
Frustration ATCOs	34 %	18 %	22 %	22 %	4 %
Frustration Engineers	24 %	10 %	24 %	30 %	12 %
Effort ATCOs	30 %	40 %	16 %	9 %	5 %
Effort Engineers	3 %	4 %	38 %	41 %	11 %
Self-rated Performance ATCOs	5 %	5 %	32 %	29 %	29 %
Self-rated Performance Engineers	2 %	13 %	27 %	34 %	24 %
Physical Demand ATCOs	52 %	29 %	13 %	5 %	1 %
Physical Demand Engineers	22 %	20 %	37 %	18 %	3 %
Mental Demand ATCOs	39 %	27 %	24 %	10 %	0 %
Mental Demand Engineers	3 %	7 %	38 %	43 %	9 %

Section 2 revealed key human performance indicators as seen above in Table 1. 66% of the ATCO mental workload is low or very low. This is an indicator of the 'underloading' condition. ATCO workload should be kept at an optimum level to maintain safety. 81% of the physical workload is also considered too low for ATCOs. Conversely, the engineers reported an increase in workload – where over 52% of respondents indicated that their workload had increased. This suggests a shift in the task loading of staff across the company. This can be explained by the fact that even when there is little air traffic, all operational services such as radars, radios and weather equipment had to be available and maintained. With social distancing and remote working, this increased the workload on engineers. It is crucial to note that although average ATCO workload was low, many smaller regional airports reported increases in workload due to increased VFR traffic. This resulted in some near-overload occurrences even when the aggregate picture appeared to be much reduced air traffic.

Despite this, 89.25% of controllers and 75% of Engineers believed that they continued to perform highly and safely throughout the period. This indicates that Safety Culture remained strong and actively engaged in by staff. Frustration was low amongst air staff despite suggestions of monotony and anxieties. Furthermore, perceived personal performance was high which indicates successful measures taken such as alternating team compositions and individual efforts to cope were appropriate. However, results showed that engineers who traditionally do not work in such large teams or operations rooms may have been left more exposed and isolated (i.e. unsupported) due to remote working.

In section 3, comments from operators indicated that they felt reasonably well supported at the operational level (i.e., on the front lines). However, going beyond the operations room, frequency analysis reveals consistent themes and drivers of worry. There are two major themes that occupy the Operator's daily concerns: a slow breakdown in intra-team communication (25%) due to home office and isolated working conditions, and confusion around leadership and crisis messages (19%). And when asked about the biggest fears and worries affecting their wellbeing, operators overwhelming report having anxiety about the future of Aviation and of having a job (58%) followed by worry about society and the economy (10%) and communication from media and government (10%).

Other comments revealed that the operation is still performing relatively well with few major occurrences, good team spirit and performance, with individual measures and professionalism remaining strong and trust in each other. However, as lockdowns persist, remote working is having an increasingly isolating effect, and staff report struggling sometimes just to get work done because of the unavailability of colleagues and tele-working barriers.

Discussion

The data from both surveys as well as evidence from occurrences has led to the conclusion that *technical skills* have not degraded. This is most likely because of proactive and professional behaviours taken by all personnel as well as support for these personnel including simulator training, briefings, communication and team resource management exercises to keep people sharp and practicing busier situations.

What has become evident though is a marked degradation in *non-technical skills*. Of particular concern is the confidence levels of personnel. This is directly linked to their level of anxiety and worry about the future. In aviation, there has been a growing threat of automation replacing much of the hands-on tactical air traffic controlling or flying of aircraft. Autopilots and auto-controllers have already replaced large sections of the skills traditionally used (Wickens, et al., 2015). When the entire industry is threatened and demand for air travel is low, it is unsurprising that the future might seem less certain. Subjective comments reveal that this anxiety or worry is contributing to reduced confidence and strain on resilience.

Taken together; confidence, resilience and trust are decreased and worry and anxiety are increased across the operational staff. This indicates an overall degradation in non-technical skills. Between July and December 2020, occurrence data reveals that human factors contribution to occurrences has remained relatively normal. However, the types of human errors have shifted. Whereas it is usually found that technology and procedures contribute to sub-conscious lapses and slips (Reason, 1990) the types of human errors observed during the period have reflected this degradation in non-technical skills. Pilots and Air Traffic controllers reported that their concentration and attention to procedures are degraded because of distraction due to worry. Corporate measures such as sending staff on leave to clear down their leave balances has also meant that many are coming back after longer periods away and some occurrences have cited being 'rusty' or 'complacent to changes' as reasons for human errors.

The data from these surveys is extensive and revealing. Furthermore, it is a snapshot of a highly unusual situation. There were many positives. Human Performance despite showing signs of significant changes to working roles, remained safe and delivered services throughout the period, with relatively few major occurrences. Frustration was generally low, and camaraderie at an operational level was high. People came through for each other and support was to be found. However, much of this professional behaviour is to be expected in highly safety focused systems such as air traffic management. The key indication though is that non-technical skills have degraded which might not be able to be sustained long term – especially if traffic is to rebound quickly once lock downs and other national measures are relaxed. Therefore, ANSPs in Europe have taken measures to implement non-technical skills training for all staff (not just operational staff). Because of the ability for non-technical skills to mediate human performance and according to the swiss-cheese concept of safety (Reason, 1990) and the Human Performance pyramid, it is vital to help staff boost-up their confidence alongside maintaining their technical day to day skills.

This concept can be applied more widely. New techniques for the teaching and practice of confidence and resilience training which includes teaching the neuropsychology concepts, human error causes and forgiveness of mistakes, acceptance techniques including elements of positive psychology and mindfulness/wellbeing and finally confidence building which includes elements of sports psychology and positive goal setting are now being taught. This method is showing positive effects on staff (although data is limited currently). But this approach of treating professionals with the idea of *'elite professional development'* is a much more positive message than treating people as if they are broken due to lack of confidence. This approach is also a proactive technique for engineering non-technical skills back into the operation.

Conclusion

The pandemic may be causing long term degradation in non-technical skills for personnel and society more generally. By reclassifying traditionally mental health or wellbeing phenomena as skills that can be taught, practiced, and lifted back up this can have a positive impact on human performance. The goal needs to be for individuals to achieve *sustainable* well-being *and* human performance. This is because of the direct impact that wellbeing (non-technical skills) has on human performance. As lockdowns continue and the future of work (e.g. working at home) changes, we have a chance to radically redesign the need for wellbeing to play a more engineered role in the required human performance. If organisations, society and individuals take a proactive approach to deigning their own wellbeing requirements against the human performance requirements this will provide a much greater benefit to outcomes in the long term. It is also much more successful than waiting until after skill degradation or negative mental health effects to try and repair them.

The aviation industry has demonstrated that proactive steps were taken based on risk assessments to maintain both the technical skills and non-technical skills of their staff. As research has emerged, these programs have been rolled out to all staff in the company, not just front-line operators. The same approach can be applied in all industries. The key is to remove the stigma and negative public relations messaging around non-technical skills and to treat them as mediating skills for total human performance that can be proactively trained and developed across the lifetime and career. By training confidence and resilience scientifically and practically this can offset the impact of reduced human performance and allow people to become more adaptive to novel and unusual situations. As is so often pointed out by human factors specialists, the price of investing in these requirements early is significantly less than investing when it is too late.

References

Analysis of Daily Traffic Variation, Eurocontrol (2020, Mar to December). Retrieved January 31, 2020. Daily Traffic Variation - States (eurocontrol.int).

Edwards, E., 1972. Man and machine: systems for safety. In: Proceedings of BritishAirline Pilots Associations Technical Symposium. British Airline Pilots Associations, London, pp. 21–36

Eurocontrol (2020). Network Operations Plan 2020 Recovery Plan. Ed 1.13. Brussels. 30 April 2020

Hart, S. G., & Staveland, L. E. (1988). Development of NASA-TLX (Task Load Index): Results of empirical and theoretical research. In *Advances in psychology* (Vol. 52, pp. 139-183). North-Holland.

Murden, F., Bailey, D., Mackenzie, F., Oeppen, R. S., & Brennan, P. A. (2018). The impact and effect of emotional resilience on performance: an overview for surgeons and other healthcare professionals. *British Journal of Oral and Maxillofacial Surgery*, *56*(9), 786-790.

Nagoski, E., & Nagoski, A. (2020). *Burnout: the secret to unlocking the stress cycle.* Ballantine Books.

Patel, V., Saxena, S., Lund, C., Thornicroft, G., Baingana, F., Bolton, P., ... & UnÜtzer, J. (2018). The Lancet Commission on global mental health and sustainable development. *The Lancet*, *392*(10157), 1553-1598.

Pereira-Sanchez, V., Adiukwu, F., El Hayek, S., Bytyçi, D. G., Gonzalez-Diaz, J. M., Kundadak, G. K., ... & da Costa, M. P. (2020). COVID-19 effect on mental health: patients and workforce. *The Lancet Psychiatry*, *7*(6), e29-e30.

Polizzi, C., Lynn, S. J., & Perry, A. (2020). Stress and coping in the time of covid-19: pathways to resilience and recovery. *Clinical Neuropsychiatry*, *17*(2).

Reason, J. (1990). *Human error*. Cambridge university press.

Schieman, S., Badawy, P. J., A. Milkie, M., & Bierman, A. (2021). Work-Life Conflict During the COVID-19 Pandemic. *Socius*, *7*, 2378023120982856.

Vink, N. (2020a). *The Risk of Skill Degradation during the Covid-19 Crisis for Austro Control Operational and Technical Staff.* Presented to Eurocontrol Human Performance Working Group. 31 March 2020.

Vink, N. (2020b). Development of an Ontology for Objective Cognitive Workload from Human Machine Interface data in Air Traffic Management. *2020 Proceedings EAAP.*

Wickens, C. D., Hollands, J. G., Banbury, S., & Parasuraman, R. (2015). *Engineering psychology and human performance*. Psychology Press.

Zacher, H., & Rudolph, C. W. (2020). Individual differences and changes in subjective wellbeing during the early stages of the COVID-19 pandemic. *American Psychologist*.

DIMENSIONS OF PILOT EXPERIENCE AND THEIR CONTRIBUTING VARIABLES

Nsikak E. Udo-Imeh
Exponent Engineering and Scientific Consulting
Bellevue, WA

Steven J. Landry
Pennsylvania State University
University Park, PA

Pilot experience is generally recognized as an insulating factor against erroneous weather-related decision making in General Aviation (GA). A pilot's level of experience is traditionally taken to correspond to the total flight hours accrued. However, there is some evidence from aviation accident databases and research that total flight hours on its own, may be an inadequate measure of pilot experience. Indeed, pilot experience may be viewed as a multidimensional attribute, with each dimension made up of several elements or variables. How individual elements align with different dimensions, or the extent to which each dimension or the elements thereof contribute to good judgement and aeronautical decision making during adverse weather encounters is unclear. This paper reports initial results from research work carried out to evaluate the extent to which total flight hours and other flight hour related experience variables are associated with the outcome of pilots' in-flight encounters with adverse weather.

Weather is a critical consideration for flight and is often cited as a causal or contributory factor in aircraft accidents (AOPA, 2009; Knecht and Lenz, 2010). Weather related GA accidents consistently involve the highest rate of fatalities of all GA accident causes (AOPA, 2009; Knecht, 2008). In 2011 for instance, 40 out of 54 weather related accidents in the non-commercial fixed-wing GA flights were fatal and 28 out of 43 were fatal in 2010 (AOPA 2011; 2012)

Most accidents caused by adverse weather generally give reasonable warning to the pilot (AOPA, 2011). Therefore, some have suggested most accidents and incidents in weather are preventable (Weener, 2014). However, adverse weather presents pilots with a dynamic, safety critical situation in which time is often limited and information uncertain. Decision making under such contexts has been described as "Naturalistic Decision Making" (Klein, Orasanu, Calderwood & Zsambok, 1993). Decision making within naturalistic contexts has been the subject of much research and our current understanding is that experience plays an important role in them (Klein, 2008). There is some consensus across different fields of endeavor that operators with high levels of experience make more accurate decisions under conditions with severe time pressure and information uncertainty compared to inexperienced operators (Adams and Ericsson, 1992).

In aviation, studies indicate pilot experience is an insulating factor against erroneous decision-making during encounters with adverse weather (Wiegmann, Goh and O'Hare, 2002). Some researchers have suggested expertise results from the experiences accumulated from time spent practicing within a domain (Ericsson, 2004). In aviation, this is tacitly understood to correspond to the total flight hours accrued. Pilot experience is typically evaluated on the basis of the total number of flight hours accumulated (Wiegmann, Goh and O'Hare, 2002; Wiggins and O'Hare, 2003; Johnson and Wiegmann, 2011). Indeed, several studies have found that pilots with higher total flying hours (more experienced) make better judgements and decisions about hazardous weather situations than pilots with lower total flying hours (Johnson and Wiegmann, 2011; Goh and Wiegmann, 2002).

Erroneous decisions made by pilots during encounters with adverse weather is often cited as a cause of GA accidents (O'Hare and Smitheram, 1995; Goh and Wiegmann, 2002; Wiggins and O'Hare, 1995; 2003). Such findings highlight two of the challenges associated with the use of total flight hours as a measure of pilot experience. First, a review of NTSB reports for related accidents reveals many involve pilots with a high number of total flying hours (Landsberg, 2004; NASA, 2007). Accidents which involve such experienced pilots suggest total flight hours may be an inadequate measure of experience. Indeed, Kochan, Jensen and Chubb (1997) have noted that more than total flying hours is required to make an expert pilot and suggested other dimensions such as the relevance, meaningfulness, recency, number and variety of the experience are also important. However, so far, no studies have been carried out to investigate the impact, if any, these dimensions may have in pilot decision making.

Secondly, some researchers have reported finding experience had no positive effect on decision making during adverse weather encounters. For instance, Goh and Wiegmann (2001) as well as the NTSB (2005) have found experience in terms of total flight hours had no positive effect on decision making during encounters with adverse weather. Instead, researchers have found other measures of experience to be more appropriate in determining superior decision-making performance in certain adverse weather situations. For instance, Wiggins and O'Hare (1995) found that a proximal measure of experience, such as cross-country flight hours was a better predictor of differences between the weather-related decision-making performance of experienced and inexperienced pilots than a global measure of experience such as total flight hours. Similarly, Wiegmann, Goh and O'Hare (2002) found that recent flight experience (hours flown in the last 90 days) was a better indicator of the accuracy of pilots' weather-related decision making than total flight hours.

The foregoing suggests a one-dimensional definition of experience, based on total flight hours may lack the resolution or discriminatory power required to fully elucidate the nature of experience that supports accurate decision-making during adverse weather encounters in GA. If that is the case, our ability to better understand and take advantage of any positive effects of experience to influence the outcome of such encounters is limited.

Experience as a Multidimensional Attribute

There is some research as well as anecdotal evidence to suggest experience is a multidimensional attribute, with each dimension made up of several elements or variables. Apart from the number of total flight hours a pilot may have accumulated, several other elements such as the number of hours flown in the aircraft make/model, total hours flown in the last 90 days, cross-country hours flown, instrument rating, certificate type and airplane rating have also been mentioned as important variables that determine the accuracy of decision making during encounters with adverse weather (Kochan, Jensen and Chubb, 1997; Wiggins and O'Hare, 1995; NTSB, 2005; Wiegmann, Goh & O'Hare, 2002).

However, not much work has been done to empirically investigate and ascertain the efficacy of the variables in helping pilots avoid accidents during encounters with adverse weather, or the relationship between the dimensions and variables, Therefore, we do not know whether any of the elements of experience alone or in combinations, reflect or are predictive of the likelihood of an accident. This paper presents the first results from a series of studies carried out to investigate the extent to which of some of the experience dimensions and variables identified in previous research, contribute to decision making during adverse weather encounters.

Approach to the Study and Data Collection

This study is predicated on the understanding that adverse weather encounters occur randomly, so nothing prevents a pilot from encountering one during a flight. Weather related incidents and accidents may be viewed as two distinct states with the potential for a unidirectional transition. An incident is an encounter with adverse weather that was resolved and did not transition to an accident state, while an accident refers to one that was not resolved and transitioned to an undesirable state, an accident. Viewed in this way, it then becomes possible to consider and investigate the key variables that that prevent a transition from incidents to accidents, since that is the preferred outcome. The general belief is that what prevents these randomly occurring incident involving adverse weather encounters from transitioning into accidents is the pilot's experience. So, if experience truly makes a difference to the outcome of adverse weather encounters, we should see significant differences between the operational experience profile of pilots who had accidents during adverse weather encounters and those who did not. Any operational experience variable that does not differ significantly between both sets of pilots may be viewed as having no effect on the outcome of adverse weather encounters.

To explore this conceptualization and thus, address the questions posed in this study, the experience profiles of a sample of pilots who had accidents from encounters with adverse weather was compared to that for a comparable sample of pilots whose encounters with adverse weather did not result in accidents. Queries were run on both the NTSB and ASRS databases to identify reports of General Aviation (Part 91) fixed wing accidents and incidents respectively, between January 1, 2005 and December 31, 2015, in which experience or decision making during adverse weather encounters was determined to be a cause or factor. Each report identified by the query was subsequently reviewed to ensure it met the criteria specified in advance for the study. Accidents and incidents during the take-off and landing phases of flights were excluded, since they could be indicative of short comings in airmanship, rather than decision making mediated by experience. Similarly, accidents and incidents during adverse weather encounters involving student pilots and those in which equipment failure was deemed a cause or factor were also excluded. Reports with incomplete date were also excluded from the study out of concern that the nature of the missing data may not be random. A total of 595 reports, comprising 218 accident and 377 non-accident flights between January 1, 2005 to December 31, 2015 satisfied criteria for inclusion in the study. Pilot experience data was then extracted from the reports and collated for analysis.

Data Analysis

Data analysis started with exploration of the data using descriptive statistics, to summarize and gain some insight into the composition and nature of each experience variable and their distribution for the two groups of pilots in the study. Standard measures of central tendency including mean, median and mode as well as measures of dispersion such as standard deviation, minimum and maximum values were computed along with the frequency distribution for each variable. Individual experience variables were analyzed to determine whether they had any relationship with the outcome of adverse weather encounters. Specifically, Chi-square tests were used to determine the extent to which each element of experience or different levels of multi-level experience variables was associated with accidents. This was followed by a determination of the strength of any such associations in terms of odds ratios.

Three pilot experience variables were considered in this first part of the study; total flight hours, hours flown in the last 90 days and hours flown in airplane make and model. Since these are expressed as continuous variables, they were categorized for the Chi-square tests. Total flight hours was broken into three categories based on Federal Aviation Regulations eligibility requirements for pilot licensure. Accordingly, the first total flight hour category included pilots with 51-250 total flight hours, the next was made up of pilots with 251 – 1500 total flight hours, while the last category included pilots with more than 1500 total flight hours. Both hours flown in the last 90 days and hours flown in airplane make and model were broken into upper and lower median categories.

Results

Descriptive statistics

Descriptive statistics for the data collected indicates accident and non-accident pilots had mean total flight hours of 2223.54 and 6093.14 hours respectively. Similar values for the median total flight hours were 760.00 and 3900.00 flight hours respectively. The mean for hours flown in the last 90 days was 48.49 hours for pilots in the accident group and 75.21 hours for those in the non-accident group, while the median hours flown in the last 90 days were 30.00 and 60.00 hours respectively. The mean hours flown in make and model for accident and non-accident pilots were 610.01 and 972.21 hours respectively, while the median values for accident and incident pilots were 174.00 and 453.00 hours respectively. Details of the descriptive statistics are contained in Table 1 below.

Variable		*Total Flight Hours*					*Hours in Last 90 days*					*Hours in Make and Model*				
	N	*Mean*	*SD*	*Med*	*Min*	*Max*	*Mean*	*SD*	*Med*	*Min*	*Max*	*Mean*	*SD*	*Med*	*Min*	*Max*
Total	595	4675.37	8879.98	2500	50	178000	65.42	64.7	50	0	680	300	1453.13	300	1	18300
Accident Pilots	218	2223.54	3528.57	760	50	22228	48.49	51.5	30	0	250	610.1	1580.37	174	2	9200
Incident Pilots	377	6093.14	10577.6	3900	57	178000	75.21	69.4	60	1	680	972.2	1358.76	453	2	9200

Table 1: Descriptive Statistics

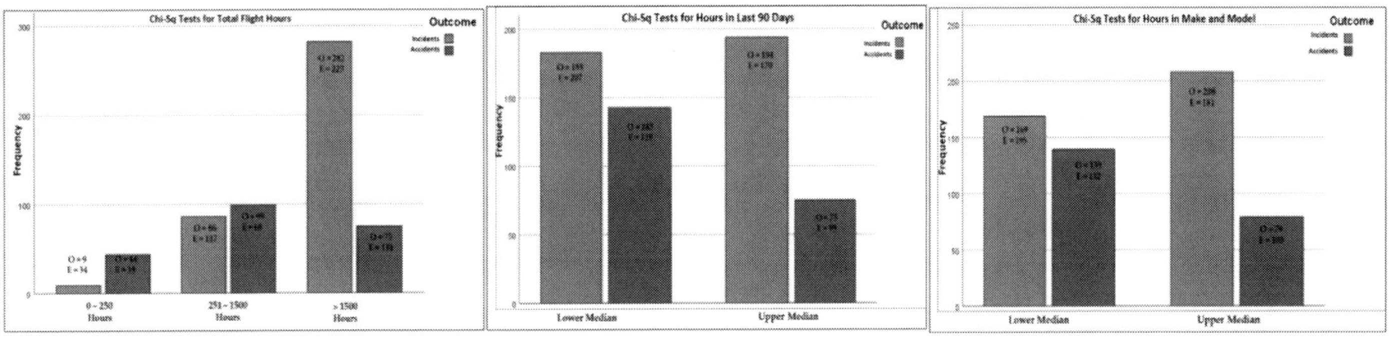

Figure 1: Chi-Square Test Results

Chi-Square Test Results

There were significant associations between total flight hours, ($\chi^2 = 109.37$, $p < 0.01$), hours in the last 90 days ($\chi^2 = 16.22$, $p<0.01$), hours in airplane make and model ($\chi^2 = 19.83$, $p < 0.01$) and the outcome of adverse weather accidents. For total flight hours, the largest differences existed between pilots within the lowest and highest categories. Pilots with 250 total flight hours or less accounted for 8.9% of the total number of accidents pilots but were associated with 20.20% of the accidents during adverse weather encounters. At the other end, pilots with more than 1500 total flight hours accounted for 60% of the total number of accidents pilots in the study and were associated with 34.4% of the accidents. Pilots that had between 251 and 1500 total flight hours made up 31.1% of accident pilots and were associated with 45.4% of the total accidents. Pilots with 250 total flight hours or less were much more associated with accidents (83%) than incidents (17%). Those with between 251 to 1500 total flight hours were more evenly spread (53.50% and 46.50% for accidents and incidents respectively). The percentages for accidents and incidents were 21% and 79% respectively for pilots who had more than 1500 total flight hours. The chart on the left of Figure 1 displays the results of the Chi-square tests for total flight hours.

For hours flown in the last 90 days, 54.8% of all the pilots studied were in the lower median, while 45.2% were in the upper median. However, 65.6% of pilots in the lower median were associated with accidents, while only 34.4% of those in the upper category were. A larger percentage of pilots in the lower category were associated with accidents (43.9%), compared to those in the upper median (27.9%). The chart on the middle of Figure 1 displays the results of the Chi-square tests for hours flown in the last 90 days.

Hours in airplane make and model followed the same trend as hours in the last 90 days. A total of 51.8% of all the pilots studied were in the lower median, while 48.2% were in the upper median. However, 63.8% of pilots in the lower median were associated with accidents, while only 36.2% of those in the upper category were. A larger percentage of pilots within the lower median were associated with accidents (45.1%), compared to the percentage in the upper median (27.5%). The chart on the right of Figure 1 displays the results of the Chi-square tests for hours flown in airplane make and model.

Discussion

This study sought to determine pilot experience variables most associated with an accident during encounters with adverse weather in GA. Much of the previous research carried out in this area have involved the use of simulation and surveys to identify risk factors associated with accidents during adverse weather encounters or the decision-making processes that contribute to such accidents (Lanicci et al., 2012). In this study, experience data for pilots involved in actual, rather than simulated encounters with adverse weather is used to determine which elements of experience are significantly associated with accidents. One advantage of this approach is that the results have a higher level of ecological validity.

Experience is believed to enable more accurate situation assessment and decision making during dynamic, safety critical encounters in which time pressure exists. How different aspects of pilot experience facilitate this during encounters with adverse weather, or which specific elements of experience more significantly impact the likelihood of accidents is not quite clear.

Associations Between Length/Duration of Experience and Accidents.

The elements of experience considered in this first part of the study were flight hour-based and delineated experience in terms of its length/duration. There were significant differences between accident and non-accident pilots on each of the length/duration experience variables evaluated. Lower levels of each experience element were significantly associated with accidents during adverse weather encounters compared to higher levels for each. This result agrees with those from several previous simulation-based studies on the subject, which also found similar associations between the elements considered here and accidents (Sawyer & Shappell, 2009; Wiegmann, et al., 2002). Chi-square tests on categorized levels of each length/duration experience variable showed statistically significant and increasing associations between increasing levels of each variable and accidents during adverse weather encounters. It is not clear whether any of the elements of experience alone or in combinations, reflect or are predictive of the likelihood of an accident. This is one of the questions to be investigated in the rest of the study.

References:

Adams, R. J., & Ericsson, A. E. (1992). Introduction to cognitive processes of expert pilots. Advanced Aviation Concepts Inc Jupiter Fl.

Aircraft Owners and Pilots Association (2011). The Nall Report: Accident trends and factors for 2008. AOPA Air Safety Foundation Publication.

Aircraft Owners and Pilots Association (2012). 2010 Nall Report. The Joseph T. Nall Report of Accident Trends and Factors for 2009. AOPA Air Safety Foundation Publication.

Burian, B., Orasanu, J., & Hitt, J. (2000). Weather-related decision errors: differences across flight types. Proceedings of the 1fourth IEA Triennial Congress of the International Ergonomics Association. Fourth Annual Meeting of the HFES (pp. 22-24). Santa Monica, CA: Human Factors and Ergonomics Society

Ericsson, K. A. (2004). Deliberate practice and the acquisition and maintenance of expert performance in medicine and related domains. Academic medicine, 79(10), S70-S81.

Ericsson, K. A., & Charness, N. (1994). Expert performance: Its structure and acquisition. American psychologist, 49(8), 725.

Goh, J., & Wiegmann, D. (2002). Human factors analysis of accidents involving visual flight rules flight into adverse weather. Aviation, space, and environmental medicine, 73(8), 817-822.Johnson and Wiegmann, 2011)

Klein, G. (2008). Naturalistic decision making. Human Factors, 50, 456-460.

Orasanu, J., Calderwood, R., & Zsambok, C. E. (1993). Decision making in action: Models and methods (pp. 138-47). G. A. Klein (Ed.). Norwood, NJ: Ablex.

Kochan, J. A., Jensen, R. S., Chubb, G. P., & Hunter, D. R. (1997). A New Approach to Aeronautical Decision-Making: The Expertise Method. Ohio State Univ Columbus.

Knecht, W. R., & Lenz, M. (2010). Causes of General Aviation Weather-Related, Non-Fatal Incidents: Analysis Using NASA Aviation Safety Reporting System Data (No. DOT/FAA/AM-10/13

Knecht, W. R. (2008). Use of Weather Information by General Aviation Pilots. Part 1. Quantitative: Reported Use and Value of Providers and Products. (No. DOT-FAA-AM-08-6).

Lanicci, J., Halperin, D., Shappell, S., Hackworth, C., Holcomb, K., Bazargan, M., ... & Iden, R. (2012). General Aviation Weather Encounter Case Studies. FAA Oklahoma City OK Civil Aerospace Medical Inst.

National Aeronautics and Space Administration (2007). General Aviation Weather Encounters. NASA ASRS (Pub. 63).

Orasanu, J., Calderwood, R., & Zsambok, C. E. (1993). Decision making in action: Models and methods (pp. 138-47). G. A. Klein (Ed.). Norwood, NJ: Ablex.

Sawyer, M. W., & Shappell, S. A. (2009, October). Eye tracking analysis of the effects of experience and training on pilots' ability to identify adverse weather conditions. In Proceedings of the Human Factors and Ergonomics Society Annual Meeting (Vol. 53, No. 1, pp. 46-50). Sage CA: Los Angeles, CA: SAGE Publications.

Lanicci, J., Halperin, D., Shappell, S., Hackworth, C., Holcomb, K., Bazargan, M., ... & Iden, R. (2012). General aviation weather encounter case studies. FAA Civil Aerospace Medical Inst. Oklahoma City OK

Weener, E.F. (2014). Presentation at 2014 SUN 'n FUN Forum, Lakeland, Florida. April 5, 2014.

Wiegmann, D. A., Goh, J., & O'Hare, D. (2002). The Role of Situation Assessment and Flight Experience in Pilots' Decisions to Continue Visual Flight Rules Flight into Adverse Weather. Human Factors: The Journal of the Human Factors and Ergonomics Society, 44(2), 189-197.

Wiggins, M. W., & O'Hare, D. (2003). Expert and novice pilot perceptions of static in-flight images of weather. The International Journal of Aviation Psychology, 13(2), 173-187.

Wiggins, M., & O'Hare, D. (1995). Expertise in aeronautical weather-related decision making: A cross-sectional analysis of general aviation pilots. Journal of Experimental Psychology: Applied, 1(4), 305.

INFORMING AVIATION SECURITY WORKFORCE ASSESSMENT AND SELECTION AMIDST THE COVID-19 PANDEMIC

Stephen R. Mitroff
Kedlin Screening International
Bethesda MD, USA
Benjamin Sharpe
Kedlin Screening International
Seattle WA, USA

The COVID-19 pandemic has presented the aviation security industry with short- and long-term challenges relating to workforce assessment that require thoughtful responses. In the short-term, the pandemic has made it difficult to administer typical assessment methods. In the long-term, as the pandemic's impact lessens and travel regains pre-pandemic levels, organizations will need to decide how best to allocate current employees and onboard new employees. While the pandemic has created both selection and assessment challenges, it now opens the door for innovations to support organizations to be better prepared to support the traveling public. The current paper discusses a tool, XRAY Screener, that may offer a way to achieve such goals. XRAY Screener has been shown to effectively and efficiently identify individuals who are best-suited to conduct X-ray screenings. The tool offers a flexible way to assess screeners, making it a viable means to assess performance during and after the pandemic.

Nearly everyone was impacted by the COVID-19 pandemic that effectively shut down non-essential services worldwide beginning in March 2020. As public and private organizations rushed to contain the spread of the novel coronavirus, each industry was forced to evaluate how their operations should or could be modified. The overall impact of COVID-19 was devastating, with global economies suffering and panic setting in for many people. This perfect storm of a crisis (Mitroff, 2020) forced difficult, if not impossible, decisions. The aviation industry was hit especially hard (e.g., Dube, 2021, Mhalla, 2020; Rimmer, 2020) as air travel had the potential to be a key means of spreading the virus. As government and airport authorities restricted travel and passengers stopped flying, nearly every aspect of normalcy was affected for the industry. It is estimated that the reduction in air travel will result in a loss of over 1% of world GDP and job losses over 25 million (Iacus et al., 2020). It is anticipated that the impact on the airline industry could be severe, with long-term financial fallout and potential consolidation of the industry (Suau-Sanchez, 2020). Therefore, as sectors of the aviation industry reopen and/or return to pre-pandemic levels of operation, steps must be taken to do so responsibly to protect passengers, reduce costs, and increase efficiency (Dube, 2021).

The impact of reduced passenger travel has a trickle-down effect that impacts all aspects of the aviation industry. For example, security is one of the most vital components of the process, and unfortunately, fewer travelers results in reduced funding for security operations as they are, at least in part, often funded through aviation travel (e.g., taxes and fees on airline ticket purchases). Moreover, having fewer passengers disrupts standard operating procedures for implementing security measures and for assessing the efficacy of the workforce to enact the measures. To pile on the potential problems, the pandemic also created specific concerns for how to have the security workforce safely interact with one another and the (limited) traveling public. With all of these hurdles, each organization has had to adjust operations to maintain security as a critical step, but to do so with a reduced traveling public and with reduced funding.

The goal of this paper is to discuss the challenges, and unforeseen benefits, the COVID-19 pandemic has presented to the aviation security industry for both during the pandemic and once the impact of pandemic eventually lessens and travel returns to pre-pandemic levels. Following discussions of the current and future states of aviation security in light of the pandemic, a tool is presented as a potential solution for many of the novel challenges that COVID-19 has created.

Challenges, and Unforeseen Benefits, of the COVID-19 Pandemic

Challenges amidst the COVID-19 Pandemic

The challenges to the aviation security industry amidst the pandemic are, unfortunately, numerous and varied. For example, there is a reduced ability to do face-to-face security measures, there is a need for a way to safely conduct security measures in a sanitary manner, and there needs to be a way to ensure the safety of the workforce. Moreover, there is the reality that some members of the security workforce will themselves get sick or need extended time off to deal with sick family members and/or to deal with new challenges in their home life (e.g., childcare). From an organizational standpoint, the pandemic has created a less obvious but highly important challenge—it is more difficult to adequately assess the capability of the workforce to ensure the security screeners are able to appropriately administer the established security measures. For example, there is a reduced capability to administer covert testing (i.e., red team, systems checks), wherein a member of the organization surreptitiously tests the security operations by attempting to bring prohibited items through the security checkpoint. Moreover, standard procedures for ongoing training efforts are potentially affected as it may not be possible to gather groups of security officers in a room for a classroom-style training session.

Once the effects of the pandemic lessen and the aviation industry begins its return to pre-pandemic passenger rates, each organization will need to find a way to "build back better" (Dube, 2021). This will present a myriad of new challenges, including the need to hire (or rehire) employees, a need to train both new employees and an out-of-practice workforce, and a need to create operations that promote the future safety of the workforce and the traveling public. Each step could require a significant financial investment, which may place a heavy burden on organizations that were hindered by the pandemic. As such, it will be critical to be thoughtful in how to be optimally efficient in returning to "normal" levels of security performance. For example, organizations will need to better assess new hires as they start to rebuild their workforce and need to make sure current employees are best-suited for their roles in conducting the necessary security operations. There will likely be numerous changes to the security procedures and each individual screener has likely been affected by the pandemic (i.e., physical, mental, or economic changes due to the pandemic), so it is critical to ensure each employee is ready to perform to the best of their ability. Just because someone was a high performer before the pandemic does not guarantee they will still be after.

Unforeseen Benefits of the Pandemic for aviation security

While the downsides of the pandemic to the aviation industry are clear, it is interesting to consider that there are also some potential unforeseen benefits. For example, if steps can be taken to ensure it is done safely (virtually or by some other means) there could be additional time for assessing and training the workforce; since there are fewer travelers, the workforce may have more flexibility. Likewise, the reduced demands on organizations to process high volumes of travelers could afford the opportunity to reevaluate standard operations; for example, whether it makes sense to have members of the workforce specialize in specific roles (Kedlin Screening International, 2018). Likewise, it may be useful to reevaluate hiring, assessment, and training processes. As the industry emerges from the pandemic, organizations will have an opportunity to explore how current procedures can be done more effectively and efficiently, how to establish new standard operating procedures that can be more resilient to future pandemics, and how to hire selectively with new criteria and measures that will better ensure the workforce is aligned with the security needs.

Innovation to Help Aviation Security Organizations During and After the Pandemic

As laid out above, there is a need in the aviation security industry to leverage innovation to more effectively and efficiently assess performance and provide training within a security workforce. XRAY Screener, an app-based selection and assessment tool, (www.kedlinscreening.com) is an example of leveraging innovation as an opportunity to be better post-pandemic. XRAY Screener has been assessed across two TSA-funded projects and has been used by security organizations in Australia since 2016. It is a flexible and portable tool that is built on a dynamic platform.

XRAY Screener was developed through a partnership that combined scientifically-backed information from academia with a successful gaming app platform. Kedlin Co. (www.airportscannergame.com) developed and published a mobile app called Airport Scanner that has players serve as X-ray Operators to detect prohibited items in simulated bags at virtual airport security checkpoints. Players use a mobile device (e.g., iPad, smartphone) and view simulated bags that may contain prohibited items (Figure 1). They use their finger to tap on any detected prohibited items. The game gained a high-level of popularity with over 20 million downloads. Dr. Stephen Mitroff and his academic research team worked with Kedlin Co. to save player-generated data, which has created a massive dataset of over 3.8 billion trials from over 15 million devices (as of March 2021). This unique dataset has served as the basis for multiple federal grants from the US Army and has provided the foundation for a number of academic publications (e.g., Mitroff et al., 2015).

Creation and Validation of XRAY Screener Tool

XRAY Screener is a modified version of a successful gaming app that combines the successful underlying mechanics of the *Airport Scanner* mobile game with a carefully controlled structure that can appropriately assess human performance in aviation security populations (Mitroff, Ericson, & Sharpe, 2018). XRAY Screener uses simulated images (Figure 1) and a well-controlled algorithm to present users with test images and then record their speed and accuracy of response. The core of XRAY Screener is an "assessment" module that presents trials to users and records their accuracy and speed at correctly identifying prohibited items and correctly identifying bags that do not contain a prohibited item. Through two TSA-funded projects, XRAY Screener was administered to over 3,000 Transportation Security Officers and performance was compared to their on-job metrics of performance (e.g., accuracy at covert tests and speed to process bags at the checkpoint). The main results are presented in Mitroff et al., 2018, which revealed that TSA Officers who performed better in the XRAY Screener assessment were both more accurate and faster when performing at an actual security checkpoint with real passengers and real bags (Figure 2).

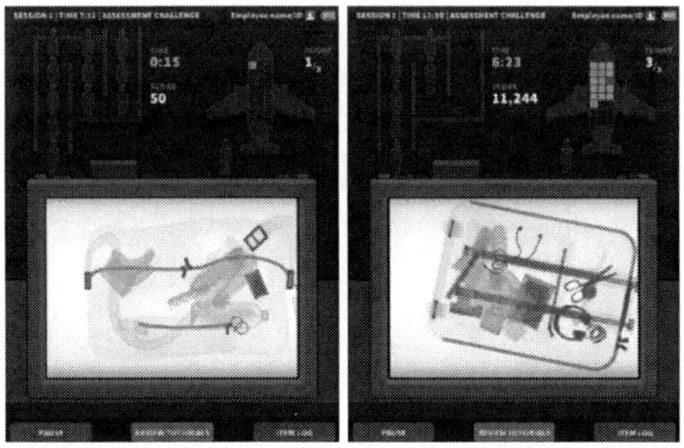

Figure 1. Sample images from XRAY Screener tool. The display on the left has a prohibited item of a liquid container and the display on the right has a prohibited item of a pair of scissors.

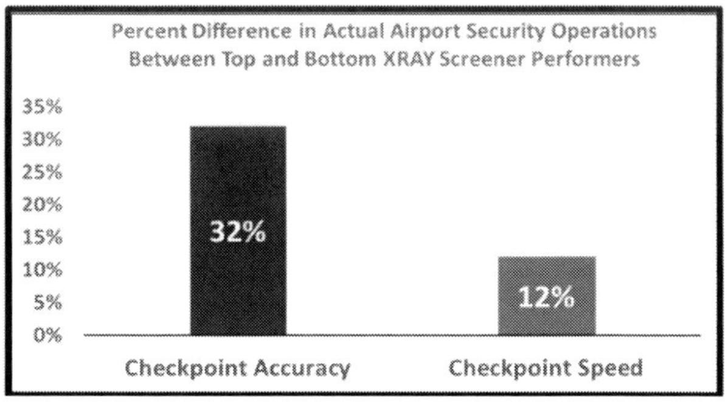

Figure 2. Percent difference in covert miss rate and checkpoint throughput (speed) between the top & bottom 25% of XRAY Screener performers. The top 25% were 32% more accurate and 12% quicker to process bags at the checkpoint (Mitroff et al., 2018).

Current Use of XRAY Screener Tool for Aviation Security Operations

Two different aviation security companies in Australia have used XRAY Screener as part of their normal operations. XRAY Screener has been used to inform new hire selection—new applicants complete the assessment and then the hiring team uses the applicant's performance to inform their decision on whether to hire the individual or not. There are many factors that go into the hiring decisions, and the XRAY Screener assessment provides insights into the individual's likelihood of being successful at the X-ray Operator role—one of the most, if not the most, critical roles in the security process. After the applicant completes the assessment, a PDF report (Figure 3) is immediately available to the organization to help them with the decision process.

XRAY Screener has also been used by these organizations to help with assessing the current workforce. After an adverse event (e.g., failing a covert test), the screeners can be rerun on the assessment and performance can be compared over time. Likewise a new XRAY Screener module is being used that involves just a very brief test (e.g., 2 minutes) administered frequently (daily to a few times a week). This provides an ongoing assessment of each individual with the possibility of identifying trends in performance.

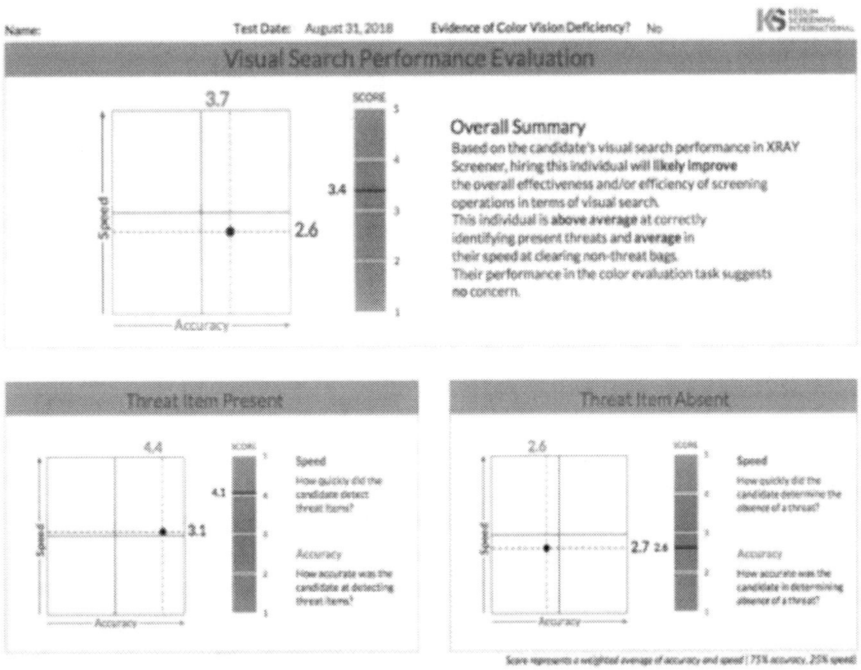

Figure 3. Sample new hire assessment report.

Use of XRAY Screener to Inform Hiring and Assessment During and After the Pandemic

As security organizations onboard new screening officers, they will need to maximize the likelihood of their success. There is a good chance that the workforce will be, at least temporarily, smaller, which makes it critical to have a streamlined workforce that is best suited to do the difficult task of X-ray screening. XRAY Screener can provide objective data for the hiring process, as is already done at a variety of airports in Australia. Moreover, as aviation security organizations consider workforce specialization, XRAY Screener can help identify those who are best suited for the X-ray Operator role. While groups have discussed specialization for years, COVID-19 might be the final push the industry needs to seriously adopt the practice.

Covert testing is the gold standard for assessment, but it is challenging to execute in a pandemic. More to the point, it is also difficult to execute in the best of times as it is time consuming and costly. As well, it is nearly impossible to get enough data on each individual screener to have a reliable estimate of their specific capabilities. XRAY Screener significantly predicts covert testing performance, which means it can serve as a safe and effective way to gather data both during and after the pandemic that can complement covert testing. It can provide officer-level data so organizations can assess both the system and individual level readiness.

Finally, the XRAY Screener "Check-in" module provides a means to gather data regularly (e.g., every shift), which can offer important insights for operational success. In good times, people vary from day to day in their readiness (e.g., due to changes in sleep, emotional experiences, fatigue, caffeine). During a world-wide pandemic, such changes are likely to be exaggerated. XRAY Screener can be used each shift to provide data to on-floor supervisors about which officers are most ready to perform right now.

Conclusions

The COVID-19 pandemic has presented the aviation industry with unprecedented challenges, but it has been inspiring to see how organizations around the world have quickly adjusted to ensure the safety of the traveling public. Aviation security is a vital component of the industry and it is critical that all steps be taken to maintain high-levels of performance through the remainder of the pandemic and to be prepared to come out stronger when the impacts lessen.

Acknowledgments

XRAY Screener is a product of Kedlin Screening International, which is a start-up company co-owned by the two authors. Any opinions presented here are those of the authors and may or may not reflect the views of government or industry organizations. XRAY Screener was validated through two TSA contracts, one to Kedlin Co. (Contract #HSTS04-15-C-CT7031) and one to Kedlin Screening International (Contract #70T04018C9NORC103).

References

Dube, K., Nhamo, G., & Chikodzi, D. (2021). COVID-19 pandemic and prospects for recovery of the global aviation industry. *Journal of Air Transport Management, 92*, 102022.

Iacus, S. M., Natale, F., Santamaria, C., Spyratos, S., & Vespe, M. (2020). Estimating and projecting air passenger traffic during the COVID-19 coronavirus outbreak and its socio-economic impact. *Safety Science, 129*, 104791.

Kedlin Screening International. (2018, October 18). To Specialize or Not? The Pros and Cons of Workforce Specialization in Aviation Security. https://www.kedlinscreening.com/to-specialize-or-not-the-pros-and-cons-of-workforce-specialization-in-aviation-security/

Mhalla, M. (2020). The impact of novel coronavirus (COVID-19) on the global oil and aviation markets. *Journal of Asian Scientific Research, 10*(2), 96.

Mitroff, I. I. (2020). Corona Virus: A Prime Example Of A Wicked Mess. *Technological forecasting and social change.*

Mitroff, S. R., Biggs, A. T., Adamo, S. H., Dowd, E. W., Winkle, J., & Clark, K. (2015). What can 1 billion trials tell us about visual search?. *Journal of experimental psychology: human perception and performance, 41*(1), 1.

Mitroff, S. R., Ericson, J. M., & Sharpe, B. (2018). Predicting airport screening officers' visual search competency with a rapid assessment. *Human factors, 60*(2), 201-211.

Rimmer, P. J. (2020). Aviation and the COVID-19 Pandmic: Flying to the'Next Normal'. *Journal of International Trade, Logistics and Law, 6*(2), 119-136.

Suau-Sanchez, P., Voltes-Dorta, A., & Cugueró-Escofet, N. (2020). An early assessment of the impact of COVID-19 on air transport: Just another crisis or the end of aviation as we know it?. *Journal of Transport Geography.*

SITUATIONAL AWARENESS TRAINING FOR OPERATORS OF UNMANNED AERIAL VEHICLES

Dr. Tarryn Kille
University of Southern Queensland
Springfield, Queensland, Australia
Prof. Paul R. Bates
University of Southern Queensland
Springfield, Queensland, Australia
Dr. Seung Young Lee
University of Southern Queensland
Springfield, Queensland, Australia
Capt. David Murray Kille
University of Southern Queensland
Springfield, Queensland, Australia

In civilian operations, the utilisation of Unmanned Aerial Vehicles (UAVs) is diverse and the application needs and performance characteristics also vary widely. To this end, the growing opportunities for UAV operations have generated an urgent need for trained operators to ensure these systems are used effectively and safely. This paper discusses the importance and integration of appropriate non-technical skills (NTS) training with a focus on situation awareness (SA) to further improve UAV mission effectiveness. The paper explores technical design and human factors challenges impacting on UAV operations. While technical design solutions to UAV systems and interfaces are examined, the authors contend that specific training strategies, which focus on the human UAV operator, should also be considered.

In recent years, civil UAV application studies have centred on the use of algorithms and unique hardware features that enable the UAV to function independently or perform more efficiently (De la Torre et al., 2016). However, this research frequently fails to stress that there remains a need for substantial human involvement in the operation of UAVs, despite the use of artificial intelligence.

As the ubiquitous use of UAVs continues, transport managers, engineers and UAV operators need to understand the key human factors issues to improve safety, usability, and human operator performance. For example, the optimum blend of automation and human interaction should consider the strengths of humans (e.g., flexibility, and decision making), and strengths of machines (e.g., accuracy, and rapid computation) (Mouloua, Gilson, Daskarolis-Kring, Kring, & Hancock, 2001).

The UAV operator plays an important role in successful UAV missions and further research is needed in the field of human integration into automated UAV systems. Therefore, the aim of this paper is to highlight the importance and integration of adequate training in non-technical skills (NTS) with a focus on situation awareness (SA) in order to further enhance the effectiveness of the UAV mission.

Literature Review

The common cause specified for an aircraft accident is human error. However, human error is often linked to a latent condition hidden within the entire operation. Such conditions may include high (or low) workload, fatigue, and limited knowledge of the situation or inadequate training. In the study of human factors, it is evident that greater success in efficiency arises when processes and facilities are built to account for people rather than fully exclude humans from the system (Abbott, Slotte, & Stimson, 1996). The benefits and strengths of combining humans and robots to achieve cooperative tasks has become widely acknowledged (Crandall & Cummings, 2007). Manipulating the levels of autonomy of the robot to cater for human input provides a good opportunity to achieve an optimal combination in mixed human/robot teams. At the heart of this assumption is the notion that robot performance is improved with human input (Kaupp & Makerenko, 2008). Thus, if human input is critical to UAV operations, then it is necessary to understand the variables that affect human performance, specifically in tasks where an operator interacts with a UAV.

The study of aviation human factors extends back some seven decades ago. Since then, a large body of theoretical and empirical research has been dedicated to human performance aspects of manned flight. While this body of research is useful in helping to address some of the impending challenges facing UAV operations, it is apparent that there are major differences between the key human factors issues associated with manned flight and remotely piloted flight. This paper contributes to the UAV body of literature as it attempts to highlight these key differences. It is hoped that it will be instructive and helpful for future UAV policy makers, designers, regulators and training providers to gain an understanding of one of the main non-technical skills training criteria associated with UAV operations.

Methods

While there is much to glean from the perspectives of manned aircraft flight operations, these findings have been considered by the authors in this research with special application to relationship with UAV operations. As such, a narrative literature review has been applied in this part of the study as it provides a valuable theory building technique, and it may also serve functions which assist in hypothesis generation (Baumeister, 1997).

Discussion

The Challenge of Situational Awareness (SA) as a Non-Technical Skill for UAV Operators

Endsley (2000) suggested a definition of SA as "the (1) perception [noticing] of the elements in the environment within a volume of time and space, the (2) comprehension of their meaning, and the (3) projection of their status in the near future" (p. 5). In the design and operation of UAVs, the notion of SA has real implications. In the context of UAVs, SA is broadly defined as the operator's awareness of the status and alterations in a machine's operation (Mouloua, Gilson, Kring, & Hancock, 2001). This awareness should provide the operator with the ability to react quickly and appropriately to unexpected events (Weimer, 1995). High levels of SA will support positive UAV mission performance (Mouloua, Gilson, Kring, et al., 2001) while poor SA is often linked to operator errors (Barnes and Matz (1998).

Themes Relating to Challenges Impacting on Situational Awareness for UAV Operators

Maintaining SA is essential for aviation safety. For the operation of highly autonomous UAVs, there are a number of factors that can collectively be extremely challenging to maintain SA. These factors include:

1. Display design that may not be ideal for maintaining SA.
2. The removal of the pilot from the aircraft, resulting in sensory isolation.
3. Delays in data links, low-grade quality of images from onboard sensors.
4. Lengthy periods of monitoring highly automated systems, leading to the operator feeling 'out of the loop'.

Themes Relating to Training Situation Awareness for UAV Operators

Improved UAV designs targeting flight training are critical processes that help pilots develop their SA capability. Bolstad, Endsley, Costello, and Howell (2010) investigated the effectiveness of six modules of training for developing and maintaining SA by using the general aviation version of the Situation Awareness Global Assessment Technique (SAGAT) installed on a computer located next to the simulator to measure SA. The research revealed that the training modules enhanced the performance of participants on these targeted skills. Results also provided promising support for the effect of the training modules in improving situation awareness.

Sorenson, Stanton, and Banks (2011) compared three theoretical frameworks covering psychology, engineering and systems ergonomics for further understanding and improving pilot SA. Although engineering and psychology provide considerable knowledge of our understanding of SA, the relationship between the individual, the artefact and the context in which they work is rarely considered by both disciplines. However, the systems ergonomics perspective offers a more holistic framework to investigate SA by exploring the complex interaction between the individual operator, the artefact and their environment. Matthews, Eid, Johnsen, and Boe (2011) also explored SA assessment utilizing an observer and self-rating methods under highly stressful and challenging training conditions. The findings revealed that subjective SA measures would not likely produce defensible estimates of SA in extreme conditions. Another study found that following a malfunction in a flight simulator, the eye movements of an experienced pilot significantly differed from a novice pilot, suggesting a different disruption to SA (van de Merwe, van Dijk, & Zon, 2012).

Throughout a mission, a pilot often alters their levels of supervisory control between a full auto pilot and other modes. This process of alternation is also common in a UAV operating environment, and depends greatly on the level of autonomy and supervision required in various stages of flight. These transitions will involve some risks and potential reduction in flight performance or an unacceptable change in workload or SA (Nguyen, Lim, Duy Nguyen, Gordon-Brown, & Nahavandi, 2018). Hainley, Duda, Oman, and Natapoff (2013) examined a pilot's efficiency, SA and work load over a number of automation mode transitions, in an attempt to establish objective measurements of gracefulness during mode transition. The experiments demonstrated that mental workload increases, and SA decrease in a monotonic fashion, with relation to the number of manual control loops the pilot is required to close as a result of the flight mode transition. The research also highlighted the reduced attention of the pilot to fuel status, terrain and altitude during times of high workload due to the attentional demands of manual control tasks.

Cuevas and Aguiar (2017) evaluated a behavioural measure to assess SA and understand how specific operator characteristics (knowledge, skills, and abilities (KSAs)) impact on the success of the mission in UAV operations. The results revealed that participants with greater manned flight experience performed better with respect to SA elements because, as expected, pilots of manned aircraft typically undergo rigorous CRM and/or human factors training throughout their flight training (Cuevas & Aguiar, 2017). The study also demonstrated a statistically significant positive correlation between gaming experience with First-Person Shooter (FPS) games and indicators for spatial orientation (Cuevas & Aguiar, 2017). The researchers explain that this finding is most likely due to the need for spatial awareness in these kinds of games where the player is an avatar in a virtual world. For the player to succeed, they must possess the skill to receive and comprehend all the available information to correctly assess their situation (Cuevas & Aguiar, 2017).

Although a significant proportion of these studies investigated manned flight operations, there are very few studies that investigate effective assessment and training programs to strengthen SA in UAV operations. Yet, many of the findings are relevant to UAV operations and provide a solid platform from which appropriate UAV operator training programs to enhance the non-technical skill of SA may be constructed. Perhaps the most striking findings, that best inform improvements to SA training for UAV operators are summarised below:

- UAV operators with greater manned flight experience performed better with respect to SA elements (Cuevas & Aguiar, 2017).
- UAV operators with gaming experience associated with First Person Shooter games showed improved performance in spatial orientation (Cuevas & Aguiar, 2017).
- UAV operators trained to effectively manage workload in highly stressful flight situations are likely to demonstrate positive skills of social cognition (social SA).
- UAV operator training needs to include pilot appreciation of the increased risk associated with transitions and changes in automation modes depending on the stages of flight. Such transitions will involve some risks and potential reduction in flight performance or an unacceptable change in workload or SA (Nguyen et al., 2018). As such, UAV operators need to be provided with guidance and procedures to assist in SA management during transitions.

Conclusion

The advent and growth of UAV operations has underscored the critical need for trained operators to ensure these systems are used effectively and safely. We have explored the technical design and human factors challenges impacting on UAV operations. Technological solutions to improving SA for UAV operators include consideration of: the principles of UAV display; the removal of the pilot from the aircraft resulting in sensory isolation; data links and sensor imagery; lengthy periods of monitoring highly automated systems which lead to the operator feeling 'out of the loop'. However, the paper also highlighted important considerations to inform improvements to training SA in UAV operators including: UAV operators with greater manned flight experience perform better with respect to SA elements; UAV operators with gaming experience show improvements in spatial orientation; procedures and the design of systems should consider supporting crew to better manage workload during stressful events; and UAV operator training should include pilot appreciation of increased risk associated with transitions in automation levels.

References

Abbott, K. A., Slotte, S. M., & Stimson, D. K. (1996). Federal Aviation Administration Human Factors Team Report on: The Interfaces Between Flightcrews and Modern Flight Deck Systems. Technical Report Retrieved August, 2018, from http://www.tc.faa.gov/its/worldpac/techrpt/hffaces.pdf

Bolstad Barnes, M. J., & Matz, M. F. (1998). Crew simulations for Unmanned Aerial Vehicles (UAV) applications: Sustained effects, shift factors, interface issues, and crew size. Paper presented at the Human Factors and Ergonomics Society Annual Meeting, Santa Monica, CA.

Baumeister, R. F. L., Mark R. (1997). Writing Narrative Literature Reviews. *Review of General Psychology, 1*(13), 311-320.

Crandall, J. W., & Cummings, M. L. (2007). *Developing performance metrics for the supervisory control of multiple robots.* Paper presented at the ACM/IEEE international conference on human-robot interaction.

Cuevas, H. M., & Aguiar, M. (2017). Assessing Situation Awareness in Unmanned Aircraft Systems Operations. *International Journal of Aviation, Aeronautics, and Aerospace, 4*(4).

De la Torre, G., Ramallo, M., & Cervantes, E. (2016). Workload perception in drone flight training simulators. *Computers in Human Behavior, 64*, 449-454.

Endsley, M. R. (2000). Theoretical Underpinnings of Situation Awareness: a Critical Review. In M. R. Endsley & D. J. Garland (Eds.), *Situation Awareness Analysis and Measurement.* Mahwah, NJ: Lawrence Erlbaum Associates.

Hainley, J. C. J., Duda, K. R., Oman, c. M., & Natapoff, A. (2013). Pilot performance, workload, and situation awareness during lunar landing mode transitions. *Journal of Spacecraft and Rockets, 50*, 793-801.

Kaupp, T., & Makerenko, A. (2008). *Measuing human-robot team effectiveness to determine an appropriate autonomy level.* Paper presented at the Robotics and automation, 2008, ICRA 2008, IEEE international conference.

Matthews, M. D., Eid, J., Johnsen, B. H., & Boe, O. C. (2011). A comparison of expert ratings and self-assessments of situation awareness during a combat fatigue course. *Military Psychology, 23*(2), 125.

Mouloua, M., Gilson, R., Kring, J., & Hancock, P. (2001). *Workload, Situation Awareness, and Teaming Issues for UAV/UCAV Operations.* Paper presented at the Human Factors and Ergonomics Society 45th Annual Meeting.

Nguyen, T., Lim, C. P., Duy Nguyen, N., Gordon-Brown, L., & Nahavandi, S. (2018). A Review of Situation Awareness Assessment Approaches in Aviation Environments. *Eprint ArXiv 1803 (08067)* Retrieved 29 December, 2018, from https://arxiv.org/ftp/arxiv/papers/1803/1803.08067.pdf

Sorenson, L. J., Stanton, N. A., & Banks, A. P. (2011). Back to SA school: contrasing three approaches to situation awareness in the cockpit. *Theoretical Issues in Ergonomics Science, 12*(6), 451-471.

van de Merwe, K., van Dijk, H., & Zon, R. (2012). Eye movements as an indicator of situation awareness in a flight simulator experiment. . *The International Journal of Aviation Psychology, 22*(1), 78-95.

Weimer, J. (1995). *Research techniques in human engineering*. Englewood Cliffs, NJ: Prentice Hall.

RECOMMENDED TRAINING PRACTICES TO PREPARE PILOTS TO COPE WITH INFORMATION CONFLICTS

Meredith Carroll, Paige Sanchez, Donna Wilt
Florida Institute of Technology, College of Aeronautics (Melbourne, FL)

As the next generation of flight deck information systems are being utilized on the flight deck, pilots now have greater amounts of information at their fingertips. Although redundant sources of information allow pilots to crosscheck, they also introduce the potential for information conflicts. There is a need to ensure pilots are trained to effectively evaluate, integrate and make decisions based on information from redundant, and potentially conflicting information. Based on findings from a literature, we present several best practice guidelines for preparing pilots to effectively respond to conflicting information. Based on data collected during a questionnaire study administered to a large sample of pilots, and a simulation-based study with B737 pilots, we transformed these guidelines into training recommendations for the pilot training community and provide use case examples of how these recommendations could be implemented.

Aeronautical decision making on the flight deck requires the integration of information from a range of different technological, environmental and human sources. Traditionally, aeronautical decision making relied on pilots experiencing and interpreting informational cues, such as instrument gauges, to assess the situation and diagnose the state of the aircraft and aircraft systems, in order to decide how to respond. As more technology becomes available on the flight deck, the piloting task is changing. Pilots are no longer purely experiencing cues and utilizing their own internal cognitive processes to transform data into information. Instead, technology is performing many of these cognitive processes and pilots are merely consuming information provided by the technology. The result is a change from a correspondence task (i.e., determining how cues correspond to past experiences) to a coherence task (i.e., assessing if there is coherence and consistency in the information being presented) that requires pilots to determine which pieces of information are accurate and relevant, and integrate the information to form an accurate representation of the situation (Mosier & Fischer, 2010; Mosier, 2002). The resulting coherence task is challenging as there are often multiple sources of redundant information that may have slightly different (a) methods by which the information is obtained and synthesized, (b) accuracy, (c) reliability and (d) security. As a result, these sources could provide conflicting information. Pilots must reconcile, make sense of, and make decisions based on these sources of information, sometimes with limited knowledge of, or access to, how the information is obtained and synthesized, and associated levels of accuracy.

Conflicting information has been shown to have deleterious effects on decision making including reduction in accuracy of decisions, longer decision times and less confidence that the decision was correct (Mosier, Sethi, McCauley, Khoo, & Orasanu, 2007; Chen and Li, 2015). In order to prepare pilots to perform effectively on the flight deck, pilots must be armed with the knowledge and skills to support them in assimilating the information, accurately assessing the situation, and making effective decisions. This effort identified best practices from the literature for preparing individuals to effectively respond to situations with conflicting information, and operationalized these into recommendations for the pilot training community.

Methods

First, we conducted a literature review to examine what empirical research has revealed about how individuals make decisions when faced with conflicting information and what training methods can help mitigate the negative impacts of conflicting information. Throughout the duration of the literature review, we reviewed approximately 300 abstracts to determine if the articles focused on the target areas. Of the initially reviewed abstracts, we selected 98 publications for a more thorough review and performed an analysis of 51 relevant publications. Thirty- six empirical studies and 15 theoretical publications were reviewed in detail and information from each article was extracted and input into an MS Excel database.

Based on the level of corroboration across publications, we quantified, prioritized, and summarized findings from the studies, resulting in two primary outcomes: 1) a framework of factors that influence decision making with conflicting information, and 2) recommended mitigations for supporting effective decision making under these circumstances. Full details of these findings are included in Carroll and Sanchez (2020). Several of these recommended mitigations focused on training, with the literature suggesting that training practitioners should teach: (a) information integration skills, (b) system knowledge, (c) metacognitive skills; as well as (d) increase trainees experience with information conflicts and decision-making biases.

Next, empirical data was collected to examine the types of information conflicts that pilots are experiencing on the flight deck, and therefore need to be trained, as well as how pilots respond to these information conflicts. We administered a questionnaire regarding pilot experiences with conflicting sources of weather, traffic, and navigation information on the flight deck to 108 pilots and conducted a simulation study in which thirty six B737 pilots were exposed to flight deck information conflicts (See Carroll, et al., 2021). The results provided a snapshot of the range of different information conflicts that pilots are experiencing on the flight deck and allowed us to marry findings from the literature review with operational knowledge regarding pilots' experience and response to information conflict. This, along with knowledge of current aviation training practices, allowed us to transform the best practices from the literature into implementable recommendations for aviation training practitioners.

Results and Discussion

This section provides best practices from the literature, recommendations for implementation, and use-case examples for classroom, simulation, and live training.

Functional System Knowledge

In preparing pilots to respond to information conflicts, a key first step is ensuring that they have knowledge of how the systems involved are supposed to work. In order to determine which information source is accurate when faced with an information conflict, performers need to know how their systems work at a functional level, including the ability to distinguish true and false alarms, (Gilson, Deaton, & Mouloua, 1996) and to know source strengths and weaknesses (Richter and Maier 2017). Performers must gain enough system knowledge to facilitate the development of accurate mental models of why systems respond in particular ways across various situations (Gilson, Deaton, & Mouloua, 1996). This knowledge will allow an understanding of times when information from a particular source is more or less trustworthy.

Pilots should be taught enough system knowledge to allow them to: (a) understand causes of false information and how to distinguish a false alarm from a true alarm, (b) understand the systems well enough to recognize when things do not go as expected and how to figure out what is happening from the information provided, and (c) recognize the strengths and weaknesses of information provided by the system. An example, with respect to distinguishing false alarms, is a situation in which the Traffic Information System - Broadcast (TIS-B) suddenly issues a traffic alert for an aircraft 100 feet directly below a pilot's aircraft. No aircraft was anywhere nearby and ATC did not advise of any traffic. Pilots need to understand that this is likely a ghost aircraft, (i.e., an artifact of their own aircraft presented as traffic). An example, with respect to recognizing strengths and weaknesses, is when the NextRad display shows moderate precipitation well left of a pilot's route, but straight ahead is clear on the NextRad display. The pilot is in the clouds and can see that the weather ahead is darker than to the left. Pilots need to understand, and most currently do, that a weakness of NextRad is its slow update rate, and the weather shown on the left may actually be directly ahead.

Techniques to Utilize in Response to Information Conflicts

Pilot should also be trained in techniques for conducting a thorough information search, evaluating conflicting cues, and inductive conflict resolution, such as envisioning missing alerts (Mosier, Sethi, McCauley, Khoo & Orasanu, 2007). Research examining pilot response to information conflicts on the flight deck showed that while exposure to information conflicts led to increased crosschecking behaviour, it often was associated with reduced confidence in their decision (Mosier, Sethi, McCauley, Khoo & Orasanu, 2007). The authors suggest that this may be due to pilots not realizing the risks associated with failing to perform a complete information search and recommend that pilots receive training focused on thorough information search and integration. Further, research has revealed that in a situation in which multiple sources of information are provided and can be in conflict, congruent, or a piece missing, performers' response to missing information was very similar to their response to congruent information, suggesting performers assumed the missing information was consistent with other sources (Chen & Li, 2015). Althought there is currently a heavy focus in aviation training on information search skills, referred to as cross-check; there is an opportunity to bolster this process by systematically training pilots how to determine which piece of information is more accurate using inductive conflict resolution skills. For example, when pilots recognize that a piece of information is missing, they could be encouraged to play devil's advocate, assume the piece of information is in conflict with other sources, and consider how they would respond in this situation. An example scenario is at a non-towered airport, a pilot hears another pilot on the radio call 5 miles west of the airport. The pilot's traffic display shows an aircraft 5 miles east of the airport, and nobody to the west. The pilot should consider both options, including that (a) the other pilot made a mistake and is actually east of the airport, or (b) the pilot really is 5 miles west but not showing on the traffic display and there is another aircraft to the east not talking.

Exposure to Information Conflicts

Once trainees understand the system at a functional level, and learn techniques for responding to information conflicts, trainees could then be exposed to unexpected, but plausible, information conflicts. Past research has shown that exposing performers to systems failures (e.g., that could result in conflicting information) can result in reduced trust in this information and reduced utilization of the information to make decisions (Dzindolet, Peterson, Pomranky, Pierce, & Beck, 2003). This is beneficial for systems which have rare false alarms, as it allows proper calibration of trust that may result in increased crosschecking behaviours. Research has also shown that exposing performers to rare false alarms resulted in increased cross-checking (Bahner, Huper, & Manzey, 2008). Currently in aviation training, little is done to train interpreting differences between NexRad radar, METARs, ASOS, and onboard weather radar information in the cockpit that could lead to information conflicts. This is because, in part, current training devices do not realistically simulate weather information in the cockpit. However, training could expose trainees to situations that could cause information conflicts via a variety of platforms, including simple low-fidelity training solutions, such as integrating mocks-ups of displays with conflicting information into PowerPoint slides, providing an opportunity to both illustrate how the information conflicts might present themselves and discuss appropriate ways to respond. Simulation-based training can provide an opportunity to present information conflicts to trainees in realistic scenarios and provide them practice in detecting and responding to information conflicts. Simulation can be used in conjunction with a debrief containing what information conflict occurred, why it occurred, and how the trainee should have responded, to help prepare trainees for responding to such occurrences in the future. Although most aviation training simulators are not currently equipped with the capabilities to introduce information conflicts between systems, there are creative ways to accomplish this. For example, the instructor could alter information in a preflight package, such as NOTAMs or preflight weather briefing so they do not agree with products such as weather on ATIS, or ATC could provide information that conflicts with information provided on a traffic or navigation display. Further, editable functions within EFB applications also provide an opportunity to alter Temporary Flight Restriction (TFR) locations and sizes, which can be in conflict with information from ATC.

Decision-Making Biases

Another area in which pilots would benefit from education and training is with respect to decision-making biases. Human decision making can be driven by biases such as our flawed assessment of how likely an event is to occur based on past experiences (availability bias) and potential outcomes based on past experiences we believe are similar (representativeness bias; Mosier et al., 2002). In the literature, biases have been shown to have a significant impact on how performers respond to information conflicts. For example, when presented with conflicting information, performers tend to choose the option that recommends action over inaction (Mosier, Keyes, & Bernhard, 2000; Skitka, 1999). Performers should be educated on these biases and associated mitigation strategies and be given the opportunity to experience how these biases impact their decision making (Parasurman and Riley, 1997). For example, in aviation training pilots learn to anticipate ATC instructions and be ready to execute them. Pilots also need to learn to verify the instructions and not be biased by expectations. Awareness of biases such as (a) take-action tendency bias (the tendency to choose action over inaction), (b) saliency bias (the most prominent piece of information is likely to carry the most weight), (c) anchoring bias (the first piece of information encountered is likely to carry the most weight) and (d) sunken-cost bias (the tendency to persist along an unfavourable course due to the amount of resources already committed) could provide pilots access to knowledge that will assist them in effectively managing conflicting information. Pilots could then be given opportunities to practice performing in situations in which these biases are likely to emerge. For instance, simulation scenarios could be designed to play into biases, such as suddenly cancelling a landing clearance when the aircraft is below the decision altitude and the pilot is expecting to land.

Self-Reflection

A final skill which could benefit pilots in preparing to respond to information conflicts is the skill of self-reflection. Also known as metacognition, this is the ability to monitor and control ones thought process (Martinez, 2006). This is a skill that can be incredibly important when integrating information from multiple sources with varying levels of integrity and reliability. Self-reflection allows a performer to be aware of what information they have collected, if any information sources are in conflict, whether they have considered why they are in conflict, and the implications for associated decisions. Pilots could be trained to use self-reflection skills such as the use of mental simulation during performance, in which a potential solution is played through in one's head to identify critical risks and relevant situational factors (Martinez 2006; Mosier and Fischer, 2010). With respect to information conflicts, pilots could be taught to mentally simulate how the scenario would play out if they were to trust and act on each of the information sources in conflict, including identifying the potential risks and negative impacts. Instructors could also be trained to use self-reflection in debriefings, specifically to encourage trainees to reflect on the information search and integration steps they performed when experiencing the information conflict, and how they should change performance in the future.

Conclusion

There is an opportunity to leverage best practices derived from the literature to prepare pilots to operate in today's information-rich cockpits, in which there are redundant sources of information. There are several techniques presented herein which can be leveraged to (a) increase pilot knowledge related to information conflicts, why they occur, and strategies to handle them, and (b) provide opportunities for pilots in training to practice responding to information conflicts. Such training practices can be utilized to arm pilots with the knowledge and skills they need to manage information conflicts on the flight deck.

Acknowledgements

This research was sponsored by the Federal Aviation Administration NextGen Human Factors Division (ANG-C1) under Contract # DTFAWA-16-D-00003.

References

Bahner, J. E., A. D. Huper, and D. Manzey. (2008). "Misuse of automated decision aids: Complacency, automation bias and the impact of training experience." *International Journal of Human-Computer Studies 66* (9): 688-699. doi:10.1016/j.ijhcs.2008.06.001

Carroll, M. & Sanchez, P. (2020). Decision making with conflicting information: Influencing factors and best practice guidelines. *Theoretical Issues in Ergonomic Science.* doi.org/10.1080/1463922X.2020.1764660

Meredith Carroll, Summer Rebensky, Donna Wilt, Warren Pittorie, Leonia Hunt, Maria Chaparro & Paige Sanchez (2021) Integrating Uncertified Information from the Electronic Flight Bag into the Aircraft Panel: Impacts on Pilot Response, *International Journal of Human–Computer Interaction, 37*:7, 630-641, DOI: 10.1080/10447318.2020.1854001

Chen, K., & Li, Z. (2015). How does information congruence influence diagnosis performance?. *Ergonomics, 58*(6), 924-934.

Dzindolet, M.T., Peterson, S.A., Pomranky, R.A., & Pierce, L.G., Beck, H.P. (2003). The role of trust in automation reliance. *International Journal of Human-Computer Studies 58*(6), 6.

Gilson, R. D., J. E. Deaton, and M. Mouloua. (1996). "Coping with complex alarms: Sophisticated aircraft cockpit alarm systems demand a shift in training strategies." *Ergonomics in Design 4* (4): 12-18. doi: 10.1177/106480469600400404

Martinez, M. E. (2006). "What is metacognition?." *Phi Delta Kappan 87* (9), 696-699. doi:10.1177/003172170608700916

Mosier, K. L., Keyes, J., & Bernhard, R. (2000). Dealing with conflicting information: Will crews rely on automation? In G. Edkins & P. Pfister (Eds.), Innovation and consolidation in aviation: Selected contributions to the Australian Aviation Psychology Symposium 2000 (pp. 243–254). Aldershot, Hampshire, UK: Ashgate Publishing Ltd.

Mosier, K. L. (2002). Automation and cognition: Maintaining coherence in the electronic cockpit. In Advances in Human Performance and Cognitive Engineering Research (pp. 93-121). Emerald Group Publishing Limited.

Mosier, K. L., & Fischer, U. M. (2010). Judgment and decision making by individuals and teams: issues, models, and applications. *Reviews of Human Factors and Ergonomics, 6*(1), 198-256.

Mosier, K. L., Sethi, N., McCauley, S., Khoo, L., & Orasanu, J. M. (2007). What you don't know can hurt you: Factors impacting diagnosis in the automated cockpit. *Human Factors, 49*(2), 300-310.

Parasuraman, R., & Riley, V. (1997). Humans and automation: Use, misuse, disuse, abuse. *Human Factors, 39*, 230–253.

Richter, T., and J. Maier. (2017). "Comprehension of multiple documents with conflicting information: A two-step model of validation." *Educational Psychologist, 52* (3): 148-166. doi:10.1080/00461520.2017.1322968

Skitka, L. J. (1999). Automation: Decision aid or decision maker? (No. NASA/CR-1998-207767.). Retrieved from NASA Technical Reports Server: https://ntrs.nasa.gov/search.jsp?R=19980048379

TRAINING AIRLINE PILOTS FOR IMPROVED FLIGHT PATH MONITORING: THE SENSEMAKING MODEL FRAMEWORK

Dorrit Billman,[1,2] Peter Zaal,[3,2] Randall Mumaw,[4,2] Thomas Lombaerts,[5,2] Isabel Torron,[4,2]
Saad Jamal,[4,2] Michael Feary[1,2]

NASA Ames Research Center[1]
Metis Technology Solutions[3]
San José State University Research Foundation[4]
KBR Wyle Services[5]
Moffett Field, California[2]

The importance and benefit of improved monitoring is increasingly recognized. Improved training may be a valuable intervention. Our study (conducted 2019) assessed and trained airline First Officers on flight path monitoring skills. The exploratory study assessed monitoring pre-training in a simulator session that included monitoring challenges (8 or 7 events). A 1-hour interactive training followed, based on the Sensemaking Model of Monitoring; it presented concepts and examples using a slide deck, discussion, and simple activities. Post-training assessment used scenarios with analogous monitoring challenges (7 or 8 events) but a different setting. Performance showed significant and relatively consistent improvement. Training monitoring as sensemaking merits further investigation.

The importance of monitoring performance is gaining increased attention. Monitoring gaps are a pervasive contributor to accidents & incidents (e.g., CAST, 2014) and is found at high rates in both line (Dismukes & Berman, 2010) and simulated flight (Mumaw, et al, 2010). Designation of the non-flying pilot as Pilot Monitoring (PM) and increased prominence in NOTECH (NonTechnical) and CRM (Crew/Cockpit Resource Management) training are other indicators of its importance. Despite the recognized need for improved monitoring training, there is not a standard approach for how best to achieve this.

We have proposed the Sensemaking Model of effective monitoring, which emphasizes that monitoring is far more than pointing one's eyes and detecting stimuli (Billman, Mumaw, & Feary, 2020; Mumaw, Billman, & Feary, 2020). Rather it is an active process of building and maintaining a relevant, accurate model of the dynamically unfolding situation. This depends both on activating an accurate mental models of "how things work" from long term knowledge and on using that understanding to guide collecting and assessing relevant information. The process of updating the situation model may frequently contain important gaps or errors (e.g., about modes, Sarter & Woods, 1995) and may contain errors of understanding, not just a failure to notice, that lead to accidents (AAIASB, 2006, cited in Dismukes & Berman, 2010). Thus, improving the overall process from noticing to understanding the situation may be a critical target of training. Further, the active, structured nature of this type of monitoring may also help the pilot stay engaged and interested. Monitoring for flight path management maybe particularly critical for safety and is the focus of our work. We characterize the monitoring process as initiating a situation model by drawing on relevant models in memory, followed by a three-phase cycle of updating this model. The pilot 1) identifies a key, relevant question that needs to be answered in the current situation, such as a gap or inconsistency in the situation model; 2) gathers relevant evidence, which requires identifying the sources of information, analyzing the information, and comparing current versus expected values; and 3) identifies what actions are needed. The central role of the Situation Model and the 3-phase cycle are illustrated in Figure 1. The monitoring cycle also provides anchor points for when and what to communicate with the other pilot. This model simplifies monitoring, highlighting certain aspects over others.

We conducted an exploratory training study that suggests the sensemaking model provides a useful framework for training monitoring. A collaborating US carrier provided us use of a training flight simulator and the pilot union supported recruitment of First Officer pilots to participate. The primary goal was to assess whether providing airline pilots with foundational training based on the Sensemaking Model of Monitoring would produce measurable improvement in flight path monitoring. To do this, we needed to develop sensitive measures of flight path monitoring that are capable of detecting change and a training intervention to produce change.

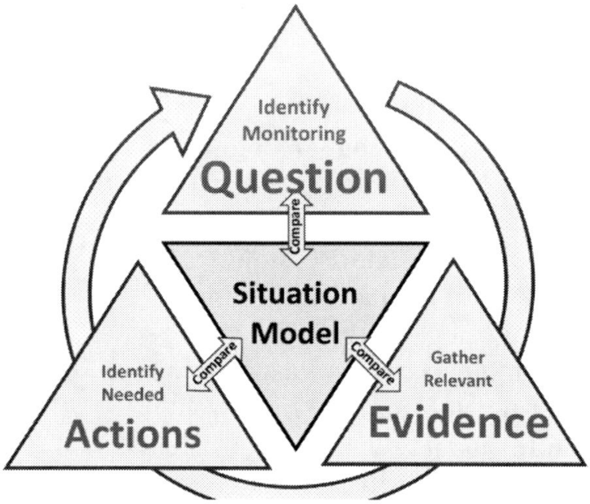

Figure 1. The Situation Model and the processes for drawing on and updating it.

Study Method

Design

We used a 2 Training (pre- vs post-training, within subject) x 2 Scenario Order (Scenarios 1&2 first vs Scenarios 3&4 first, between subject) x 2 Display Configuration (with vs without Flight Director, within subject) design. The pre- versus post-training variable assessed the impact of training, while Scenario Order was a counterbalancing factor. In Scenario Order 1, participants flew Scenarios 1 & 2 before training; in Scenario Order 2, they flew Scenarios 3&4 before training. Performance of each individual and on each item was assessed pre- and post-training. Pragmatic factors of scheduling and simulator availability constrained feasible designs, as discussed in Limitations. The study also asked whether presence versus absence of the Flight Director affected pilot monitoring. Eye tracking data was also collected. See Zaal et al, 2021.

Participants

The participants were 19 First Officers (FOs) who were active and current on the 737 NG. Flight hours ranged from 4100 to 14000 with a median of 7000. Participants were recruited through the union and were offered $100 and NASA stickers as an honorarium. We sought pilots in their first five years at the airline, as they might benefit more from additional training than more experienced pilots; 17 met this criterion.

Procedures and Equipment

The study took 3.5 hours and had five phases: the orientation and demographics interview; simulator session one training tutorial; simulator session two; and the study feedback interview. The simulator was a CAE 737-700 full-flight simulator used in the airline's standard configuration, equipped with Seeing Machine eye-tracking. Participants always served as FO and PM; unknown to the FO, the Captain was a confederate who introduced scripted errors. Sessions were also staffed with an Instructor Pilot introducing ATC clearances, and two experimenters.

Training was conducted as a tutorial structured by a slide deck that standardized coverage. The slide deck included reading information based on the sensemaking model of monitoring, answering both comprehension and examples-from-experience questions, debriefing exercises both of their own simulator flight and of a video snippet, and follow-up discussions of the participant's response.

Key Materials

Approach Scenarios 1&2 (Airport A) versus 3&4 (Airport B) were counterbalanced between Pre- and Post-training sessions. To measure monitoring performance 15 challenging events (see Table 1) were designed so that noticing and understanding the event would lead to specific, identifiable behaviors, and enable objective scoring. Behaviors were typically talking to the Pilot Flying (PF), but some were control actions. Integrating so many issues for the PM to catch while maintaining plausibility relied critically on collaboration with senior pilots, drawing on reported safety events and their own line experience. The challenging events in Scenarios 1&2 versus 3&4 were designed in pairs to pose challenges with similar difficulty, but in different airports and conditions (see Table 1). Matched pairs proved possible for 14 of the 15 events.

Table 1. *Challenge Event Descriptions, by Matched Pair (where possible)*

Challenge Type	Scenario 1	Scenario 3
High on Path (ATC)	**#1** Slowed by ATC	**#9** Held high by ATC
Inappropriate mode	**#2** PF remains in VNAV	**#10** PF selects HDG SEL
Instrument issues	**#3** Given wrong altimeter setting	**#11** False glideslope
Did not enter value	**#4** Field elevation not set on MCP	
	Scenario 2	Scenario 4
Inappropriate mode	**#5** Auto-flight/PF interaction VS	**#12** PF engages LVL CHG
Shortened lateral path	**#6** ATC gives direct-to	**#13** ATC gives direct-to
Inappropriate mode	**#7** PF selects LNAV	**#15** PF fails to arm APP
Airspeed error	**#8** PF calls flaps 25 when too fast	**#14** PF fails to call for flaps 5

The training materials introduced the key concepts of monitoring as active inquiry and understanding as characterized by the Sensemaking Model and included activities to support understanding and integration. The training session was guided by an experimenter and was intended to maximize participant understanding of the key concepts and their application. The importance and centrality of a situation model were explained and illustrated. Training addressed: 1) identifying what question about the situation is the priority to answer, 2) how to gather evidence and assess it against expected values to answer the priority question, and 3) how to identify whether and what actions need to be taken by the pilots. The importance of talking to share information and align the situation models of each pilot was addressed. Pilots applied these concepts in creating a short self-debriefing of an event they had just flown and, later, of a video of another pilot's monitoring the same situation. Examples of monitoring situations were used to illustrate the concepts, and pilots were asked to produce examples from their experience as well. The training materials were focused on information content and were not designed to resemble official training materials; delivery was supported by flexible interaction with an experimenter. Individualized feedback was provided as our goal was to maximize learning, not to control or identify specific training components.

Results

We scored performance on each monitoring event based on the crew videos and simulator data streams.The intended scoring was to code 4 operationally distinct performance levels, two passing, two unsuccessful. A 3- level scale was used for 6 events, because behavior provided only 3 operationally consequential levels of performance .

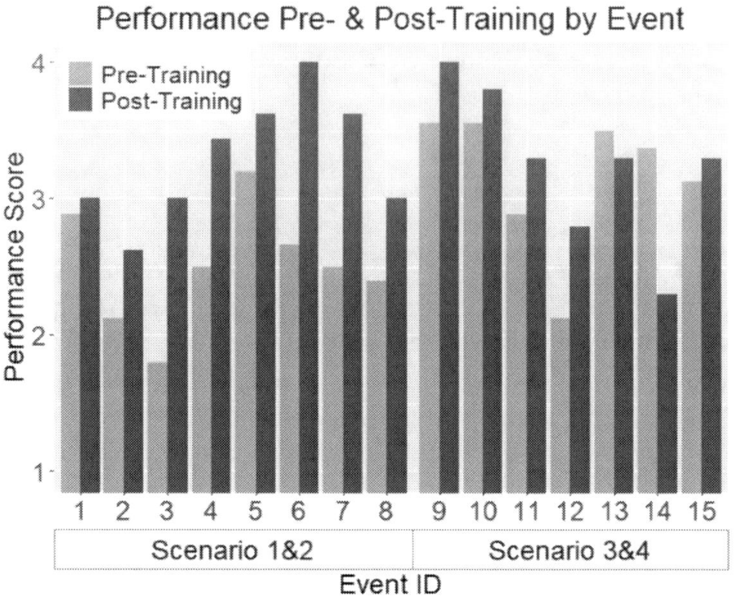

Figure 2. Pilot performance per event.

Overall mean performance score pre- training was 2.8 and post-training was 3.23, showing better performance after training. Figure 2 shows that performance on 13 of the 15 events was higher after training than before. Considered by individual, 13 of the 19 participants improved, 1 had identical pre- and post-training scores, while 5 scores declined. Table 2 shows pre- and post- training performance broken down by the counterbalancing factor of presentation order. To test the significance of training, we modeled Performance Score (ordinal) as predicted by Training (fixed factor), by Scenario Order (fixed factor), by the Training X Scenario Order interaction, by Participant (categorical random factor), and by Event (categorical random factor) using the Gamma distribution family in cumulative link mixed models (clmm routines in R). We compared this to a model that did not include the Training factor. Comparison using a likelihood ratio test showed that including the Training factor significantly improved fit ($\chi2(2)=10.868$, $p=.00437$). While an interaction between presentation order and training is suggested in Table 2, neither the effect of Scenario Order, nor the Scenario Order X Training interaction was significant (for Scenario Order, $\chi2(2)= 2.066$, $p=.356$, or for Scenario Order X Training, $\chi2(1)=.624$, $p=.430$).

Table 2. *Mean Performance Score Before and After Training*

Subject Group	Pre-training	Post-training
Grp1:Scenarios 1&2 first	2.51 *Items 1-8*	3.19 *Items 9-15*
Grp 2:Scenarios 3&4 first	3.17 *Items 9-15*	3.28 *Items 1-8*
Overall	2.80	3.23

An analogous assessment of the pass/fail score using binomial tests also found a significant effect of training ($\chi2(2)=10.341$, $p=.00568$) but not of Scenario Order or of the

Scenario Order X Training, though these became marginally significant, Scenario Order, $\chi2(2)= 4.999$, p=.0821, or for Scenario Order*Training, $\chi2(1)=3.066$, p=.07999. Improvement on the pass/fail score as well as the multi-point scale suggests that training did not simply shift subtle levels of performance within pass or fail categories, but was associated with shifts from unacceptable to satisfactory performance.

Interpretation of time-of-successful-completion is complicated because different events contribute to this measure before and after training and the role of time differed across events. However, for all events that were successfully performed, there was no difference in time to successfully complete the events before (mean = 103 s) versus after (mean = 97 s) training.

Conclusions

Summary

The study found significant improvement in monitoring simulator events designed to require active monitoring, after a brief training and a pre-test experience simulating approaches on a different airport. Our two before- and two after-training scenarios in an airline simulator, assessed 15 challenging events in total. Our design allowed the variation from participants and from events to be modeled, thus increasing statistical sensitivity. The design of the scenarios was challenging, and the fact that they were sensitive to changes after training suggests they can provide a useful resource for future research. The tutorial introducing the sensemaking concepts of monitoring was brief and had not been intensively developed. Nevertheless, this exploratory study suggests that monitoring can be improved with a modest intervention.

Limitations

Our design has two particular limitations, concerning diagnosticity and transfer. First, our study lacked a control group who flew both our pre- and post- training scenarios but had no training. We used performance in the two simulator sessions as our *measurement* of training, but experience in a simulator is itself a powerful learning opportunity. Without a no-training control, we cannot tell whether changes after training are due to the tutorial or due to learning from the first simulator session. Of course, both learning from the simulator sessions and from the tutorial would be valuable if that learning was retained and transferred. However, turning to the second limitation, we did not measure performance after a delay or in a new context. At short delay, memory for the specific cues, such as the confederate pilot or memory of the pilot's own self-debriefing may have produced a brief, context-specific benefit. Our study cannot diagnose whether change in performance is due to the tutorial, the initial simulator session, or the combination. Further, we do not know if performance would transfer to different or delayed simulator sessions, let alone to line operations. Our study uses a limited number of participants and events, and it lacked sensitive measures of pilot skill or experience and of item difficulty.

Future Directions

Monitoring continues to gain importance as system supervision becomes an increasing part of pilots' roles .Our study contributes a valuable initial finding about the ability to improve monitoring performance. This suggests that further efforts to use the Sensemaking Model of

Monitoring to guide training may be worthwhile. A basic replication with a new set of participants would be useful. Very recently a partial replication using a more polished training session but without performance measurement was undertaken (Baron, 2021). Even more informative would be the ability to test performance over a delay and in a setting which did not as strongly cue the training experiences. Rather than aiming to separate the effects of direct training about monitoring from the effects of additional time in the simulator, it may be more valuable to address how simulator time and targeted, out-of-simulator training can be integrated. In particular, linking concepts to simulator exercises through prebriefs and debriefs, and delayed assessment may provide a powerful integrative structure. This strategy for improved monitoring training is ambitious and may require incremental development, but we are excited about the prospects.

References

Baron, B. (April 2021) The Sensemaking Model for Airline Pilot Training: Building and Maintaining Expert Flight Path Management. accepted at Teaching, Colleges & Community. tcchawaii.org.

Billman, D., Mumaw, R.J., & Feary, M. (2020). A model of monitoring as sense-making: Application to flight path management and pilot training. Proceedings of the Human Factors and Ergonomics Society Annual Meeting, vol. 64.

CAST (2014). Airplane state awareness joint safety analysis team. https://www.skybrary.aero/bookshelf/books/2999.pdf

Dismukes, R. K., & Berman, B. A. (2010). Checklists and monitoring in the cockpit: Why crucial defenses sometimes fail, NASA Technical Memorandum (NASA TM-2010–216396). MoffettField, CA: NASA Ames Research Center.

Mumaw, R.J, Billman, D. & Feary, M. (2020) Analysis of Pilot Monitoring Skills and a Review of Training Effectiveness. NASA Technical Memorandum (NASA/TM-20210000047). MoffettField, CA: NASA Ames Research Center.

Mumaw, R.J.,Roth,E.,Vicente, K.J.,Burns, C.M. (2000) There Is More to Monitoring a Nuclear Power Plant than Meets the Eye. *Human Factors, 42*, 36–55.

Sarter, N. B., & Woods, D. D. (1995). How in the World Did We Ever Get into That Mode? Mode Error and Awareness in Supervisory Control. *Human Factors*, *37*(1), 5–19.

Zaal, P., Lamberts, T. et al (2021). An Eye-Tracking Study into the Effect of Training on Pilot Monitoring, to be presented at the AIAA 2021 AVIATION Forum.

Acknowledgements

The research was supported in part by the FAA, "Training Methods for Improved Pilot Monitoring Performance." We thank the airline's pilots and simulator engineers who helped develop the experiment, instructor pilot who recorded two training snippets, the pilots who served as confederates and instructors during the simulator sessions, and the pilots who participated in the experiment. Thanks to Megan Shyer for help with graphs, to Kathryn Ballard for statistical discussions, and to Immanuel Barshi for comments on the paper.

WHAT WE KNOW ABOUT TEAMWORK AND MULTITEAM COORDINATION IN AVIATION: OVERVIEW MODEL

Glenn E. Littlepage
Middle Tennessee State University
Murfreesboro, TN, USA
Michael B. Hein
Middle Tennessee State University
Murfreesboro, TN, USA
Andrea M. Georgiou
Middle Tennessee State University
Murfreesboro, TN, USA

Much of the work of pilots, flight attendants, air traffic controllers, aircraft mechanics, and flight operations center personnel is done in teams and coordination within and between teams is required. This is the first in a five-article series discussing theory and research relating to teamwork in aviation. This article presents a comprehensive model of teamwork in aviation. It builds on leading teamwork theories and integrates other aviation-relevant constructs such as decision making, technology, and culture. All components of the model have been extensively supported in the general team literature, but the extent of aviation-specific research varies considerably across constructs. Additional articles in this series examine the various components in greater detail.

In this article, we discuss the importance of effective teamwork within and across the multiple facets of commercial aviation. We provide a broad framework of factors affecting teamwork including teamwork processes, factors supporting teamwork, and contextual features affecting teamwork. In the other articles in this series, we review the extant literature on teamwork in aviation. We also identify gaps in research, and provide conclusions and suggestions for research and practice.

Airline operations require coordinated action and information flow among multiple components including airline flight operations, maintenance, ground operations, airport management, air traffic control (ATC), pilots, and cabin crew (Loukopoulos et al., 2009). Teamwork is required within each of these components, but coordination is also needed between components. Thus, the airline industry is composed of multiteam systems (Cahil et al., 2014; Shuffler et al. 2015).

Operating as an effective team is vital for safe airline operations (Helmreich, & Foushee, 2019). Within aviation, this is often referred to as crew resource management (CRM), a term developed to describe effective team interaction, decision-making, and safety management. CRM training has emphasized team-related factors such as leadership, climate, communication, and decision-making (Kanki et al., 2019). Teamwork failures have been identified as major proximal causes of mishaps among both pilot (Miranda, 2018) and ATC teams (Read & Charles, 2018). Although the importance of teamwork in aviation is well-recognized, a comprehensive model of teamwork in aviation is lacking and researchers have expressed the need for a multifactor model of teamwork in aviation (Edwards et al., 2012).

Although teamwork is critical in many contexts, high-risk organizations such as aviation present some challenges to effective teamwork that are not generally found in most other contexts. These include challenges related to safety, culture, technology, decision requirements, need for adaptation, and aviation-specific organizational policies. These challenges suggest the need for an analysis of teamwork within and among the various aviation specializations.

Figure 1 provides an organizational framework for the presentation of the specific aviation-related research on teamwork. It is not intended to supplant extant teamwork models. Rather it is meant to present the teamwork constructs that are discussed and to illustrate the relationships between these teamwork constructs. These constructs apply to teamwork within each of the teams that operate within the aviation industry and also to the multiteam systems. For simplicity, the figure is presented as a path model, but in actuality, complex recursive patterns exist.

Although numerous models or teamwork have been identified (Rousseau et al. (2009), we draw heavily on the work of Salas et al. (2005) and Marks et al. (2001). The Marks and Salas models are among the most influential models of teamwork. We also draw heavily on the work of Klein (2008) as he provides a perspective on decision-making especially relevant to aviation.

Marks et al. (2001) proposed a hierarchical model of teamwork processes including three major categories of teamwork processes: transition processes, action processes, and interpersonal processes. Marks and colleagues emphasize the sequential nature of teamwork processes by conceptualizing teamwork as consisting of recurring patterns of transition and action phases. Transition processes involve planning activities that occur before or between active performance-episodes and provide the basis for coordinated goal directed team behavior. Transition performance includes teamwork behaviors related to mission analysis, goal specification, and strategy formulation. Action processes consist of behaviors occurring while the team is actively seeking to accomplish the task. Action processes include monitoring progress toward goals, monitoring resources, monitoring the performance of team members to provide assistance as needed, and coordination (sequencing and timing of actions). Interpersonal processes involve proactive and reactive conflict management, maintaining confidence and motivation, and managing member emotions and cohesion. Interpersonal processes are conceptualized as occurring during both transition and action phases. Meta-analytic results support the construct validity of the Marks model, including the relationship between effective team performance and both overall teamwork and each of the teamwork processes (LePine et al., 2008).

Salas et al. (2005) proposed a teamwork model with five teamwork processes and three coordinating mechanisms that support effective teamwork. Some of the teamwork processes are similar to those proposed by Marks and colleagues, but three additional teamwork processes were proposed: team orientation, adaptability, and team leadership. An additional facet of teamwork is communication. Communication plays a major role in effective team performance (Mesmer-Magnus & DeChurch, 2009). Communication is recognized as critical for effective teamwork in aviation (Kanki, 2019).

The third theoretical perspective incorporated in analysis is decision-making. Although decision-making has been identified as an important teamwork competency (Cannon-Bowers et al., 1995), it is implicit, but not prominent, in most teamwork models including the Marks et al (2001) and Salas et al. (2005) models. Despite its limited emphasis in most teamwork models, the importance of group decision-making is well established (Castellan, 1993; Forsyth, 2019; Janis, 1989) and is prominent in aviation research. Because some decision situations are routine while others require rapid response, effective team performance in aviation involves both vigilant (e.g., Forsyth, 2019; Janis, 1989) and naturalistic decision-making (e.g., Klein, 2008).

Following Marks et al. (2001), transition and action processes are conceptualized as *sequential processes*. The quality of interpersonal processes, leadership, decision making, and communication, affect both planning and implementation and are conceptualized as *permeating processes*.

Salas and colleagues (2005) also proposed conditions that support teamwork: mutual trust, effective communication, and shared mental models. We expand on these coordinating mechanisms by including the additional emergent states of situation awareness, psychological safety, transactive memory, and collective efficacy.

Finally, two *exogenous influences* are included: technology and culture. Both factors represent contextual features that affect teamwork and performance in aviation. Within aviation, technology has major impact on both individual task performance and teamwork. Many types of culture (international, organizational, and professional) can affect teamwork and team and multiteam performance. (Merritt, 2000; Strauch, 2010).

Figure 1.

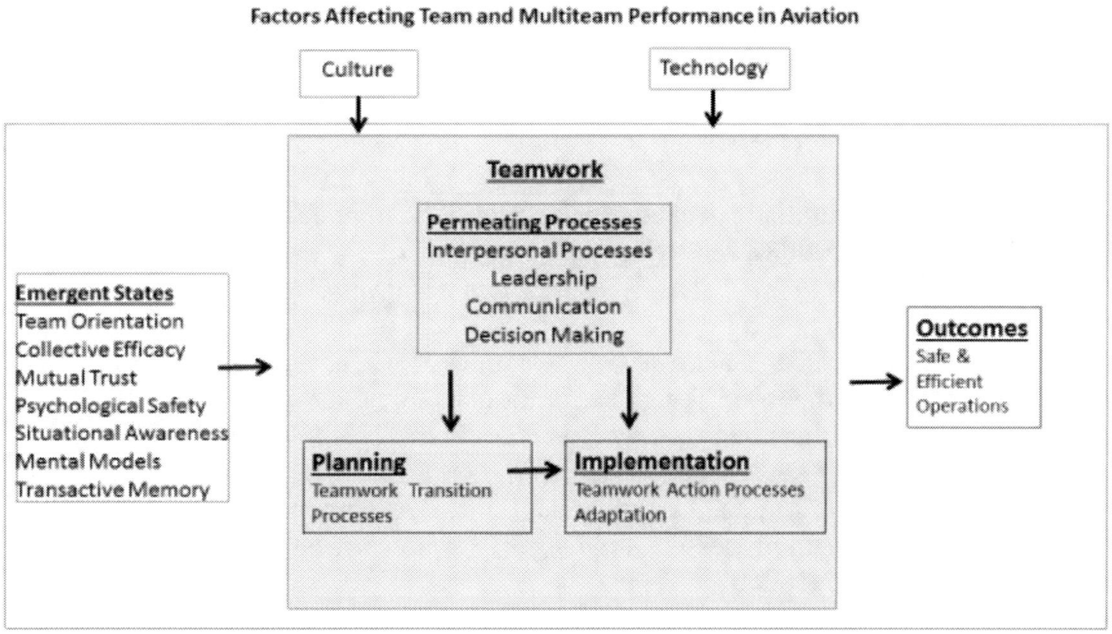

Method

We searched for relevant articles using the PsycINFO database by entering the term aviation paired with various teamwork related search terms (e.g., teamwork, decision-making, communication, etc.). We also examined conference proceedings of the International Symposium on Aviation Psychology, FAA resources, reference sections of relevant articles, and other articles and conference papers of which we were aware. The search yielded 116 articles dealing specifically with teamwork in aviation. While it is unlikely that the search identified all relevant articles, it provides a relatively comprehensive picture of literature relating to teamwork in aviation.

References

Cahill, J., McDonald, N., & Losa, G. (2014). A sociotechnical model of the flight crew task. *Human Factors, 56,* 1337-1363. doi: 10.1177/0018720814532684

Cannon-Bowers J, Salas E, & Converse S. (1993). Shared mental models in expert team decision making. In N. John Castellan, Jr. (Ed.), *Individual and group decision making: Current issues,* (pp. 221-246). Hillsdale, NJ, US: Erlbaum.

Casellan, N. J., Jr. (1993). *Individual and group decision making: Current issues.* Lawrence Erlbaum Associates Publishers.

Edwards, T., Sharples, S., Wilson, J. R., & Kirwan, B. (2102). Factor interaction influences on human performance in air traffic control: The need for a multifactorial model. *Work 41*, 159-166. DOI: 10.3233/WOR-2012-0151-159

Forsyth, D. R. (2019). *Group dynamics*, 7th ed. Boston, MA: Cengage.

Helmreich R, Foushee H. (2019). Why CRM? Empirical and Theoretical Bases of Human Factors Training. In B. G. Kanki, R. L. Helmreich, & J. Anca (Eds.), *Crew Resource Management* (pp. 3-52). London, UK: Academic Press.

Kanki, B. G. (2019). Communication and crew resource management. In B. G. Kanki, J. Anca, & T. R. Chidester (Eds.), *Crew Resource Management,3rd ed.* (pp. 103-138). London, UK: Academic Press.

Klein, G. (2008). Naturalistic decision making, *Human Factors, 50*, 456-60. doi: 10.1518/001872008X288385

Janis, I.L. (1989). *Crucial decisions: Leadership in policymaking and crisis management:* NY: The Free Press.

LePine, J. A., Piccolo, R. F., Jackson, C. L., Mathieu, J. E., & Saul, J. R. (2008). A meta-analysis of teamwork processes: Tests of a multidimensional model and relationships with team effectiveness criteria. *Personnel Psychology, 61*, 273-307. doi:10.1111/j.1744-6570.2008.00114.x

Marks, M. A., Mathieu, J. E., & Zaccaro, S. J. (2001). A temporally based framework and taxonomy of team processes. *Academy of Management Review, 26*, 356-376. doi:10.2307/259182607

Mesmer-Magnus, J. R., & DeChurch, L. A. (2009). Information sharing and team performance: A meta-analysis. *Journal of Applied Psychology, 94*, 535-546. doi:10.1037/a0013773

Merritt, A. (2000). Culture in the cockpit: Do Hofstede's dimensions replicate? *Journal of Cross-Cultural Psychology, 31*(3), 283-301.

Miranda, A. T. (2018). Understanding human error in naval aviation mishaps. *Human Factors*, *60*(6), 763–777. doi.org/10.1177/0018720818771904

Read, K., & Charles, R. (2018). Understanding teamwork errors in royal air force air traffic control. *Safety Science*, *109*, 36–45. https://doi.org/10.1016/j.ssci.2018.04.030

Rousseau, V., Aubé, C., & Savoie, A. (2006). Teamwork Behaviors: A Review and an Integration of Frameworks. *Small Group Research*, *37*(5), 540–570. https://doi-org.ezproxy.mtsu.edu/10.1177/1046496406293125

Salas, E., Sims, D. E., & Burke, C. S. (2005). Is there a "big five" in teamwork? *Small Group Research, 36*, 555-599. doi:10.1177/1046496405277134

Shuffler, M. L., Jiménez-Rodríguez, M., & Kramer, W. S. (2015). The science of multiteam systems: A review and future research agenda. *Small Group Research*, *46*(6), 659-699. doi:10.1177/1046496415603455

Strauch, B. (2010). Can cultural differences lead to accidents? Team cultural differences and sociotechnical system operations. *Human Factors, 52*(2), 246-263. Doi:10.1177/0018720810362238

WHAT WE KNOW ABOUT TEAMWORK AND MULTITEAM COORDINATION IN AVIATION: EMERGENT STATES SUPPORTING TEAMWORK IN AVIATION

Glenn E. Littlepage
Middle Tennessee State University
Murfreesboro, TN, USA
Michael B. Hein
Middle Tennessee State University
Murfreesboro, TN, USA
Andrea M. Georgiou
Middle Tennessee State University
Murfreesboro, TN, USA

This paper describes a variety of factors that can facilitate teamwork. These include team orientation, collective efficacy, mutual trust, psychological safety, shared situational awareness, shared mental models, and transactive memory. Aviation-specific research on each of these states is reviewed.

A number of factors have been identified that facilitate effective teamwork. These include emergent cognitive and affective states can serve as coordinating mechanisms to support effective teamwork and team performance (Salas et al., 2005). These states are developed or refined during team interaction and impact subsequent team processes. The importance of each of these states has been documented in the general team literature. Aviation-specific research has focused on some states, while other states have received little research attention.

Team Orientation

Team orientation is an attitude that team performance can be improved by coordination and cooperation with other team members (Salas et al., 2005). A meta-analysis (Bell, 2007) indicated that preference for teamwork was positively related to team performance. Although team orientation is generally conceptualized as an individual characteristic, one's orientation toward working with teammates can be shaped by interactions within the team. While aviation research has not focused directly on team orientation, Cahill et al.'s (2014) series of interview and observational studies of flight operations suggested that shared task responsibility and the need for coordination across disciplines are essential for safe and efficient airline operations.

Collective Efficacy

Collective efficacy is the shared belief that the team can perform its tasks. Meta-analytic evidence indicates that collective efficacy is related to effective team performance (Stajkovic, Lee et al., 2009). We found few studies of collective efficacy in aviation, and those were limited to ATC teams. Studies of ATC teams indicate that collective efficacy is related to backup behavior (Smith-Jentsch et al., 2009) and effective team performance (Mathieu et al., 2010).

Mutual Trust

Mutual trust involves a shared belief that team members will properly perform their duties and protect the interests of other team members (Salas et al., 2005). Trust promotes cooperation, information sharing, and willingness to rely on information provided by others. Although we are not aware of studies of trust among aviation professionals, a study of occupational stereotypes among aviation students suggests that a lack of professional trust is not a major issue (Lillard et al., 2015).

Psychological Safety

Psychological safety refers to the belief that it is safe to take interpersonal risks such as suggesting changes, raising doubts and objections, or admitting mistakes or a lack of knowledge or expertise. Edmondson & Lei (2014) review extensive evidence indicating that a climate of psychological safety can facilitate the discussion of problems leading to error correction and improved work practices.

Creation of a psychologically safe climate facilitates team error prevention and management, and this is one of the major goals of CRM (Tullo, 2010; Velazquez & Bier, 2015). Surveys of first officers and flight attendants suggested that psychological safety facilitates questioning or challenging of actions and decisions of superiors in both groups. Psychological safety was related to flight attendants speaking up to the lead flight attendant and also mitigated the chilling effects of status on first officers' speaking up to the captain. Feelings of psychological safety within the flight attendant group facilitated boundary spanning and was associated with lead flight attendants speaking up to pilots (Bienefeld & Grote, 2014).

Situation Awareness and Assessment

One of the most critical emergent cognitive states supporting teamwork and multiteam coordination is a shared awareness and assessment of the situation (Endsley, 2015). Aircrews need to have a shared understanding of weather conditions, terrain, altitude, location, flight traffic, airport conditions, flight plan deviations, and the mechanical condition of the aircraft. Situation assessment requires not only an awareness of the situation, but also an accurate interpretation of its meaning and implications. Analysis of a national accident database indicted that about 62% of accidents involved failures of situation awareness (Endsley, 2010). Examples include fatal crashes that have occurred where distracted cockpit crews failed to monitor basic situational factors such as fuel or altitude. Results of flight simulator studies and analysis of incident reports provide additional evidence indicating that situation awareness among pilot teams is related to effective teamwork processes and team performance (e.g., Brannick et al., 1995; Nullmeyer, & Spiker, 2003). Ineffective aircrews displayed situation awareness deficiencies such as lack of vigilance and lack of awareness of the environment and of aircraft systems (Hausler et al., 2004). Examination of ATC incident reports revealed that the lack of situation awareness was related to the frequency and severity of errors (Rodgers et al., 2000).

Maintenance is often performed by teams and involves initial inspection, diagnosis, repair, and final inspection. Typically, these activities are performed by different individuals and frequently multiple systems are serviced simultaneously by different technicians. It is important to maintain shared situation awareness about the status of the airplane and the maintenance activities, including assessments and reasons for actions (Endsley & Robertson, 2000).

Awareness of risks (e.g., severe weather, mechanical issues) provides a foundation for threat detection and effective decision making, and is critical to mission success (Helmreich et al., 1999). The National Transportation Safety Board (NTSB) has identified distraction as one factor that can undermine situation awareness. Examples include accidents and issues that occurred when pilots failed to monitor flight conditions while attending to a minor problem, used portable electronic devices, or engaged in social conversations with a flight attendant. (Chute & Wiener, 1996; Endsley, 2010; NTSB, 2017).

Mental Models

Shared mental models provide shared expectations that allow for more efficient coordination and reduce the need for explicit communication. This is especially important under time-sensitive and high workload conditions. Task mental models focus on procedures, strategies, and cue-response associations. Teamwork mental models reflect roles, interdependencies, and interaction requirements. Equipment mental models involve understanding of equipment operation, and technology (Cannon-Bowers et al., 1993). A meta-analysis indicated that both task mental models and teamwork mental models were related to teamwork and to team performance (DeChurch & Mesmer-Magnus, 2010).

Importance of mental models has been demonstrated for pilots, ATC, and for multiteam operations. While there have been some conflicting results, overall patterns have emerged. Performance is highest when both task and teamwork mental model are shared and accurate (e.g., Mathieu et al., 2005; Mathieu et al., 2010; Smith-Jentsch et al., 2005). An accurate and shared task model allows team members to have a common and appropriate understanding of actions that need to be taken. An accurate and shared teamwork model allows team members to allocate tasks and coordinate activities to effectively implement actions deriving from the task mental model.

Not only are shared mental models important within teams, they are important in multiteam contexts as well. Lack of shared mental models can lead to disconnects. Bearman et al., (2010) identify three types of disconnects common to aviation: informational, evaluative, and operational. Informational disconnects are when the two team members do not have the same information. Evaluative disconnects occur when both parties have a different interpretation or give different weights to the information. For example, pilots and air traffic controllers have different framing and cue utilization for risk assessment (Mosier & Fischer, 2015). ATC personnel tend to base risk assessments on distance between aircraft, but pilot's risk assessments are largely based on time to respond and options to control the situation (Fischer et al., 2003). These evaluative disconnects can lead to operational disconnects (mismatches between different team members plans and/ or actions) such as a pilot choosing to avoid challenging weather rather than adhere to ATC directives (Bearman et al., 2010).

Although there is limited research on equipment or technology mental models in the general team literature, there is evidence that shared mental models of technology are important within aviation. One area where an inadequate technology mental model is evident is mode errors (Sarter, 2008; Sarter et al., 2007; Sarter & Woods, 1994). Mode errors occur when the pilots do not understand the current state of a system, the permitted actions, and the future actions taken by an automated system. Mode errors result in inappropriate or ineffective actions or failure to take action when needed. Other studies indicate that differences in experience and comfort level with aviation technology may create different perceptions of individual workload and confidence and can undermine shared situation awareness (Fernandes & Smith, 2011; Martin et al., 2011).

Transactive Memory

Transactive memory refers to a shared understanding of the areas of expertise held by specific team members. An effective transactive memory system allows for specialization and coordination among team members. Meta-analysis revealed a strong relationship between transactive memory and both team processes and team performance (DeChurch & Mesmer-Magnus, 2010). Aviation requires coordination among various specializations, therefore a shared understanding of the types of knowledge possessed by each specialization is needed. Little research has examined the role of transactive memory in aviation, but a study of senior aviation students indicated that they showed relatively high levels of transactive memory (Littlepage e al., 2016). In a related study, transactive memory was found to predict teamwork and to have an indirect effect on both routine and adaptive performance (Wertheimer & Littlepage, 2017). A study of ATC teams found that transactive memory was related to requesting and accepting backup behavior (Smith-Jentsch et al., 2009).

Conclusions

These studies illustrate the importance of emergent states. They also indicate that while aviation research has addressed some emergent states, others are in need of additional aviation-specific research. The next paper in this series describes research on teamwork processes.

References

Bell, S. (2007). Deep-level composition variables as predictors of team performance: A meta-analysis. *Journal of Applied Psychology, 92*(3), 595-615. doi: 10.1037/0021-9010.92.3.595

Bearman, C., Paletz, S. B. F., Orasanu, J., & Thomas, M. J. W. (2010). The breakdown of coordinated decision making in distributed systems. *Human Factors, 52,* 173-188. doi: 10.1177/0018720810372104

Bienefeld, N., & Grote, G. (2014). Speaking up in ad hoc multiteam systems: Individual-level effects of psychological safety, status, and leadership within and across teams. *European Journal of Work and Organizational Psychology, 23*(6), 930-945. doi:10.1080/1359432X.2013.808398

Brannick, M.T., Prince, A., Prince, C., & Salas, E. (1995). The measurement of team process. *Human Factors, 37,* 641-653. doi: 10.1518/001872095779049372

Cannon-Bowers J, Salas E, & Converse S. (1993). Shared mental models in expert team decision making. In N. John Castellan, Jr. (Ed.), *Individual and group decision making: Current issues,* (pp. 221-246). Hillsdale, NJ, US: Erlbaum.

Cahill, J., McDonald, N., & Losa, G. (2014). A sociotechnical model of the flight crew task. *Human Factors, 56,* 1337-1363. doi: 10.1177/0018720814532684

Chute, R. D. & Wiener, E. L. (1996). Cockpit-cabin communication: II. Shall we tell the pilots? *The International Journal of Aviation Psychology, 6*(3), 211-231. doi: 10.1207/s15327108ijap0603_1

DeChurch, L. A., & Mesmer-Magnus, J. R. (2010). The cognitive underpinnings of effective teamwork: A meta-analysis. *Journal of Applied Psychology, 95*, 32-53. doi:10.1037/a0017328

Edmondson, A. C., & Lei, Z. (2014). Psychological safety: The history, renaissance, and future of an interpersonal construct. *Annual Review Of Organizational Psychology And Organizational Behavior, 1*23-43. doi:10.1146/annurev-orgpsych-031413-091305

Endsley, M. R. (2010). Situation awareness in aviation systems. In J. A. Wise, V. D. H. Hopkin, & D. J. Garland (Eds.). *Handbook of aviation human factors*. Boca Raton, FL: CRC. doi: 10.1201/b10401-15

Endsley M. (2015) Final reflections: Situation awareness models and measures. *Journal of cognitive engineering and decision making, 9*(1):101-111. doi: 10.1177/1555343415573911

Endsley, M. R., & Robertson, M. B. (2000). Situation awareness in aircraft maintenance teams. *International Journal of Industrial Ergonomics, 26,* 301-325. doi: 10.1016/S0169-8141(99)00073-6

Fernandes A. B., Smith P. J. (2011). Human-centered design to support flexibility and adaptability in airport surface management. *Proceedings of the 16th International Symposium on Aviation Psychology 16, 106-111*. Dayton, OH.

Fischer, U., Davison, J., & Orasanu, J. (2003). What makes flight situations risky? Examining commercial and general aviation pilots' concepts of risk. *Proceedings of the 12th International Symposium on Aviation Psychology 11,* 380-384. Dayton, OH.

Hausler, R., Klampfer, B, Amacher, A., & Naef, W. (2004). Behavioral markers in analyzing team performance of cockpit crews. In R. Dietrich & T. Childress (Eds.) *Group interaction in high risk environments*. (pp. 25-37). Burlington, VT: Ashgate.

Helmreich, R. L., Klinect, J. R., & Wilhelm, J. A. (1999). Models of threat, error, and CRM in flight operations. *Proceedings of the tenth international symposium on aviation psychology,* 677-682, Columbus, OH: The Ohio State University.

Lillard, R. M., Littlepage, G. E., Moffett, R. G. III, Hein, M. B., Pearn, M., & Beaufore, A. (April, 2015). *Occupational bias in aviation: Explicit vs. implicit attitudes*. Poster presented at the annual conference of the Society of Industrial and Organizational Psychology, Philadelphia, PA.

Littlepage, G. E., Hein, M. B., Moffett, R. G. III, Craig, P. A., Georgiou, A. M. (2016). Team training for dynamic cross-functional teams in aviation: behavioral, cognitive, and performance outcomes. *Human Factors, 58*(8), 1275-1288. doi: 10.1177/0018720816665200

Martin, W.L., Murray, P.S., & Bates, P.R. (2011). Reducing pathological stress effects and increasing pilot performance during unexpected in-flight events. *Proceedings of the 16th International Symposium on Aviation Psychology 16,* 381-386. Dayton, OH.

Mathieu, J.E., Heffner, T. S., Goodwin, G. F., Cannon-Bowers, J.A., & Salas, E. (2005). Scaling the quality of teammates' mental models: Equifinality and normative comparisons. *Journal of Organizational Behavior, 26*, 37-56. doi:10.1002/job.296

Mathieu, J. E., Rapp, T. L., Maynard, M. T., & Mangos, P. M. (2010). Interactive effects of team and task shared mental models as related to air traffic controllers' collective efficacy and effectiveness. *Human Performance, 23*(1), 22-40. doi: 10.1080/08959280903400150

Mosier, K. L., & Fischer, U. M. (2015). Judgment and decision making by individuals and teams. In D. Harris & W. Li (Eds.), *Decision making in aviation* (pp. 139-197). Burlington, VT: Ashgate.

National Transportation Safety Board. (2017) NTSB 2017-2018 most wanted list of transportation safety improvements. Eliminate distractions. Retrieved from https://www.ntsb.gov/safety/mwl/Documents/2017-18/2017MWL-FctSht-EliminateDistraction-A.pdf

Nullmeyer, R. T., & Spiker, V. A. (2003). The importance of crew resource management behaviors in mission performance: Implications for training evaluation. *Military Psychology, 15,* 77-96. doi: 10.1207/S15327876MP1501_06

Rogers, M. D., Mogford, R. H., & Strauch, B. (2000). Post-hoc assessment of situation awareness in air traffic control incidents and major aircraft accidents. In R. Endsley, & D. J. Garland (Eds.), *Situation awareness analysis and measurement* (pp. 73-112). Mahwah, NJ: Erlbaum.

Salas, E., Sims, D. E., & Burke, C. S. (2005). Is there a "big five" in teamwork? *Small Group Research, 36*, 555-599. doi:10.1177/1046496405277134

Smith-Jentsch, Kraiger, K., Cannon-Bowers, J., & Salas, E. (2009). Do familiar teammates request and accept more backup? Transactive memory in air traffic control. *Human Factors, 51,* 181-192. doi: 10.1177/0018720809335367

Smith-Jentsch, K. A., Mathieu, J. E., & Kraiger, K. (2005). Investigating linear and interactive effects of shared mental models on safety and efficiency in a field setting. *Human Factors, 51*, 181-192. doi: 10.1037/0021-9010.90.3.523

Sarter, N. (2008). Investigating mode errors on automated flight decks: Illustrating the problem-driven, cumulative, and interdisciplinary nature of human factors research. *Human Factors*, *50*(3), 506–510. https://doi-org.ezproxy.mtsu.edu/10.1518/001872008X312233

Sarter, N. B., Mumaw, R. J., & Wickens, C. D. (2007). Pilots' Monitoring Strategies and Performance on Automated Flight Decks: An Empirical Study Combining Behavioral and Eye-Tracking Data. *Human Factors*, *49*(3), 347–357. https://doi-org.ezproxy.mtsu.edu/10.1518/001872007X196685

Sarter, N. B., & Woods, D. D. (1994). Pilot Interaction With Cockpit Automation II: An Experimental Study of Pilots' Model and Awareness of the Flight Management System. *International Journal of Aviation Psychology*, *4*(1), 1. https://doi-org.ezproxy.mtsu.edu/10.1207/s15327108ijap0401_1

Stajkovic, A. D., Lee, D., & Nyberg, A. (2009). Collective efficacy, group potency, and group performance: Meta-analysis of their relationships and test of a mediation model. *Journal of Applied Psychology, 94*, 814-828. Doi: 10.1037/a0015659

Tullo, F. J. (2010). Teamwork and organizational factors. In B. G. Kanki, R. L. Helmreich, & J Anca (Eds.). *Crew resource management*. San Diego, CA: Academic Press.

Velazquez, J., & Bier, N. (2015). SMS and CRM: Parallels and Opposites in their Evolution. *Journal of Aviation / Aerospace Education & Research*, *24*(2), 55–78. https://doi-org.ezproxy.mtsu.edu/10.15394/jaaer.2015.1616

Wertheimer, M. & Littlepage, G. E. (2017). *Mapping predictors of team effectiveness in an airline simulation.* Poster presented at the annual conference of the Society for Industrial and Organizational Psychology, Orlando, FL.

WHAT WE KNOW ABOUT TEAMWORK AND MULTITEAM COORDINATION IN AVIATION: TEAMWORK PROCESSES IN AVIATION

Michael B. Hein
Middle Tennessee State University
Murfreesboro, TN, USA
Glenn E. Littlepage
Middle Tennessee State University
Murfreesboro, TN, USA
Andrea M. Georgiou
Middle Tennessee State University
Murfreesboro, TN, USA

Much of the work of pilots, flight attendants, air traffic controllers, aircraft mechanics, and flight operations center personnel is done in teams and coordination within and between teams is required. This is the third in a five-article series discussing theory and research relating to teamwork in aviation. This article presents a core piece of the comprehensive model of teamwork in aviation.

The airline industry involves complex interdependent tasks where planning and implementation are required and adaptation is needed. Under such conditions, teamwork is especially important (e.g., LePine et al., 2008). Teamwork has been studied extensively with respect to cockpit crews, but is important in other teams and across the entire multiteam system.

Sequential Teamwork Processes

Planning (Transition Processes)

In team research, planning activities have been discussed as transition processes (Marks et al., 2001). Flight crews are frequently composed of members with little experience working together. Flight crews were more effective when the captain used the initial preflight briefing to establish norms of safety, effective communication, and cooperation (Ginnett, 2019). Other studies demonstrated the importance of initial planning (Lei et al., 2016), contingency planning (Thomas, 2004), and workload assignment (Hausler et al., 2004). Zijlstra, et al., (2012) found that effective crews had more consistent and more reciprocal communication patterns during initial planning. These studies indicate that transition processes of mission analysis and strategy formulation are related to aircrew performance.

Implementation

Implementation involves attempts to carry out the plans and decisions that have been made Implementation includes action processes needed to carry out the work. Under non-routine situations adaptation may be needed as well.

Action processes. Four teamwork action processes were proposed by Marks and colleagues (2001): goal monitoring, systems monitoring, team monitoring and backup, and coordination. All are important components of teamwork in aviation. Two simulator studies provide evidence of the importance of monitoring progress toward goals and systems monitoring. Compared to less effective teams, more effective teams maintained standards, managed contingencies more effectively, and were more aware of time and the status of aircraft systems (Hausler et al., 2004). Likewise, effective time management was correlated with more effective performance of cockpit teams (Nullmeyer et al., 2003).

Another important aspect of teamwork in aviation is team backup behavior. An observational study of pilots conducting scheduled flight operations indicated that monitoring and cross-checking was related to effective error management (Thomas et al., 2006). Backup behavior by the flight attendants was associated with more effective performance during a simulated emergency (Bienefeld et al., 2014). Interviews with ATC personnel indicated the frequent use of backup behavior (Owen, 2004). Experienced ATC teams displayed teamwork behaviors such as team monitoring and backup behavior, workload balancing, contingency planning, and proactive communication of information to a larger degree than novice controllers (Malakis et al., 2010). Coordination involves the proper timing and sequencing and compatibility of interdependent actions (Marks et al., 2001). Clear assignment of responsibilities among cockpit crews facilitates coordinated action and is related to effective performance (Nullmeyer et al., 2003) and to managing errors (Thomas et al., 2006). Coordination between pilots and ATC can be an issue because of differential access to information, differing risk assessments, and differences in the preferred actions and timing of actions to prevent traffic conflicts (Davison et al., 1999). Coordination issues between pilots and ATC have led to violations of clearances, unnecessary weather encounters, and near collisions (Bearman et al., 2010).

Adaptation. Hatano and Inagaki (1986) distinguish between routine expertise and adaptive expertise and Kozlowski (1998) extended the concepts to the team level. Routine expertise allows for effective team performance under familiar conditions, but adaptive expertise is needed when teams face unfamiliar situations. Adaptive expertise allows for team adaptation and adjustments in team processes in response to non-routine events (Burke,et al, 2006). In aviation, the need to adapt can be triggered by a wide range of circumstances such as changes in weather, mechanical issues, events aboard the aircraft, air traffic, or human error (Loukopoulos et al., 2009). Tschan, et al., (2018) found nonsignificant or modest relations between performance on routine and non-routine tasks. In an airline simulation study, Littlepage and Wertheimer (2017) found that routine and adaptive performance were unrelated. These findings suggest the importance of separate analysis of performance in routine and non-routine situations. In a flight simulation study, Chen, et al., (2005) found effects of both transition and action phase teamwork processes on adaptive performance. Nullmeyer et al., (2003) found that effective performance was related to situation awareness, clear allocation of responsibilities, use of sound tactics, time management, and willingness to change plans. Waller (1999) indicated that the adaptation of cockpit crews was enhanced when they reactively and proactively acquired and shared information in an attempt to establish shared situation awareness, quickly reassessed task priorities, and assigned tasks. While longer, more complex, and more interactive communication was associated with routine performance, Waller observed a different pattern for situations requiring adaptation. Under non-routine situations, adaptive performance was associated with shorter, simpler communications with less discussion. This pattern allows the team to quickly assess a situation and take corrective action (Lei et al., 2016). Under routine conditions, initial planning and contingency planning can facilitate team performance. When unanticipated events occur, however, in-process planning is needed and it may need to be done very quickly (Lei et al., 2016). Across studies, a pattern of results emerges suggesting that many of the factors that facilitate performance in routine situations also apply when adaption is needed. However, non-routine situations require greater flexibility, more rapid response, and simpler communication patterns. Next, we describe research on teamwork processes that impact both the transition and action phases.

Permeating Teamwork Processes

Four overriding teamwork processes are involved in both of the sequential processes of teamwork (planning and implementation) and in the development and maintenance of emergent states. The permeating processes of interpersonal teamwork processes, leadership, communication and decision-making are necessary to effectively accomplish both collaborative planning and implementation.

Interpersonal processes

The quality of interpersonal processes impacts each of the other teamwork categories. The Marks et al. teamwork model (2001) includes three categories of interpersonal processes: conflict management, motivation and confidence building, and affect management. Standards for CRM training emphasize the importance of interpersonal processes, including positive interpersonal relations, conflict resolution, and a climate supporting assertiveness. Occasionally, incident reports indicate safety issues resulting from conflict or poor interpersonal relations. For example, in a classic article, Foushee (1984) reported an incident where a first officer was reprimanded by the captain for making legitimate safety warnings. The captain instructed the first officer to "just look out the damn window" (p. 888). Although interpersonal skills are implied in the discussion of CRM training, and research in other areas (e.g., communication, leadership) and is relevant, there is very little aviation research that directly examines the impact of specific interpersonal processes such as proactive and reactive conflict management, maintenance of motivation, and affect management.

Leadership

Salas et al., (2005) identified leadership as an important component of teamwork. Leadership has been found to be related to team performance in a variety of aviation contexts including pilots during normal conditions (Brannick et al., 1995) and flight crews in emergency situations, (Bienefeld et al., 2014). Leadership also impacts the quality of error management (Thomas et al., 2006), and pre-flight planning (Cahill et al., 2013). Three leadership theories seem to have special relevance to aviation: LMX theory, shared leadership, and functional leadership. Both LMX and shared leadership emphasize the importance of trust, respect, and open communication. LMX theory suggests the quality of leader-member relationships is based on perceptions of competence, dependability, and interpersonal compatibility (Graen et al., 1995). Wilson et al., (2010) suggest that LMX theory is especially relevant to leadership in the cockpit.

Shared leadership is also beneficial in aviation. A typical cockpit crew consists of a captain and a first officer who alternate primary control of the aircraft. While the captain has final authority, open discussion and collaborative problem solving are expected between both cockpit crewmembers. The captain is expected to create a culture of psychological safety where crewmembers feel free to raise questions, suggest alternative courses of action, and engage in mutual monitoring. Mandatory CRM training is designed, in part, to emphasize shared leadership, but findings that errors are less likely to be corrected if they are made by the captain suggest that shared leadership is not fully embraced (Thomas, 2004). Brannick and colleagues (1995) found that shared leadership in the cockpit was related to effective team performance. Directive facets of leadership such as clearly assigning task responsibilities (Bowers et al., 1998; Foushee & Manos, 1981) and establishing norms (Ginnett, 2019) are related to effective performance of cockpit crews. Thus, a balance between formal and shared leadership may be appropriate (Grote, 2016). The shared nature of leadership is apparent in multiteam situations involving pilots and dispatchers and pilots and ATC. When a plane is preparing for flight or in flight, the dispatcher and the pilot have joint responsibility for making the best decisions possible (e.g., fuel load, alternate airports). While the final authority lies with the captain, dispatchers are expected to assertively advocate their preferred course of action (Federal Aviation Administration, 2005). Likewise, ATC personnel provide altitude and course instructions, but the pilot can make requests, discuss options, or even fail to comply if he or she feels safety is threatened. Bienefield & Grote (2014) observed leadership of pilots and flight attendants during simulated emergency situations. Both formal leadership and shared leadership strongly correlated with the quality of the decision and crew performance. Likewise, perception of leader inclusiveness predicted speaking up among both pilot crews and cabin crews. Functional leadership involves the leader assessment of the situation and actions to correct deficiencies (Hackman & Walton, 1986). These actions can include addressing teamwork-related issues such as goals, procedures, and responsibilities. Studies of leadership in coordinated aviation combat simulations revealed that functional leader behaviors increase coordination and performance (DeChurch & Marks, 2006; Murase et al., 2014). In a recent description of leadership in cockpit teams, Grote (2016) indicated that a large portion of the research is based on a functional approach, stressing leadership processes rather than the formal leadership role.

References

Bearman, C., Paletz, S. B. F., Orasanu, J., & Thomas, M. J. W. (2010). The breakdown of coordinated decision making in distributed systems. *Human Factors, 52,* 173-188. doi: 10.1177/0018720810372104

Bienefeld, N., & Grote, G. (2014). Shared leadership in multiteam systems: How cockpit and cabin crews lead each other to safety. *Human Factors, 56,* 270-286. doi: 10.1177/0018720813488137

Burke, C. S., Stagl, K. C., Salas, E., Pierce, L., & Kendall, D. (2006). Understanding team adaptation: A conceptual analysis and model. *Journal of Applied Psychology, 91,* 1189-1207. doi: 10.1037/0021-9010.91.6.1189

Cahill, J., McDonald, N., & Losa, G. (2013). Understanding and improving flight crew performance of preflight flight planning, and briefing task. *The International Journal of Aviation Psychology, 23,* 27-48. doi: 10508414.2013.746158

Chen, G., Thomas, B., & Wallace, J. C. (2005). A multilevel examination of the relationships among training outcomes, mediating regulatory processes, and adaptive performance. *Journal of Applied Psychology, 90,* 827-841. doi: 10.1037/0021-9010.90.5.827

Davison, J., & Orasanu, J. (1999). Alternative perspectives on traffic risk. In R. Jensen (Ed.), *Proceedings of the Tenth International Symposium on Aviation Psychology* (pp. 483-488). Columbus: Ohio State University.

DeChurch, L. A., & Marks, M. A. (2006). Leadership in multiteam systems. *Journal of Applied Psychology*, *91*(2), 311-329. doi:10.1037/0021-9010.91.2.311

Ginnett, R. C. (2019). Crews as groups: Their formation and their leadership. In B. G. Kanki, J. Anca, & T. R. Chidester (Eds.) Cockpit Resource Management. (pp. 73-102). London, UK: Academic Press.

Graen, G. B., & Uhl-Bien, M. (1995). Relationship-based approach to leadership: Development of leader-member exchange (LMX) theory of leadership over 25 years: Applying a multi-level multi-domain perspective. *The Leadership Quarterly*, *6*, 219-247. doi:10.1016/1048-9843(95)90036-5

Grote, G. (2016). Leading high-risk teams in aviation. In C.Peus, S. Vrann, & B. Sehyns (Eds.), *Leadership lessons from compelling contexts: Monographs in leadership and management, Volume 8*, 189-208. Bingley, WY, UK: Emerald Publishing Group.

Hackman, J. R., & Walton, R. E. (1986). Leading groups in organizations. In P. S. Goodman, & Associates (Eds.), *Designing effective work groups*. (pp. 72–119). San Francisco, CA: Jossey-Bass.

Hatano, G., & Inagaki, K. (1986). Two courses of expertise. In H. W. Stevenson, H. Azuma, & K. Hakuta (Eds.), *Child development and education in Japan.* (pp. 262–272). New York, NY: W H Freeman/Times Books/ Henry Holt & Co.

Hausler, R., Klampfer, B, Amacher, A., & Naef, W. (2004). Behavioral markers in analyzing team performance of cockpit crews. In R. Dietrich & T. Childress (Eds.) *Group interaction in high risk environments*. (pp. 25-37). Burlington, VT: Ashgate.

Kozlowski, S. W. J. (1998). Training and developing adaptive teams: Theory, principles, and research. In J. A. Cannon-Bowers & E. Salas (Eds.), *Making decisions under stress: Implications for individual and team training.* (pp. 115–153). Washington, DC: American Psychological Association. https://doi-org.ezproxy.mtsu.edu/10.1037/10278-005

LePine, J. A., Piccolo, R. F., Jackson, C. L., Mathieu, J. E., & Saul, J. R. (2008). A meta-analysis of teamwork processes: Tests of a multidimensional model and relationships with team effectiveness criteria. *Personnel Psychology, 61*, 273-307. doi:10.1111/j.1744-6570.2008.00114.x

Loukopoulos L, Dismukes K, Barshi I. (2009). *The multitasking myth: Handling complexity in real-world operations.* Burlington, VT: Ashgate Pub. Ltd.

Marks, M. A., Mathieu, J. E., & Zaccaro, S. J. (2001). A temporally based framework and taxonomy of team processes. *Academy of Management Review, 26*, 356-376. doi:10.2307/259182607.

Nullmeyer, R. T., & Spiker, V. A. (2003). The importance of crew resource management behaviors in mission performance: Implications for training evaluation. *Military Psychology, 15,* 77-96. doi: 10.1207/S15327876MP1501_06

Owen, C. (2004). Beyond teamwork! Reconceptualising communication, coordination, and collaboration in air traffic control. *Human Factors Aerospace Safety*, *4*(4): 289-306.

Salas, E., Sims, D. E., & Burke, C. S. (2005). Is there a "big five" in teamwork? *Small Group Research, 36*, 555-599. doi:10.1177/1046496405277134

Tschan, F, Semmer, N. K., Vetterli , M., Hunziker, P. R., & Marsch, S. C. (2018). Predicting team-performance and leadership in emergency situations by observing standardized operational procedures: a prospective single-blind simulator-based trial. BMJ Stel Epub ahead of print: August, 10, 2018. doi:10.1136/bmjstel-2018-000342

Wertheimer, M. & Littlepage, G. E. (2017). *Mapping predictors of team effectiveness in an airline simulation.* Poster presented at the annual conference of the Society for Industrial and Organizational Psychology, Orlando, FL.

Wilson, K. A., Guthrie, J. W., Salas, E., & Howse, W. R. (2010). Team process. In J. A. Wise, V. D. Hopkin, & G. J. Garland (Eds.). *Handbook of aviation human factors*, 2nd ed. Boca Raton, FL: CRC. doi: 10.1201/b10401-12

WHAT WE KNOW ABOUT TEAMWORK AND MULTITEAM COORDINATION IN AVIATION: TEAMWORK COMMUNICATION AND DECISION MAKING IN AVIATION

Michael B. Hein
Middle Tennessee State University
Murfreesboro, TN, USA
Glenn E. Littlepage
Middle Tennessee State University
Murfreesboro, TN, USA
Andrea M. Georgiou
Middle Tennessee State University
Murfreesboro, TN, USA

A variety of teams operate within aviation and decisions are made both within individual teams and with multiteam collaboration as well. As a result, multiple decision contexts exist and communication issues are ubiquitous. Two different approaches to decision making are described. The utility of each approach may vary across situational factors such as time pressure and attentional capacity. This is the fourth in a five-article series discussing theory and research relating to teamwork in aviation. This article presents a core piece of the comprehensive model of teamwork in aviation,

Permeating Teamwork Processes

Four overriding teamwork processes are involved in both of the sequential processes of teamwork (planning and implementation) and in the development and maintenance of emergent states. The permeating processes of interpersonal teamwork processes, leadership, communication and decision-making are necessary to effectively accomplish both collaborative planning and implementation. In this article we will cover communication and decision making.

Communication

Communication is characterized as a permeating process because it is intertwined with all the processes and emergent states. Meta-analysis indicated that both the sharing of relevant information and the openness of communication were related to team performance (Mesmer-Magnus et al., 2009). Communication provides a mechanism to share individual situation awareness, impact collective efficacy, serve as a vehicle for planning, or backup behavior, coordination, and so forth.

Communication is needed to convey relevant information (Waller, 1999), but more communication may not always be better (e.g., Zijlstra et al., 2012). Consistent patterns are reported concerning the nature of communication in effective cockpit crews. Communications about the environment and flight status promote shared situation awareness and are related to performance (Bowers et al., 1998; Foushee & Manos, 1981). Compared to less effective cockpit crews, effective ones quickly settled into stable patterns of reciprocal communication (Zijlstra et al., 2012). In effective cockpit crews, commands were likely followed by acknowledgements and questions were likely to be followed by answers (Kanki et al., 2019).

Other communication factors relating to effective flight crew performance include assertive communications to question decisions and point out problems (Bowers et al., 1998) and the use of directives (Bowers et al., 1998; Foushee & Manos, 1981). Directives are more likely to be associated with high performance when they are explicit, provide reasons, and are framed in terms of shared goals rather than status (Orasanu-Engel & Mosier, 2019, 2010; Mosier & Fischer, 2015). These studies indicate the importance of clear, assertive, respectful, and proactive two-way communication among the cockpit crew. Despite this consistent pattern, some evidence suggests a more task-contingent approach to communication. A simulation study with experienced pilots found that relatively long dialogues with frequent speaker switches were associated with high performance under routine conditions, but were negatively related to performance under difficult, non-routine flight segments (Lei et al., 2016). This may also suggest that frequent communication is appropriate under routine conditions, but may detract from effective adaptation, especially under time pressure.

Observation of experienced ATC teams indicated that they tend to use clear, concise, meaningful, and timely communications and proactively communicate within the team and with other ATC teams (Malakis et al., 2010). ATC communication problems can lead to altitude and lateral displacement (course heading) errors that can result in serious air traffic issues (Grayson, 1981). ATC communication problems are most common under two conditions: shift change and sector handoffs. Both of these conditions involve passing control of aircraft to another controller team.

Communications between ATC and pilots can be problematic. Although text-based communication systems are sometimes used, much of the communication between ATC and pilots is verbal contact via radio. Both accident investigations and incident reports indicate that errors in communications between ATC and pilots can cause serious safety risks (Billings & Cheaney, 1981; Kanki, 2019). Another issue is script-based anticipation errors which involve pilots or controllers hearing what they expect to hear (National Aeronautics and Space Administration, 2009).

Communication issues between flight attendants and pilots have been cited as a factor in aircraft accidents (Chute & Wiener, 1996; Ford, Henderson, & O'Hare, 2013). Communications to pilots are sometimes less effective because attendants are unaware of proper terminology for airplane components or the functional significance of issues they observe (Chute & Wiener, 1996). Attendants sometimes notice abnormal conditions (e.g., vibrations, unusual noises, ice on wings), but fail to report their concerns to pilots (Bienefeld & Grote, 2012).

In aviation maintenance, communication is needed between technicians working on the same aircraft (either concurrently or across shifts), between technicians and the lead technician, with other maintenance facilities concerning deferred maintenance, and with other organizational units. A high number of maintenance errors occur following shift transfer. Shift change errors tend to involve more critical aircraft systems and are more likely to result in serious consequences (Endsley & Robertson, 2000). For example, one fatal accident was caused because a technician removed stabilizer screws, but did not inform the incoming technician (Flight Safety Foundation, 1991). Following a flight with mechanical issues, best practice is for the pilots and a maintenance technician to meet and discuss the issue. When this does not occur, the technician must use the pilot's log notes to diagnose the problem. Frequently the log notes provide only a cursory description of the problem, making accurate diagnosis difficult (Munro et al., 2008).

Decision-Making

Vigilant decision-making. Many decisions are relatively routine and made without extensive time pressure (e.g., calculating fuel load or planning for potential diversions based on weather forecasts). In these situations, vigilant decision models may apply (e.g., Forsyth, 2019; Janis, 1989). These models suggest specific sequential steps such as analysis of the nature of the problem, generation of multiple alternatives, evaluation of positive and negative consequences of various alternatives, choice, and implementation. Pilots face varying degrees of time pressure. Frequently, during the preflight phase (where transition processes occur) there is time for vigilant processing. But, during some action phases of the flight, particularly during the takeoff and approach/landing phases, decisions may need to be made quickly (Thomas, 2004).

Naturalistic decision-making. An important component of aviation decision-making is understanding and managing threats. Decisions are often made in reaction to threats (e.g., severe weather, mechanical malfunction) or errors made by the crew or other parties. Thomas (2004) found that threat and error management was a critical component of decision-making across all phases of flight. Because aviation is often a dynamic environment where decisions have to be made under time pressure, with incomplete information and competing goals, the naturalistic decision-making model (NDM) is appropriate for many team decisions (Lipshitz, Klein, Oransu & Salas, 2015). Expertise plays a strong role in NDM as experts often make decisions by pattern matching the current situation to past experiences or recognition-primed decision making (Klein, 2008).

When issues arise, there are often regulations, decision rules and standard operating procedures (SOPs) that apply; in these cases, situation assessment is the critical process. In other cases, teams may face situations where they need to develop novel solutions to novel problems (Canas, Antoli, Fajardo, & Salmeron, 2005). Good decision-making under these conditions requires not only accurate situation awareness and risk assessment, but also metacognitive processes, shared mental models, and efficient resource management. Team decision making is also facilitated by an open communication climate and high levels of trust (Oransu-Engel & Mosier, 2019).

As Salas et al. (2005) noted, the specific ways team concepts manifest are often dependent on context. In aviation, task contexts vary in many ways. For pilots, initial flight planning involves transition teamwork processes while in-flight operations represent action processes. Thus, flight planning and in-flight operations are two distinct phases of flight that provide different contexts with potential to moderate the relationship between communication, decision-making, and leadership with outcome measures (Cahill et al, 2014; Thomas, 2004). Furthermore, task requirements vary across the in-flight operations phases of takeoff, cruise, and approach/landing. Likewise, task demands differ between routine and non-routine tasks. Greater appreciation of the moderating effects of task conditions is likely to result in a more refined understanding of emergent states and processes affecting team performance in aviation. It appears that the most appropriate communication strategies, leader behavior, and decision-making approach may be task contingent. Simpler and directive communication may be needed when non-routine events are experienced while more explanation and solicitation of input may be appropriate when conditions are routine. Vigilant decision-making may be effective for important decisions when time is available, but naturalistic decision making seems to be more appropriate under time pressure. Shared leadership may be appropriate in many situations, but directive leadership may be more appropriate when time demands are extreme. Thus, the ability to adapt leadership and team behavior to situational demands is critical.

References

Billings, C. E., & Cheaney, E. S. (Eds.). (1981). *Information transfer problems in the aviation system* (NASA Technical Paper 1875). Retrieved from https://ntrs.nasa.gov/archive/nasa/casi.ntrs.nasa.gov/19810022620.pdf

Bienefeld, N., & Grote, G. (2012). Silence that may kill: When aircrew members don't speak up and why. *Aviation Psychology and Applied Human Factors, 2*(1), 1-10. doi: 10.1027/2192-0923/a00021

Bowers, C. A., Jentsch, F., Salas, E., & Braun, C. C. (1998). Analyzing communication sequences for team training needs assessment. *Human Factors, 40*, 672-679. doi: 10.1518/001872098779649265

Cahill, J., McDonald, N., & Losa, G. (2013). Understanding and improving flight crew performance of preflight flight planning, and briefing task. *The International Journal of Aviation Psychology, 23,* 27-48. doi: 10508414.2013.746158

Cañas, J. J., Antolí, A., Fajardo, I., & Salmerón, L. (2005). Cognitive inflexibility and the development and use of strategies for solving complex dynamic problems: Effects of different types of training. *Theoretical Issues in Ergonomics Science*, *6*(1), 95–108. https://doi-org.ezproxy.mtsu.edu/10.1080/14639220512331311599

Chute, R. D., & Weiner, E. L. (1995). Cockpit-Cabin communication: I. A tale of two cultures. *International Journal of Aviation Psychology*, *5*(3), 257. doi-10.1207/s15327108ijap0503_2

Endsley, M. R., & Robertson, M. B. (2000). Situation awareness in aircraft maintenance teams. *International Journal of Industrial Ergonomics, 26,* 301-325. doi: 10.1016/S0169-8141(99)00073-6

Flight Safety Foundation (1991). *Aviation Safety Network accident description for September 11, 1991.* https://aviation-safety.net/database/record.php?id=19910911-0.

Foushee, H. C. (1984). Dyads and triads at 35,000 feet: Factors affecting group process and aircrew performance. *American Psychologist, 39,* 885-893. doi: 10.1037/0003-066X.39.8.885

Ford, J., Henderson, R., & O'Hare, D. (2013). Barriers to intra-aircraft communication and safety: The perspective of flight attendants. *The International Journal of Aviation Psychology, 23*(4), 368-387. doi: 10.1080/10508414.2013.834167

Forsyth, D. R. (2019). *Group Dynamics*, 7th ed. Boston, MA: Cengage.

Grayson, R. L. (1981). Information transfer in the surface component of the system: Problems associate with briefing of relief controllers. In C. E. Billings, & E. S. Cheaney (Eds.), *Information transfer problems in the aviation system* (NASA Tech Paper 1875).

Janis, I.L. (1989). *Crucial decisions: Leadership in policymaking and crisis management:* NY: The Free Press.

Kanki, B. G. (2019). Communication and crew resource management. In B. G. Kanki, J. Anca, & T. R. Chidester (Eds.), *Crew Resource Management,3rd ed.* (pp. 103-138). London, UK: Academic Press.

Klein, G. (2008). Naturalistic decision making, *Human Factors, 50*, 456-60. doi: 10.1518/001872008X288385

Lei, Z., Waller, M. J., Hagen, J., & Kaplan, S. (2016). Team adaptiveness in dynamic contexts: Contextualizing the roles of interaction patterns and in-process planning. *Group & Organization Management, 4,* 491-525. doi: 10.1177/1059601115615246

Lipshitz R, Klein G, Orasanu J, Salas E. (2015) Taking stock of naturalistic decision making. In D. Harris & W. Li (Eds.), *Decision making in aviation* (pp. 3-24). Burlington, VT: Ashgate.

Malakis, S., Kontogiannis, T., & Kirwan, B. (2010). Managing emergencies and abnormal situations in air traffic control (part II): Teamwork strategies. *Applied Ergonomics, 41*, 628-635. doi: 10.1016/j.apergo.2009.12.018

Mesmer-Magnus, J. R., & DeChurch, L. A. (2009). Information sharing and team performance: A meta-analysis. *Journal of Applied Psychology*, *94*, 535-546. doi:10.1037/a0013773

Mosier, K. L., & Fischer, U. M. (2015). Judgment and decision making by individuals and teams. In D. Harris & W. Li (Eds.), *Decision making in aviation* (pp. 139-197). Burlington, VT: Ashgate.

Munro, P. A., Kanki, B. G., & Jordan, K. (2008). Beyond "inop": Logbook communication between airline mechanics and pilots. *The International Journal of Aviation Psychology*, *18*, 86-103. doi: 10508410701749563

Orasanu-Engle J. & Mosier, K. L., (2019). Flight crew decision-making. In B. G. Kanki, & J. Anca, & T. R. Chidester (Eds.) *Crew resource management* , 3rd ed. (pp. 139-184). San Diego, CA: Academic Press.

Salas, E., Sims, D. E., & Burke, C. S. (2005). Is there a "big five" in teamwork? *Small Group Research, 36*, 555-599. doi:10.1177/1046496405277134

Thomas, M. J. W. (2004). Predictors of threat and error management: Identification of core nontechnical skills and implications for training systems design. *International Journal of Aviation Psychology, 14,* 207-231. doi:10.1207/s15327108ijap1402_6

Waller, M. J. (1999). The timing of adaptive group responses to nonroutine events. *Academy of Management Journal,* 42, 127-137. doi: 10.2307/257088

Zijlstra, R. H., Waller, M. J., & Phillips, S. I. (2012). Setting the tone: Early interaction patterns in swift-starting teams as a predictor of effectiveness. *European Journal of Work and Organizational Psychology*, *21,* 749-777. doi: 10.1080/1359432X.2012.690399

WHAT WE KNOW ABOUT TEAMWORK AND MULTITEAM COORDINATION IN AVIATION: CONTEXTUAL FACTORS AFFECTING TEAMWORK IN AVIATION

Andrea M. Georgiou
Middle Tennessee State University
Murfreesboro, TN, USA
Glenn E. Littlepage
Middle Tennessee State University
Murfreesboro, TN, USA
Michael B. Hein
Middle Tennessee State University
Murfreesboro, TN, USA

Technology and culture are two major influences that play a role in team processes and performance. Technology can lessen the cognitive workload; however, the use of technology has the potential to undermine shared cognition and teamwork. Aviation technology has been developed to support individual performance, without sufficient analysis of the impact on team performance. Due to the international nature of aviation, cultural differences can play a role in teamwork. Dimensions of national culture, such as power distance and gender roles, along with inadequate English proficiency may impact the effectiveness of communication and teamwork. Status barriers and physical barriers such as the closed cockpit door can compromise communication and coordination between pilots and flight attendants. Teamwork issues related to status and power differences within the cockpit were a primary motivation for the development of CRM. These findings suggest that a wide range of contextual factors can affect teamwork attitudes and behaviors.

Technology and Teams

Aviation is heavily based in technology and technological advances continue to be introduced. Some view technology as functioning as a teammate (Fiore & Wiltshire, 2016; Hoeft et al., 2006). Technology can lessen the cognitive load by allowing for the offloading of duties (i.e., the use of autopilots and automated airspace management tools) and by providing shared displays and memory archives (such as the onboard Flight Management System-FMS with navigation, communication, performance, etc.) Despite the benefits of technology, the relationship between technology and performance is complex. Aviation technology is highly reliable, but not infallible. For example, highly sophisticated Air Traffic Control (ATC) technology can identify airspace conflicts, but the algorithms do not fully account for factors such as winds and airplane handling characteristics. In an ATC simulation study, Rovira & Parasuraman (2010) found that when an automated airspace conflict management system was perfectly reliable, airspace conflicts were noticed and handled more quickly. However, when the technology occasionally provided false alarms or failed to note airspace conflicts, performance declined markedly. These findings highlight the issue of trust in technology.

If teams underestimate the reliability and accuracy of information provided by technology, they may discount accurate information. If they overestimate the accuracy and reliability of technology, they may discount information from other sources. Reports of aviation accidents and near misses indicate that either too much or too little trust in technology can have serious consequences (Rovira & Parasuraman, 2010). Under time pressure, decisions suggested by technology tend to have very high salience and lead to the under-utilization of other sources of information. Another issue is that an overreliance on technology may result in the erosion of skills and expertise (Mosier & Fischer, 2015).

Communication technology can affect communication effectiveness and team performance. While text-based communication, such as ACARS, is asynchronous and less rich, this digital data can be stored for future reference. Comparison of voice and text chat communication among teams operating unmanned aircraft systems revealed that text-based communication involved greater communication lag and different patterns of coordination. There is also concern that the use of text-based communication can impede situational awareness. Agent-based modeling predicted only small performance differences between voice communication and text communication utilizing a single display window. However, performance degraded quickly when multiple windows were used (Cooke et al., 2014).

While technology often reduces the workload and cognitive load of human operators, this is not always the case. Monitoring of automated systems is a critical component of the jobs of some aviation personnel, such as flight crewmembers and ATC tower operators (Sarter & Woods, 1994). This creates demands for vigilance, a task that is difficult to consistently maintain. Eye tracking data indicate that pilots monitored basic flight indicators, such as altitude and airspeed, but often failed to monitor information related to automated flight modes. In addition, they sometimes failed to understand the meaning or significance of annunciations from automated systems (Sarter et al., 2007).

A great deal of the technology in aviation has been developed to support individual performance, often without an analysis of how to implement the technology to maximize performance at the team level (Maynard & Rantanen, 2005). Wright and Kaber (2005) found that differing levels of automation to support decision-making affected team processes in different ways. They suggest using a team-based task analysis that includes the impact of technology on individual work and team processes.

Emerging Issues with Technology and Automation

Dismukes et al.(2007) suggested that accidents should be viewed as a failure of the sociotechnical system. This is especially relevant as new technologies are introduced. NextGen and SESAR (Single European Sky ATM Research Joint Undertaking) are two major platforms for the modernization of aviation with extensive sets of technological changes planned for aviation. The implementation of these programs is designed to accommodate greater air traffic and they will involve the introduction of several interdependent technological changes (FAA, 2016). These changes will result in new navigation and communication systems, including a greater use of text communication between ATC and pilots and greater autonomous action by technological systems. Technology and automation have great potential to improve functioning in aviation and will be relied on more and more as technological innovations are implemented. However, there are potential threats to team processes and performance in technological change and we need a much better understanding of how technology and teams in aviation interact as we move forward. Especially needed is research on how proposed technological changes impact teamwork and multiteam systems.

Cooke and colleagues (2014) examined one aspect of technology (communication technology) that affects team and multiteam performance. The multiteam nature of aviation requires communication across teams, such as between ATC and pilots. The expanding use of text-based communication technology raises interesting issues. Text-based communication has both advantages and disadvantages. Examination of nonessential text messages can be delayed so as to avoid conflict with high workload demands. Text messages can provide a more lasting record that can serve as a memory aid. However, asynchronous text-based communications can impede performance in time critical situations. In addition, text-based communication lacks the richness of verbal communication (Cooke et al., 2014). Teamwork issues can arise from a greater use of text-based communication. For example, in the cockpit environment both pilots are generally aware that voice (radio) communication is occurring. This shared awareness allows the pilot not actively involved in the conversation to monitor the conversation and/or ask the other pilot to relay information. If a single text display is provided, only one crewmember may be aware of the message and the second member may not be aware that a message was received. On the other hand, if each crew member has a display, there may be uncertainty about whether each member has noticed the information. Thus, both information sharing and the maintenance of shared situation awareness may be more difficult with text communication.

National and Professional Culture

Culture can exert strong, but often unrecognized influences on behavior, including team behavior (Markus & Conner, 2013). Cultural influences exist within the aviation industry on the professional, organizational, and national levels and can positively and negatively affect operations (Cookson, 2015; Helmreich et al., 2001). Cultural differences can negatively affect team and multiteam performance, especially during high workload and high stress periods (Strauch, 2010). Based on data from 9,400 pilots in 19 countries, Merritt (2000) found that national culture had a greater influence than professional culture in the cockpit. One facet of national culture that affects safety is language. The International Civil Aviation Organization (ICAO) has cited inadequate English language proficiency as a contributing factor in several past accidents, yet Cookson et al. (2011) concluded that language proficiency was rarely the sole issue. Dimensions of national culture such as power distance, gender roles, individualism/collectivism, and uncertainty avoidance affect teamwork in aviation (Wilson et al., 2010). For example, Avianca flight 052 crashed due to the pilots' failure to request immediate landing clearance due to extreme fuel shortage. Helmerich (1994) identified cultural norms as the cause of the Columbian pilots' mitigated communication with ATC.

Mjos (2004) examined dimensions of professional culture among pilots and found that, with increasing experience, captains became more individualistic and focused more on individual needs and less on team cooperation. Pilot perceptions of higher captain dominance coupled with higher individualistic tendencies in the cockpit resulted in first officers feeling less encouragement, conflict tolerance, and reward. These negative professional and social perceptions are dangerous as they create an unsafe atmosphere that restricts communication.

Merritt and Helmreich (1996) discovered that an organization's culture can affect the professional culture in that a positive organizational climate was correlated with more positive CRM attitudes and behaviors. Likewise, the perception of a negative organizational culture that does not support positive CRM attitudes was correlated with a more negative professional culture. A positive professional culture amongst pilots consists of professional pride and motivation, which results in a higher probability of safe flights. On the other hand, a negative professional culture can increase the chances of an accident and is associated with unrealistic pilot perceptions of vulnerabilities, disregard for approved safety procedures, and disregard for teamwork (Helmreich, et al., 2001). Accident rates vary greatly across air carriers and these differences have been attributed to differences in organizational safety culture (Mjos, 2004). A study of small commercial air services in Alaska illustrates how organizational culture can create pressures to take risks. Pilots reported flying in challenging conditions due to explicit or implicit norms, pressure from other pilots, and in some cases policies that paid pilots only if flights were completed (Bearman et al., 2009). These findings suggest that the organizational and professional culture of an airline can affect teamwork attitudes and behaviors.

Expanding outside the cockpit for safety culture issues, culture barriers exist between pilots and flight attendants, which can compromise their abilities to effectively communicate and coordinate. In addition to differences in roles and responsibilities between pilots and flight attendants, there are status differences and social categorization differences in their uniforms, scheduling, travel, and hotel accommodations (Ford et al., 2013). The closed cockpit door creates a physical and cultural barrier between the two entities, which can cause a reduction in pilot and flight attendant communications. It is vital that pilots communicate with flight attendants prior to encountering turbulence to avoid serious injuries to crewmembers and passengers. Chute and Weiner (1995) collected surveys from 177 United States line pilots and 125 flight attendants and discovered 87% of the flight attendants said they sometimes experience turbulence without warning from the pilots. Likewise, as previously noted, flight attendants are frequently reluctant to communicate with pilots. This breakdown in communication could be related to the culture barriers that exist between the two teams.

Conclusion

With the increasing efforts and research of national and international groups, such as NextGen, ICAO, and SESAR, it is important we continue to research how technology and culture influences individual and team performance. Technological advances provide a vast range of benefits with automation and safety, but users must be aware of the potential negative effects such as overreliance, loss of situational awareness, and complacency. Dimensions of national culture and professional culture impacts communication, coordination, and teamwork. A review of the teamwork literature suggests the need for future quantitative and qualitative research in the areas of technology and culture in aviation.

References

Bearman, C., Paletz, S. B. F., Orasanu, J., & Thomas, M. J. W. (2010). The breakdown of coordinated decision making in distributed systems. *Human Factors, 52,* 173-188. doi: 10.1177/0018720810372104

Chute, R. D. & Wiener, E. L. (1996). Cockpit-cabin communication: II. Shall we tell the pilots? *The International Journal of Aviation Psychology, 6*(3), 211-231. doi: 10.1207/s15327108ijap0603_1

Cooke, N. J., Myers, C. W., & Rajivan, P. (2014). Implications of text chat for air crew communication and coordination. In M. Vidulich, P. S. Tsana, & J. Flach (Eds.), *Advances in aviation psychology: Volume 1* (pp. 125-136). Farnham, Surrey, England: Routledge.

Cookson, S. (2015). 'We need priority please' Mitigated speech in the crash of Avianca flight 052. *Proceedings of the International Association of Aviation Psychology, 18,* 578-583. Dayton, OH: ISAP.

Cookson, S., Zagreb, T., & Cove, N. (2011). Revisiting the assumptions underlying ICAO's language proficiency program. *Proceedings of the International Association of Aviation Psychology, 16,* 26-31. Dayton, OH: ISAP.

Dismukes, K., Bermn, B. A., & Loukopoulos, L. D. (2007). *The limits of expertise: Rethinking pilot error and the causes of airline accidents.* Aldershot, England: Ashgate.

Federal Aviation Administration [FAA]. (2016). *Delivering NextGen.* Retrieved from https://www.faa.gov/nextgen/delivering

Fiore, S. M., & Wiltshire, T. J. (2016). Technology as teammate: Examining the role of external cognition in support of team cognitive processes. *Frontiers in Psychology, 7.*

Ford, J., Henderson, R., & O'Hare, D. (2013). Barriers to intra-aircraft communication and safety: The perspective of flight attendants. *The International Journal of Aviation Psychology, 23*(4), 368-387. doi: 10.1080/10508414.2013.834167

Helmreich, R. L. (1994). Anatomy of a system accident: The crash of Avianca Flight 052. *International Journal of Aviation Psychology, 4*, 265-284. doi:10.1207/s15327108ijap0403_4

Helmreich, R.L., Wilhelm, J.A., Klinect, J.R., & Merritt, A.C. (2001). Culture, error and Crew Resource Management. In E. Salas, C.A. Bowers, & E. Edens (Eds.), *Applying resource management in organizations: A guide for professionals.* Hillsdale, NJ: Erlbaum.

Hoeft, R. M., Kochan, J. A., & Jentsch, F. (2006). Automated systems in the cockpit: Is the autopilot, "George," a team member? In C. Bowers, E. Salas, & F. Jentsch (Eds.), *Creating high-tech teams.* Washington, DC: APA.

Markus, H. R., & Conner, A. (2013). *Clash!: 8 cultural conflicts that make us who we are*. New York: Hudson Street Press.

Maynard, P. W. & Rantanen. E. M. (2005). Teams, teamwork, and automation in air traffic control. *Proceedings of the International Symposium on Aviation Psychology,* 484-489. Dayton, OH: ISAP.

Merritt, A. (2000). Culture in the cockpit: Do Hofstede's dimensions replicate? *Journal of Cross-Cultural Psychology, 31*(3), 283-301.10.1177/0022022100031003001

Merritt A, & Helmreich R. (1996) Human factors on the flight deck: The influence of national culture. *Journal of Cross-Cultural Psychology, 27*(1):5-24. doi: 10.1177/0022022100031003001

Mjos, K. (2004). Basic cultural elements affecting the team function on the flight deck. *The International Journal of Aviation Psychology, 14*(2), 151-169. doi: 10.1207/s15327108ijap1402_3

Mosier, K. L., & Fischer, U. M. (2015). Judgment and decision making by individuals and teams. In D. Harris & W. Li (Eds.), *Decision making in aviation* (pp. 139-197). Burlington, VT: Ashgate.

Rovira, E., & Parasuraman, R. (2010). Transitioning to future air traffic management: effects of imperfect automation on controller attention and performance. *Human Factors*, *52*(3), 411–425. doi- 10.1177/0018720810375692

Sarter, N. B. & Woods, D. D. (1994). Pilot interaction with cockpit automation II: An experimental study of pilots' model and awareness of the flight management system. *The International Journal of Aviation Psychology (4)*, 1-28. Doi: 10.1207/s15327108ijap0401_1 *50*, 506-510.

Sarter, N. B., Mumaw, R. J., & Wickens, C. D. (2007). Pilots' monitoring strategies and performance on automated flight decks: An empirical study combining behavioral and eye-tracking data. *Human Factors, 49*, 347-357. doi: 10.1518/001872007X196685

Strauch, B. (2010). Can cultural differences lead to accidents? Team cultural differences and sociotechnical system operations. *Human Factors, 52*(2), 246-263. doi:10.1177/0018720810362238

Wilson, K. A., Guthrie, J. W., Salas, E., & Howse, W. R. (2010). Team process. In J. A. Wise, V. D. Hopkin, & G. J. Garland (Eds.). *Handbook of aviation human factors*, 2nd ed. Boca Raton, FL: CRC. doi: 10.1201/b10401-12

Wright, M. C., & Kaber, D. B. (2005). Effects of automation of information-processing functions on teamwork. *Human Factors*, *47*(1), 50-66. doi:10.1518/0018720053653776

A PASSIVE ELECTROENCEPHALOGRAPHY BRAIN-COMPUTER INTERFACE PREDICTS MENTAL WORKLOAD DURING FLIGHT SIMULATION

Adam Fraser, Kathleen Van Benthem and Chris M. Herdman
Visualization and Simulation (VSIM) Centre
Carleton University
Ottawa, Canada

The objective of the present research was to investigate an electroencephalography (EEG) brain-computer interface (BCI) for monitoring realistic variations in mental workload during virtual reality (VR) flight simulation. Many aviation accidents are related to pilot cognition and a mismatch between task demands and cognitive resources. Real-time neurophysiological monitoring offers an approach to identifying high-workload mental states by obtaining continuous, objective measurements without adding to the workload of the pilot. Workload was manipulated by varying navigational difficulty and communication tasks during VR flight simulation. EEG data collected during simulated flight was analyzed to evaluate performance of passive BCI for classification of workload level. BCI approaches were guided by EEG workload literature. A classification rate of 75.9% was obtained, with Alpha and Beta frequency bands being most informative. The results indicate that a passive EEG-BCI may be an effective strategy for monitoring workload and enhancing flight safety.

Electroencephalography (EEG) is a relatively non-invasive and temporally precise neuroimaging device and has become a popular instrument for monitoring mental states (Abreu et al., 2018). In recent decades EEG has been applied to indexing mental workload states (e.g., Berka et al., 2007) and more recently has been included in research relating to monitoring mental workload during flight activities (e.g., Dehais et al., 2019; Harja et al., 2020).

The motivation for incorporating EEG in pilot workload monitoring is that EEG provides an opportunity for objective and continuous measurements of workload level. Achieving reliable EEG measurements of workload has the potential to facilitate prevention of frequent workload-related accidents in aviation and contribute to aviation psychology research. The non-physiological standard for workload evaluation is subjective reporting, wherein pilots rank their level of workload (e.g., NASA Task Load Index) *after* performing a flight operation. However, this method has its limitations. For example, perceived workload does not always correlate well with task performance (see Matthews et al., 2020 for a review). Assessing workload through subjective questionnaires also requires stopping the primary task (or directing attention away from it) which restricts use of this method in real-world settings.

Efforts are being made to establish EEG into a passive BCI (pBCI) for the purpose of classifying high-workload mental states during flight. A pBCI system employs a neuroimaging device to acquire a signal that then gets fed to an analysis program for the purpose of classifying neural activity as relating to a certain mental state. pBCI are distinct from conventional 'active'

BCI which incorporate a response or action such as controlled movement over a robotic limb (e.g., Hochberg et al., 2012). pBCI is now being explored in pilot mental workload research. For example, Dehais et al. (2019) employed an off-line EEG pBCI to classify high- and low-workload periods during flight and obtained 71% accuracy. Although promising, 71% likely illustrates the low-end of potential for pBCI as Dehais et al. employed a 6 dry-electrode system in an actual aircraft which encompasses many engineering and signal acquisition limitations.

The purpose of the present research was to evaluate the efficacy of pBCI as a pilot workload monitoring tool under dynamic flight environments. Workload was manipulated through changes in pilot related tasks. Detecting changes from a 'medium' to high level of workload is most critical for flight safety, therefore participants were continuously loaded with tasks even when not in the high-workload condition. EEG was collected during flight and analyzed off-line to determine the predictive power of EEG-pBCI on workload level. The EEG workload literature guided selection of specific EEG features and scalp locations. Alpha, Beta, and Theta EEG oscillations were hypothesized to reflect workload level, particularly at frontal and parietal electrode sites.

Method

Participants

Fifteen participants with no flying experience were recruited for the present study. All participants were briefed on task requirements, and experiment materials before providing written consent. Ethics were approved by the Carleton University Research Ethics Board (CUREB). Participants were reimbursed for their participation with refreshments and course-credit.

Procedure

Participants 'flew' three practice circuits and four test circuits in a VR flight environment. Half of the test circuits contained a radio-message call-sign memorization task. As shown in Figure 1, participants were instructed to navigate through a series of large rectangular hoops which outlined the oval path of the circuit. Circuits were initiated at altitude of the first hoop at the end of the downwind leg of the circuit. Each circuit took approximately six minutes to complete. Participants were paused by the experimenter when they returned to the starting point of the circuit. After each circuit, participants were presented with questionnaires. Participants were queried about their comfort relating to the VR system and asked to recall the call signs after high-workload circuits.

Mental workload was manipulated between circuits via the presentation of the call sign task, and within circuits by the segment of flight. The high-workload (HWL) condition included all the flight time that occurred during the crosswind and base legs in the circuits that included the call sign tasks. The crosswind segments were relatively challenging for participants as it involved rapid change in heading and altitude. During HWL circuits participants were instructed to listen for and remember the aircraft call signs mentioned in pre-recorded air-to-air communication messages (e.g., "Pendleton Traffic, this is *Delta Echo Foxtrot*, Cessna 150, Five

Miles to the Northeast, Inbound for touch and gos"). The medium-workload (MWL) condition was all flight that occurred during the runway segments in the circuits that did not contain the call sign task. This segment contained straight flight without curves or changes in altitude, heading, or airspeed.

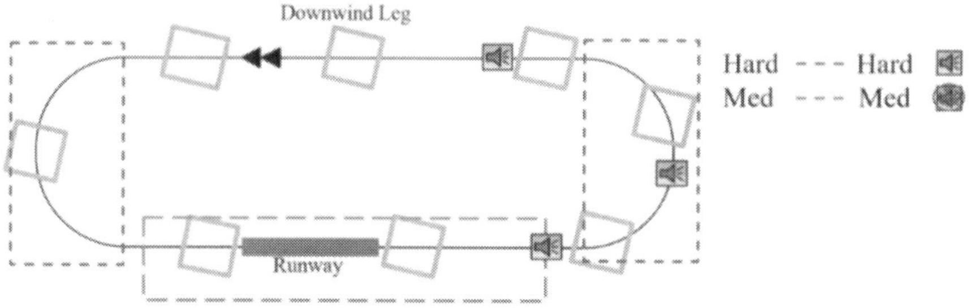

Figure 1. Illustration of flight circuit. Participants began each circuit at the location of the double arrows and at altitude of the first hoop (in green). The red dashed line outlines the navigationally challenging portions of the flight path (Base and Crosswind legs) and the green corresponds to the easiest section of flight (runway leg). Each curve took approximately 40 seconds to complete and each straight leg approximately 140 seconds. The speaker symbols represent the locations where pre-recorded messages were played.

Equipment

Flight simulation apparatus: An HTC Vive VR headset (2016) was used to graphically display the 3D flight simulation, including a full Cessna 172 model aircraft and all exterior terrain and airspace. The flight simulation was produced by Lockheed Martin's Prepar3d software. The location was geo-specific terrain consisting of coastal and mountainous regions surrounding an aerodrome in Hong Kong. The VR headset provided a 360-degree virtual environment. Flight instruments were made visible in the simulation and corresponded to the physical locations of the yoke, throttle, and flaps in the flight control unit (See Figure 2). The simulation produced aircraft realistic visuals and engine noise. Weather conditions were clear with no experience of turbulence.

Electroencephalography: Electrophysiological data was collected using an EMOTIV EPOC+ 14 channel wireless EEG system with electrodes located at AF3, F7, F3, FC5, T7, P7, O1, O2, P8, T8, FC6, F4, F8, AF4. The channel placements follow the international 10-20 system and were referenced online to electrodes P3 and P4. Channels AF3 and AF4 were positioned underneath the top of the VR headset to accommodate the simultaneous use of the two devices (see Figure 2). The EEG recordings were collected at 2048 Hz, and then down-sampled to 256 Hz and were transmitted wirelessly via Bluetooth to an iMac desktop computer.

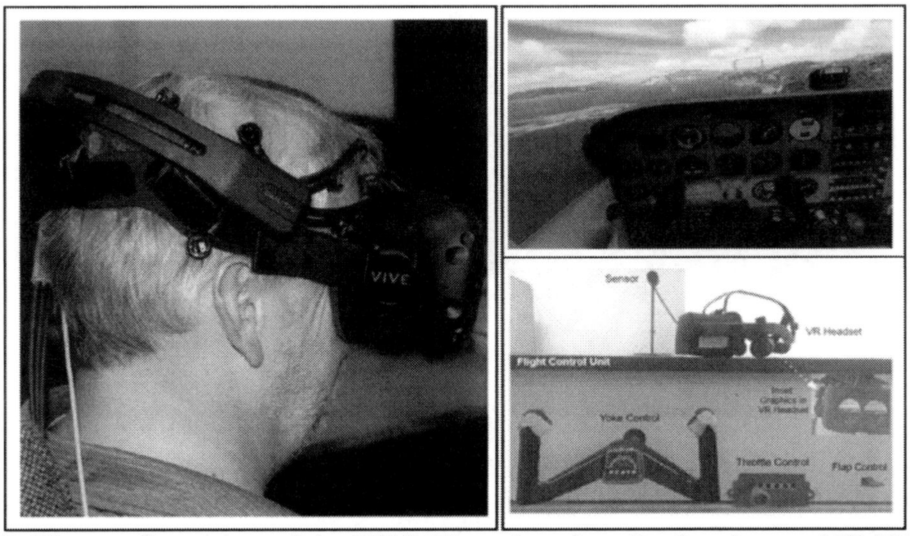

Figure II. The configuration of the HTC Vive virtual reality headset and EMOTIV EPOC+ EEG headset on the left. The participants' view of the simulation environment is displayed on the top right, and the physical instrument layout is shown on the bottom right.

Measures

Continuous EEG measures were transformed into power spectral densities via Hamming windowed sinc FIR filter using the MATLAB plugin EEGLab. Frequency ranges were defined to correspond to conventional EEG 'frequency bands'. The frequency bands were defined as: Delta (1-4 Hz), Theta (4-8 Hz), Alpha (8-12 Hz), and Beta (12-32 Hz).

Analysis

EEG spectral power densities were used as predictors in a classification scheme using a linear discriminant analysis (LDA) algorithm via BCILab. LDA has been recommended as a favorable machine learning algorithm for EEG-BCI as its relative simplicity is favorable for sampling limitations of most human EEG research paradigms (Lotte at al., 2007). Spectral power densities were computed for each 1-second window in high- and medium-workload conditions. There were 120 data points for each condition for each participant. A k-fold cross-validation scheme was used, where 200 of the total data points were used for training and 40 were used for testing. This classification scheme was applied to various approaches including reducing electrodes and reducing frequency band inclusion with the aim of reducing complexity of the BCI system.

Results

Analysis of power spectral densities and classification scores revealed that classification performance was enhanced by evaluating only the Theta, Beta, and Alpha bands. Figure 3 shows the distributions of classification scores from best performing to worst performing across participants. The model including only the Theta, Beta, and Alpha bands are shown in red.

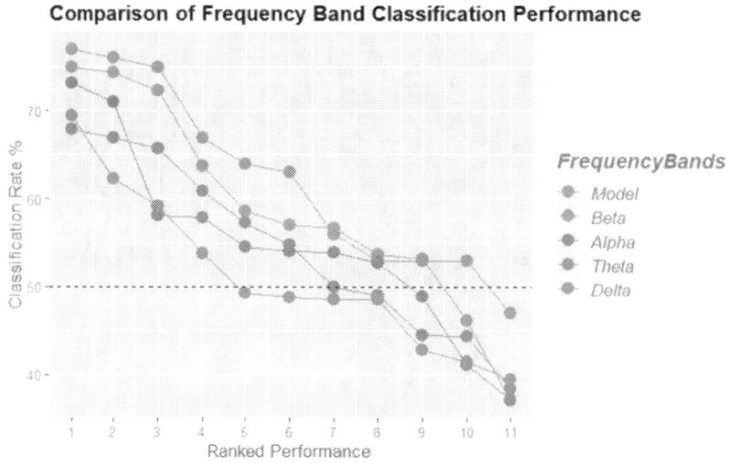

Figure III. Comparison of classification performance for spectral filtering approach. The full model (red) contains oscillatory information between 4 and 32 Hz. Scores from left to right are ordered from better to worse performance separately for each approach (i.e., participant order is varied for each line graph, and y columns do not necessarily correspond to the same participant).

Theta, Beta, and Alpha bands were employed as the spectral filtering model for the following analyses. First all electrodes were included which resulted in a mean classification rate of 56.5% (SD = 13.5%). Classification was improved to a mean of 61.4% (SD = 11.5) with electrode reduction only the primary electrodes that have been related to workload in previous research (AF3, AF4, F3, F4, FC5, FC6, P7, P8, O1, & O2). The third analysis removed the two occipital electrodes due to implications of noise related to eye movements and visual inconsistencies (63%, SD = 11.9). The fourth analysis involved removing four participants with poor BCI performance and may be related to the phenomena of BCI 'illiteracy'. BCI illiteracy occurs in about 20% of subjects where the necessary detection of brain signals is unsuccessful and likely related to neuroanatomical properties (Allison & Neuper, 2010). Lastly, classification approaches were divided into two separate classification approaches for sequential circuits to eliminate temporal effects on EEG signal quality. The classification score for the first two circuits was averaged with the classification score of the last two circuits for each participant and resulted in a classification accuracy of 75.9% (SD = 7.5%).

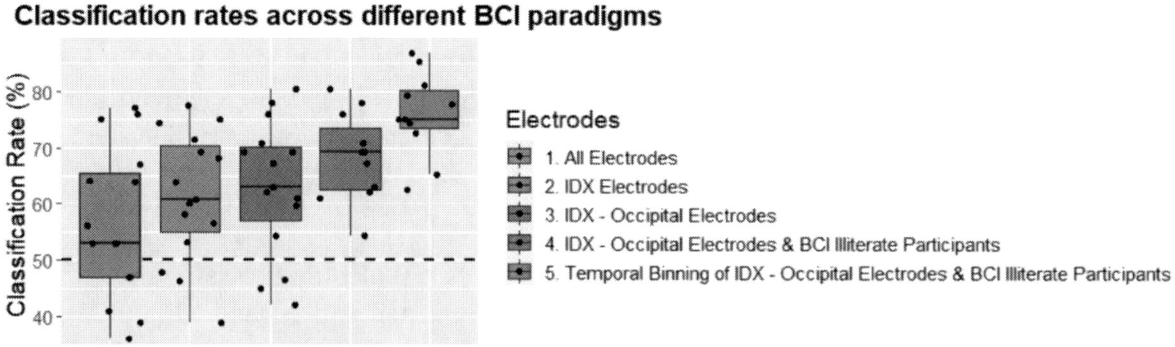

Figure IV. Distributions of participant classification rates as electrode selection was refined. The 'index electrodes' are described in Figure 4. Note: IDX = index electrodes.

Discussion

The present research investigated an EEG-pBCI for monitoring mental workload during VR flight simulation. Workload was manipulated by varying navigational difficulty and performing communication tasks. The workload manipulations were selected for enhancing ecological validity by corresponding with workload variations experienced in regular flight. EEG data was collected and used to classify periods of flight as medium- or high-workload.

Several pBCI approaches were used. Each modification reduced complexity and increased pBCI accuracy, and was grounded in the literature. Similarly, the predictive EEG oscillations and the relevant brain regions matched the hypotheses. Particularly that oscillations within the Theta, Alpha, and Beta range and at parietal and frontal regions were most predictive of workload levels. The final pBCI scheme was successful in classifying medium- versus high-workload conditions 75.9% of the time.

The final classification accuracy is estimated to be a conservative approximation of the potential of pBCI. Longer training phases, individual customization, and training over repeated uses may be feasible strategies to enhance classification. We conclude that, with further development, a passive EEG-BCI may be an effective tool for monitoring pilot workload and enhancing flight safety.

References

Abreu, R., Leal, A., & Figueiredo, P. (2018). EEG-Informed fMRI: A Review of Data Analysis Methods. *Frontiers in Human Neuroscience, 12*.

Allison, B., & Neuper, C. (2010). Could anyone use a BCI? In Brain-Computer Interfaces: HumanComputer Interaction Series (pp. 35–54).

Berka, C., Levendowski, D. J., Lumicao, M. N., Yau, A., Davis, G., Zivkovic, V. T., Olmstead, R. E., Tremoulet, P. D., & Craven, P. L. (2007). EEG correlates of task engagement and mental workload in vigilance, learning, and memory tasks. *Aviation, Space, and Environmental Medicine*, *78*(5), B231–B244.

Dehais, F., Duprès, A., Blum, S., Drougard, N., Scannella, S., Roy, R. N., & Lotte, F. (2019). Monitoring Pilot's Mental Workload Using ERPs and Spectral Power with a Six Dry-Electrode EEG System in Real Flight Conditions. *Sensors (Basel, Switzerland)*, *19*(6).

Hajra, S. G., Xi, P., & Law, A. (2020, October). A comparison of ECG and EEG metrics for in-flight monitoring of helicopter pilot workload. In *2020 IEEE International Conference on Systems, Man, and Cybernetics (SMC)* (pp. 4012-4019). IEEE.

Lotte, F., Congedo, M., Lécuyer, A., Lamarche, F., & Arnaldi, B. (2007). A review of classification algorithms for EEG-based brain-computer interfaces. *Journal of Neural Engineering*, *4*(2), R1–R13. https://doi.org/10.1088/1741-2560/4/2/R01

Matthews, G., De Winter, J., & Hancock, P. A. (2020). What do subjective workload scales really measure? Operational and representational solutions to divergence of workload measures. Theoretical Issues in Ergonomics Science, 1-28

Identifying Opportunities for Augmented Cognition During Live Flight Scenario: An Analysis of Pilot Mental Workload using EEG

Nicholas Wilson[1], Hamed Taheri Gorji[2], Jessica VanBree[4], Bradley Hoffmann[3], Kouhyar Tavakolian[3], Thomas Petros[4]

University of North Dakota; Departments of Aviation[1], Electrical Engineering[2], Biomedical Engineering[3], and Psychology[4]

Augmented cognition is a form of human-systems interaction in which physiological sensing of a user's cognitive state is used to precisely invoke system automations when needed. The present study monitored the in-flight physiological state of the pilot to determine the optimal combination of EEG indices to predict variations in workload, or opportunities for augmented cognition.The participants were 10 collegiate aviation students with FAA commercial pilot certificates and current medical certificates. Each participant performed a uniform flight scenario that included procedures that varied in workload demands. All maneuvers were performed while simultaneously acquiring EEG data in flight. The EEG data were divided into periods of high and low workload. Power spectral density values were computed and subjected to several machine learning methods to distinguish high and low workload periods. The results indicate excellent classification accuracy for distinguishing low and high workload. The present results further demonstrate the potential of augmented cognition.

A growing body of research focusing pilot, driver or operator physiological and cognitive state monitoring during operations of air or ground vehicles facilitates our understandings of the role of human and machine operations (Dussault et al., 2004; Wilson et al., 2019; Guragain et al., 2019; Wang et al., 2019). Within this domain of research, we can apply observed changes in physiological and cognitive state to invoke augmented cognition, or system adaptation based on the condition of the operator. One promising avenue of research in augmented cognition involves developing the capability to continuously monitor an individual's level of fatigue, stress, attention, task engagement, and mental workload in operational environments using physiological parameters (Berka et al., 2007). These physio-cognitive monitoring systems have a wide range of potential applications that could significantly enhance performance, productivity, and safety in military, industrial, and educational settings, including evaluating alternative interface designs, enhancing skill acquisition, and optimizing the ways humans interact with technology (Berka et al., 2007).

Monitoring of the operator functional state can determine if or when the operator is task-saturated, stressed or disengaged and allow the introduction of adaptive aiding by implementation of some form of automation. Attempts to implement adaptive aiding have utilized physiological triggering of adaptive aiding (Wilson & Russell, 2007). One of the challenges for those engaged in operator state monitoring is to utilize the most sensitive set of sensors that are the least intrusive and most practical for the operator.

Blanco et al. (2018) examined the utility of dry electrode EEG measures for distinguishing workload during simulated flight. All participants had previously experienced basic flight training and tested under three different flight scenarios of differential levels of difficulty (easy, medium difficulty, difficult). Each participant flew each scenario once for 10 minutes in counterbalanced order. Scalp EEG dry electrode signals were recorded from Fz, FCz, Cz and Pz using the International 10-20 system. The authors reported that a strong negative correlation between behavioral performance and EEG workload measures. However, a subset of subjects demonstrated increased cognitive workload without any decrement in flight performance. Perhaps physiologically based workload measures can be used to assess learning proficiency during pilot training to identify pilots who are cognitively saturated and at a higher risk to perform poorly as new cognitive challenges emerge.

The present study collected physiological data from pilots while they executed flight patterns that varied in their workload. The purpose of this research was to demonstrate the validity of EEG measures for distinguishing workload in flight. Some of the higher workload flight maneuvers are executing a missed approach at minimums and performing consecutive steep turns. Whereas maneuvers that were classified as low workload included flying straight and level and taxiing at an un-towered airport. To cross-validate perceived workload differentiation between maneuvers, elements of the flight profile were individually ranked by experienced faculty and/or flight instructors at the University of North Dakota's John D. Odegard School of Aerospace Sciences.

Methods

Participants

Ten undergraduate aviation students participated in this study. Study participants held a Federal Aviation Administration (FAA) commercial pilot certificate and either a FAA Class I or Class II medical certificate. The average self-reported flight hours of each participant were 323.6 at the time of the study, with a range of 170 to 840. Each participant was current in the aircraft type flown and all had experience with the Garmin G1000 avionics system. Study participants were informed and provided consent through the approved Institutional Review Board (IRB) protocol. Participants were provided a monetary incentive for their participation.

Experimental Procedure

Informed consent was first obtained from each participant in advance of the meeting time at the airport. Upon arrival at the airport, the participant completed a demographic, recent sleep, and recent stimulant (e.g. caffeine) intake questionnaire and was subsequently connected to the ABM-B-Alert X24 data collection system (Advanced Brain Monitoring, Inc). A baseline recording was collected while the participant was in a quiet, closed office space. Once the baseline recording was completed, the participant boarded a common four-seat single engine trainer aircraft equipped with Garmin G1000 avionics. The participant then performed a pre-determined flight sequence while at the control of the aircraft at the direction of a safety-pilot (the PI) with support of a research assistant sitting in the back seat of the aircraft. During the collection of physiological data, the safety pilot and research assistant noted times of maneuvers. Later the aircraft flight data was downloaded from the avionics to cross-reference against the performed maneuvers. To add a second cross-reference of workload, the PI collected survey data from "experts" aviation faculty or airport leadership to classify maneuvers included in the data collection flight profile as "high" "medium" or "low" workload. This information was used to determine time periods where changes in workload are anticipated to occur.

Image 1
ABM B-Alert

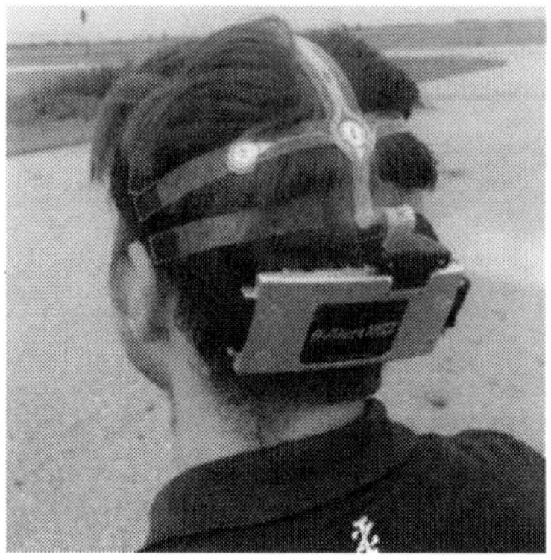

Image 2
Data Collection

Figure 1
EEG 10/20 Placement

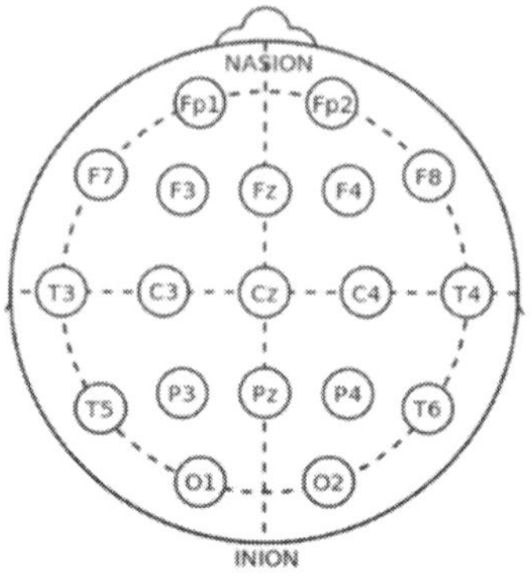

Note. Image 1 showing the ABM B-Alert X24 system. Image 2 showing the data collection flight environment. Figure 1 showing the international 10/20 electrode placement.

EEG recording was accomplished using the Advanced Brain Monitoring (ABM) X-24 system (ABM, 2020). The ABM system includes 20 electrodes placed in the standard international 10-20 system (Fp1, Fp2, F7, F3, Fz, F4, F8, T3, C3, Cz, C4, T4, T5, P3, Pz, P4, T6, O1, POz, and O2) with a sampling rate of 256 Hz. Data is transferred via Bluetooth signal to a laptop with corresponding ABM EEG recording software.

EEG Pre-processing

The recorded EEG data were pre-processed using EEGLAB, an open-source interactive MATLAB toolbox (Delorme & Makeig, 2004) (http://sccn.ucsd.edu/eeglab). The data was filtered using a high-pass filter (1 Hz) followed by a low-pass filter (45 Hz) to remove low-frequency drifts and high-frequency artifacts. Subsequently, the filtered data were visually inspected, and noisy channels, dead channels (channel data indicated no activity over longer periods), muscle activity, mechanical artifacts in the time domain were removed manually, and using EEGLAB "clean_rawdata" plugin. On average, 19.5 EEG channels remained for further analyses (range: 18–20; SD = 0.67). Then, all missing channels were interpolated by spherical algorithm to minimize the potential bias toward a hemisphere. In the next step, Independent Component Analysis (ICA) was computed using the EEGLAB runica function in order to extract independent components (ICs) from signals in scalp level that represent maximum statistical independent sources (Gramann, Ferris, Gwin, & Makeig, 2014; Gramann et al., 2011). Using the ICLabel toolbox (Pion-Tonachini, Kreutz-Delgado, & Makeig, 2019), which is an automatic independent component (IC) classification algorithm, source descriptions including brain, non-brain, eye, muscle, heart, and other sources were automatically assigned to each IC. Consequently, the artifactual ICs with an assigned probability of higher than 0.8 were selected and eliminated from the data, and cleaned EEG signals were used for further processing.

Feature Extraction and Selection

The feature extraction step was performed using power spectral analysis. Fast Fourier Transform (FFT) using one-second hamming windows with 50% overlap was used to transform the EEG into power spectral density (PSD). Each EEG channel based on its frequency was divided into four sub-bands, namely Delta (1-4Hz), Theta (4-8Hz), Alpha (8-13Hz), and Beta (13-30Hz). Since each EEG sub-band has a different frequency range, the average power spectrum for each channel and sub-band was calculated and used for further analysis. Moreover, we calculated the ratios of average spectral powers for theta of each electrode in the frontal area divided by the alpha of the electrodes in the parietal and occipital region; theta divided by beta for each EEG electrode, and beta divided by alpha plus theta for each EEG channel resulting in 148 features. Lasso cross-validation (LassoCV) algorithm was employed to select the most important features.

Classification

Classification refers to a supervised method in which algorithms aim to learn from one portion of already labeled data called training data and uses the learned pattern and information to classify the new unseen portion of data into a proper class. The main goal of this study is to determine the level of the cognitive load of pilots in two classes of high workload and low workload, and this task is defined under the binary classification category. We used the support vector machine (SVM) algorithm as the binary classifier because it is considered one of the most widely used technique in the field of brain signal analysis due to the robust approach for recognition of the complex pattern, good generalization performance, and its efficient computational cost (Wei et al., 2018). To achieve a more accurate estimate of the SVM performance on unseen data and prevent our model from overfitting, we used k-fold cross-validation (k=5), in which all the data were split into 5 subsets. The k-fold cross-validation is an iterative process (k times), and each time the model is evaluated by one of the k subsets while the k-1 subsets used for training and the final results will be the average of all k time evaluations. Moreover, we used accuracy, precision, recall and F score to evaluate the performance of the proposed model to differentiate between high and low workloads.

Result

The top ten features out of 148 features were selected using the LassoCV based on the highest absolute coefficient value as follows: frontal (F4) theta/parietal, occipital (Poz) alpha, frontal (F8) theta/parietal, occipital (Poz) alpha, frontal (F7) theta/parietal (P3) alpha, frontal (F4) theta/parietal (P3) alpha, temporal (T4) theta, temporal (T3) delta, central (C3) theta, temporal (T6) beta/T6 alpha + T6 theta, temporal (T3) beta/T3 alpha + T3 theta, and central (C3) beta/C3 alpha + C3 theta. At the next step, to classify the high and low workloads using the features mentioned above, we trained and tested our SVM classifier with 5-fold cross-validation, and 95.00 $\pm$ 0.30 percent was the highest accuracy achieved. Moreover, the SVM classifier resulted in 100% precision, 90.00 $\pm$ 0.20 recall, and 93.33 $\pm$ 0.13 F-score. The results show that the selected feature can successfully be used as an indicator for the level of the pilots' cognitive workloads.

Figure 2
EEG Data Collection and Analysis Process

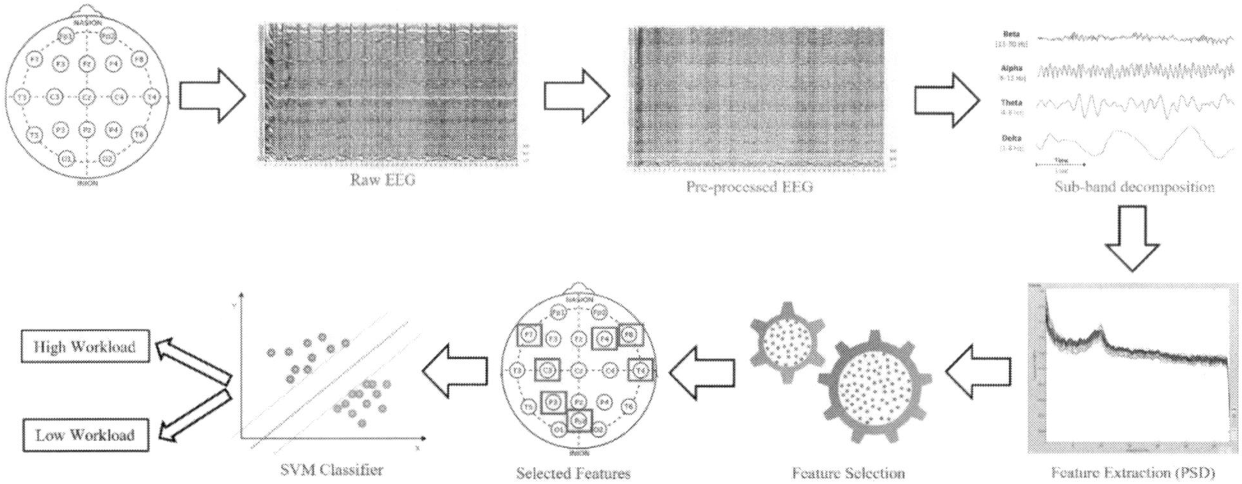

Note. Process above shows collection of EEG data from participant through pre-processing, decomposition to EEG sub-bands, feature extraction and selection through SVM classifier. Ultimately, periods of high and low cognitive workload are determined.

Limitations and Future Directions

Results of this study are consistent with Dussault et al. (2004) in showing cognitive workload fluctuations during flight scenarios. These results will enhance our confidence in establishing reliability during active monitoring of pilot cognitive workload during periods of high workload. The results of this study also support technical feasibility of continued development of advanced headset technology designed to improve pilot situational awareness and monitoring physiological measures underway by Wilson and Tavakolian (2019). Earlier EEG research within a flight simulator also showed promise of EEG and other external measures such as eye tracking to detect periods of drowsiness and fatigue with pilots in a collegiate aviation environment (Guragain et al., 2019; Wang et al., 2019).

The nature of the data collection was in a "live" flight environment. As such, there is greater possibility of motion artifact as a result of aircraft vibration or typical pilot activity which could influence data quality on individual electrodes. Additionally, environmental conditions such as temperature, wind or turbulence may influence certain workload or stress indicators from one flight to another, however, the flight sequence was nearly identical from one flight sequence to another, as such, changes in workload were expected between maneuvers, regardless of outside environmental conditions. Also, the dataset used in this analysis included only 10 participants. A larger dataset could improve data validity and generalizability.

This research provides a foundation for understanding changes in pilot cognitive workload during live flight of aircraft. Such research allows us to establish benchmarks where augmentation of a pilot's available tools or increases in automation may serve to improve the safety of flight. Examples of changes in automation could include changes in density of displayed flight information during high workload conditions or increases in the control exerted by autopilot(s) on relevant flight control surfaces to aid in aircraft control and stability. Future opportunity exists to establish a more formal link between human and machine (referred to as human-machine interface), within the aviation and aerospace domain.

Acknowledgments

The authors would like to thank Mr. Kyle Bernhardt, Ms. Sydney Kramer and Mr. Ryan Krebs for their support during data collection. This research effort received financial support from the ND EPSCoR STEM grants program and the John D. Odegard School of Aerospace Sciences.

References

Advanced Brain Monitoring (ABM). (2021). Carlsbad, CA.

Berka, C., Levendowski, D. J., Lumicao, M. N., Yau, A., Davis, G., Zivkovic, V. T., ... & Craven, P. L. (2007). EEG correlates of task engagement and mental workload in vigilance, learning, and memory tasks. *Aviation, space, and environmental medicine*, *78*(5), B231-B244.

Blanco, J.A.; Johnson, M.K.; Jaquess, K.J.; Oh, H.; Lo, L.-C.; Gentili, R.J.; Hatfield, B.D. Quantifying cognitive workload in simulated flight using passive, dry EEG measurements. IEEE Trans. Cogn. Dev. Syst. 2018, 10, 373–383.

Delorme, A., & Makeig, S. (2004). EEGLAB: an open source toolbox for analysis of single-trial EEG dynamics including independent component analysis. *Journal of neuroscience* methods, 134(1), 9-21.

Dussault, C., Jouanin, J. C., & Guezennec, C. Y. (2004). EEG and ECG changes during selected flight sequences. Aviation, space, and environmental medicine, 75(10), 889-897.

Gramann, K., Ferris, D. P., Gwin, J., & Makeig, S. (2014). Imaging natural cognition in action. *International Journal of Psychophysiology, 91*(1), 22-29.

Gramann, K., Gwin, J. T., Ferris, D. P., Oie, K., Jung, T.-P., Lin, C.-T., . . . Makeig, S. (2011). Cognition in action: imaging brain/body dynamics in mobile humans. *Reviews in the Neurosciences, 22*(6), 593-608.

Guragain, B., Rad, A. B., Wang, C., Verma, A. K., Archer, L., Wilson, N., & Tavakolian, K. (2019, July). EEG-based Classification of Microsleep by Means of Feature Selection: An Application in Aviation. In *2019 41st Annual International Conference of the IEEE Engineering in Medicine and Biology Society (EMBC)* (pp. 4060-4063). IEEE.

Pion-Tonachini, L., Kreutz-Delgado, K., & Makeig, S. (2019). ICLabel: An automated electroencephalographic independent component classifier, dataset, and website. *NeuroImage, 198*, 181-197.

Wei, Z., Wu, C., Wang, X., Supratak, A., Wang, P., & Guo, Y. (2018). Using support vector machine on EEG for advertisement impact assessment. *Frontiers in neuroscience, 12*, 76.

Wang, C., Guragain, B., Verma, A. K., Archer, L., Majumder, S., Mohamud, A., ... & Tavakolian, K. (2019). Spectral Analysis of EEG During Microsleep Events Annotated via Driver Monitoring System to Characterize Drowsiness. *IEEE Transactions on Aerospace and Electronic Systems*, *56*(2), 1346-1356.

Wilson, G. F., & Russell, C. A. (2007). Performance enhancement in an uninhabited air vehicle task using psychophysiologically determined adaptive aiding. *Human factors*, *49*(6), 1005-1018.

Wilson, N. D., & Tavakolian, K. (2019). *U.S. Patent No. 10,328,852*. Washington, DC: U.S. Patent and Trademark Office.

METACOGNITIVE SITUATION AWARENESS IN AVIATION

Haydee M. Cuevas
Embry-Riddle Aeronautical University
Daytona Beach, Florida
Marisa D. Aguiar
Embry-Riddle Aeronautical University
Daytona Beach, Florida

This exploratory study was aimed at gaining a better understanding of metacognitive situation awareness. Seven subject matter experts, two each for commercial aviation and aviation maintenance and three for air traffic control, were asked to define 'situation awareness' as it relates to their job and identify the knowledge, skills, and strategies enabling them to effectively monitor, evaluate, and regulate their situation awareness as they perform their job. Findings from this line of research can guide the design, development, and evaluation of approaches for enhancing and assessing metacognitive situation awareness.

Metacognitive situation awareness refers to the operator's ability to monitor, evaluate, and regulate their situation awareness. Metacognitive monitoring of one's situation awareness has been shown to influence performance in both safety-critical roles, such as air traffic control (McNally et al., 2017; Sethumadhavan, 2011) and command and control (Rousseau et al., 2009), as well as more mundane tasks such as driving (Soliman & Mathna, 2009). In essence, metacognitive situation awareness is a higher order cognitive skill bridging the cognitive processes of situation awareness and metacognition, as described next.

Endsley (1995, p. 36) formally defined situation awareness (SA) as "…the perception of the elements in the environment within a volume of time and space, the comprehension of their meaning and the projection of their status in the near future." Simply stated, SA involves being aware of what is happening around you to understand how information, events, and your own actions will impact your goals and objectives, both now and in the near future. Although alone it cannot guarantee successful decision making, SA does support the necessary input processes (e.g., cue recognition, situation assessment, prediction) upon which good decisions are based. Numerous studies have highlighted the vital role of SA to ensure successful performance in complex domains (e.g., Artman, 2000; Endsley, 1993; Furniss & Blandford, 2006; Sharma et al., 2019; Skrypchuk et al., 2020).

Metacognition has been defined as the awareness of one's own cognitive processes and the ability to understand, control, and manipulate these processes (Davidson et al., 1994, Osman & Hannafin, 1992). Metacognition, therefore, involves two distinct dimensions: knowledge of one's cognitions and regulation of these cognitions (Schraw, 1998). Knowledge of cognition refers to one's awareness and understanding of one's own thoughts and cognitive processes. Regulation of cognition refers to the behaviors one enacts to control and manipulate these processes, such as seeking new information and self-testing one's knowledge. Metacognition has a long established history of research demonstrating its importance for numerous cognitive outcomes. Metacognition plays an essential role in communication and comprehension (both oral and written; see Flavell, 1979), problem solving (e.g., Davidson et al., 1994; Davidson & Sternberg, 1998; Mayer, 1998), memory (e.g., Bjork, 1994; Brown, 1978), and self-regulated learning (e.g., Gourgey, 1998; Hofer et al., 1998; Winne & Hadwin, 1998; Winne & Stockley,

1998). Metacognition has also been shown to be critical to the development of expertise (Glaser, 1989; MacIntyre et al., 2014; Osman & Hannafin, 1992; Smith et al., 1997; Sternberg, 1998).

Method

The aviation domain involves completion of dynamic, highly technology-dependent operations and affords different aviation settings for exploring metacognitive situation awareness. This initial study focused on three settings: commercial aviation, air traffic control, and aviation maintenance. Seven subject matter experts (SMEs), two each for commercial aviation and aviation maintenance and three for air traffic control, were individually asked to respond to the following two questions: (1) Define 'situation awareness' as it relates to your job, and (2) What knowledge, skills, and strategies enable you to monitor, evaluate, and regulate your situation awareness as you perform your job? Below are brief summaries of each SME's background, organized by domain.

Commercial Aviation (CA)

CA-SME-1 holds FAA ATP, CFI, CFII, MEI and Advanced Ground Instructor licenses and ratings and currently has just under 5,000 hours of flight time logged over a period of 40 years. CA-SME-1 has been employed as a regional airline captain for the past three years. Prior to this, CA-SME-1 worked as a flight simulator instructor for one and half years. CA-SME-2 has experience in multiple aircraft as Captain (A-320 / B-737-200 / 300 / 500 / 700 /800 / 900) and as First Officer (B-777 / B-747 / B-767 / 757 / B-737). CA-SME-2 is currently employed as a Captain on the B-737 and as a Line Check Airman and has been working in this occupation for 30 years. CA-SME-2's previous occupation was as a U. S. Air Force pilot (T-38 Instructor, F16 Fighter Pilot; HC-130 Combat Rescue Operations), for 13 years active duty and then 8 years in the U. S. Air Force Guard/Reserve.

Air Traffic Control (ATC)

ATC-SME-1 received training in the U. S. Air Force and then transitioned to the civil air traffic domain, spending eight years in the FAA. ATC-SME-1's experience includes controlling in the tower, approach control, and area control. Currently, ATC-SME-1 works as an ATC Instructor in both tower and radar and has been in this occupation for 29 years. ATC-SME-2 began in air traffic control in the U. S. Air Force and then worked civilian ATC before transitioning back to the military. ATC-SME-2 is currently employed as an Air Traffic Supervisor and has been in this occupation for 21 years. ATC-SME-3 controlled aircraft both in a tower and radar environment at six different facilities, ranging from a VFR tower to a major international airport, then retired from the FAA and started teaching air traffic control. ATC-SME-3 is currently employed as an Associate Professor in Air Traffic Management and has been in this position for 14 years. Prior to this, ATC-SME-3 work for 27 years as an Air Traffic Controller and Supervisor.

Aviation Maintenance (AM)

AM-SME-1 is a U. S. Army trained CH47 rotary wing mechanic, A&P certified. AM-SME-1 worked in the Army for seven years and then transitioned to industry, working in aviation safety, hazmat, tool control, and maintenance. AM-SME-1 currently is employed in quality control as a Technical Inspector and has worked in the aviation industry for 19 years. Prior to this, AM-SME-1 worked in automotive maintenance for five years. AM-SME-2 is employed as a technician, supervisor and Chief Inspector working in a 14 CFR Part 145 Repair

Station for aircraft operated by general aviation operators, 135 operators and air carriers. AM-SME-2 has been in this occupation for 43 years. Prior to this, AM-SME-2 worked as a motorcycle technician for seven years.

Results and Discussion

Results are organized by the three aviation settings: commercial aviation, air traffic control, and aviation maintenance. Common themes across settings are also discussed.

Commercial Aviation

CA-SME-1 defined SA as one's innate, learned, and practiced ability to evaluate the operating environment, while considering various environmental inputs, to identify errors and threats as they arise to ensure appropriate actions can be taken to maintain safe operations. CA-SME-2 described SA as the ability to see and comprehend the 'big picture,' while simultaneously conducting other relevant tasks. The CA-SMEs collectively agreed the knowledge, skills, and abilities required for commercial pilots to effectively monitor their SA are drawn from both crew resource management and recurrent training. According to CA-SME-1, commercial airline pilots use the 'Prepare, Repair, Recover' strategy drawn from crew resource management model of situation awareness to monitor the SA of the flight team. To achieve this, briefs and debriefs are used to assist in the SA monitoring process. CA-SME-2 also highlighted the importance of recurrent training and the use of checklists as additional tools and strategies commercial airline pilots use to monitor their SA. According to CA-SMEs, the knowledge, skills, and strategies enabling commercial airline pilots to evaluate their SA are drawn from training. CA-SME-1 identified the crew resource management model as an effective tool to identify errors and threats at an early stage to ensure a quick return to safe operating conditions. Additionally, CA-SME-1 emphasized the importance of recognizing the following identifiable barriers to good SA: poor workload management, complacency, failure to share information, distractions, fixation, ineffective communication, slang and acronyms, stress and fatigue, and poor briefings. The CA-SMEs agreed effective regulation of one's SA is achieved through the knowledge, skills, and strategies drawn from the experience of training and recurrent training. Additionally, CA-SME-1 emphasized the importance of developing a deeper understanding of one's own cognitive biases. CA-SME-1 described the '3D' strategy as a useful SA regulation tool: pay attention to every *detail*, practice *diligence* consistently, and maintain *discipline* to resist the temptation to deviate in real time.

Air Traffic Control

The ATC-SMEs collectively defined SA as the process of acquiring and maintaining an accurate mental picture of the managed airspace in terms of ongoing traffic, while anticipating the potential for unexpected changes. Per the ATC-SMEs, the knowledge, skills, and abilities required for an air traffic controller to effectively monitor their SA include good listening skills, scanning techniques, and background knowledge drawn from the experience of working in the field. Both ATC-SME-1 and ATC-SME-2 highlighted the importance of using the past experience of having managed various types of airspace traffic as the foundation for the knowledge, skills, and abilities required to assist in the monitoring process. The ATC-SMEs collectively agreed prior experience and previous knowledge of airspace traffic flow are required to evaluate one's SA while controlling the airspace. Drawing from the predictability of experience and a keen understanding of how pilots behave and handle their aircraft provides air traffic controllers with the premise to evaluate their own SA. The ATC-SMEs identified the importance and utility of having a foundational understanding of federal rules and regulations encompassing air traffic control operations aids in the process of regulating one's SA as an air traffic controller. According to the ATC-SMEs, additional strategies could be implemented to regulate one's SA, such as minimizing extraneous discussions with fellow controllers and tuning in to other frequencies to stay up-to-date on what is happening within their managed airspace.

Aviation Maintenance

AM-SME-1 defined SA as paying attention to the paperwork, the environment in which you are performing maintenance, the items being inspected, and measurement and mitigation of any risks associated with any of the above. In terms of knowledge, skills, and strategies to support their M-SA, AM-SME-1 highlighted the importance of experience, teamwork, constant vigilance, and carefully assessing and continually reassessing the situation, especially with regard to risk assessment. AM-SME-2 defined SA as a comprehensive analysis of all of the aspects of aircraft maintenance, operations, crew operations and how they are interdependent. AM-SME-2 also highlighted the importance of risk assessment, relying on historical information, trend analysis, accident analysis and predictive techniques. The goal is to eliminate repetitive operation discrepancies, reduce Time Between Failures (TBF), analyze dispatch rate success, and meet industry standards for operational safety and efficiency.

Common Themes across Settings

In defining SA, the SMEs all highlighted the ability to formulate and maintain an accurate picture, with consideration for the interdependence of multiple elements in the operational environment as well as other relevant tasks. When asked to identify the knowledge, skills, and abilities required to effectively monitor, evaluate, and regulate their SA, the SMEs all emphasized the importance of risk management, including diligence and measurement and mitigation of any risks. Other common themes across the three operational settings included training, background knowledge and experience, communication skills, teamwork, constant vigilance, and avoiding distractions. Elements of crew resource management were explicitly identified by the CA-SMEs and implied in the responses from the ATC-SMEs and AM-SMEs.

Study Limitations and Implications for Future Research

Findings in this exploratory study are promising but limited by the small number of SMEs. Also, the depth of the responses were varied, with some SMEs providing greater details and others less details. To address these limitations, future research is warranted with a larger number of SMEs and in-depth questions designed to elicit more detailed responses. For instance, the SMEs could be asked to provide real-world examples demonstrating the application of the knowledge, skills, and strategies they identified. Future research could also solicit input from SMEs in other aviation settings such as unmanned systems, ground and ramp operations, and airport operations. With a richer understanding of the knowledge, skills, and strategies underlying metacognitive situation awareness, a conceptual framework can be proposed to inform a quantifiable operationalization of this construct. In turn, this quantifiable operationalization would enable translating the three components of metacognitive situation awareness (monitor, evaluate, and regulate) into observable behaviors. To illustrate, knowledge important for metacognitive situation awareness could be demonstrated by answering knowledge questions. Essential skills could be demonstrated by executing tasks for which the skill is needed. Strategies supporting effective metacognitive situation awareness could be demonstrated in realistic simulated scenarios and evaluated by trained observers.

Conclusion

Findings from this line of research can guide the design, development, and evaluation of approaches for enhancing and assessing metacognitive situation awareness. Insights gained from a broader range of SMEs can inform the development of training programs targeting key knowledge, skills, and strategies underlying metacognitive situation awareness. A quantifiable operationalization of metacognitive situation awareness can be used to develop valid and reliable measures to evaluate the effectiveness of such programs as well as the utility of decision aids aimed at supporting operator metacognitive situation awareness.

References

Bjork, R. A. (1994). Memory and metamemory considerations in the training of human beings. In J. Metcalfe & A. P. Shimamura (Eds.), *Metacognition: Knowing about knowing* (pp. 185-205). Cambridge, MA: The MIT Press.

Brown, A. L. (1978). Knowing when, where, and how to remember: A problem of metacognition. In R. Glaser (Ed.), *Advances in instructional psychology* (Vol. 1) (pp. 77-165). Hillsdale, NJ: LEA.

Crandall, B., Klein, G., & Hoffman, R. R. (2006). *Working minds: A practitioner's guide to cognitive task analysis*. MIT Press.

Davidson, J. E., Deuser, R., Sternberg, R. J. (1994). The role of metacognition in problem solving. In J. Metcalfe & A. P. Shimamura (Eds.), *Metacognition: Knowing about knowing* (pp. 207-226). Cambridge, MA: The MIT Press.

Davidson, J. E. & Sternberg, R. J. (1998). Smart problem solving: How metacognition helps. In D. J. Hacker, J. Dunlosky, & A. C. Graesser (Eds.), *Metacognition in educational theory and practice* (pp. 47-68). Mahwah, NJ: LEA.

Endsley, M. R. (1995). Toward a theory of situation awareness in dynamic systems. *Human Factors, 37* (1), 32-64.

Flavell, J. H. (1979). Metacognition and cognitive monitoring: A new area of cognitive-developmental inquiry. *American Psychologist, 34*, 906-911.

Furniss, D., & Blandford, A. (2006). Understanding emergency medical dispatch in terms of distributed cognition: A case study. *Ergonomics, 49*(12-13), 1174–1203. https://doi.org/10.1080/00140130600612663

Glaser, R. (1989). Expertise and learning: How do we think about instructional processes now that we have discovered knowledge structures? In D. Klahr & K. Kotovsky (Eds.), *Complex information processing: The impact of Herbert A. Simon* (pp. 269-282). Hillsdale, NJ: LEA.

Gourgey, A. E. (1998). Metacognition in basic skills instruction. *Instructional Science, 26*, 81-96.

Hofer, B. K., Yu, S. L. & Pintrich, P. R. (1998). Teaching college students to be self-regulated learners. In D. H. Schunk, & B. J. Zimmerman (Eds.), *Self-regulated learning: From teaching to self-reflective practice* (pp. 57-85). New York, NY: Guilford Publications.

MacIntyre, T. E., Igou, E. R., Campbell, M. J., Moran, A. P., & Matthews, J. (2014). Metacognition and action: A new pathway to understanding social and cognitive aspects of expertise in sport. *Frontiers in Psychology, 5*, Article 1155, 1-12. ISSN: 1664-1078.

Mayer, R. E. (1999). Instructional technology. In F. T. Durso, R. S. Nickerson, R. W. Schvaneveldt, S. T. Dumais, D. S. Lindsay, & M. T. H. Chi (Eds.), *Handbook of applied cognition* (pp. 551-569). Chichester, England: John Wiley & Sons.

McNally, K. I., Morris, A. P. & Best, C. (2017). Metacognitive monitoring and control in visual change detection: Implications for situation awareness and cognitive control. Plos one. 2017; 12 (9):e0176032. https://doi.org/10.1371/journal.pone.0176032

Novacek, P. F. (2017). Exploration of a confidence-based assessment tool within an aviation training program. *Journal of Aviation/Aerospace Education & Research, 26* (1), Article 3. https://doi.org/10.15394/jaaer.2017.1717

Osman, M. E., & Hannafin, M. J. (1992). Metacognition research and theory: Analysis and implications for instructional design. *Educational Technology Research & Development, 40*, 83-99.

Rousseau, R. Tremblay, S., Banbury, S., Breton, R. & Guitouni, A. (2010). The role of metacognition in the relationship between objective and subjective measures of situation awareness. *Theoretical Issues in Ergonomics Science, 11* (1-2), 119-130. https://doi.org/10.1080/14639220903010076

Schraw, G. (1998). Promoting general metacognitive awareness. *Instructional Science, 26*, 113-125.

Sethumadhavan, A. (2011). Knowing what you know: The role of meta-situation awareness in predicting situation awareness. *Proceedings of the Human Factors and Ergonomics Society Annual Meeting, 55* (1), 360-364. https://doi.org/10.1177/1071181311551074

Sharma, A., Nazir, S., & Ernstsen, J. (2019). Situation awareness information requirements for maritime navigation: A goal directed task analysis. *Safety Science, 120*, 745-752. https://doi.org/10.1016/j.ssci.2019.08.016

Skrypchuk, L., Langdon, P., Sawyer, B. D. & Clarkson, P. J. (2020) Unconstrained design: improving multitasking with in-vehicle information systems through enhanced situation awareness. *Theoretical Issues in Ergonomics Science, 21* (2), 183-219, https://doi.org/10.1080/1463922X.2019.1680763

Smith, E. M., Ford, J. K., & Kozlowski, S. W. J. (1997). Building adaptive expertise: Implications for training design strategies. In M. A. Quinones & A. Ehrenstein (Eds.), *Training for a rapidly changing workplace: Applications of psychological research* (pp. 89-118). Washington, D. C.: American Psychological Association.

Soliman, A. M. & Mathna, E. K. (2009). Metacognitive strategy training improves driving situation awareness. *Social Behavior and Personality, 37*(9), 1161-1170. https://doi.org/10.2224/sbp.2009.37.9.1161

Sternberg, R. J. (1998). Metacognition, abilities, and developing expertise: What makes an expert student? *Instructional Science, 26*, 127-140.

Winne, P. H. & Hadwin, A. F. (1998). Studying as self-regulated learning. In D. J. Hacker, J. Dunlosky, & A. C. Graesser (Eds.), *Metacognition in educational theory and practice* (pp. 277-304). Mahwah, NJ: LEA.

Winne, P. H. & Stockley, D. B. (1998). Computing technologies as sites for developing self-regulated learning. In D. H. Schunk, & B. J. Zimmerman (Eds.), *Self-regulated learning: From teaching to self-reflective practice* (pp. 106-136). New York, NY: Guilford Publications.

PILOT SITUATION AWARENESS AND RISK OF CRITICAL INCIDENTS USING A NOVEL ONLINE FLIGHT SIMULATION TOOL

Kirsten Brightman[1]
Kathleen Van Benthem[1]
Chris M. Herdman[1]
1. Carleton University
Ottawa, Canada

Pilot situation awareness (SA) is a critical influence on decision making and an important element in maintaining the safe control of an aircraft. The present research investigated whether measures of pilot SA, gathered via an online computer-based cognitive screening tool for pilots, could be used to predict pilot's likelihood of real-world critical incidents. A risk score for each pilot was developed based on their self-reported critical incidents from their actual flight history. It was hypothesized that individuals with lower SA scores would have higher risk scores. The impact of age and pilot experience were also considered, as these factors are known to influence achievement of SA. Results report on comprehensive models of flight performance that quantify the effects of three levels of SA on risk.

Accident and fatality rates in general aviation (GA) have remained consistently high (Kenny, 2020) despite targeted safety strategies. Analysis of general aviation incidents indicates that approximately 70% of accidents are due to pilot error (Kenny, 2020). Specifically, SA errors are frequently linked to pilot-related accidents in general aviation (Bolstad et al., 2010, Jones & Endsley, 1996). Successful performance and safety outcomes rely heavily on pilot SA. Screening pilots for declines or deficits in SA may represent a potential approach to mitigate aviation accident and fatality rates. A current gap in the aviation domain are ecologically valid screening tools that identify pilots who may be at risk for SA failures.

SA is a well-studied element of pilot cognition and is recognized as a fundamental component in performance and safety outcomes in aviation. SA is thought to be a salient causal factor in aviation accidents. Notably, of major air carrier incidents, 88% of those involving human error could be attributable to problems with SA (Endsley, 1995a). SA has been identified as the most significant human factor causation in commercial air transport accidents (Kharoufah et al., 2018). In simulated flight SA is a significant contributor to performance (Bolstad et al., 2010) and predictive of safety outcomes (Van Benthem & Herdman, 2020). Endsley (1995a, p. 36) describes SA as "the perception of the elements in the environment within a volume of time and space, the comprehension of their meaning, and the projection of their status in the near future." The safe operation of an aircraft depends upon the accuracy and completeness of the pilot's SA. SA is thought to build on three hierarchical levels of cognitive processing (Endsley,1995a). The first level involves the pilot's perception of relevant situational elements, their status and their characteristics without interpretation to the larger picture. Approximately 76% of SA errors can be attributed to failures in Level I (Jones & Endsley, 1996). Level II SA involves comprehending the significance of situation elements to current goals and circumstances. At this level, a pilot develops a mental model of the environment, in which elements and events are related. The third and final level of SA involves using knowledge regarding the status and significance of elements to predict future states of the environment.

Aging affects multiple elements of cognition necessary for generating and maintaining SA, such as working memory and attention. (Finnigan et al., 2011). As such, SA is thought to be negatively affected by age-related cognitive decline (Bolstad & Hess, 200). Pilots are not immune to typical cognitive aging (Hardy & Parasuraman, 1997) and there is a growing body of literature which suggests age has a significant negative impact on pilot performance. Older GA pilots have been shown to perform worse than younger pilots in simulated flight (Taylor et al., 2007, Van Benthem & Herdman, 2016). Some of the negative impacts of aging might be alleviated by experience as it has the potential to extend a pilot's working memory capacity through increased automaticity of relevant skills. However, there have been mixed conclusions regarding if expertise can compensate for age-related decrements in performance.

The present research explored whether measures of SA from an online cognitive screening tool for aviators can predict self-reported critical incident risk. Structural equation modeling (SEM) was used to model the direct effects of pilot age and experience on their ability to achieve SA at the three levels, as well as the subsequent direct effects of SA on self-reported critical incident data. SA measures were based on Endsley's three-level characterization. A risk score for each pilot was developed based on their self-reported critical incidents from their actual flight history. Three main hypotheses were investigated. First, increased age was hypothesized to negatively affect SA at all levels. Second, increased experience level was hypothesized to positively affect SA at all levels. Finally, decreased SA at all levels was hypothesized to be associated with higher risk scores.

Method

Participants

The sample was composed of 65 pilots, with ages ranging from 18 to 80 years (M = 48.78, SD = 12.47). Admission criteria include being a licensed/permitted pilot and holding a Canadian medical certificate. Pilots had between 46 and 26,500 logged flight hours (M=3376.01, SD=5570.31), and held an active license for a range of 1 to 54 years (M=20.59, SD=12.10). Certification level ranged from Student to Airline Transport, but the majority were Private VFR with 1 or more additional ratings. All participants provided informed consent to participate in the study in accordance with the Carleton University Research Ethics Board

Materials

Participants completed the study online using a personal electronic device. During the flight exercise, participants watched five short videos from the view of the left pilot's seat of a Cessna 172 Skyhawk. The virtual flight videos were delivered by a screen display and included the interior/exterior of the Cessna 172 Skyhawk and exterior terrain details. Underneath the flight video, two sliders were displayed which signified either a flight instrument or a mental state (e.g. mental workload, SA).

Procedure

Participants completed a pre-flight questionnaire regarding their expertise, demographics, and history of critical incidents. A list of seventeen possible pilot-caused critical incidents was provided. Participants selected all of those which they had personally experienced while acting as pilot in command. Next, participants completed one practice flight leg followed by four test legs. Throughout each video, participants were instructed to monitor and adjust the two sliders so that they accurately reflected the indicated flight instrument or the pilot's self-rated mental state. As pilots were not able to control the simulated aircraft, the slider task (e.g., matching slider values to actual altitude) was included as an alternative visual motor task. After completing each leg, pilots were asked questions relating to Level 1 SA (details regarding other aircraft heard in the radio call messages, and instrument panel monitoring) and Level 2 SA (Ownship and Conflict Detection [derived from radio call messages]). Responses to the questions were used to generate the SA variables.

Conceptual Framework

The current research presents a conceptual framework for understanding SA and pilot risk. Age, certification level, SA at each level were incorporated in the proposed model to examine their influences on risk for critical incidents. Level 1 SA variables related to recall of static information. This information came either from radio calls, such as other aircraft call signs, type, location and intention (SA Others & Intention Others) or was derived from pilot's own instruments (Instrument Error). Level 2/3 SA variables involved awareness of dynamic information (Ownship and Conflict Detection). The number of hours flown was included in the model as a control variable. The main outcome variable of the model was risk score. A risk score was generated for each participant using their responses to the critical incident questionnaire. Each critical incident was assigned a grade from 1 to 5. Grading of critical incidents was primarily based on fatality rates associated with the incident, with 5 representing the highest risk of a fatality. Grades were established and assigned based on expert advice and accident data from the Transportation Safety Board of Canada (2018), International Civil Aviation Organization (2020) and Joseph T. Nall Report (Kenny, 2020). To generate the risk score, the corresponding grades for each critical incident selected by a participant were summed.

Results

Regarding the evaluation of the latent constructs (outer models), the final average variance extracted for all variables was above the 0.5 threshold, representing acceptable convergent validity. Using Cronbach's alpha and composite reliability as guides, all latent variables were found to have acceptable reliability (above 0.7). All final indicator loadings were above the threshold of 0.5 and considered acceptable. The inner model defines the relationship between the latent constructs and directly measures variables. All indices for assessing general fit and quality of the model were acceptable, and the structural model demonstrated a very good fit to the data. The r^2 associated with the outcome variable (Risk Score) and the path coefficients are essential measures for assessing the inner model. *r-Squared* is the standard method used to examine the predictive power of the structural model. It can be seen from Figure 1 that the model has a high predictive power, and accounts for 54% of the variance in the risk score. Figure 1 also demonstrates the path coefficients and *p*-values for each hypothesis, and it can be noted that most hypotheses were supported (non-significant paths were removed from the model).

As predicted, age significantly influenced all SA outcomes, such that increased age resulted in poorer SA scores. Certification level had a significant influence on Instrument Error (β= -0.228, p=.026) and SA Others (β= 0.223, p=.028) such that higher certification level resulted in better SA scores: this is consistent with the hypothesis that experience would positively affect SA. However, certification did not significantly influence Level 2 or 3 SA. Intention (β= -0.247, p=.017) and Conflict Detection (β= -0.308, p=.004) significantly affected Risk Score, such that individuals who performed better on these measures of SA had lower risk scores; partially supporting hypothesis 3. SA Others was found to significantly affect Risk Score (β= 0.208, p=.038), however the direction of the relationship was not as hypothesized. The results indicated an unexpected effect that those who performed better on the SA Others tasks also had higher risk scores. Instrument Error and Ownship were not significant influences on Risk score.

Figure 1.

Hypotheses Testing Results

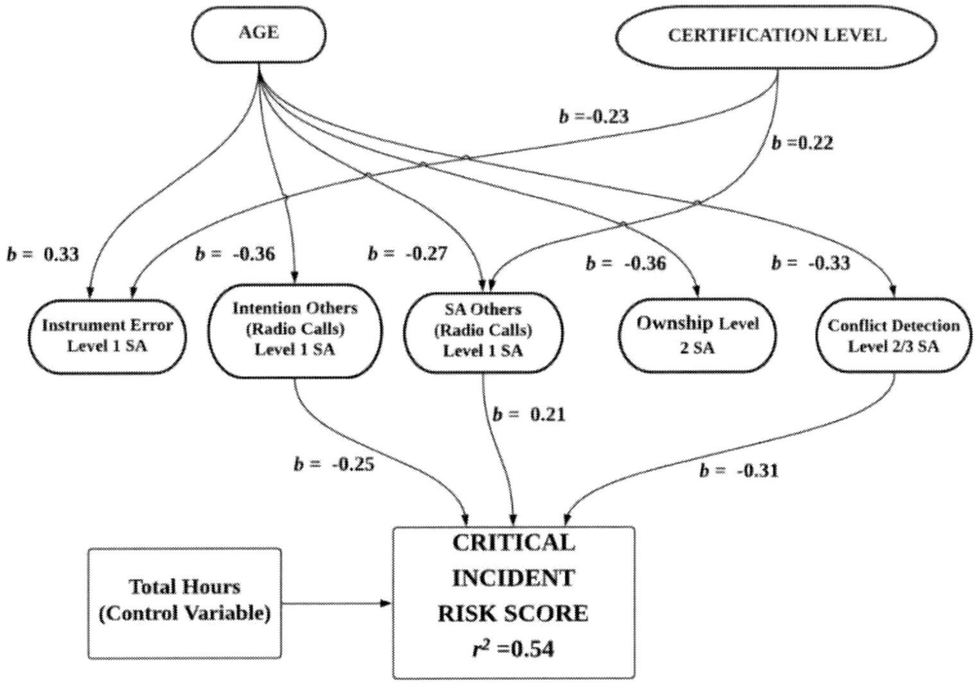

Discussion

The present research aimed to model the direct and indirect effects of pilot age and experience on their ability to achieve SA at all three levels. In the present study age was shown to have a significant effect on all levels of SA. Overall as the age of the pilot increased their performance on SA tasks suffered. The finding that older age was associated with poorer SA outcomes is consistent with findings from the literature, which suggests that SA may be negatively affected by age-related cognitive decline. Age-related cognitive decline may affect Level 1 SA by limiting the amount of information which can be processed and the efficiency of retrieval processes (Bolstad & Hess, 2000). These effects may in turn diminish the development of accurate mental models (Level 2 SA) and future projections (Level 3 SA). The present research provides support for the account that cognitive-aging negatively affects pilot SA.

The effects of experience on the three levels of SA were also examined. As hypothesized, experience had a significant effect on Instrument Error and SA Others in the prediction direction.

The analysis showed there to be no significant relationship between experience and Ownship, Intention or Conflict. The finding that experience had significant effects on Level 1 SA but not Level 2 or Level 3 is consistent with results from Endsley et al. (2002) who reported that differences in SA between experience levels was most pronounced in Level 1.

The primary goal of the present study was to evaluate whether pilot's SA abilities are related to their history of critical incidents. The results of the SEM analysis partially supported the hypothesis that pilots who performed better on SA tasks would have lower Risk Scores. Of the five SA factors, three had significant effects on Risk Score; SA Others, Intention and Conflict. Intention and Conflict demonstrated directionality effects which were consistent with hypotheses, such that pilots who scored higher on these SA tasks had lower risk scores. These findings indicate that SA abilities could be a significant factor contributing to accidents. Of all SA factors, Conflict (Level 3 SA) was found to have the greatest influence on Risk Score. Mental projection is considered a demanding task which people struggle to perform well (Jones & Endsley, 1996). However, it seems that individuals who can perform this task accurately may have better safety outcomes. Jones and Endsley (1996) reported high numbers of SA incidents at the first and second level of SA. Individuals who can successfully generate and uphold Level 3 SA may have effectively avoided making common errors seen in Level 1 and 2. Based on our results, the ability to generate Level 3 SA is a significant and influential predictor of a pilot's history of critical incidents.

Unexpectedly, the direction of the relationship between SA Others and pilot risk showed that pilots who performed better on these SA tasks had higher Risk Scores. The opposite direction of effects for SA Others and pilot risk may represent a paradox with self-reporting critical incidents. For a pilot to report that they've experienced an incident involving a SA failure, the pilot has to be consciously aware that this failure has occurred. A pilot cannot report, for example, that they've had a loss of SA or a near-miss if they don't realize that this has happened. To identify that an SA incident occurred, the pilot has to be aware enough of themselves and their surroundings to recognize that there has been an error. Following this logic, we suggest that individuals with superior awareness may perform better on the SA tasks and also report more critical incidents.

The main purpose of this study was to model the relationship between SA, individual factors and pilot's history of real-life critical flights incidents. The implications of our results are applicable to the development of cognitive assessment tools in the aviation domain which may use SA abilities as a main predictor of safety. Virtual assessment tools which can accurately predict pilot risk have the potential to meaningfully improve safety outcomes in aviation. Knowledge of the precise cognitive processes underlying SA, individual differences predicting SA, and the influence of SA on end performance, will be of utmost importance in the development of cognitive assessment tools for aviators.

References

Bolstad, C. A., Endsley, M. R., Costello, A. M., & Howell, C. D. (2010). Evaluation of Computer-Based Situation Awareness Training for General Aviation Pilots. *The International Journal of Aviation Psychology*, *20*(3), 269–294. https://doi.org/10.1080/10508414.2010.487013

Bolstad, C. A., & Hess, T. M. (2000). Situation awareness and aging. In *Situation awareness analysis and measurement* (277-302).

Endsley, M. (1995a). A taxonomy of situation awareness errors, human factors in aviation operations; *Proceedings of the 21st Conference of the European Association for Aviation Psychology (EAAP)*, *3*, 287–292.

Endsley, M. R. (1995b). Toward a Theory of Situation Awareness in Dynamic Systems. *Human Factors*, *37*(1), 32–64. https://doi.org/10.1518/001872095779049543

Finnigan, S., O'Connell, R. G., Cummins, T. D. R., Broughton, M., & Robertson, I. H. (2011). ERP measures indicate both attention and working memory encoding decrements in aging. *Psychophysiology*, *48*(5), 601–611. https://doi.org/10.1111/j.1469-8986.2010.01128.x

Hardy, D. J., & Parasuraman, R. (1997). Cognition and flight performance in older pilots. *Journal of Experimental Psychology: Applied*, *3*(4), 313–348. http://dx.doi.org.proxy.library.carleton.ca/10.1037/1076-898X.3.4.313

International Civil Aviation Organization. (2020). Safety Report 2020. Retrieved from: https://www.icao.int/safety/Documents/ICAO_SR_2020_final_web.pdf.

Jones, D. G., & Endsley, M. R. (1996). Sources of situation awareness errors in aviation. *Aviation, Space, and Environmental Medicine*, *67*(6), 507–512.

Kharoufah, H., Murray, J., Baxter, G., & Wild, G. (2018). A review of human factors causations in commercial air transport accidents and incidents: From to 2000–2016. *Progress in Aerospace Sciences*, *99*, 1–13. https://doi.org/10.1016/j.paerosci.2018.03.002

Kenny, D. J. (2020). *30th Joseph T. Nall report-general aviation accidents in 2018*. Frederick, MD: AOPA.

Taylor, J. L., Kennedy, Q., Noda, A., & Yesavage, J. A. (2007). Pilot age and expertise predict flight simulator performance. *Neurology*, *68*(9), 648–654. https://doi.org/10.1212/01.wnl.0000255943.10045.c0

Transportation Safety Board of Canada. (2018). Statistical Summary: Air Transportation Occurrences in 2018. Retrieved from: https://www.bst-tsb.gc.ca/eng/stats/aviation/2018/ssea-ssao-2018.html

Van Benthem, K., & Herdman, C. M. (2020). *A Virtual Reality Cognitive Health Screening Tool for Aviation: Managing Accident Risk for Older Pilots.* Manuscript submitted for publication.

Van Benthem, K., & Herdman, C. M. (2016). Cognitive factors mediate the relation between age and flight path maintenance in general aviation. *Aviation Psychology and Applied Human Factors*, *6*(2), 81–90. http://dx.doi.org.proxy.library.carleton.ca/10.1027/2192-0923/a000102

THE EFFECTS OF INCREASED VISUAL INFORMATION ON COGNITIVE WORKLOAD IN A HELICOPTER SIMULATOR

Reilly Innes[1], Zachary Howard[2], Alexander Thorpe[1], Scott Brown[1] & Ami Eidels[1]

[1]University of Newcastle, Newcastle, Australia
[2]University of Western Australia, Perth, Australia

Workload in highly demanding environments can be influenced by the amount of information given to an operator, and consequently, it is important to limit the potential overload. In the current study, we used the Detection Response Task (DRT) to assess the effects of enhanced heads-up display information ("symbology") on cognitive workload in a simulated helicopter environment. Participants (highly trained military pilots) completed simulated helicopter flights, which varied visual conditions and the amount of information given. During these flights participants completed a DRT. With increased heads-up display information, pilots landing accuracy improved across visual conditions. The DRT captured the increased workload resulting from the varying environmental conditions, and provided evidence for heads-up display information having negligible effects on workload. Our study shows that the DRT is a useful workload measure in simulated helicopter settings. We also show that the increased level of symbology appeared to assist pilots flight behaviour and landing ability, without compromising safety. This research highlights that a) the DRT is an easily implemented and effective measure of cognitive workload in a variety of settings and b) the potential for further cognitive workload evaluative methods in similar aviation and applied settings.

In modern society, we have seemingly unlimited sources of information available, however, processing, integrating, and using this information is difficult. For example, in helicopters, pilots have constant access to an array of flight metrics such as speed, altitude, roll and more, all intended to assist the pilot. The primary difficulty in using information is the limits of human attention and perception, which are commonly overlooked.

Driver distraction research has been used to highlight the limits of human attention and the potential consequences of such distraction on cognitive workload (Strayer & Johnston, 2001). Cognitive workload refers to the overall level of cognitive demand placed on an individual from a task/s (Lee et al., 2008; Innes et al., 2020). Cognitive workload includes demands related to the number of tasks at hand, the difficulty of those tasks, associated time pressures and the overall mental and physical effort exerted (Hart & Staveland, 1988).

The detection response task (DRT) is a behavioural measure of cognitive workload, which requires individuals to detect a salient stimulus (light or vibration) and respond as quickly as possible, whilst concurrently completing another task (such as driving; Engstrom et al., 2005; Strayer et al., 2013). There are various methods of measuring cognitive workload, such as subjective questionnaires, eye tracking, and heart rate variability monitoring, however, these fail to provide performance based results, which are important if main task performance is difficult to quantify (Innes et al., 2020). The DRT is easily applied to simple psychological task designs or more complex real world designs. For example, the DRT has been used extensively for simulated and real-world driver distraction studies to assess the level of workload induced from mobile phones, conversations with passengers and smart assistants (Strayer et al., 2013; Strayer et al., 2019). Response times from the DRT give an indication of cognitive workload, as fast responses indicate more available cognitive resources.

Previous cognitive workload studies have shown that errors and inferior task performance are more likely when cognitive workload is high. In a multitasking environment, cognitive demands can be difficult to assess, as adding items to process or increasing task difficulty can lead to a depletion of cognitive resources. When resources are low (i.e. workload is high), errors are more likely. Yet additional sources of information could also lead to redundacy gains, so performance may increase. In helicopter piloting and interface design, maximising informative assistance to aid performance whilst minimising distraction is vital. In this scenario, poor performance has critical consequences, and so needs to be optimised without overloading pilots.

Recently Airbus and Hensoldt have developed state-of-the-art sensor systems, which allow more information to be available to pilots. This information includes 3D mapping of the environment, clearly identified hazards and landing guides – all of which can be displayed in a pilots heads-up display (HUD). In the current study, we aimed to assess the cognitive workload induced under varying amounts of HUD information using the DRT. Highly trained helicopter pilots completed a simple flight path and landing in differing visual conditions, and with differing levels of HUD information. Both flight results (landing execution and flight metrics) and DRT results were assessed to evaluate the impact of HUD information on flight performance and workload. It was hypothesized that increased HUD information would lead to greater flight perfomance, however, we also hypothesized that cognitive workload would similarly increase.

Method

Participants

The participants' were limited to three pilots due to the highly specific requirements of the task. Thus, we ran a small-n study (Smith & Little, 2018) with high repetition to maximise availability of data. All three participants were highly trained helicopter pilots, each with more than 4,000 flying hours experience and extensive simulator experience. The three pilots completed the design seven, five and three times respectively. The uneven number of trials per participant was due to pilot availability, although no pilot varied greatly in DRT results or flight results between iterations.

Materials and Design

Data was collected in an Airbus MRH90 Taipan Multi Role helicopter simulator. The simulator incorporated three partially overlapping screens which made up 200◦ x 40◦ field of vision. The participant sat at a distance of approximately two metres from the screen. Controls in the simulator included a collective shaft, cyclic shaft and two foot pedals. The participants were shown an electronic map and a multi-function display, which indicated altitude, ground speed, collective power and helicopter roll. Participants were also fitted with a headpiece which was placed over the participants eyes. The headpiece acted as goggles, so that the participant could still see the simulator. In conditions where symbology was added, additional information was overlaid in their visual field. The location and angle of the headpiece was tracked at high rate so that information projected into the visual field mapped accurately and dynamically onto the visual environment.

Cogntive load was assessed via a DRT device, closely adhering to ISO 17488 (2016). The DRT device included a vibrating pad, which was taped to the participant's skin near their shoulder, and a response button, which was attached to the collective shaft nearest to where the pilots thumb sat. With an already crowded visual environment, we proposed the use of the tactile DRT to limit visual competition.

Each participant completed two simultaneous tasks – the flight simulation and DRT. For the DRT, a short stimulus was elicited via a vibration. The participant was required to respond via the response button to each iteration of the stimulus. The stimulus lasted for one second (or until the response button was pressed, whichever came first). The DRT stimulus was elicited at an interval of 3 - 5 seconds and occurred for the duration of each simulated flight. For the full DRT method, see ISO 17488 (2016).

	NO SYMBOLOGY (∅) Headpiece off	**2D MINIMAL (2D)** *LIDAR off	**3D MAX SYMBOLOGY (3D)** *LIDAR on, All symbology.
DAY VIS = 12000, TIME = 1600 DUST = OFF	DAY ∅ **A**	DAY 2D **D**	DAY 3D **G**
NIGHT VIS = 12000, TIME = 2000 DUST = OFF, FLIR = (ON) FLIRTIME =2000, FLIRVIS= 2400	NIGHT ∅ **B**	NIGHT 2D **E**	NIGHT 3D **H**
DUST VIS = 1200 TIME = 1600 DUST = ON	DUST ∅ **C**	DUST 2D **F**	DUST 3D **I**

Figure 1: Details of each tactical approach flown, with conditions of symbology and environment shown. There were two additional conditions where the DRT was not present – these were the 3D and no symbology Night conditions.

The flight simulation involved participants undertaking a short predetermined flight path and subsequent landing. There were three conditions of visual environment: High Visibility (Day), Low Visibility (Dust) and Night. There were three conditions of symbology: no symbology, 2D symbology and 3D. A full summary of conditions can be seen in Figure 1. The 2D symbology condition was made as similar as possible to the standard heads-up display used by military helicopter pilots in modern large-platform helicopters. The 3D symbology condition contained extra information, and the no symbology condition contained less. For an example of the three symbology conditions, see Innes et al., (2020). Two baseline conditions were added where participants completed the flight without the DRT in the Night condition *without* symbology and the Night condition *with* 3D symbology. The experiment was thus a 3x3+2 within subjects design (see Figure 1).

Procedure

All three participants were familiar with the simulator environment, and were given instructions about the DRT. The DRT commenced as soon as the pilot lifted the collective shaft for each condition. The flight path was identical for all 11 conditions. Participants were instructed to take off and fly for around one minute towards a mountain, where they would then complete a horseshoe turn at a deignted gate point. Following this the participants were instructed to begin descending to the Landing Zone, which was in the centre of a sports field.

Participants completed all 11 conditions. The order of the nine DRT conditions was randomized and the remaining two baseline conditions were completed following the corresponding DRT-active conditions. If, during a flight trial, the participant crashed or there were any technical issues, the run was restarted. Responses in these trials were recorded separately. Participants were given short breaks between flight trials and long breaks between blocks of flight trials. All flight data was recorded. DRT response times and misses were recorded.

Results

Several flight metrics were used to evaluate the quality of flight for each trial. These metrics included flight path variability measures and landing data. For brevity we report here only the latter. The main reason to evaluate landing/flight quality parameters was to ensure there was no task trade-off between the flight and the DRT. DRT response time and misses were analysed. Flights where the participant crashed were removed. For each metric we completed Bayesian repeated measures ANOVAs to measure effects of environmental conditions, symbology conditions and the interaction. All analysis was completed using the statistical program JASP (JASP Team, 2019).

Flight Metrics. We assessed the accuracy of landing data by borrowing appropriate precision measures from ballistic sciences. Participants were instructed to land at a specified and marked point in the virtual environment (centre of a football field). We measured the absolute distance from this landing zone (LZ) to the actual landing location ("landing error") and the "circular error probable" (CEP), which is the median error radius (Nelson, 1988, p.1). CEP results are shown in Figure 2.

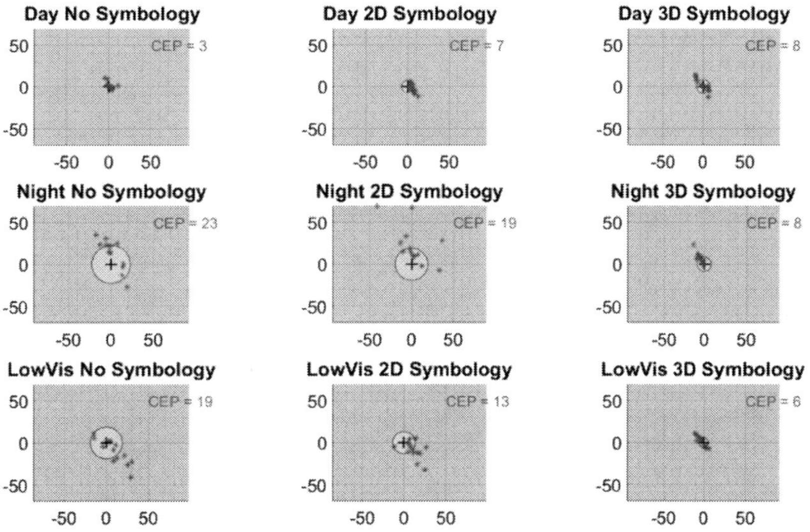

Figure 2: CEP plots for eah flight condition. The grid shows visual conditions as rows and symbology levels as columns. The yellow circle (which is quite small in some plots), shows the CEP for the landings. Each blue dot represents an individual flight landing.

A Bayesian repeated measures ANOVA on distance from the LZ (for each block of landings) revealed strong evidence for main effects of environment and symbology, and for their interaction effect (all $BF_{10}>1000$). Post-hoc analysis showed that landings in the night and dust conditions were significantly worse than the day condition (Night vs Day; $BF_{10}= 137.78$, Dust vs

Day; BF_{10}= 62.95). Landings completed in the 3D symbology condition were also much closer to the designated LZ than with no symbology (BF_{10}= 65.32) and 2D symbology (BF_{10} = 14.39). An interaction effect was also shown, such that the 3D symbology appears to be unaffected by the environmental condition, whereas no symbology and 2D symbology landed closer to the LZ in day conditions, but did much worse in the night and dust conditions. Figure 2 and the ANOVA results suggest that landings completed in the Day condition were most accurate. Furthermore, symbology was shown to have the greatest effect on landings, with 3D symbology landings more consistently closer to the designted landing spot than other symbology conditions.

DRT. Figure 3 shows mean DRT response times across flight and symbology conditions. A two-way repeated measures Bayesian ANOVA of log-transformed RT showed a preference for the model that included the effect of visual condition when observing differences across symbology and time of day (BF_{10}= 4.909). A comparison of visual conditions showed some evidence for a difference between the Day and Night conditions (BF_{10}= 3.335), and good evidence for a difference between Day and Dust conditions (BF_{10}= 6.842). A comparison of symbology conditions showed ambiguity for a difference between the conditions. A two-way repeated measures Bayesian ANOVA of misses showed a preference for the model which included symbology, environment and the interaction, with positive evidence in favour of the null (i.e. evidence for no difference between conditions of symbology and visual conditions; BF_{01}= 3.844). These results should be interpreted with caution, as the small sample size prevents reliable statistical testing.

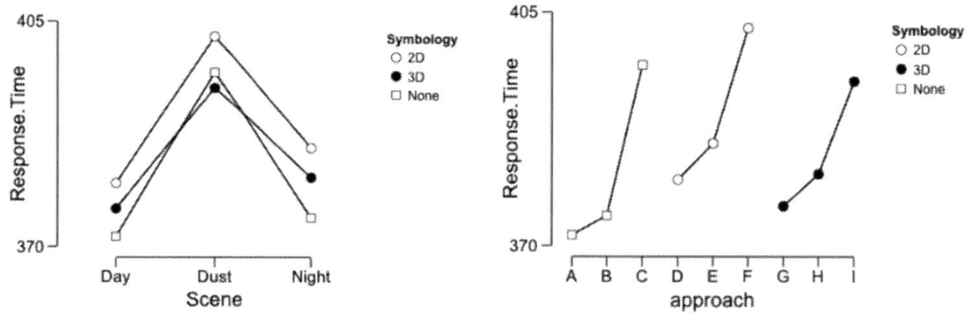

Figure 3: Left; Mean DRT response times for the visual conditions. Right; Mean DRT response times for the individual approaches. Data points in both panels are grouped across symbology conditions (2D, 3D, none) for ease of interpretation.

Discussion

Results indicated that landing performance declined in more difficult conditions (i.e. conditions with lower visibility and conditions with higher symbology), as expected. This was clear in the analysis of flight landings, but performance was difficult to distinguish through variability metrics. The variability metrics did however provide a descriptive analysis of the flight, where in low visibility and low symbology conditions, pilots tended to fly lower and slower. Importantly, it was shown that despite a difficulty increase from the degraded visual environment, workload was relatively unaffected by the symbology. There was no difference in workload, as measured by DRT responses, across the three symbology levels (none, 2D, 3D). However, this is not due to lack of senstivty as the DRT was sensitive enough to detect a change in workload between the Day and Dust conditions. This scenario appears ecologically valid, as pilots in brown-out experience degraded visual conditions and higher amounts of flight errors.

These results highlight the importance of workload evaluation, the sensitivity of the DRT to change in workload and the usability of 3D symbology for flight assitance in night and dust conditions. 3D symbology was especially useful in degraded visual conditions, but added little to performance in high visibility conditions. This is an important finding that highlights the need for adaptive user interface, where extra information may be useful in some conditions, but not others. Further, the additional information appeared to come at no extra cost to the pilots workload. There was a difference between 2D and 3D symbology, which could indicate that the 2D symbology was less intuitive than the 3D symbology, potentially using extra attentional resources or obscuring the field of view (in comparison to No symbology and 3D symbology). This finding is promising for implentation of 3D symbology to deliver increased information to assist pilots.

There were some practical limitations to the current experiment. First, each flight path was relatively short (around 3 minutes), which limited the DRT to around 40 trials. Further, the difficult part of the flight - the landing - only lasted around 30 seconds of the total time. Whilst this highlights the sensitivity of the DRT across such limited number of trials, more subjects and trials over a greater and more variable flight path (where data from specific sections was collated) could provide an avenue for future research.

Acknowledgements

This research was supported by a University of Newcastle Industry Linkage Pilot Grant to SB and AE. We thank Airbus and Hensoldt for their in-kind support.

References

Engström, J., Aberg, N., Johansson, E., & Hammärback, J. (2005). Comparison between visual and tactile signal detection tasks applied to the safety assessment of in-vehicle information systems. University of Iowa

Hart, S. G., & Staveland, L. E. (1988). Development of nasa-tlx (task load index): Results of empirical and theoretical research. In *Advances in psychology* (Vol. 52, pp. 139–183). Elsevier.

Innes, R. J., Evans, N. J., Howard, Z. L., Eidels, A., & Brown, S. D. (2020). A broader application of the detection response task to cognitive tasks and online environments. *Human factors.*

JASP Team. (2019). JASP (Version 0.11.0)[Computer software]. Retrieved from https:// jasp-stats.org/

Lee, J. D., Young, K. L., & Regan, M. A. (2008). Defining driver distraction. *Driver distraction: Theory, effects, and mitigation*, 13 (4), 31–40.

Strayer, D. L., Cooper, J. M., McCarty, M. M., Getty, D. J., Wheatley, C. L., Motzkus, C. J., . . . Horrey, W. J. (2019). Visual and cognitive demands of carplay, android auto, and five native infotainment systems. *Human Factors*, 61 (8), 1371–1386.

Strayer, D. L., Cooper, J. M., Turrill, J., Coleman, J., Medeiros-Ward, N., & Biondi, F. (2013). Measuring cognitive distraction in the automobile. AAA Foundation for Traffic Safety.

Strayer, D. L., & Johnston, W. A. (2001). Driven to distraction: Dual-task studies of simulated driving and conversing on a cellular telephone. Psychological science, 12 (6), 462–466.

Wickens, C. D., Hooey, B., L., Gore, B., F., Sebok, A., & Koenicke, C. S. (2009). Identifying black swans in NextGen: Predicting human performance in off-nominal conditions. *Human Factors, 51*, 638-651. doi: 10.1080/10508410802597382

DETECTING A LOSS OF SITUATIONAL AWARENESS

Stephen R. Pickard, Ami Eidels, Eric J. Beh
University of Newcastle, Australia

Leslie M. Blaha
711th Human Performance Wing, Air Force Research Laboratory
Pittsburgh, PA, USA

Based within the aviation context, this research is attempting to detect a loss of situational awareness (SA) using Detection Response Tasks (DRT). The research is in two phases: an initial survey and then an experimentation phase. A survey has been distributed attempting to gain an understanding of what current aviation personnel understand of the construct of SA and how it is currently taught and assessed, and what are the observed indicators of a loss of SA. The survey results will be used to validate some previously identified elements of SA around the link between SA and cognitive capacity and workload, and the cognitive and physical indicators of SA loss. The survey results have provided input into the experimentation phase of the research project. In the experimentation phase, a DRT-style experiment using Modifiable Multitasking Environment (ModME) software will simulate pilot workloads and behaviours, and through gradual increases in workload, use the DRT responses as a cognitive performance indicator of a loss of SA.

Situation awareness (SA) is loosely defined as the perception of events and elements in time and space (Endsley, 1990). Though the term is commonly used in aviation, transportation, and other critical industries, there is still debate around the most appropriate formal and operational definitions (Endsley, 1993). In this paper, we investigated pilots' understanding of SA, their SA experiences through training and assessment, and the signs of SA loss via an online survey. We analysed the data using a specialised statistical technique, correspondence analysis, to identify the structure of the association between categorical variables using data visualization. We explain how survey outcomes are used to develop a methodology to investigate SA in the laboratory.

THE SITUATION AWARENESS SURVEY

The survey, containing 76 items in 6 sections, was implemented online through LimeSurvey. Five sections were multiple choice questions; the sixth section allowed free-form comment. Section 1 identified demographic data around an individual's current role, measures of experience (years of flying and number of flying hours), the types of aircraft flown, and whether their experience was civil, military or both. Section 2 questions pertained to the participant's background of SA training and assessment through their initial pilot training. Section 3 asked about the participant's current understanding of SA. The questions were modelled on descriptions identified in a literature review of contemporary, scientific descriptions of SA. Section 4 contained similar questions to Section 2, but relating to training and assessment in the respondents current environment. Section 5 asked questions related to the respondent's observations of when someone is losing situational awareness. Section 6 was a free-form question to allow the respondents to include any additional pertinent information. A four point Likert scale was used for most questions with response options: Strongly Disagree, Disagree, Agree, Strongly Agree. A 'Not Applicable' option was provided for some questions. A neutral or middle option was not provided; respondents had to answer in either a negative or positive way. At points in the survey, negative questions were used to discourage response bias. These negative items' responses were reversed during analysis.

The target survey participants were pilots and aircrew members who were either: currently employed as a pilot, recreational pilot, flying instructors, or undergoing pilot training. The target population also covered fixed wing, rotary wing (Helicopter) and operators of unmanned aircraft systems (UAS). To simplify the approach to potential respondents, organisations with access to pilots were approached to have them distribute the survey to employees or members. The organisations approached included: Regular Public Transport (RPT) operators, The Australian Defence organisation, civil flying clubs and associations, Defence Industry organisations, other commercial aircraft operators in the general aviation space, and Emergency Services. These organisations potentially provided access to approximately 12,000-15,000 respondents. Over the 12 months the survey was open, 15 organisations distributed the survey and we received 26 responses, and of those, only 24 were complete.

Analysis of Survey Results

Demographics showed that all respondents had civil flying backgrounds (no military); respondents were predominantly pilot or pilot instructors with only two registering as student pilots; all flew fixed wing with two having also flown rotary wing; experience levels ranged from >3 years to >21 years and from >50 to >5,000 flyer hours.

We used correspondence analysis (CA) to explore the survey results and visualize the associations between the outcomes. The technique is described in detail in a number of textbooks (Beh & Lombardo, 2014), whereas here we provide only a concise description. Pearson's χ^2 statistic is typically used to assess statistical significance of the relationship between two or more categorical variables; when such an association is statistically significant, this measure is used in CA to examine its structure in a low (two, or three) dimensional graphical format using a correspondence plot. To perform the CA, the Likert scales were converted into numerical values, and a table of response frequencies was constructed for each question. Correspondence plots were then constructed using R code on the table of counts for each section of data. Each correspondence plot provided an excellent visual summary of the association, explaining at least 80% of the association in the table as measured using Pearson's χ^2 statistic. The survey results validated the use of the negative questions to identify participants completing the survey correctly and not just ticking the same response. The participants generally answered as expected for positive and negative questions. Key results are reported below.

Training of Situational Awareness

Of the 24 respondents, 21 reported that they received SA training and assessment during their initial pilot training. Results showed a strong alignment with questions relating to the focus on SA training in contemporary training environments. Respondents agreed or strongly agreed that SA is presented as a critical aviator skill and was taught throughout elements of the course. Similar responses were received for the questions relating to SA assessment in that, SA assessment was conducted throughout the course and was assessed against established criteria to provide objective assessment.

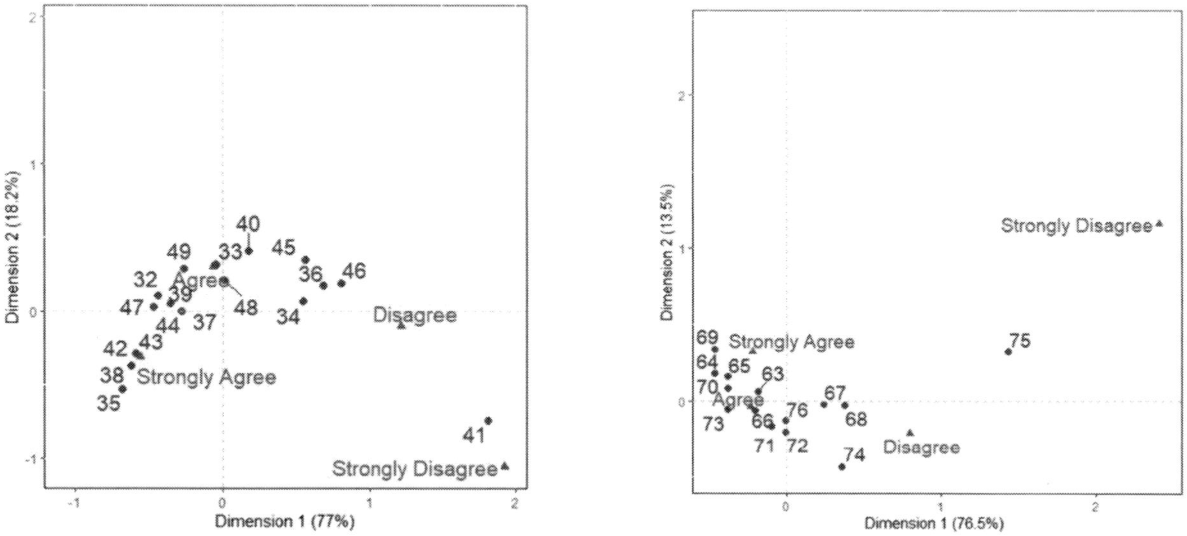

(a) Q32-49 Understanding of SA

(b) Q63-76 Detecting a loss of SA

Figure 1: Correspondence Bi-Plots

Understanding of Situational Awareness

The first 3 questions in this section referred to the respondents' assessment of their own understanding of SA and their evaluation of their colleagues' understanding of SA. There was a high level of agreement that respondents (24 of 24 Agree or Strongly Agree) and their colleagues (21 of 24 Agree or Strongly Agree) had a good understanding of SA, but then the responses to a check question (Q36) are somewhat contradictory and more than half indicated that, 'it was not a well understood concept'. The remaining questions in this section were developed around the contemporary description of SA from the literature. A correspondence plot was developed for the remaining questions in this section (Q32-49), as shown in Figure 1(a). Each point on the plot corresponds to a survey question and the proximity of the dot to the four options shows their association. As can be seen in the CA plot, there is a strong association across the responses to the 'Agree' and 'Strongly Agree' options. For example Q49, 'Situational Awareness is making decisions on available information', responses were Agree (16), Strongly Agree (8) and Disagree (1), hence the location of the dot in relation to 'Agree' on the plot. The responses to Q41 were more closely associated with 'Strongly Disagree'. Overall, the CA plot shows that the respondents generally Agree or Strongly Agree with contemporary construct of SA as identified in the literature, for example Endsley's 3-stage definition of SA (Endsley, 1995) and the link to cognitive processes (Durso et al., 2006; Flach & Rasmussen, 2017).

Detecting a loss of Situational Awareness

The questions in this section referred to physical and cognitive indicators of a loss of SA. Figure 1(b) shows the CA plot for this section. Again, there was strong association between the responses and indicators of loss of SA identified in the literature. The respondents indicated Agree or Strongly Agree to the questions; it was easy to detect a loss of SA (Q63) the cognitive indicators questions; not detecting threats (Q65), loss of scanning technique (Q66), poor decision making (Q69), not following procedures (Q71) and inability to recall basic information (Q72). This cognitive element of SA is most interesting and steers toward the concept that experiments that measure cognitive performance and workload can also be used to measure SA, or a loss of SA as identified by various authors (Endsley, 1990; O'hare, 1997; Wickens, 1999).

EXPERIMENTAL PLATFORM FOR ASSESSING SA

From the analysis of the survey and the literature review the nexus between cognitive capacity, workload, and the indicators of a loss of SA should be explored through experimentation. Such an experiment has two main requirements: First, an experimental platform that allows to systematically change the complexity, or load, of the task (Howard et al., 2020) and second, an objective way for measuring operators' workload. The Detection Response Task (DRT) is a well established methodology for observing changes in cognitive workload and capacity, especially in the automotive industry (Strayer et al., 2006; Conti et al., 2012). Yet, there has been limited use of the DRT methods to measure or detect a loss of SA as experienced by an aircraft pilot.

The planned DRT experiment aims to replicate pilot behaviours in operating an aircraft as much as possible in a simple interface. The types of behaviours would cover the fine and gross motor-skills (i.e., use of a joystick), use visual scanning techniques to switch between stimuli, mimic some of the stimuli types commonly seen inside and outside the cockpit, and use cognition to detect error or threat conditions and project future potential outcomes. The DRT should also use an inexpensive and easily accessible, simple to use interface that minimises variables, reduces or minimises practice effect, allows a primary task that employs fine and gross motor skills that can be easily automated by the participant, and contains a secondary visual/cognitive task to measure workload.

For experimental testing, we will use the ModME platform, illustrated in Figure 2, which provides a suitable interface and allows manipulation of task complexity. According to Jones, as task complexity increases, performance will be relatively linear to a threshold where SA will be impacted and there will be a relatively rapid drop-off in performance (Jones & Endsley, 1996). It is anticipated that this drop in performance will be observable as a rapid increase in response times and error rate in the DRT, and serve as an indicator of a loss of SA. By using the stair-casing technique to increase the task complexity within the DRT, it should be possible to identify the threshold where performance degrades for individuals and use this as a baseline for further experiments.

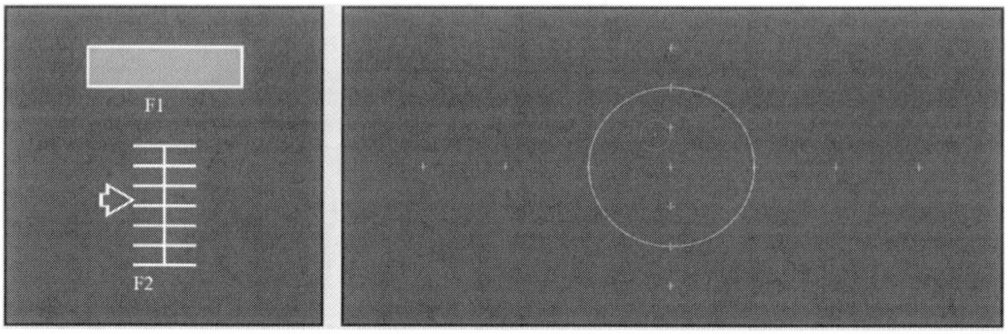

(a) Low complexity SA task configuration

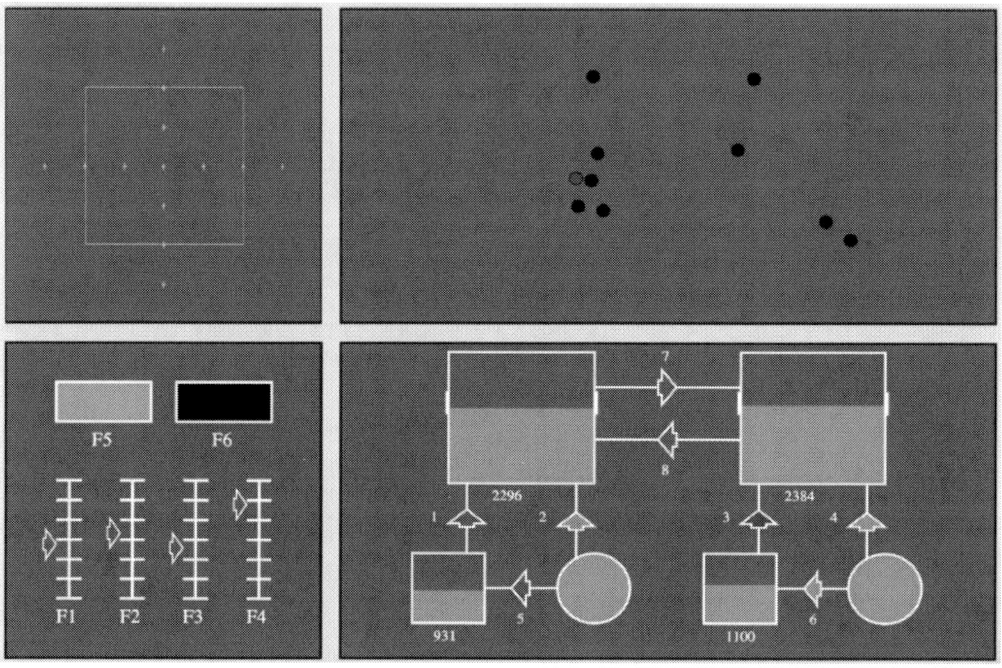

(b) High complexity SA task configuration

Figure 2: ModME interface configurations for a low and high complexity SA task concepts. (a) uses a low-load Monitoring (left) and Cross-hair Tracking tasks (right). (b) uses (clockwise from upper left) Cross-hair Tracking, Multi-object Tracking, high-load Monitoring, and Resource Management tasks.

For the initial experimentation phase, a lower complexity ModME configuration, shown in Figure 2(a), will be used. This configuration, with relatively slow-moving stimuli, would be relatively easy to complete and provide a baseline in which participants can complete the task with minimal errors and fast RTs. Over subsequent trials, the speed and complexity of the stimuli will increase in a stair-cased manner to track how RT and errors change as indicators of overload and loss of SA. This would validate the first level of SA as identified by Endsley (1995). Later experiments would use a similar methodology but a more complex set of tasks to measure a loss of SA at the 2nd and 3rd level of SA. For the 2nd and 3rd level of SA experiments, we will use increasingly complex ModME configurations containing up to four subtasks, similar to Figure 2(b). This will require scanning of frequent events occurring in multiple screen locations and monitoring to identify threats and potential future states.

References

Beh, E. J., & Lombardo, R. (2014). *Correspondence analysis: Theory, practice and new strategies.* John Wiley & Sons.

Conti, A., Dlugosch, C., Vilimek, R., Keinath, A., & Bengler, K. (2012). An assessment of cognitive workload using detection response tasks. In *Proceedings of the fourth international conference on applied human factors and ergonomics.*

Durso, F. T., Bleckley, M. K., & Dattel, A. R. (2006). Does situation awareness add to the validity of cognitive tests? *Human Factors*, *48*(4), 721–733.

Endsley, M. R. (1990). *Situation awareness in dynamic human decision making: Theory and measurement* (Unpublished doctoral dissertation). University of Southern California Los Angeles, CA.

Endsley, M. R. (1993). A survey of situation awareness requirements in air-to-air combat fighters. *The International Journal of Aviation Psychology*, *3*(2), 157–168.

Endsley, M. R. (1995). Toward a theory of situation awareness in dynamic systems. *Human factors*, *37*(1), 32–64.

Flach, J. M., & Rasmussen, J. (2017). Cognitive engineering: Designing for situation awareness. In *Situational awareness* (pp. 333–360). Routledge.

Howard, Z. L., Evans, N. J., Innes, R. J., Brown, S. D., & Eidels, A. (2020). How is multi-tasking different from increased difficulty? *Psychonomic Bulletin & Review*, *27*, 937–951.

Jones, D. G., & Endsley, M. R. (1996). Sources of situation awareness errors in aviation. *Aviation, space, and environmental medicine.*

O'hare, D. (1997). Cognitive ability determinants of elite pilot performance. *Human factors*, *39*(4), 540–552.

Strayer, D. L., Drews, F. A., & Crouch, D. J. (2006). A comparison of the cell phone driver and the drunk driver. *Human factors*, *48*(2), 381–391.

Wickens, C. (1999). Cognitive factors in aviation. *Handbook of applied cognition*, 361–389.